Quantum Mechanics
A Modern Introduction

Quantum Mechanics
A Modern Introduction

Ashok Das and Adrian C. Melissinos
University of Rochester
Rochester, New York

GORDON AND BREACH SCIENCE PUBLISHERS
New York · Philadelphia · London · Paris · Montreux · Tokyo · Melbourne

Published 1986
Fourth Printing 1990

Gordon and Breach Science Publishers

Post Office Box 786
Cooper Station
New York, New York 10276
United States of America

5301 Tacony Street, Drawer 330
Philadelphia, Pennsylvania 19137
United States of America

Post Office Box 197
London WC2E 9PX
United Kingdom

58, rue Lhomond
75005 Paris
France

3-14-9, Okubo
Shinjuku-ku, Tokyo 169
Japan

Private Bag 8
Camberwell, Victoria 3124
Australia

Library of Congress Cataloging in Publication Data
Das, Ashok, 1953–
Quantum mechanics.

Includes index.
1. Quantum theory. I. Melissinos, Adrian C.
(Adrian Constantin), 1929– . II. Title.
QC174.12.D36 1986 530.1′2 85-17531

'Αφιερωμένο στήν Ναταλία καί τόν

'Αυτώνη Λιβιερᾶτο

To

B.K.D., J.R.D.
and Ammani

Contents

Preface xi

Introduction xiii

1. WAVE-PARTICLE DUALITY 1

1.1 Wave-particle Duality 1
1.2 Interference 10
1.3 Description of Particles by Waves 13
1.4 Process of Measurement 23
1.5 The Uncertainty Principle 27
1.6 Wave Packets 33
1.7 Correspondence with Classical Physics 40
1.8 Applications: Particle in a Box 43
1.9 Summary 48
Problems 50

2. STATES, AMPLITUDES, AND OPERATORS 56

2.1 States of a Quantum-Mechanical System 56
2.2 Representation of Quantum-Mechanical States 61
2.3 Properties of Quantum-Mechanical Amplitudes 69
2.4 Operators and Change of State 77
2.5 Language of Quantum Mechanics 84
Problems 88

3. OBSERVABLES AND THE DESCRIPTION OF QUANTUM SYSTEMS 92

3.1 Process of Measurement 93
3.2 Time Dependence of Quantum-Mechanical Amplitudes 102
3.3 Observables with No Classical Analogue: Spin 106
3.4 Dependence of Quantum-Mechanical Amplitudes on Position: The Wavefunction 113
3.5 Superposition of Amplitudes: Interference 117
3.6 Identical Particles 122
3.7 Central Concepts of Quantum Mechanics 126
Problems 129

4. STATIONARY STATES OF A QUANTUM SYSTEM 133
4.1 The Hamiltonian Matrix and the Time Evolution of Quantum-Mechanical States 134
4.2 Time-Independent Perturbation of an Arbitrary System 140
4.3 Simple Matrix Examples of Time-Independent Perturbation 146
4.4 Energy Eigenstates of a Two-State System 151
4.5 Time-Independent Perturbation of a Two-State System 159
4.6 General Description of Two-State Systems: Pauli Matrices 167
4.7 Summary 174
Problems 175

5. TRANSITIONS BETWEEN STATIONARY STATES 179
5.1 Transitions in a Two-State System 180
5.2 Time-Dependent Perturbations: The ''Golden Rule'' of Quantum Mechanics 185
5.3 Phase Space 190
5.4 Emission and Absorption of Radiation 193
5.5 Scattering of a Particle from a Static Potential 199
5.6 Nuclear Magnetic Resonance 204
5.7 Energy Width of Quasi-Stationary States 211
Problems 215

6. THE COORDINATE REPRESENTATION 218
6.1 Compatible Observables 220
6.2 Quantum Conditions and the Uncertainty Relation 223
6.3 Coordinate Representation of Operators 226
6.4 Time Dependence of Expectation Values: Ehrenfest's Theorem 231
6.5 The Schrödinger Equation 237
6.6 The Periodic Potential 244
6.7 Barrier Penetration 248
6.8 Summary 253
Problems 254

7. SYMMETRIES AND CONSTANTS OF THE MOTION 260
7.1 Compatible Observables and Constants of the Motion: A Review 261
7.2 Symmetry and Conservation Laws in Quantum Mechanics 264
7.3 Symmetry Transformations and Their Generators 272

7.4 Angular Momentum Operators and Their Eigenvalues 280
7.5 Representations of the Angular Momentum Operators and Their Eigenstates 286
7.6 Composition of Angular Momenta 293
Problems 298

8. BOUND STATES: PART I 305
8.1 Separation of the Center-of-Mass Motion 306
8.2 Spherically Symmetric Potentials 309
8.3 Solution of the Radial Equation for the Coulomb Potential 314
8.4 Further Discussion of the Solutions of the Hydrogen Atom 319
8.5 Effect of Spin 327
8.6 Fine and Hyperfine Structure of Atomic Spectra 332
8.7 Variational Method: Application to the Linear Potential 339
8.8 Quarkonium 346
8.9 Summary 353
Problems 355

9. BOUND STATES: PART II 363
9.1 The Simple Harmonic Oscillator in One Dimension 364
9.2 Representations of the One-Dimensional s.h.o. 371
9.3 Molecular Band Spectra 380
9.4 The s.h.o. in Three Dimensions 388
9.5 The Shell Model of the Nucleus 395
9.6 The Square-Well Potential 399
9.7 The Deuteron 407
9.8 Summary 410
Problems 411

10. SYSTEMS WITH IDENTICAL PARTICLES 421
10.1 Indistinguishability and Exchange Symmetry 422
10.2 The Periodic Table 430
10.3 Raman Spectra of Homonuclear Molecules 436
10.4 The Helium Atom 442
10.5 The Hydrogen Molecule 452
10.6 Systems with Identical Bosons 462
Problems 472

11. SCATTERING: PART I 477
11.1 Differential and Total Cross Section 478
11.2 Solutions of the Scattering Equation: The Method of Partial Waves 486
11.3 Applications of Partial-Wave Analysis 497
11.4 Low-Energy *n-p* Scattering 504
11.5 Angular Distribution 510
11.6 Conclusion 517
Problems 519

12. SCATTERING: PART II 521
12.1 Scattering of Identical Particles 521
12.2 Energy Dependence and Resonance Scattering 526
12.3 The Lippman–Schwinger Equation 535
12.4 Born Approximation 540
12.5 Inelastic Scattering 547
12.6 Conclusion 554
Problems 556

Appendix 1: The Fourier Transform 560
Appendix 2: The Dirac Delta Function 564
Appendix 3: Normalization of Plane Waves and Wave Packets 570
Appendix 4: Review of Matrix Algebra 577
Appendix 5: Quantum Conditions and the Poisson Brackets 590
Appendix 6: Composition of Angular Momenta 592
Appendix 7: Matrix Elements of Vector Operators 596
Appendix 8: Operators in Spherical Coordinates 600
Appendix 9: Legendre Polynomials and Spherical Harmonics 606
Appendix 10: Special Functions: Laguerre and Hermite Polynomials 613
Appendix 11: Expectation Values of $\hat{r}^{-s}$ for Hydrogen 620
Appendix 12: Spherical Bessel Functions 626
Appendix 13: Evaluation of Overlap Integrals 629
Appendix 14: Units and Physical Constants 633
Index 635

Preface

Quantum mechanics was invented as the culmination of the effort to understand atomic and radiation phenomena. It began in 1900 with Planck's attempt to derive his celebrated radiation formula. Then, in 1913, Bohr gave his theory of Balmer's series which eventually led Heisenberg, Schrödinger, Dirac, and others to construct the entire edifice of quantum mechanics in the mid-twenties. Since then, quantum mechanics has been applied to a vast number of areas: nuclear physics, solid state physics, statistical mechanics, chemical reactions, particle physics, etc. It ranks with the greatest of human creations in its philosophical conception and its successful applications.

Because of the immensity of its achievement, almost all books on quantum mechanics have followed more or less the historical route. However, the period from 1925 to the present has been characterized by an unprecedented burst of energy and advances in physics. The result is that the sheer number and variety of accurate and sensitive experiments supporting quantum mechanics have proliferated enormously. With this has come a much deeper understanding and appreciation of the subject. It is therefore both logical and proper to present quantum mechanics in a way not restricted to its historical sequence, but instead to concentrate on the most appropriate experiments, and from these to build up its foundation.

This new book by A. Das and A. C. Melissinos, two of our leading physicists, does just that. By a careful selection of topics in particle and nuclear physics, as well as in atomic and molecular physics, they are able to bring college and first-year graduate students to a comprehensive understanding of the subject and its richness. They succeed in a natural and effortless way in making the excitement of these new developments an integral part of a rigorous training in quantum mechanics.

T. D. Lee
Enrico Fermi Professor of Physics
Columbia University
New York, New York

Introduction

Physics is the science that describes physical systems in terms of the few basic laws that, we believe, govern the universe. The discovery of these laws has progressed hand-in-hand with the experimental evidence that could be obtained at any particular time; this knowledge has, in turn, made possible the technology on which so much of our present society is based. While the physical laws stand by themselves, their ramifications and their connection with observable effects are derived through the language of mathematics with its concise and powerful forms.

The physical laws that describe nature at distances and momenta much smaller than those encountered in direct human experience, are the subject of quantum mechanics. The same laws also apply to macroscopic objects, of course, but in that case it is much preferable, and easier, to use an appropriate approximation to quantum mechanics: classical mechanics. The macroscopic theories of classical physics were a well-established science long before 1925 when quantum mechanics was put on a firm foundation. Thus, it has been customary in the teaching of physics to introduce the classical laws first and to address quantum mechanics only later. This choice is motivated by tradition and by the fact that direct human experience is at the macroscopic level.

Modern science, on the other hand, has progressed well beyond macroscopic systems; any practicing scientist, whether chemist, electrical engineer, physicist or biologist must be freely conversant—at a quantitative level—with the laws governing molecules and atoms. This text is an outgrowth of this point of view. We have tried to present quantum mechanics in a self-consistent manner. Thus, we do not rely on classical arguments to make quantum mechanics plausible but instead introduce the principles directly. In the spirit of this approach, the wavefunction in coordinate space, in spite of its great utility, is not discussed until the concepts of quantum-mechanical states and amplitudes have been clearly established.

Throughout this text we have used physical examples to illustrate the principles and concepts that are introduced. For instance, electron diffraction plays the role of the inclined plane of classical mechanics. The examples lead to numerical answers and thus should provide the reader with a feeling for the order of magnitude of the physical observables that are encountered in quantum problems. They also show the simplicity of quantum-mechanical calculations and the remarkable agreement between theory and experimental observation.

The first chapter serves as an introduction and stresses the dual description of natural phenomena in terms of a particle or of a wave picture. In Chapters 2 and 3 we introduce the principles of quantum mechanics and the concept of state vectors, their transformations and representations. Dynamics, that is, the equation of motion, is introduced in Chapter 4. Here we use the simple two-level system as an example and find its stationary states under a variety of conditions. This leads to a discussion of perturbation theory and to the introduction of the Pauli matrices at an early point in the text. Transitions between stationary states are introduced in Chapter 5 and make possible the discussion of several important physical applications. The emission and absorption of radiation, Coulomb scattering and nuclear magnetic resonance are treated as examples.

The coordinate representation is introduced in Chapter 6 and the connection with classical physics is discussed in a formal fashion that includes applications of the one-dimensional Schrödinger equation. In Chapter 7 we discuss the relation between symmetries and constants of motion. This is a fundamental relationship in nature, one that became more clearly evident as quantum mechanics was understood better. In particular, we derive the spectrum of the eigenstates of the angular momentum, which is essential for any application of quantum mechanics to three-dimensional systems. Chapters 8 and 9 deal with bound states where we try to stress the similarity of all such systems as well as the universality of the techniques used to describe them. An extensive account of the hydrogen atom, including spin-orbit coupling, is given. The linear potential applicable to quarkonium is chosen as an example of the variational method. The one-dimensional simple harmonic oscillator is solved by the operator technique in Chapter 9. The square well and oscillator potentials are treated in three dimensions and the results are applied to the nuclear shell model and to the deuteron.

Systems with identical particles are considered in Chapter 10. The consequences of the fundamental notion of indistinguishability are analyzed by using the properties of the permutation operator and are applied to both Fermi and Bose systems. We use this opportunity to discuss multielectron systems, the helium atom and the molecular bond. The two concluding chapters are devoted to a discussion of scattering. In Chapter 11 we introduce general concepts about scattering and the method of partial waves; this is followed by several examples of low-energy scattering. In

Chapter 12 we examine the energy dependence of scattering and the connection with the bound state spectrum. We indicate the general solution of the scattering problem and consider the Born approximation as an application.

Several topics which can be considered mathematically technical and therefore could distract from the continuity of the text have been placed in thirteen appendices. The reader may already be familiar with some of this material or may need to consult the appendices only occasionally. Nevertheless, the appendices are written at the same level as the main text and, we hope, serve to make the book complete and self-contained. A set of problems is provided after each chapter. The problems can be solved with the material discussed in the text and we urge the reader to do so: there is no better way of mastering a subject than by working out problems. Within each chapter the problems are arranged in order of increasing difficulty.

The text is addressed to students who have completed an introductory sequence of physics and calculus. No particular preparation in advanced calculus or linear algebra is needed, even though familiarity with these subjects can only be helpful to the reader. No knowledge of atomic phenomena is required, the examples serving to illustrate these effects. However, the text is quantitative and is meant to provide the reader with the foundations of quantum mechanics. It can be used either at the junior/senior level or at the first-year graduate level, depending on the emphasis and choice of topics.

There are various paths that can be followed through the text. For advanced students Chapter 1 could be omitted and Chapters 2 and 3 treated as review material; similarly, Chapter 6 could be covered rapidly if the reader is familiar with simple applications of the Schrödinger equation. Conversely, for beginning students Chapter 5 could be postponed until after Chapters 8 and 9 have been covered. In general the final sections of Chapters 8–12 can be omitted without loss of continuity, as they deal with specific applications. For a one-term course we would recommend Chapters 1–4, 6, and parts of 8. In any case, we feel that the organization of the material allows the instructor a variety of choices according to his own preference and the characteristics of his class.

This book is an outgrowth of courses in quantum mechanics that we have given to seniors and to first-year graduate students at the University of Rochester. We are grateful to our students, who have provided not only the reason for engaging in this project but also suggestions, criticism, and enthusiasm. While this is primarily a textbook, we hope that it will also be useful as a reference. We have in mind, in particular, scientists who have completed their formal education some time ago yet need to use quantum mechanics in their current work.

Our own outlook on the subject has been shaped by many influences, including the extensive literature on quantum mechanics at all levels. We mention below some of the most prominent texts that, we feel, would be excellent sources for further reading:

L. I. Shiff, *Quantum Mechanics*, Third edit., McGraw Hill, New York (1968). This is an excellent and complete reference. The presentation is somewhat compact and relatively advanced.

R. P. Feynman, R. E. Leighton, and M. Sands, *The Feynman Lectures*, Vol. III, Adisson-Wesley, Reading, Mass. (1965). A book of very deep physical insight that makes for wonderful reading. However, for the first-time student of quantum mechanics it is difficult to use as a text.

C. Cohen-Tannoudji, B. Diu, and F. Laloe, *Quantum Mechanics*, English trans., Wiley, New York (1977). A very complete but lengthy description of the fundamentals of quantum mechanics with applications to atomic systems.

P. A. M. Dirac, *The Principles of Quantum Mechanics,* Oxford Univ. Press, Clarendon (1958). The classical text by one of the inventors of quantum mechanics. The notation, style and subjects covered reflect the date of first publication (1930). Nevertheless this is the work of a great master.

L. Landau and E. M. Lifshitz, *Quantum Mechanics*, English trans., Pergamon (1968). A classical text with a wealth of physical applications; well organized and at the intermediate level.

Our choice of units is that found in the research literature rather than the MKS system. This is because atomic phenomena are best described in terms of their natural units. In such units equations can be easily interpreted and the magnitude of experimental quantities can be quickly predicted and understood. We have kept $\hbar$ and c in all equations, but use dimensionless (and other) combinations of the fundamental constants as often as possible.

In the preparation of the text we have benefited from the perspective and suggestions of many colleagues. In particular we wish to mention Drs. J. B. French, V. S. Mathur, and S. Okubo and Mr. P. Panigrahi. Drs. H. L. Helfer and T. G. Castner were kind enough to use parts of the text in its preliminary form in their own classes, providing us with student feedback. We are indebted to Mrs. Edna Hughes-McLain and to Mrs. Judith Mack for their patient and skillful typing of the many versions of the manuscript and to Mr. J. Sheedy for the preparation of the illustrations. Finally, we are grateful to our families for their encouragement and understanding during the course of this work.

We are aware that in spite of the publisher's and our own best efforts some errors and misprints may still be found in the text. We would greatly appreciate it if the readers would kindly bring them to our attention. But above all we hope that this book will bring some inspiration and joy to its readers.

A. Das
A. C. Melissinos

Chapter 1

WAVE-PARTICLE DUALITY

In this chapter we develop the basic concepts of quantum mechanics. We emphasize the fact that particles can exhibit wave properties much in the same way light waves can display particle-like behavior. Furthermore, we discuss the essential feature of quantum mechanics—that measurement at the microscopic level leads to a probabilistic description of systems. The wave entities of a particle are given in the De Broglie hypothesis, and we describe these relations in the context of various examples including the scattering of high-energy protons.

The process of measurement affects the system, of course, and hence leads to uncertainties in the measured quantities. We discuss Heisenberg's uncertainty relation and apply it to simple physical phenomena. We introduce the concept of a wave packet as it describes a localized particle, and we discuss the various properties of wave packets exposing the correspondence principle. We then use wave packets to derive the energy levels of a particle confined in a box as well as the Fermi energy in a conductor. The technical discussions of Fourier transforms, the Dirac delta function, and the wave packets are given in appendices 1–3.

1.1. WAVE-PARTICLE DUALITY

Quantized Observables

The term *observable* is generally used for a physical quantity that can be determined experimentally. For example, the position of a body on a line or with reference to a three-dimensional coordinate system, the time interval between the swings of a pendulum, and the force exerted on a body or the momentum and kinetic energy of a

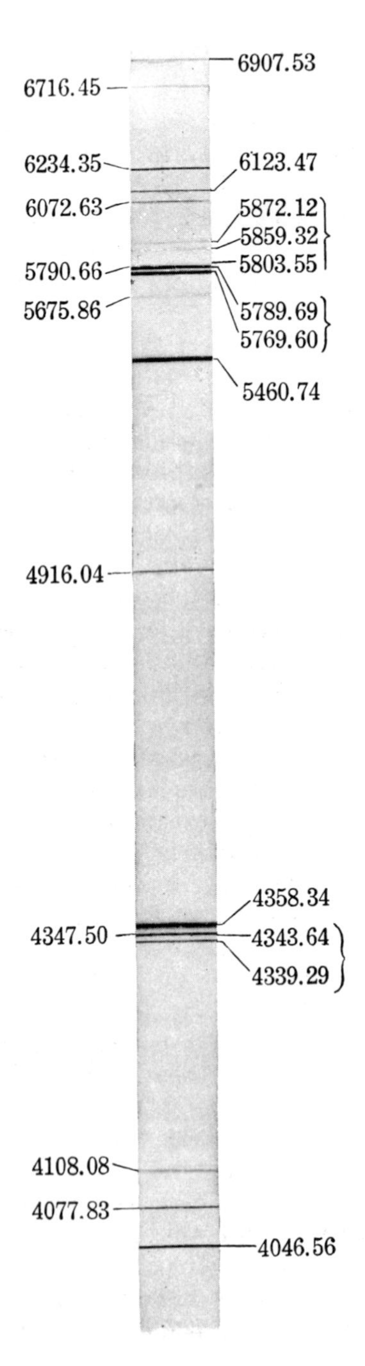

FIGURE 1.1. The spectrum of mercury in the visible. The wavelengths of the lines are given in angstroms ($1\ \text{Å} = 10^{-10}$ m). [From G. P. Harnwell and J. J. Livingood, Experimental Atomic Physics McGraw Hill, New York (1933); by permission.]

moving object—are all observables. Furthermore, the observables can either assume a continuous range of values or they can be restricted to have only discrete or quantized values. As an example, we note that the momentum or kinetic energy of a macroscopic body can have any continuous value. On the other hand, when the electric charge of a body is measured, it is always found to be a multiple of the charge of the electron

$$e = 1.602 \times 10^{-19} \text{ coul}$$

This does not seem surprising since we assume that the total charge is due to the presence (or absence) of some number of electrons with respect to the neutral state of the body. Nevertheless, one concludes that the observable electric charge is always quantized, namely, that it is a multiple of the quantum of charge and can take only discrete (and not a continuous set of) values.

A similar situation exists when we examine the frequency spectrum of the light emitted by atoms, as shown in Fig. 1.1. One observes a discrete set of spectral lines distinct from the continuous spectrum emitted by a glowing filament. Further analysis indicates that the spectral lines arise because of transitions of the configuration of the atomic electrons between different energy levels (or states) of the atom and, consequently, that the atom can be found only in discrete states of energy. Thus, the observable energy of an atom is quantized, as shown in Fig. 1.2. This is a rather surprising conclusion because experience with macroscopic bodies indicates that their energy can have any one of a continuum of values.

Another demonstration of a quantized observable occurs in the Stern–Gerlach experiment sketched in Fig. 1.3. A beam of sodium or silver atoms is produced by heating the element in an oven and letting the atoms escape through an orifice; further collimation selects atoms with almost parallel trajectories. If the beam passes through a highly inhomogeneous magnetic field, then it separates into two components along the direction of the field.

This behavior of the sodium (or silver) atoms can be understood as follows: Alkaline atoms, as they are usually called, have an orbit structure such that only one electron is in an unpaired state. The electron has an intrinsic angular momentum—referred to as spin—and associated with it a magnetic dipole moment μ. If we think of this moment as a classical vector, then for the alkaline atoms the

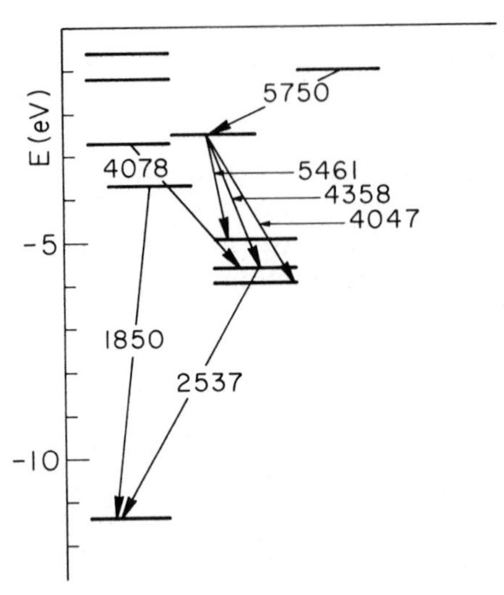

FIGURE 1.2. Some of the energy levels of the mercury atom. The most prominent transitions are indicated. The numbers give the wavelength in angstroms.

magnitude of $\boldsymbol{\mu}$ is given by $|\boldsymbol{\mu}| = [s(s+1)]^{1/2}(2\mu_B)$, where $s = \frac{1}{2}$ and μ_B is the Bohr magneton, which has the value

$$\mu_B = \frac{e\hbar}{2m_e} = 9.274 \times 10^{-24} \text{ A-m}^2$$

Since the electron charge is negative, the magnetic moment is directed opposite to the spin direction. When such a magnetic dipole is placed in a magnetic field **B**, it acquires a potential energy given by

$$U = -\boldsymbol{\mu} \cdot \mathbf{B} \tag{1.1a}$$

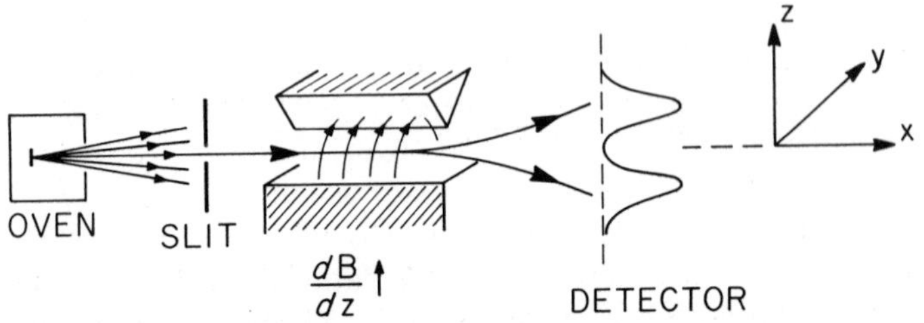

FIGURE 1.3. Schematic diagram of the Stern–Gerlach apparatus.

If the field is homogeneous, then the dipole is subjected to a torque

$$\boldsymbol{\tau} = \boldsymbol{\mu} \times \mathbf{B} \tag{1.1b}$$

that gives rise to a precession of the dipole moment around the field direction, as shown in Fig. 1.4(a). If the magnetic field were inhomogeneous (say along the z-direction) the dipole would experience an additional force given by

$$F_z = \boldsymbol{\mu} \cdot \hat{e}_z \frac{dB_z}{dz} = \mu_z \frac{dB_z}{dz} = |\boldsymbol{\mu}| \frac{dB_z}{dz} \cos\theta \tag{1.1c}$$

Here θ represents the orientation of the dipole moment with respect to the Z-axis, as shown in Fig. 1.4(b). Thus, an inhomogeneous magnetic field would cause an amount of deflection of the trajectory which depends on the orientation of the dipole moments. Furthermore, since we expect the atoms (sodium or silver) to come out of the oven with fairly random orientations of their spin, a logical expectation would be to obtain a broad distribution of atoms at the detector. Instead, two distinct spots are observed. Figures 1.5(a) and (b) show, respectively, the distributions observed at the detector when the magnetic field is turned off and on.

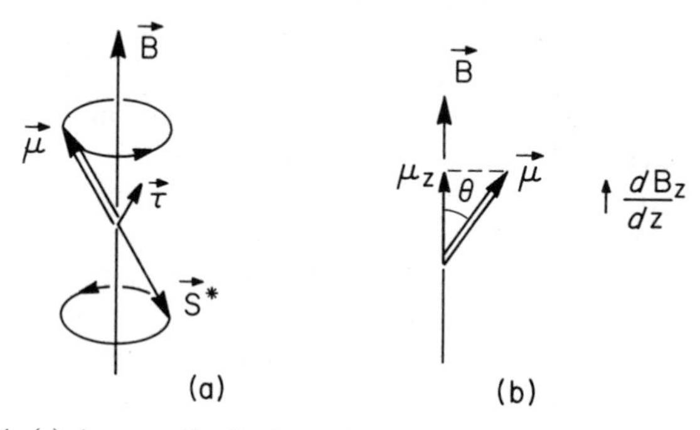

FIGURE 1.4. (a) A magnetic dipole precesses in a uniform magnetic field. (b) A magnetic dipole experiences a net force when it is placed in an inhomogeneous magnetic field.

The interpretation of these results is that the magnetic moment of these atoms is found only in one of two orientations, namely, either parallel or antiparallel to the z-direction, which is the direction of

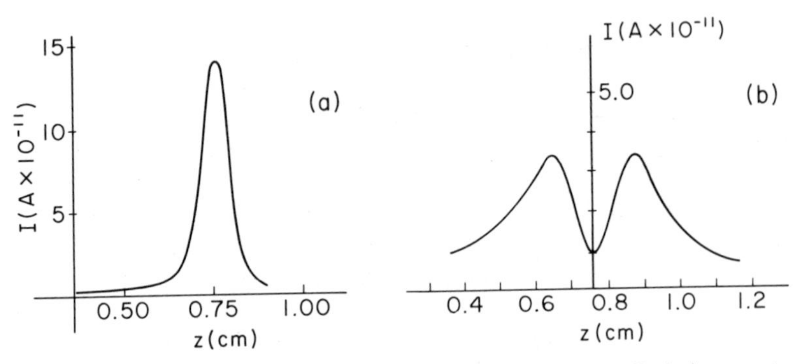

FIGURE 1.5. The beam current in the detector plane of a Stern–Gerlach apparatus (a) without the magnetic field and (b) when the magnetic field is switched on.

the field. The projection of the magnetic moment onto the z-direction is found to be exactly

$$\mu_z = \pm\tfrac{1}{2}(2\mu_B) \qquad (1.1d)$$

If the magnet is rotated so that the field and the field gradient are along the x-direction, then the splitting is also along the X-axis. One concludes that the possible orientations of the electron's magnetic moment in a magnetic field (or, more explicitly, the projections of the spin angular momentum onto any fixed axis) are quantized. The direction of quantization depends on the orientation of the measuring apparatus.

The existence of a discrete spectrum for a physical observable is also encountered in classical physics. For example, the frequencies of oscillation of a tight string or of an air column are discrete. However, what is different is that these frequencies are not affected by the measurement process. In quantum mechanics the process of measurement itself can influence the possible outcome of the measurement, as we have seen in the Stern–Gerlach experiment. We will return to the discussion of the measurement process in Section 1.4. Another class of phenomena where the expectations from macroscopic experience are contradicted in the microscopic regime are discussed below.

Most optical phenomena are explained by the wave nature of light. Among many other things, the wave nature suggests that light waves can exchange any amount of energy with the surroundings. With this assumption, if one calculates the spectrum of radiation for a blackbody, then the theoretical predictions lead to a curve, which does not agree with the experiments. For example, Wien's law leads to a spectrum that agrees with the experimental curve at shorter wavelengths but differs quite a lot at longer wavelengths. Similarly, although the Rayleigh–Jean law gives good agreement at higher wavelengths, it runs into difficulties at shorter wavelengths.

Planck assumed that electromagnetic radiation (light) of frequency ν can exchange energy with the surroundings only in units of $h\nu$ where

$$h = 2\pi\hbar = 6.62 \times 10^{-34}\ \text{J-sec}$$

This is known as Planck's constant. With this assumption, the theoretical prediction for the blackbody spectrum coincides completely with the experimental curve shown in Fig. 1.6. This implies that the energy of the radiation field is quantized.

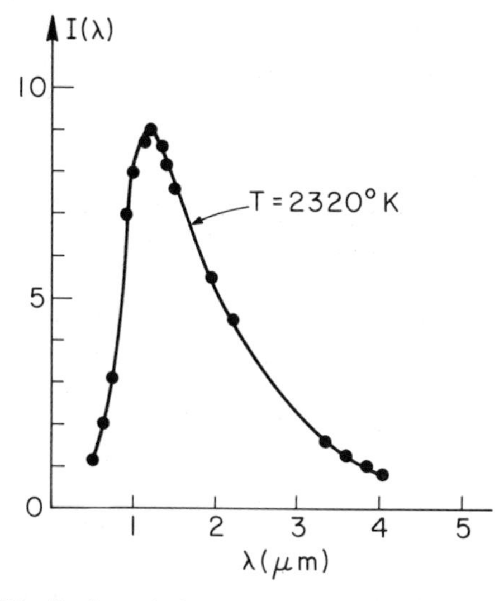

FIGURE 1.6. Distribution of the radiation emitted by a blackbody heated to $T = 2320$ K. The curve is the prediction of Planck's radiation law.

Further support of this hypothesis comes from the photoelectric effect. It was observed that electrons were released from metals when they were irradiated with light. However, electrons were emitted only if the frequency of the radiation was greater than a critical frequency which depended on the metal. Furthermore, if the frequency was less than the critical value, no amount of radiation could release the electrons from the metal. This presented a puzzle if one believed in the continuous exchange of energy by the light waves. Einstein solved this puzzle by using Planck's hypothesis. Namely, radiation of frequency ν can only exchange energy

$$E = h\nu \tag{1.2a}$$

For the electrons to be released, therefore, we must have

$$h\nu = e\phi + \mathrm{KE} \tag{1.2b}$$

where $e\phi$ is the energy with which electrons are bound to the particular metal, and KE stands for the kinetic energy of the released electron. If we denote the binding energy as $e\phi = h\nu_0$, then we have

$$h(\nu - \nu_0) = \mathrm{KE} \geqslant 0 \tag{1.2c}$$

This clearly explains why radiation below a critical frequency cannot release the electrons. It also supports the hypothesis about the particle nature of light waves.

That radiation sometimes behaves like particles, is decisively demonstrated by the Compton effect. When light is scattered by an electron the wavelength λ of the light changes. This change depends only on the angle at which radiation is detected. This phenomenon can only be explained if we assume that the light waves behave like particles with energy and momentum given by

$$E = h\nu \tag{1.3a}$$

$$p = \frac{E}{c} = \frac{h\nu}{c} = \frac{h}{\lambda} \tag{1.3b}$$

and that the scattering process is treated as that of particles with conservation of momentum as well as energy. The following example shows that light waves do exhibit particlelike behavior and that their momentum and energy are quantized.

Example: The Compton effect

Let us examine in detail the scattering of a light quantum (photon) of frequency ν_0 by an electron at rest. The electron and the photon are scattered by angles θ and θ_1, respectively, as shown in Fig. 1.7.

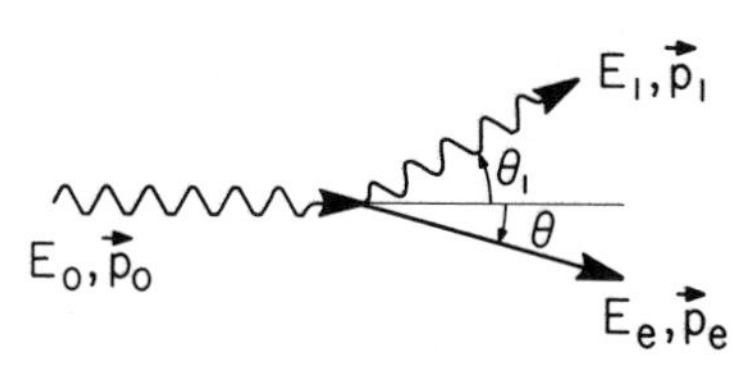

FIGURE 1.7. Scattering of a photon from a free electron.

If we think of the photon as a particle, then it has energy and momentum given by

$$E_0 = h\nu_0$$

$$p_0 = \frac{h\nu_0}{c}$$

The scattered wave at the detector then would have energy and momentum given by

$$E_1 = h\nu_1$$

$$p_1 = \frac{h\nu_1}{c}$$

If, after scattering, the electron moves away with a momentum of magnitude p_e, then conservation of momentum gives

$$p_0 = p_e \cos\theta + p_1 \cos\theta_1 \tag{1.4a}$$

$$0 = -p_e \sin\theta + p_1 \sin\theta_1 \tag{1.4b}$$

Equations (1.4*a*) and (1.4*b*) lead to the relation

$$p_e^2 = p_0^2 - 2p_0p_1 \cos\theta_1 + p_1^2 \tag{1.4c}$$

Similarly, conservation of energy gives

$$E_0 + m_e c^2 = (p_e^2c^2 + m_e^2c^4)^{1/2} + E_1 \tag{1.4d}$$

which, upon simplification, leads to

$$p_e^2 = p_0^2 + p_1^2 - 2p_0p_1 + 2m_ec(p_0 - p_1) \tag{1.4e}$$

Equating the two relations in Eqs. (1.4*c*) and (1.4*e*) we obtain

$$\frac{1}{p_1} - \frac{1}{p_0} = \frac{1 - \cos\theta_1}{m_ec} \tag{1.4f}$$

In terms of wavelengths this formula translates to

$$\lambda_1 - \lambda_0 = \frac{h}{m_ec}(1 - \cos\theta_1) \tag{1.4g}$$

This states that the change in the wavelength of the X rays depends depends on the angle at which they are detected.

1.2. INTERFERENCE

Consider a machine gun spraying bullets at an armored plate with holes at s_1 and s_2. If the hole s_2 is closed, then the bullets passing through s_1 produce a pattern at a distant wall as shown in Fig. 1.8(a). Similarly if the hole s_1 is closed, then the passage of bullets through s_2 gives rise to the pattern of Fig. 1.8(b). If both s_1 and s_2 are open we expect the resulting pattern to be the sum of the two patterns obtained previously. This is shown in Fig. 1.8(c). Such behaviour is observed for bullets, and it shows, in particular, that the passage of the bullets through $s_1(s_2)$ is independent of whether $s_2(s_1)$ is open or closed.

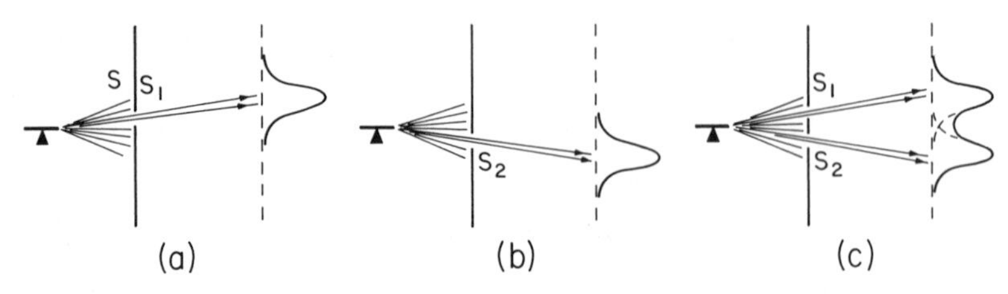

FIGURE 1.8. A hypothetical machine gun firing bullets at a steel wall: (a) with a single opening S_1 in the wall; (b) with a single opening S_2 in the wall; (c) with both openings at the same time.

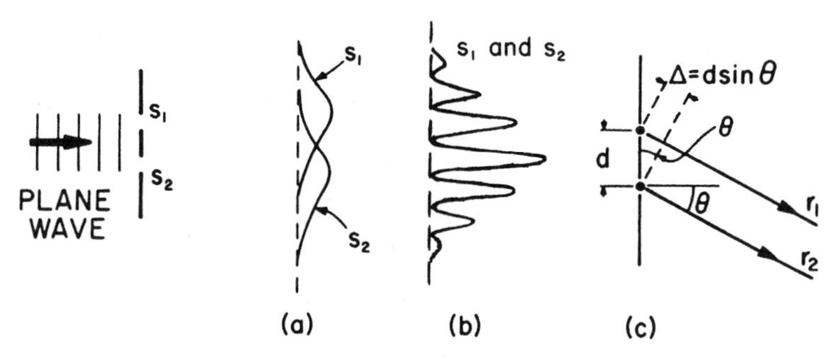

FIGURE 1.9. Monochromatic parallel light incident on two slits. (a) The pattern from each of the slits. (b) The interference pattern when both slits are open. (c) The path length of the light emitted from each slit depends on the angle θ. We assume that the screen is at infinity.

We can perform a similar experiment by shining a beam of parallel monochromatic light onto a double slit and then closing one or the other of the slits alternately. The result is shown in Fig. 1.9(a) and is similar to the pattern seen with the bullets. When both the slits are open, however, what is observed is a set of fringes, as shown in Fig. 1.9(b); i.e., there are points of maximum and minimum illumination. This clearly is not the sum of the patterns obtained when only one of the slits is open at a time. The light experiment differs from the experiment with bullets in this respect. Such a pattern in the case of light waves is understood to be due to the interference between the waves emerging from s_1 and s_2. When the angle θ [see Fig. 1.9(c)] is such that the path difference $\Delta = d \sin \theta$ of the two light waves from the two slits to the point of observation equals

$$\Delta = d \sin \theta = \frac{\lambda}{2}, \frac{3\lambda}{2}, \frac{5\lambda}{2}, \ldots \tag{1.5a}$$

the two waves interfere destructively, and we observe the minima of illumination. In contrast, when

$$\Delta = d \sin \theta = 0, \lambda, 2\lambda, 3\lambda, \ldots \tag{1.5b}$$

the two waves interfere constructively, and we observe the maxima of illumination.

As an example, for red laser light let $\lambda = 6600$ Å. If the spacing between the slits is $d = 0.04$ mm, the first minimum occurs at an angle

$$\theta \simeq \sin\theta = \frac{\lambda}{2d} = \frac{6.6\times 10^{-7}\ \text{m}}{2\times 4\times 10^{-5}\ \text{m}} = 8.25\times 10^{-3}\ \text{rad} \tag{1.6}$$

If the light wave emerging from slit s_1 is described by an electric field

$$E_1(r_1, t) = A\cos(\omega t - kr_1) \tag{1.7a}$$

and the light wave emerging from slit s_2 is described by

$$E_2(r_2, t) = A\cos(\omega t - kr_2) \tag{1.7b}$$

then when only s_1 is open the intensity is proportional to

$$I_1 = |A|^2\cos^2(\omega t - kr_1) \tag{1.8a}$$

Correspondingly, when only s_2 is open the intensity is proportional to

$$I_2 = |A|^2\cos^2(\omega t - kr_2) \tag{1.8b}$$

When both slits are open the intensity is proportional to

$$\begin{aligned} I_{1,2} &= |E_1 + E_2|^2 \\ &= |A\cos(\omega t - kr_1) + A\cos(\omega t - kr_2)|^2 \\ &= \left| 2A\cos\left[\omega t - \frac{k(r_1+r_2)}{2}\right]\cos\frac{k(r_1-r_2)}{2}\right|^2 \\ &= 4|A|^2\cos^2\left[\omega t - \frac{k(r_1+r_2)}{2}\right]\cos^2\frac{k\Delta}{2} \\ &= 4|A|^2\cos^2\frac{\pi\Delta}{\lambda}\cos^2\left[\omega t - \frac{k(r_1+r_2)}{2}\right] \\ &\neq I_1 + I_2 \end{aligned} \tag{1.8c}$$

Note that when $\theta = 0$, the intensity is twice as large as the sum of the intensities when either of the slits is open.

We can slowly decrease the intensity of the incident light wave so that only a single photon passes through the slits at any one moment. This is very much like the experiment with the bullets. A natural question is whether the resulting pattern in this case would be the same as for the bullets. The answer, surprisingly, is in the negative. That is, if one waits long enough (so that a sufficient number of photons have been accumulated), the same interference pattern would be obtained. In fact, it is not possible to tell whether any single photon arriving at the observation plane has come from s_1 or s_2. This shows, therefore, that the behavior of the photons passing through the slits differs from the analogous process when bullets are used.

If the two slits are replaced by a diffraction grating, the light waves show a diffraction pattern with a central maximum and a series of secondary maxima and minima. The positions of the maxima occur at

$$d \sin \theta = n\lambda \tag{1.9}$$

where d is the spacing of the grating, n is the order of the maxima, and λ is the wavelength of the light used. A similar experiment can be performed with electrons. It is no longer possible to make slits or holes of appropriate dimension, but one can use a crystalline material. For example, the spacing between neighboring atoms in rock salt is $d = 2.82$ Å. In the Davisson–Germer experiment, a beam of electrons is incident on the crystal, and a diffraction pattern is observed as we would expect in an experiment with light. This experiment, therefore, shows that the electrons also behave like waves under appropriate conditions. Furthermore, if we use the observed diffraction pattern to find the wavelength associated with the electron in this experiment, it turns out to be inversely proportional to their momentum. The conclusion from these experiments is that under appropriate conditions, just as light waves can display particlelike behavior, particles, like electrons, can exhibit wave nature. The "appropriate conditions" refer to the scale of the quantities involved. For example, bullets, in principle, also possess wave nature and can give rise to interference patterns. However, note from the Davisson–Germer experiment that bullets, being very massive, would have an extremely small wavelength. Hence, the separation between the maxima would be extremely small, and consequently the interference pattern would escape detection.

1.3. DESCRIPTION OF PARTICLES BY WAVES

From the discussion of the previous sections we realize that particles also have wavelike behavior. A wave in one dimension is commonly represented by a function of the form

$$f(\omega t \mp kx) \tag{1.10}$$

where the $\mp$ sign corresponds to waves travelling in the $+x$ or $-x$ directions, respectively, and

$$\begin{aligned} \omega &= 2\pi\nu = \text{angular frequency} \\ k &= \frac{2\pi}{\lambda} = \text{wave number} \end{aligned} \tag{1.11}$$

If the wave has a sine or a cosine form, then it is called a harmonic wave. An arbitrary harmonic wave can be written as

$$f(\omega t - kx) = A \sin(\omega t - kx) + B \cos(\omega t - kx) \tag{1.12}$$

One can generalize this to three space dimensions. A wave is then denoted by

$$f(\omega t - \mathbf{k} \cdot \mathbf{x}) \tag{1.13}$$

Here $\mathbf{k}$ is the wave vector, i.e., it is the wave number multiplied by the unit vector along the direction of propagation of the wave.

If we have to ascribe wave nature to particles, then we have to find the appropriate values of the parameters ω and $\mathbf{k}$ for a wave describing a particle of energy E and momentum $\mathbf{p}$. This is done in complete analogy with the case for the photons. For example, we know that for photons

$$\begin{aligned} E &= cp \\ E &= h\nu = \hbar\omega \end{aligned} \tag{1.14a}$$

or

$$\omega = \frac{E}{\hbar} \tag{1.14b}$$

Furthermore,

$$\begin{aligned} \lambda &= \frac{c}{\nu} = \frac{2\pi c}{\omega} = 2\pi c \frac{\hbar}{E} \\ &= h\frac{c}{E} = \frac{h}{p} \end{aligned} \tag{1.14c}$$

Hence, we obtain

$$k = \frac{2\pi}{\lambda} = \frac{p}{\hbar}$$

or

$$\mathbf{k} = \frac{\mathbf{p}}{\hbar} \tag{1.14d}$$

In analogy with these relations De Broglie suggested that the wave corresponding to a particle with energy E and momentum $\mathbf{p}$ also has the wave entities given by

$$\omega = \frac{E}{\hbar}$$
$$\mathbf{k} = \frac{\mathbf{p}}{\hbar} \tag{1.15}$$

These are known as De Broglie relations and have been tested in several experiments such as the Davisson–Germer experiment. We now discuss a few other experiments that support the above hypothesis.

Scattering of High-Energy Protons

Let us examine the implications of the De Broglie relations applied to the scattering of high-energy protons. When protons of energy 200 GeV are scattered (elastically) from hydrogen, a diffraction minimum is observed at transverse momentum $p \sin\theta \simeq 1.2$ GeV/c. The experimental curve is shown in Fig. 1.10 [R. Rusack *et al.*, *Phys. Rev. Lett.* **41,** 1632 (1978)].

Before analyzing the experimental result, we introduce the units used in the study of elementary particles. The masses of particles are conveniently expressed in units of energy known as electron volts (eV).

$$1\ \text{eV} = 1.6 \times 10^{-12}\ \text{erg}$$

The mass of the electron and the proton are given in eV as

$$m_e = 0.511\ \text{MeV}/c^2 = 5.11 \times 10^5\ \text{eV}/c^2$$

$$\begin{aligned} m_p &= 938.28\ \text{MeV}/c^2 = 938.28 \times 10^6\ \text{eV}/c^2 \\ &= 0.938\ \text{GeV}/c^2 \end{aligned} \tag{1.16}$$

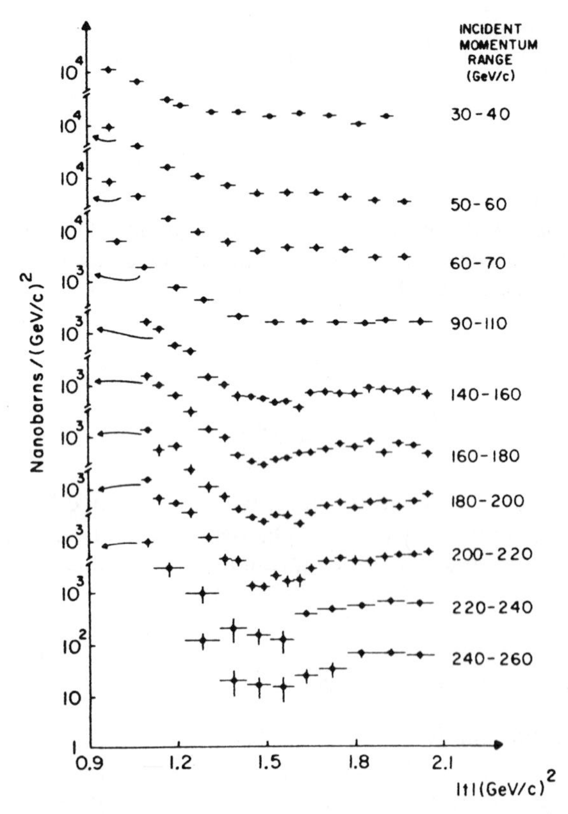

FIGURE 1.10. The differential cross section for proton–proton scattering at incident momenta from 30 to 260 GeV/c. [From R. Rusack *et al.*, *Phys. Rev. Lett.* **41,** 1632 (1978).]

In these units Planck's constant has the value

$$\hbar = \frac{h}{2\pi} = 6.58 \times 10^{-22}\ \text{MeV-sec} \tag{1.17a}$$

Furthermore, remembering that the speed of light is

$$c \simeq 3 \times 10^{10}\ \text{cm/sec}$$

we have

$$\hbar c \simeq 1.97 \times 10^{-11} \text{ MeV-cm}$$
$$\simeq 197 \text{ MeV-F} \tag{1.17b}$$

where 1 fermi (F) $= 10^{-13}$ cm $= 10^{-15}$ m $= 1$ fm.

To discover the experimental implications of proton scattering, we associate a plane wave with the incoming protons and think of the diffraction as due to scattering from the protons in the hydrogen target. This process is similar to the diffraction of light waves from a circular hole of radius R equal to the radius of the proton. The first minimum in this case occurs at an angle†

$$\theta \simeq \sin\theta = 0.61 \frac{\lambda}{R} \tag{1.18}$$

Since the proton is very relativistic we can neglect its rest mass, and hence

$$p = \frac{E}{c} = 200 \text{ GeV}/c \tag{1.19a}$$

On the other hand, we know

$$p \sin\theta = 1.2 \text{ GeV}/c \tag{1.19b}$$

Equations (1.19a) and (1.19b) yield a value for the angle

$$\theta \simeq \sin\theta = \frac{1.2}{200} = 6 \times 10^{-3} \text{ rad} \tag{1.20}$$

From the De Broglie relations, one obtains the value of the wavelength associated with the incoming protons

$$\lambda = \frac{2\pi}{k} = 2\pi \frac{\hbar}{p} = 2\pi \frac{\hbar c}{pc} = 2\pi \times \frac{197 \text{ MeV-F}}{200 \text{ GeV}}$$
$$= 2\pi \times 0.985 \times 10^{-16} \text{ cm} \tag{1.21}$$

† For a slit of width $2R$ the first diffraction minimum occurs at $\sin\theta = \lambda/2R$. The derivation for circular apertures is given, for example, in F. Lobkowicz and A. C. Melissinos, *Physics for Scientists and Engineers*, Vol. II, p. 79, W. B. Saunders, Philadelphia (1975).

Using this value of the wavelength and the value of θ from Eq. (1.20) we can determine the radius of the proton from Eq. (1.18) to be

$$\begin{aligned} R &= 0.61\frac{\lambda}{\theta} \\ &= 0.61\times\frac{2\pi\times 0.985\times 10^{-16}\text{ cm}}{6\times 10^{-3}} \\ &= 0.63\times 10^{-13}\text{ cm}\simeq 0.63\text{ F} \end{aligned} \tag{1.22}$$

This result is of the correct order of magnitude, the radius of the proton being $R\simeq 1$ F.

Diffraction of Electrons from a Thin film of Silver

Electrons of sufficient energy can traverse thin films of matter. If the film material is crystalline, it acts as a three-dimensional grating, but because of the random orientation of the crystalline domains the diffraction pattern consists of circles centered on the incident beam. Such diffraction patterns obtained with electrons of energy 24 and 37 keV are shown in Fig. 1.11.

$E = 28$ keV

$E = 37$ keV

FIGURE 1.11. Diffraction of electrons from a thin film of silver for incident kinetic energy $E = 28$ keV and $E = 37$ keV.

For a cubic lattice of spacing a, the diffraction angle θ is given by†

$$2a \sin\theta = \lambda(m^2+n^2+q^2)^{1/2} \tag{1.23}$$

where m, n, and q are integers. From the radius r of the diffraction ring and the distance s of the screen, the angle $\theta \simeq r/s$ has been plotted for 35-keV electrons in Fig. 1.12 against the various combinations of the integers m, n, and q. From the experimental data we obtain

$$\frac{\lambda}{2a} = 0.0078 \tag{1.24}$$

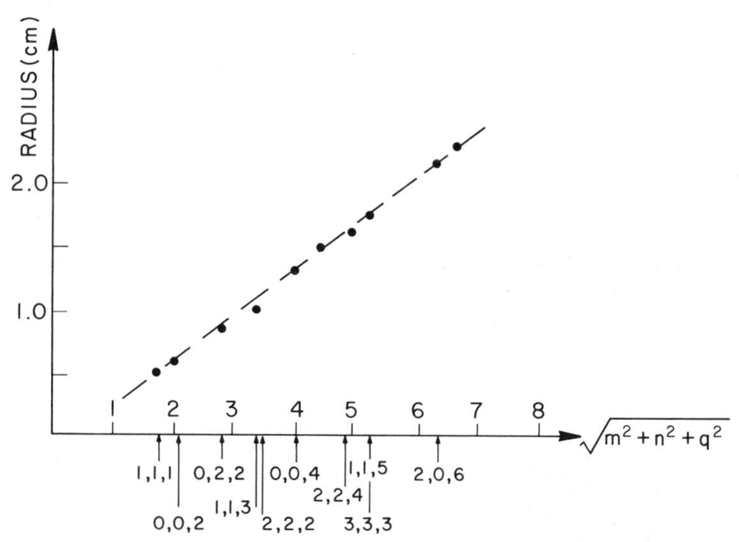

FIGURE 1.12. Plot of the diffraction ring radius versus the indices of the scattering plane for the data shown in Fig. 1.11.

To calculate the wavelength we note that the electrons are nonrelativistic. Hence, using

$$T = \frac{p^2}{2m_e} = 35 \text{ keV}$$

† See, for example, G. P. Harnwell and J. J. Livingood, *Experimental Atomic Physics*, p. 168, McGraw Hill, New York (1933).

we obtain

$$p^2 \simeq 2 \times 0.5\ \text{MeV}/c^2 \times 35\ \text{keV}$$
$$= 35 \times 10^{-3}\ (\text{MeV}/c)^2 \tag{1.25a}$$

Hence

$$p \simeq 0.19\ \text{MeV}/c \tag{1.25b}$$

Thus, the De Broglie relation gives the wavelength

$$\lambda = \frac{h}{p} = 2\pi\frac{\hbar}{p} = 2\pi\frac{\hbar c}{pc} = 2\pi \times \frac{197\ \text{MeV-F}}{0.19\ \text{MeV}}$$
$$\simeq 6.5 \times 10^3\ \text{F} = 0.065\ \text{Å} \tag{1.26}$$

Using Eq. (1.24) we can now determine the lattice spacing, which turns out to be

$$a = \frac{\lambda}{2 \times 0.0078} = \frac{0.065\ \text{Å}}{2 \times 0.0078} \simeq 4.17\ \text{Å} \tag{1.17}$$

This is in good agreement with the value obtained from X ray diffraction.

That electron diffraction patterns are similar to the diffraction patterns of waves can be seen in Fig. 1.13, which corresponds to the diffraction of X rays from a thin gold film. Since gold is crystalline the diffraction pattern (known as the Debye–Scherrer pattern) consists of circular rings centered on the beam direction. For the same accelerating voltage, X rays have longer wavelength than electrons since $\lambda = h/p = hc/E \simeq 0.35$ Å for $E = 35$ keV. Thus, the diffraction angles are correspondingly larger.

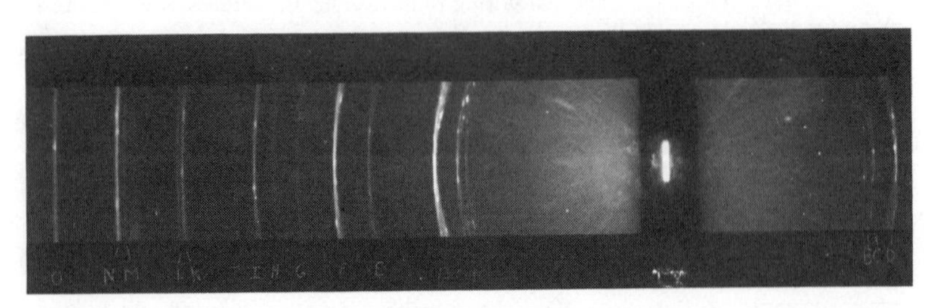

FIGURE 1.13. Debye–Scherrer pattern of X ray diffraction from a crystalline material.

Refraction of Electrons

When a light wave is incident on the interface between two regions with different refractive indices, it is bent in accordance with Snell's law [see Fig. 1.14(a)]

$$\frac{\sin \theta_1}{\sin \theta_2} = \frac{n_2}{n_1} \tag{1.28}$$

where n_1 and n_2, respectively are the refractive indices of the two media. The same phenomenon is observed with matter waves and can be demonstrated with an electron beam, as shown in Fig. 1.14(b). For light waves the refractive index is defined as the ratio of the velocity of light c to the velocity of propagation c' in the medium.

$$n = \frac{c}{c'} \tag{1.29a}$$

Furthermore, since the frequency ν does not change, the wavelength changes as

$$\lambda' = \frac{c'}{\nu} = \frac{c}{\nu}\frac{1}{n} = \frac{\lambda}{n} \tag{1.29b}$$

Thus

$$\frac{\sin \theta_1}{\sin \theta_2} = \frac{n_2}{n_1} = \frac{\lambda_1}{\lambda_2} \tag{1.30}$$

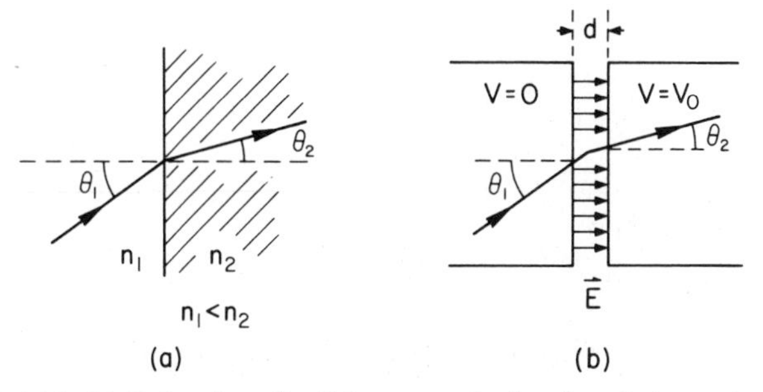

FIGURE 1.14. (a) Refraction of a light ray at the interface between two media of differing refractive index. (b) Refraction of an electron beam at the interface between two regions of space held at different potential.

For a beam of electrons, a region of differing refractive index is a region of differing electrostatic potential because

$$\lambda=\frac{h}{p}=\frac{h}{(2mT)^{1/2}}=\frac{h}{[2m(E-U)]^{1/2}}=\frac{h}{[2m(E+eV)]^{1/2}} \tag{1.31}$$

where T is the kinetic energy of the electrons and U is the potential energy (for example, in the region of an accelerating potential $U<0$). In Fig. 1.14(b) two regions with electric potential $V=0$ and $V=V_0$ are shown. If they are separated by a small distance d, the electric field between the two regions is

$$\mathscr{E}=\frac{V_0}{d} \tag{1.32}$$

and is directed as shown. This will accelerate the electrons in the x-direction and change their momentum according to

$$e\mathscr{E}=\frac{dp_x}{dt}=v_x\frac{dp_x}{dx}=\frac{p_x}{m}\frac{dp_x}{dx}$$

or

$$\int p_x\,dp_x=me\mathscr{E}\int dx$$

or

$$\frac{1}{2}((p_x')^2-p_x^2)=meV_0 \tag{1.33}$$

Here p_x and p_x' are the x-component of the momentum of the electron in the two regions. The angle of incidence and refraction are given, respectively, by

$$\sin\theta_1=\frac{p_y}{(p_x^2+p_y^2)^{1/2}}=\frac{p_y}{(2mT_1)^{1/2}} \tag{1.34a}$$

and

$$\sin\theta_2=\frac{p_y}{[(p_x')^2+p_y^2]^{1/2}}=\frac{p_y}{(p_x^2+p_y^2+2meV_0)^{1/2}}$$

$$=\frac{p_y}{[2m(T+eV_0)]^{1/2}}=\frac{p_y}{(2mT_2)^{1/2}} \tag{1.34b}$$

$$\frac{\sin\theta_1}{\sin\theta_2}=\frac{(2mT_2)^{1/2}}{(2mT_1)^{1/2}}=\frac{\lambda_1}{\lambda_2} \tag{1.35}$$

Here we have used Eq. (1.31) in the final equality. Thus, we obtain Snell's law.

What we have demonstrated here is that the introduction of matter waves to describe particles through the De Broglie relations is consistent with the expected motion of particles in a region of changing potential. The trajectories of particles are equivalent to the rays of the corresponding waves. In fact, the focusing elements (lenses) in an electron microscope consist of regions where the electrostatic potential (i.e., the refractive index) changes abruptly.

1.4. PROCESS OF MEASUREMENT

At this point we may ask what the matter waves really describe. To answer this we examine the double-slit experiment again.

We first consider the arrangement shown in Fig. 1.9 and consider, in particular, the case when the intensity of the incident light is very low. If we wait for long enough time so that a large number of photons have been accumulated, the familiar interference pattern will emerge in the observation plane. We cannot predict for any single photon where it will hit the observation plane. It certainly cannot strike at any one of the minima, but a hit at any other place is possible. In spite of our inability to predict exactly where the photon will arrive, we can calculate the probability that a photon will strike the observation plane at a particular position. The higher the probability of a hit at any point, the higher the observed intensity there. The same is true for an experiment involving electrons. The electrons arrive one at a time and are spread over the interference pattern. We designate the arrival of an electron at the point x as an event and assign a definite probability $P(x)$ that this event would occur.

In order to be able to account for the interference we further postulate that the probability $P(x)$ is the absolute square of the probability amplitude $\phi(x)$, i.e.,

$$P(x) = |\phi(x)|^2 = \phi^*(x)\phi(x) \tag{1.36}$$

The probability amplitude $\phi(x)$ is a complex function and is not an observable. It is only the probability that an event will occur that is observable. We would like to contrast this with the electric field of an electromagnetic wave or the density variations of a sound wave, etc., which are real functions and hence correspond to real physical observables. The necessity for requiring that the probability amplitude be complex will be discussed in Section 1.6.

In the experiment shown in Fig. 1.9, we assign the amplitude $\phi_1(x)$ and the probability $P_1(x)$ for the arrival of an electron at point x when only s_1 is open. Correspondingly, $\phi_2(x)$ and $P_2(x)$ refer to the case when only s_2 is open and $\phi_{12}(x)$ and $P_{12}(x)$ to the case when both slits are open. It follows then that

$$\begin{aligned} P_1(x) &= |\phi_1(x)|^2 \\ P_2(x) &= |\phi_2(x)|^2 \\ P_{12}(x) &= |\phi_{12}(x)|^2 \end{aligned} \tag{1.37}$$

In the case of the double-slit experiment it is observed that an interference pattern arises and that

$$\phi_{12}(x) = \phi_1(x) + \phi_2(x) \tag{1.38}$$

That is, the amplitude for the event is the sum (or superposition) of the amplitudes for the occurrence of the event due to all possible channels. This statement about the linear superposition of amplitudes in quantum mechanics is the first postulate of the theory. It follows from Eq. (1.38) that

$$P_{12}(x) = |\phi_{12}(x)|^2 = |\phi_1(x) + \phi_2(x)|^2 \tag{1.39}$$

which clearly indicates the possibility of interference if the complex functions $\phi(x)$ are wavelike.

The experiment can be modified by attaching a small counter next to each slit so that one can know exactly which slit the electron went through. After accumulating sufficient events we can plot all three functions $P_1'(x)$, $P_2'(x)$, and $P_{12}'(x)$, which represent the probabilities when the information about which slit the electron passed through is known. It is obvious then that

$$\begin{aligned} P_1'(x) &= P_1(x) = |\phi_1(x)|^2 \\ P_2'(x) &= P_2(x) = |\phi_2(x)|^2 \end{aligned} \tag{1.40a}$$

However,

$$P'_{12}(x) \neq P_{12}(x) \tag{1.40b}$$

In fact, $P'_{12}(x)$ must be the sum of $P'_1(x)$ and $P'_2(x)$ by definition since it represents the total sample of observed events. Thus

$$P'_{12}(x) = P'_1(x) + P'_2(x) = |\phi_1(x)|^2 + |\phi_2(x)|^2 \tag{1.40c}$$

and the interference pattern has disappeared.

Thus, we see that the process of measurement (i.e., the addition of counters next to the slits) has altered the outcome of the measurement; recall the Stern–Gerlach experiment, also. In fact, knowledge of the slit through which the electron passed has destroyed the interference pattern.

One concludes that Eq. (1.39) is valid or that, equivalently, interference can occur, only when a certain event can take place in two or more indistinguishable ways; in this case the amplitudes must be added. When a certain event occurs through distinguishable ways, it is the probabilities that are added.

From the discussion above it is clear that the wave function of a particle can be thought of as a probability amplitude describing the particle. Thus

$$P(x, t) = \phi^*(x, t)\phi(x, t) \tag{1.41}$$

describes the probability of finding the particle at point† x at time t.

We also note that the possibility of interference is related to the uncertainty Δx in the position of the electron at the plane of the slits. That is, the uncertainty in the momentum associated with the spreading of the interference or diffraction pattern arises because of the uncertainty in the position of the electron. This observation can be cast in quantitative form and is the second postulate of the theory known as the uncertainty principle

$$\Delta x \, \Delta p_x \geqslant \frac{\hbar}{2} \tag{1.42}$$

where Δp_x is the uncertainty in the x component of the momentum.

† Of course, where x is a continuous variable, it is $P(x, t)\,dx$ that represents the probability of finding the particle within the interval dx centered at x.

This inequality is an expression of Heisenberg's uncertainty relation, and it implies that the position and the momentum of a particle cannot be measured simultaneously with arbitrary accuracy.

The consequences of the uncertainty principle are not felt in our daily experience due to the small value of $\hbar$. As an example, consider a 1-mm-diam steel ball and assume that we know its position accurately up to $\Delta x = 1\ \mu\text{m} = 10^{-4}$ cm. The corresponding uncertainty in its momentum from Eq. (1.42) is

$$\begin{aligned} \Delta p_x &\geqslant \frac{\hbar}{2\,\Delta x} = \frac{\hbar c}{2\,\Delta x}\frac{1}{c} \\ &= \frac{197\ \text{MeV-F}}{2 \times 10^{-4}\ \text{cm}}\frac{1}{c} \\ &= 98.5 \times 10^4 \times 10^{-13}\ \text{MeV}/c \\ &\simeq 10^{-7}\ \text{MeV}/c \end{aligned} \tag{1.43a}$$

The density of the steel ball is $\rho \sim 8$ g/cm^3. Hence

$$\begin{aligned} m &= \frac{4\pi}{3}(0.5 \times 10^{-3}\ \text{m})^3 \times (8 \times 10^3\ \text{kg/m}^3) \\ &= 4 \times 10^{-6}\ \text{kg} \simeq 2.27 \times 10^{24}\ \text{MeV}/c^2 \end{aligned} \tag{1.43b}$$

The uncertainty in the velocity, therefore, is

$$\begin{aligned} \Delta v_x &= \frac{\Delta p_x}{m} = \frac{10^{-7}\ \text{MeV}/c}{2.27 \times 10^{24}\ \text{MeV}/c^2} \\ &\simeq 4 \times 10^{-32} c \\ &= 1.2 \times 10^{-21}\ \text{cm/sec} \end{aligned} \tag{1.43c}$$

Such a velocity would produce a displacement of the ball by 1 μm after a time interval of 10^9 years. Clearly, it is not possible to observe the effect of the uncertainty relation using a steel ball.

In contrast, for the electron in the ground state of the hydrogen atom, the kinetic energy T is of the order of 13.6 eV.

$$T = \frac{p^2}{2m_e} = 13.6\ \text{eV} \tag{1.44a}$$

Since

$$m_e \simeq 0.5\ \text{MeV}/c^2 \tag{1.44b}$$

we find

$$p \simeq 3.6 \times 10^{-3}\,\text{MeV}/c \tag{1.44c}$$

If we assume that the uncertainty in the momentum is of the order of the momentum itself, i.e., $\Delta p \simeq p$, then we obtain from Eq. (1.42)

$$\Delta x \geqslant \frac{\hbar}{2\,\Delta p} = \frac{\hbar c}{2 \times 3.6 \times 10^{-3}\,\text{MeV}}$$

$$= \frac{197\,\text{MeV-F}}{7.2 \times 10^{-3}\,\text{MeV}} \simeq 2.7 \times 10^{-9}\,\text{cm} \tag{1.44d}$$

which is of the same order of magnitude as the size of the hydrogen atom. Thus, the uncertainty principle tells us that we cannot speak with any precision of a particular position of the electron in the atom or of the orbit of the electron around the proton. It is for this reason that the methods of classical mechanics (which ignore the uncertainty principle) are applicable to macroscopic problems and that at the microscopic level one must use quantum mechanics.†

1.5. THE UNCERTAINTY PRINCIPLE

In Section 1.4 we introduced the uncertainty principle as a postulate of quantum mechanics. One form of the uncertainty relation was given in Eq. (1.42). We indicated there that the interference pattern was a consequence of the uncertainty as to which slit the photon or the electron passed through. We will show below that any wave automatically satisfies the uncertainty principle and that consequently the description of particles by probability waves is compatible with this postulate.

As an example, let us consider the diffraction of a plane wave through a slit of width a as shown in Fig. 1.15. If this is a probability wave, it is clear that the passage through the slit localizes the position of the particle along the Y-axis to within an uncertainty

† There are few cases of macroscopic problems, however, where it is necessary to use not classical but quantum theory.

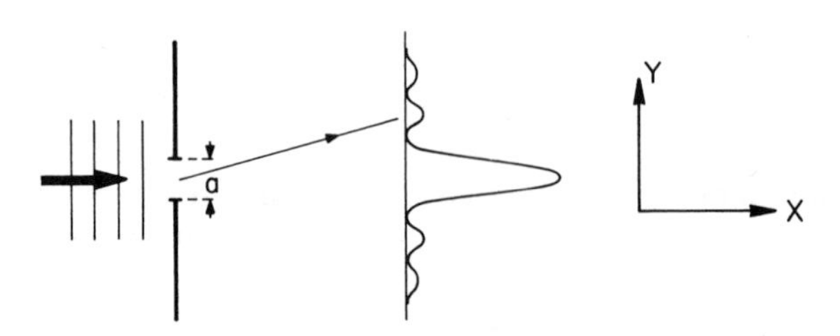

FIGURE 1.15. Diffraction of a plane wave through a single slit.

where

$$|\Delta y| = \frac{a}{2} \tag{1.45a}$$

The wave is diffracted, and the first minimum is found at an angle θ given by

$$\theta \simeq \sin\theta = \pm\frac{\lambda}{2a} \tag{1.45b}$$

Thus, the direction of motion of the particle is uncertain to within

$$|\Delta\theta| = \frac{\lambda}{a} = \frac{h}{p}\frac{1}{a} \tag{1.45c}$$

where we have used the De Broglie relations for the wavelength of the probability wave. Therefore the uncertainty in momentum in the y-direction is given by

$$|\Delta p_y| = p\,|\Delta\theta| = p\frac{h}{p}\frac{1}{a} = \frac{h}{a}$$

or

$$|\Delta y|\,|\Delta p_y| = \frac{h}{2} > \frac{\hbar}{2} \tag{1.45d}$$

Here we have used Eq. (1.45a).

As a second example, let us consider a plane wave incident on a grating of spacing d, as shown in Fig. 1.16(a). Narrow peaks will occur at angles θ which satisfy

$$d\sin\theta = m\lambda \tag{1.46a}$$

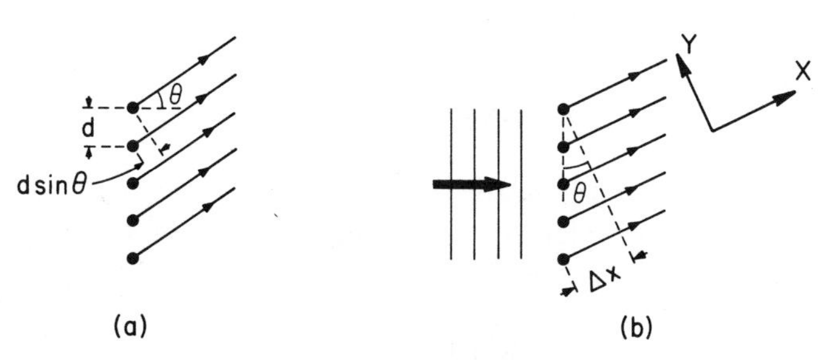

FIGURE 1.16. Diffraction of a plane wave from a grating.

where m is the order of the maximum. The resolution of a grating with N grooves is given by

$$\frac{\Delta\lambda}{\lambda}=\frac{1}{Nm} \tag{1.46b}$$

This resolution is achieved only if all the grooves in the grating participate in the diffraction process and, as can be seen from Fig. 1.16(b), this implies a path difference and hence an uncertainty Δx in the position of the wave fronts along the scattered direction given by $\Delta x = Nd \sin\theta = Nm\lambda$. Thus we have

$$\frac{\Delta\lambda}{\lambda}=\frac{1}{Nm}=\frac{\lambda}{\Delta x}$$
$$\left|\Delta\left(\frac{1}{\lambda}\right)\right|=\frac{1}{\Delta x} \tag{1.47a}$$

On the other hand, we know from the De Broglie relations that

$$\lambda=\frac{h}{p_x}$$
$$\left|\Delta\left(\frac{1}{\lambda}\right)\right|=\frac{|\Delta p_x|}{h} \tag{1.47b}$$

Combining the results of Eqs. (1.47a) and (1.47b) we have,

$$\frac{|\Delta p_x|}{h}=\frac{1}{\Delta x}$$

or

$$\Delta x\,|\Delta p_x| = h > \frac{\hbar}{2} \tag{1.47c}$$

This result is in accordance with the uncertainty principle.

We can also consider the uncertainty in the time required to establish a principal maximum. Let a wave front reach the plane of the grating at $t=0$. From Fig. 1.16(b) it is evident that the contribution from the last groove will reach the detector at a time

$$\Delta T = \frac{\Delta x}{c} = \frac{Nm\lambda}{c} \tag{1.48a}$$

after the contribution from the first groove. If we use this result in Eq. (1.47a) we have

$$\left|\Delta\left(\frac{1}{\lambda}\right)\right| = \frac{1}{\Delta x} = \frac{1}{c\,\Delta T} \tag{1.48b}$$

Remembering that

$$\lambda\nu = c$$

$$\left|\Delta\left(\frac{1}{\lambda}\right)\right| = \frac{|\Delta\nu|}{c} = \frac{1}{c\,\Delta t} \tag{1.48c}$$

$$|\Delta\nu|\,\Delta T = 1$$

leads to the result

$$|\Delta E|\,\Delta T = h > \frac{\hbar}{2} \tag{1.49}$$

which is another form of the uncertainty relation.

The relations that we have derived above are the well-known connections between a wave amplitude and its Fourier transform. For instance, a plane wave of fixed frequency cannot be localized in space. On the contrary, a sharp pulse that is well localized in space contains a broad spectrum of frequencies. A quantitative treatment of these questions is given in Appendix 1.

We would point out here the effect of the measurement process on the particle. As shown in Fig. 1.15, before the plane wave reaches the slit we know its momentum and hence its direction of propagation exactly, but it has an infinite extent in the y-direction, i.e., the position is infinitely uncertain. After passing through the slit, however, the position is known, but the information on the direction of propagation has been broadened (made uncertain) due to diffraction.

As an example of the application of the uncertainty principle we now calculate the lowest energy state of the hydrogen atom. Since the electron is bound to the proton its total energy $E = T + U$ must be negative. We assume that the electron can be found in any one of a discrete set of energy states, as shown in Fig. 1.17. Since $U < 0$ for a bound state, $T < |U|$. Let us denote by E_0 the energy of the lowest state.

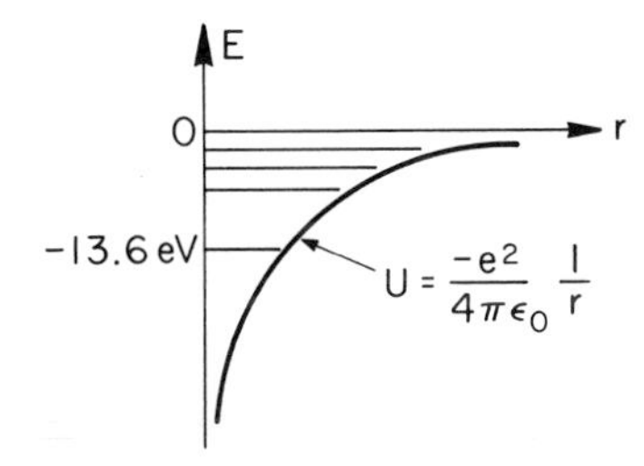

FIGURE 1.17. Discrete energy levels of an electron bound in a Coulomb potential.

If the size of the atom in this state is a we can set

$$\Delta r \simeq a \tag{1.50a}$$

and if the uncertainty in the momentum is the maximum possible

$$\Delta p \simeq p \tag{1.50b}$$

then from the uncertainty relation we have

$$\Delta p \, \Delta r \simeq pa \simeq \hbar \tag{1.50c}$$

The kinetic energy is given by

$$\frac{p^2}{2m} = T = \frac{\hbar^2}{2ma^2} \tag{1.51a}$$

The potential energy is of the order of

$$U = -\frac{e^2}{4\pi\varepsilon_0}\frac{1}{a}$$

where ε_0 is the dielectric constant of the vacuum. Thus

$$E = T + U$$

$$= \frac{\hbar^2}{2ma^2} - \frac{e^2}{4\pi\varepsilon_0}\frac{1}{a} \tag{1.51a}$$

Since we seek the state of the lowest energy we must minimize Eq. (1.51*b*). Thus

$$\frac{dE}{da} = 0 = -\frac{\hbar^2}{ma^3} + \frac{e^2}{4\pi\varepsilon_0}\frac{1}{a^2}$$

or

$$a = 4\pi\varepsilon_0 \frac{\hbar^2}{me^2} \tag{1.51c}$$

Substituting this value of the radius into Eq. (1.51*b*), the ground-state energy turns out to be

$$E = \frac{\hbar^2}{2m}\left(\frac{me^2}{4\pi\varepsilon_0\hbar^2}\right)^2 - \frac{e^2}{4\pi\varepsilon_0}\frac{me^2}{4\pi\varepsilon_0\hbar^2}$$

$$= -\frac{1}{2}\frac{e^4}{(4\pi\varepsilon_0)^2}\frac{m}{\hbar^2} = -13.6\ \text{eV} \tag{1.52}$$

The result of Eq. (1.52) is the true value of the lowest energy state of the hydrogen atom. However, the assumptions made in the calculation are only qualitative. In particular, we have ignored the fact that the position and the momentum of the electron are three-dimensional quantities. The important conclusion of this exercise is that the uncertainty principle prevents the hydrogen atom from collapsing. This is because if the value of a for the atom becomes smaller, the momentum must increase. The kinetic energy will increase as $1/a^2$, whereas the attractive potential energy will change only as $-1/a$, and soon the total energy will become positive allowing the electron to escape from the atom. The strength of the electromagnetic interaction $e^2/4\pi\varepsilon_0$, coupled with the uncertainty principle, determines the size of atoms.

1.6. WAVE PACKETS

We have seen that consideration of probability waves allows us to explain a variety of observed phenomena, in particular the diffraction of particle beams. We also have seen that a probability amplitude must be complex. A complex wave along the X-axis is written as

$$\phi(x, t) = Ae^{-i(\omega t - kx)} = Ae^{-(i/\hbar)(Et - px)} \tag{1.53a}$$

where the minus sign in the exponential is a convention adopted in quantum mechanics. Such a wave is also called a plane wave and describes a particle of momentum

$$p = \hbar k \tag{1.53b}$$

Thus, the value of the momentum is well defined.

We note that $|\phi(x, t)|^2$ determines the probability (an observable) of finding the particle at time t at position x, and for the plane wave we see that

$$|\phi(x, t)|^2 = |A|^2 \tag{1.53c}$$

which is independent of x and t. That is, the probability of finding the particle anywhere in space is always the same. In other words, the uncertainty in the position of the particle is infinite. This is reasonable because for well-defined momentum ($\Delta p = 0$) the uncertainty principle requires that $\Delta x = \infty$.

This example also clarifies the assertion made earlier that a probability amplitude must be complex. This is because, if we had chosen a real form for the probability amplitude, say

$$\chi(x, t) = A \cos(\omega t - kx) \tag{1.54a}$$

then

$$|\chi(x, t)|^2 = |A|^2 \cos^2(\omega t - kx) \tag{1.54b}$$

As shown in Fig. 1.18(a), the function $\cos^2(\omega t - kx)$ oscillates between 0 and 1, implying that there are points in space where the probability of finding the particle is zero. This violates the uncertainty principle and, in fact, no combination of real functions can give uniform probability for a wave of well-defined momentum as is required by the uncertainty principle.

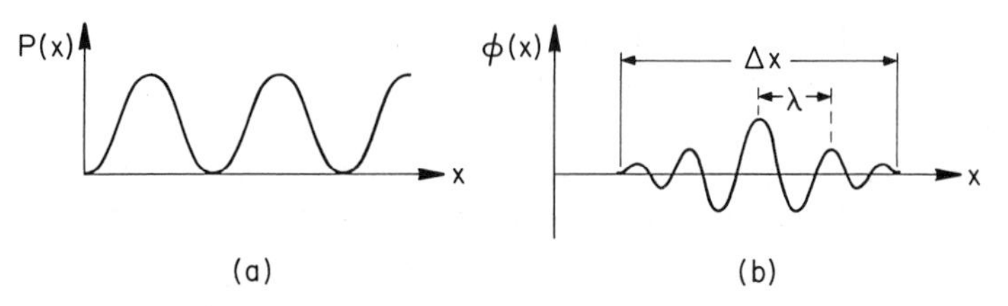

FIGURE 1.18. (a) The probability function $\cos^2(\omega t - kx)$. (b) The amplitude for a wave packet.

A particle, on the other hand, is fairly well localized in space. Therefore, our next task is to find a probability amplitude that leads to a localized probability in a particular region of space. We know that such a wave will contain several values of k since, by the uncertainty principle, p (and hence k) can no longer be well defined. Such a wave, commonly called a wave packet (or more precisely a space packet), is shown in Fig. 1.18(b) and can be constructed by superimposing several harmonic waves of different wave number k.

Let the spatial extent of the wave packet in Fig. 1.18(b) be Δx. The wavelength λ is no longer defined precisely but if the packet contains N wiggles it will hold approximately

$$\lambda = \frac{\Delta x}{N}$$

or

$$k = \frac{2\pi}{\lambda} = 2\pi \frac{N}{\Delta x} \tag{1.55a}$$

As the wiggles die out, it becomes difficult to determine N, and we can assign an uncertainty of $\Delta N = 1$ oscillations to the determination of N. Thus

$$\Delta k \simeq 2\pi \frac{\Delta N}{\Delta x} = \frac{2\pi}{\Delta x} \tag{1.55b}$$

Since $k = p/\hbar$ we have

$$\Delta k = \frac{\Delta p}{\hbar} = \frac{2\pi}{\Delta x}$$

or

$$\Delta p \, \Delta x = 2\pi\hbar = h \tag{1.55c}$$

which is again a statement of the uncertainty principle. The sharper we make the wave packet in space, the broader will be the momentum range of the particle, and vice versa.

To see how we can construct a space packet mathematically, consider the superposition of two plane waves with wavenumbers k_1, k_2 differing only slightly from k_0. Thus

$$\phi(x, t) = A_1 e^{-i(\omega_1 t - k_1 x)} + A_2 e^{-i(\omega_2 t - k_2 x)} \tag{1.56a}$$

If we consider this wave at a fixed time t and for simplicity set $t = 0$, then

$$\begin{aligned}\phi(x, t) &= A_1 e^{ik_1 x} + A_2 e^{ik_2 x} \\ &= e^{i/2(k_1+k_2)x}(A_1 e^{i/2(k_1-k_2)x} + A_2 e^{-i/2(k_1-k_2)x})\end{aligned} \tag{1.56b}$$

and if $A_1 = A_2 = A$

$$\phi(x, t) = e^{i/2(k_1+k_2)x} 2A \cos[(k_1 - k_2)x] \tag{1.56c}$$

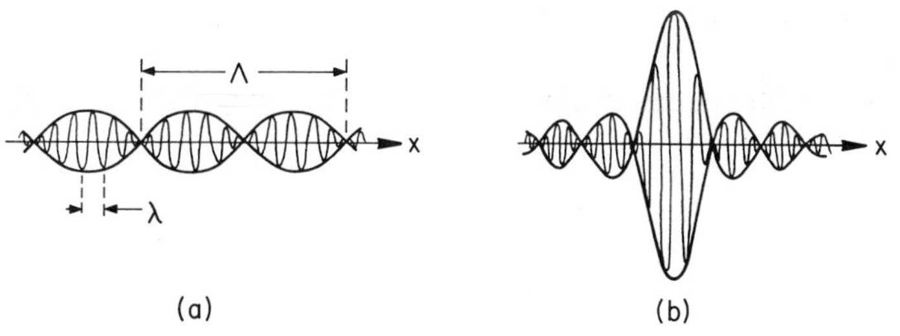

FIGURE 1.19. Superposition of waves. (a) Two waves of slightly differing frequency. (b) Several waves with a spread in frequency result in a pattern localized in space.

We see that the result of superimposing two waves is a wave modulated in space†, as shown in Fig. 1.19(a). This pattern will propagate along the X-axis as time increases. If we superimpose more and more waves with wave numbers k near k_0, and of appropriate amplitude, an isolated packet will result as shown in Fig. 1.19(b). We designate the space packet by

$$G(x, t) = \sum_n A_n e^{-i(\omega_n t - k_n x)} \tag{1.57a}$$

† When two harmonic waves of slightly differing frequencies are superimposed, then at a fixed point in space we obtain a *beating* or time packet.

where the wave number k_n and the angular frequency ω_n are near the central values k_0 and ω_0, and A_n is the amplitude of the nth wave. Equation (1.57a) represents a Fourier series, and we can construct any desired space packet by an appropriate choice of the coefficients A_n. In the limit that the wave numbers k_n are continuous, the summation in Eq. (1.57a) is replaced by an integral

$$G(x, t) = \int_{-\infty}^{\infty} dk A(k) e^{-i(\omega t - kx)} \tag{1.57b}$$

where the amplitudes $A(k)$ of the superimposed waves are a continuous function of the wave numbers as shown in Fig. 1.20 with the maximum support at the central value k_0. We note here that the angular frequency ω is related to the wave number through the De Broglie relations such that

$$\omega = f(k) \tag{1.58}$$

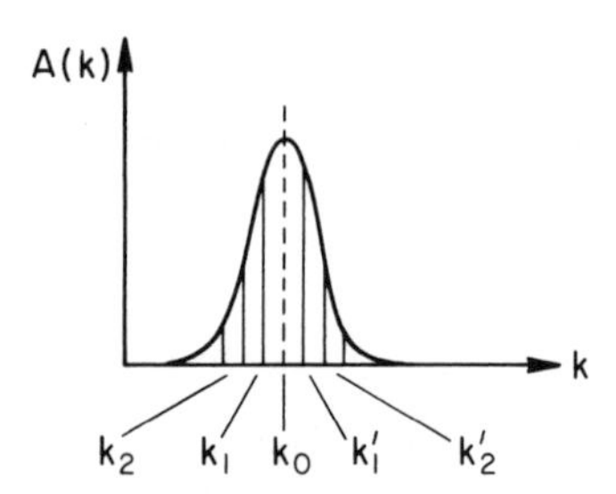

FIGURE 1.20. The distribution in frequency of a wave packet plotted versus the wavenumber k.

If $A(k)$ and $\omega = f(k)$ are given, then it is possible to determine $G(x, t)$ from Eq. (1.57b). The converse is true, namely, if $G(x, t)$ is given at all space–time points, then we can determine $A(k)$. In fact, we define a momentum wave packet as

$$\tilde{\phi}(k, t) = \frac{1}{(2\pi)^{1/2}} e^{-i\omega t} A(k) = \frac{1}{(2\pi)^{1/2}} \int dx G(x, t) e^{-ikx} \tag{1.59}$$

We recognize that this is simply the Fourier transform (see Appendix 1) of the space packet $G(x, t) = \phi(x, t)$ and serves as another mathematical representation of the probability amplitude describing the occurrence of the event; $\tilde{\phi}(k, t)$ is referred to as the "wave function in momentum space." In problems where our interest lies in states of well-defined momentum, it is easier to obtain results using $\tilde{\phi}(k, t)$ rather than $\phi(x, t)$.

Example

To better understand the formalism we consider the following space packet, which in momentum space contains all wave numbers between k_1 and k_2 with equal amplitude, as shown in Fig. 1.21(a). Thus

$$\begin{aligned} A(k) &= 1 \qquad k_1 \leq k \leq k_2 \\ &= 0 \qquad k < k_1 \quad \text{and} \quad k > k_2 \end{aligned} \tag{1.60a}$$

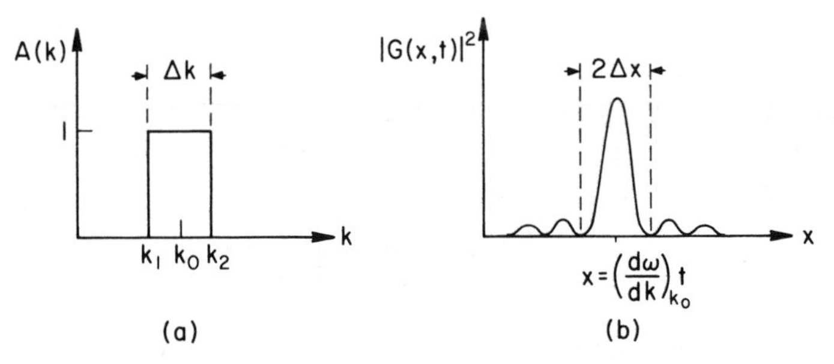

FIGURE 1.21. (a) A wave packet with a square-pulse distribution in frequency. (b) The corresponding probability of finding the packet at the position x at time t.

If k_1 and k_2 do not differ much from $k_0 = (k_1 + k_2)/2$ we can expand $\omega = f(k)$ to first order† and write

$$\omega = f(k) = \omega_0 + (k - k_0)\left(\frac{df}{dk}\right)_{k_0} + \cdots \tag{1.60b}$$

† For a more general case see Problem 1.13.

The space packet expression then becomes

$$G(x, t) = \int_{-\infty}^{\infty} dkA(k)e^{-i(\omega t - kx)}$$

$$= \int_{k_1}^{k_2} dk \exp\{-i[\omega_0 t + (k - k_0)(df/dk)_{k_0} t - kx]\}$$

$$= \exp[-i(\omega_0 t - k_0 x)] \int_{k_1}^{k_2} dk \exp\{-i(k - k_0)[(df/dk)_{k_0} t - x)]\} \tag{1.61a}$$

Defining $z = (df/dk)_{k_0} t - x$, Eq. (1.61a) gives

$$G(x, t) = e^{i(\omega_0 t - k_0 x)} \int_{k_1}^{k_2} dke^{-i(k-k_0)z}$$

$$= e^{-i(\omega_0 t - k_0 x)} \frac{1}{-iz}(e^{-i(k_2-k_0)z} - e^{-i(k_1-k_0)z})$$

$$= e^{-i(\omega_0 t - k_0 x)} \frac{\sin \Delta kz/2}{z/2} \tag{1.61b}$$

where $\Delta k = k_2 - k_1$. This result is the familiar diffraction amplitude which peaks when $z = 0$. The probability of finding the particle at position x at time t is [see Fig. 1.21(b)]

$$|G(x, t)|^2 = \frac{4 \sin^2\{\Delta k/2[(df/dk)_{k_0} t - x]\}}{\{\Delta k[(df/dk)_{k_0} t - x]\}^2} (\Delta k)^2 \tag{1.62}$$

Note that the envelope of the space packet is a wave since it is a function of the form $f(v_g t - x)$ with

$$v_g = \left(\frac{df}{dk}\right)_{k_0} = \left(\frac{d\omega}{dk}\right)_{k_0} \tag{1.63a}$$

It propagates along the x-axis with velocity v_g and is localized at any time at $x = v_g t$. (The peak is always located at $z = v_g t - x = 0$.) The expression $(d\omega/dk)_{k_0} = v_g$ is called the group velocity and determines how fast a packet (i.e., an observable signal) propagates. This differs from the phase velocity of the probability wave which is defined from the De Broglie relations to be

$$v_p = \frac{\omega}{k} = \frac{E}{p} \tag{1.63b}$$

v_p is not always precisely defined and sometimes can even be larger than c, the speed of light. However, this is all right since the phase velocity is not an observable. It is the group velocity that can be measured experimentally and which has to be less than the speed of light.

The group velocity of a wave packet can also be obtained from the following argument. The shape of the space packet is caused by the interference of waves with different k. Their amplitudes are arranged so that they interfere destructively everywhere in space except at the position $x(t)$ of the packet. If this condition is to be maintained, then the relation between the phases of waves with different k must also be maintained. In other words, the phase

$$\delta(k) = \omega(k)t - kx$$

must be stationary:

$$\begin{aligned} \frac{d\delta(k)}{dk} &= 0 = \frac{d\omega(k)}{dk}t - x \\ x &= \frac{d\omega}{dk}t = v_g t \end{aligned} \tag{1.64}$$

We note that the width of the packet in Fig. 1.21(b) is inversely proportional to Δk. Consider the first minimum in Eq. (1.62). It occurs when

$$\begin{aligned} \frac{\Delta kz}{2} &= \pi \\ v_g t - x &= \frac{2\pi}{\Delta k} \end{aligned} \tag{1.65a}$$

When Δx is defined as in Fig. 1.21(b), we have

$$\Delta x = \frac{2\pi}{\Delta k}$$

and hence

$$\Delta x\, \Delta p = 2\pi\hbar = h > \frac{\hbar}{2} \tag{1.65b}$$

which is in agreement with the uncertainty principle.

1.7. CORRESPONDENCE WITH CLASSICAL PHYSICS

We have seen that a particle can be described by a space packet that is fairly localized in space. It also contains a spread of wave numbers k around some central value k_0. Because of the uncertainty principle we can no longer talk of a well-defined coordinate and momentum for the particle. However, we can calculate the average value of the coordinate and momentum associated with the space packet or the particle denoted, respectively, by $\langle x \rangle$ and $\langle p \rangle$. The meaning of these quantities will be discussed in the following chapters. Here we simply explain how they are calculated in quantum mechanics.

We stated earlier that the absolute square of the probability amplitude $|\phi(x, t)|^2$ denotes the probability of finding a particle at point x at time t. However, since x is a continuous variable, it is more meaningful to talk about a quantity

$$P(x, t)\, dx = |\phi(x, t)|^2\, dx \tag{1.66a}$$

which simply gives the probability of finding a particle in the interval dx around point x at time t.

Since the particle must be somewhere in space at any given instant of time, the sum of all probabilities must equal one. In other words

$$\int_{-\infty}^{\infty} P(x, t)\, dx = \int_{-\infty}^{\infty} |\phi(x, t)|^2\, dx = 1 \tag{1.66b}$$

Equation (1.66) is called the *normalization* condition. It is clear that the plane waves defined as

$$\phi(x, t) = Ae^{-i(\omega t - kx)} \tag{1.67a}$$

appear to be non-normalizable since

$$\int_{-\infty}^{\infty} |\phi(x, t)|^2\, dx = \int_{-\infty}^{\infty} |A|^2\, dx \to \infty \tag{1.67b}$$

However, the simple form of these functions allows us to normalize them to the Dirac delta function, which we will discuss in Appendix 2. An alternate normalization is the box normalization where we take the physical space to have a finite but large extent. We discuss the box normalization and the Dirac normalization of the plane waves as well as the wave packet normalization in Appendix 3.

The momentum wave packet represents a probability amplitude. In fact,

$$P(k, t)\, dk = |\tilde{\phi}(k, t)|^2\, dk \tag{1.68a}$$

gives the probability of finding a particle with momentum in the interval dk around k at time t. Since the probability of finding the particle with any momentum must be unity we have

$$\int_{-\infty}^{\infty} P(k, t)\, dk = \int_{-\infty}^{\infty} |\tilde{\phi}(k, t)|^2\, dk = 1 \tag{1.68b}$$

It is only when these normalization conditions given by Eqs. (1.66*b*) and (1.68*b*) are satisfied that we can talk of the corresponding wave functions as probability amplitudes. Given any wave function we can always multiply it by a constant such that one of the above normalization conditions holds. In fact, we do this in Appendix 3 for the example of the wave packet we worked out earlier. We simply give the results here. The normalized amplitudes are

$$\phi(x, t) = \frac{1}{(2\pi\, \Delta k)^{1/2}}\, e^{-i(\omega_0 t - k_0 x)} \left\{ \frac{2 \sin[\Delta k/2(x - v_g t)]}{x - v_g t} \right\} \tag{1.69a}$$

$$\begin{aligned} \tilde{\phi}(k, t) &= \frac{1}{(\Delta k)^{1/2}}\, e^{-i\omega t} \qquad && k_1 < k < k_2 \\ &= 0 && k < k_1 \quad \text{and} \quad k > k_2 \end{aligned} \tag{1.69b}$$

Here, as defined before,

$$\begin{aligned} \Delta k &= k_2 - k_1 \\ k_0 &= \frac{k_1 + k_2}{2} \\ v_g &= \left(\frac{d\omega}{dk}\right)_{k_0} \end{aligned} \tag{1.69c}$$

Once the normalized wave amplitudes are known, then the average values of the coordinate and momenta are defined as

$$\begin{aligned} \langle x \rangle &= \int_{-\infty}^{\infty} xP(x, t)\, dx = \int_{-\infty}^{\infty} x\, |\phi(x, t)|^2\, dx \\ \langle p \rangle &= \hbar \langle k \rangle = \hbar \int_{-\infty}^{\infty} kP(k, t)\, dk = \hbar \int_{-\infty}^{\infty} k\, |\tilde{\phi}(k, t)|^2\, dk \end{aligned} \tag{1.70}$$

Armed with these ideas, we calculate $\langle x\rangle$ and $\langle p\rangle$ for the wave packet we discussed earlier:

$$\langle x\rangle = \int_{\infty}^{\infty} x \frac{1}{2\pi\,\Delta k} \frac{4\sin^2[\Delta k/2(x-v_g t)]}{(x-v_g t)^2}\,dx$$

$$\langle x\rangle - v_g t = \frac{4}{2\pi\,\Delta k}\int_{-\infty}^{\infty} \frac{\sin^2[\Delta k/2(x-v_g t)]}{(x-v_g t)^2}(x-v_g t)\,dx$$

$$= \frac{4}{2\pi\,\Delta k}\int_{-\infty}^{\infty} \frac{\sin^2[\Delta k/2(x-v_g t)]}{x-v_g t}\,d(x-v_g t)$$

$$= 0 \tag{1.71a}$$

which follows† from the fact that the integrand is odd in the integration variable $z = x - v_g t$. We therefore have

$$\langle x\rangle = v_g t \tag{1.71b}$$

which, as we have seen before, simply states that the peak of the space packet propagates with the group velocity v_g.

Similarly, the average momentum of the wave packet is calculated, and

$$\langle p\rangle = \hbar\int_{-\infty}^{\infty} k\,|\tilde{\phi}(k,t)|^2\,dk$$

$$= \hbar\int_{k_1}^{k_2} k\frac{1}{\Delta k}\,dk = \frac{\hbar}{k_2-k_1}\int_{k_1}^{k_2} k\,dk = \hbar\frac{k_1+k_2}{2}$$

$$= \hbar k_0 \tag{1.71c}$$

as expected.

Let us assume that the particle is a free particle, i.e., that it only has kinetic energy. Then

$$\omega = \frac{E}{\hbar} = \frac{1}{\hbar}\frac{p^2}{2m} = \frac{1}{\hbar}\frac{\hbar^2 k^2}{2m} = \frac{\hbar k^2}{2m} \tag{1.72a}$$

† We have used the fact that $v_g t$ is constant for a narrow space packet and therefore can be introduced under the integral as well as the fact that $\phi(x, t)$ is normalized.

and hence†

$$v_g = \left(\frac{d\omega}{dk}\right)_{k_0} = \frac{\hbar k_0}{m} = \frac{\langle p \rangle}{m} = \langle v \rangle \tag{1.72b}$$

This states simply that the group velocity is nothing other than the average velocity associated with the wave packet. If we use this fact, then Eq. (1.71*b*) becomes

$$\langle x \rangle = \langle v \rangle t \tag{1.73a}$$

This is similar to the classical laws of physics which say that

$$x = vt \tag{1.73b}$$

when no external forces are acting on the particle.

Equation (1.73*a*) represents a general connection between classical and quantum mechanics known as the *correspondence principle* which says that the average values of physical observables calculated in quantum mechanics obey the laws of classical mechanics.

This is also observed in the Stern–Gerlach experiment. The average value of the deflection in the Stern–Gerlach experiment is zero as would be expected from classical mechanics if randomly oriented magnetic dipoles were subjected to an inhomogeneous magnetic field.

1.8. APPLICATIONS: PARTICLE IN A BOX

Let us consider a physical situation where the potential energy $U = 0$ for $-L < x < L$, and $U = \infty$ for $x \leqslant -L$ and $x \geqslant L$, as shown in Fig. 1.22(a). We wish to obtain the amplitude $\psi(x, t)$ for finding a

† The same result is obtained in the relativistic case

$$\frac{d\omega}{dk} = \frac{dE}{dp} = \frac{d}{dp}(p^2c^2 + m^2c^4)^{1/2} = \frac{pc^2}{(p^2c^2 + m^2c^4)^{1/2}} = \frac{pc^2}{E} = v$$

where the last equality follows from

$$E = \frac{mc^2}{[1 - (v/c)^2]^{1/2}} \quad \text{and} \quad p = \frac{mv}{[1 - (v/c)^2]^{1/2}}$$

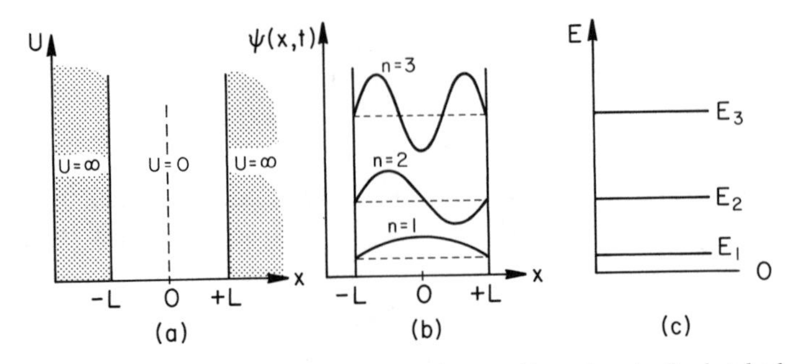

FIGURE 1.22. (a) An infinite potential well in one dimension; in the hatched area the potential is infinite, whereas inside the well it is zero. (b) The probability amplitude for different states in the well is given by standing waves. (c) The energy levels for the one-dimensional, infinitely deep potential well.

particle at the position x. In the region $-L<x<L$ the particle is free so that $U=0$ and $E=T$. Therefore, the amplitude will be a superposition of harmonic waves with ω and k given by the De Broglie relations

$$\omega=\frac{E}{\hbar} \tag{1.74a}$$

and

$$\begin{aligned} E&=\frac{p^2}{2m}=\frac{\hbar^2k^2}{2m} \\ k&=\pm\frac{1}{\hbar}(2mE)^{1/2} \end{aligned} \tag{1.74b}$$

Note that the possibility of k becoming positive or negative simply reflects the fact that the wave could be travelling either along the $+x$ or $-x$ axis.

In the region $x<-L$ or $x>L$, the potential energy is infinite and, therefore, for any given energy E the kinetic energy $T=E-U$ is infinitely negative. The particle cannot penetrate into a region of infinite potential energy and, therefore, the probability of finding the particle at $x\leqslant -L$ and $x\geqslant L$ must vanish. We conclude that

$$\psi(x,t)=0 \quad \text{for} \quad x=\pm L \tag{1.75a}$$

In other words, $\psi(x, t)$ must be a standing wave. There are two such independent waves, valid in the interval $-L \leqslant x \leqslant L$

$$\psi_e(x, t) = Ae^{-i\omega t} \cos kx = \frac{A}{2}(e^{-i(\omega t - kx)} + e^{-i(\omega t + kx)}) \quad (1.75b)$$

$$\psi_0(x, t) = Be^{-i\omega t} \sin kx = \frac{B}{2i}(e^{-i(\omega t - kx)} - e^{-i(\omega t + kx)}) \quad (1.75c)$$

As the subscripts denote, these are even and odd solutions in x. The form of the solutions reflects that standing waves result from the superposition of waves travelling toward $+x$ and $-x$.

We note that the boundary condition $\psi(\pm L, t) = 0$ can be satisfied only if k takes on discrete values. Thus, for the even solutions

$$\begin{aligned} kL &= (2n+1)\frac{\pi}{2} \\ k &= (2n+1)\frac{\pi}{2L} \qquad n = 0, 1, 2, \ldots \end{aligned} \quad (1.76a)$$

Similarly, for the odd solutions

$$\begin{aligned} kL &= n\pi \\ k &= 2n\frac{\pi}{2L} \qquad n = 1, 2, 3, \ldots \end{aligned} \quad (1.76b)$$

We can combine both results by writing

$$k_n = n\frac{\pi}{2L} \qquad n = 1, 2, 3, \ldots \quad (1.76c)$$

where we must recognize that the odd values of n correspond to the even solution, whereas the even values of n refer to the odd solution. The probability waves corresponding to $n = 1, 2$, and 3, are shown in Fig. 1.22(b).

It follows from Eq. (1.76c) that the energy of the particle inside the well also takes on discrete values given by

$$E_n = \frac{\hbar^2 k_n^2}{2m} = \frac{\hbar^2 \pi^2}{8mL^2} n^2 \quad (1.77)$$

The concept of probability waves has led us to discover that when a particle is confined in a region of space, its energy cannot have any arbitrary value but rather only certain discrete values. That is, the energy is *quantized.* Figure 1.22(c) is an energy level diagram where the energy E_n of the particle is shown for $n = 1, 2$, and 3 (note the similarities with Fig. 1.2).

A physical situation approximating the potential of Fig. 1.22(a) exists in the motion of free electrons in a conductor. Inside the conductor, the electrons are free, but they cannot leave the conductor. However, in contrast to the one-dimensional example we discussed the electrons can move in three dimensions. In this case we must specify all three components of the wave vector **k**. The probability wave is of the form

$$\psi(x, t) = A\left\{\begin{matrix}\sin\\ \cos\end{matrix}(k_x x)\right\}\left\{\begin{matrix}\sin\\ \cos\end{matrix}(k_y y)\right\}\left\{\begin{matrix}\sin\\ \cos\end{matrix}(k_z z)\right\}e^{-i\omega t} \qquad (1.78a)$$

and each component of the wave vector must satisfy a quantization condition similar to Eq. (1.76*c*). That is,

$$k_x = n_x \frac{\pi}{2L_x}$$

$$k_y = n_y \frac{\pi}{2L_y} \qquad n_x, n_y, n_z = 1, 2, 3, \ldots \qquad (1.78b)$$

$$k_z = n_z \frac{\pi}{2L_z}$$

For simplicity, we assume that the confining region is a cube of side $2L$ (i.e., $L_x = L_y = L_z = L$), and we say that the particle is in a box. It follows that the energy levels in this case are given by

$$\begin{aligned} E_{n_x,n_y,n_z} &= \frac{\hbar^2}{2m}(k_x^2 + k_y^2 + k_z^2) \\ &= \frac{\hbar^2\pi^2}{8mL^2}(n_x^2 + n_y^2 + n_z^2) \end{aligned} \qquad (1.78c)$$

Note that different combinations of the integers n_x, n_y, and n_z can give rise to the same energy. That is, in this case the probability waves can be degenerate in energy. For example, the combinations $(n_x = 1, n_y = 1, n_z = 2)$, $(n_x = 1, n_y = 2, n_z = 1)$, and $(n_x = 2, n_y = 1, n_z = 1)$ represent distinct probability waves but all with the same energy. The energy level diagram for the three-dimensional case is

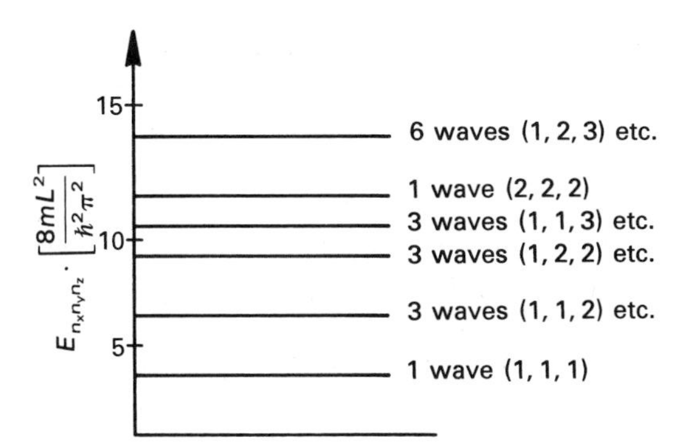

FIGURE 1.23. The energy levels for the three-dimensional, rectangular, infinitely deep potential well.

shown in Fig. 1.23, and we emphasize that the spectrum is quite different from the one-dimensional case shown in Fig. 1.22(c).

To find the energy of an electron in the conductor we must know n_x, n_y, and n_z; however, because of the Pauli principle (see Chapter 3) only two electrons can have the same combination of n_x, n_y, and n_z. Furthermore, the most stable configuration is one of lowest energy so that the lower energy states of the system are occupied first. Therefore, the highest energy of an electron (known as the Fermi energy E_F) is obtained when the sum of all combinations of n_x, n_y, and n_z equals one half the number of all free electrons present in the conductor. The integers n become very large and we will first consider n_x and n_y as shown in Fig. 1.24(a). Each dot

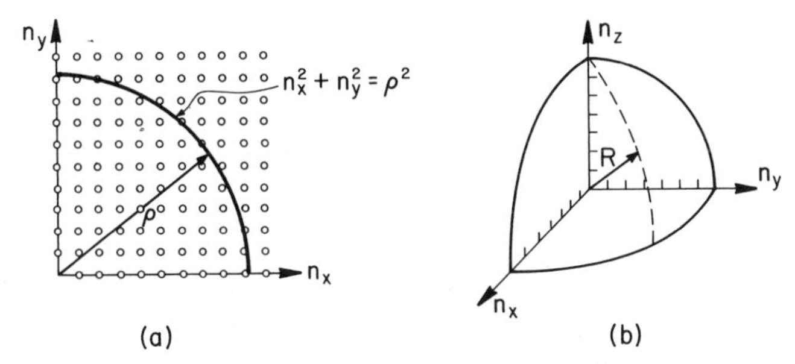

FIGURE 1.24. (a) The number of combinations of the squares of two integers. (b) Extension to three integers.

represents a particular combination of n_x and n_y and the number of combinations N_c contained in a radius $\rho = (n_x^2 + n_y^2)^{1/2}$ is

$$N_c = \tfrac{1}{4}\pi\rho^2 = \frac{\pi}{4}(n_x^2 + n_y^2) \tag{1.79a}$$

The factor of 1/4 is required because n_x and n_y are always positive. Extending this argument to three dimensions [Fig. 1.24(b)] we find that

$$N_c = \frac{1}{8}\left(\frac{4}{3}\pi R^3\right) = \frac{\pi}{6}(n_x^2 + n_y^2 + n_z^2)^{3/2} \tag{1.79b}$$

and the number of free electrons is $N_e = 2N_c$.

We can now introduce Eq. (1.79*b*) into Eq. (1.78*c*) to obtain the Fermi energy

$$E_F = \frac{\hbar^2\pi^2}{8mL^2}\left(\frac{3N_e}{\pi}\right)^{2/3} = \frac{\hbar^2}{2m_e}\left(3\pi^2\frac{N_e}{8L^3}\right)^{2/3} \tag{1.80a}$$

Note that $8L^3 = (2L)^3$ is the volume of the box so that $(N_e/8L^3) = \rho_e$ is the free-electron density in the conductor. Therefore,

$$E_F = \frac{\hbar^2}{2m_e}(3\pi^2\rho_e)^{2/3} \tag{1.80b}$$

In a typical conductor like copper, one electron is free for each atom, and E_F predicted from Eq. (1.80*b*) is $E_F = 7.1$ eV. This is very close to the experimental value, indicating that our calculation is correct. It is important to note that the result of Eq. (1.80*b*) does not depend on the dimensions of the conductor. It was convenient to introduce a box of finite size, but this was a calculational artifact on which the physical observables should not depend.

1.9. SUMMARY

We have tried to bring out in this chapter the essential features of particle behavior in the microscopic domain. The deterministic classical laws are replaced by the probabilistic description of systems. This leads, among other things, to the consequence that not all particle attributes can be measured accurately. The measurement of the position and the momentum of the particle, for example, have to be uncertain by an amount given by the Heisenberg uncertainty principle. We have associated a wave packet with a sufficiently localized particle. We have calculated the expectation values of the position and momentum of particles associated with a wave packet. We have also shown that the expectation values obey equations corresponding to the classical laws. This feature, known as the correspondence principle, will be discussed in more detail in later chapters. Finally, we have used the ideas of probability waves to show that the energy of the free electrons in a conductor is quantized, and showed how this leads to the correct value for the Fermi energy in the conductor.

Problems

PROBLEM 1

In a demonstration of the diffraction from two slits it was found that the spacing between interference minima was

$$\Delta x_I = 5.5 \text{ mm}$$

and the first *diffraction* minimum was located at

$$\Delta x_D = 14.2 \text{ mm}$$

from the beam center. The distance of the screen from the slits was $L = 1$ m and the wavelength (neon laser) was $\lambda = 6328$ Å.

(*a*) Find the width of the slits and their separation.
(*b*) Make a quantitative plot (on graph paper) of the observed pattern.

PROBLEM 2

In a Stern–Gerlach experiment the detector is 2.5 m from the center of the inhomogeneous field. The length of the inhomogeneous field is 1 m. The magnetic moment of the atoms is $\mu_0/2$, where $\mu_0 = 0.929 \times 10^{-23}$ J A-m^2 is the Bohr magneton. Find the strength of the gradient dB/dz necessary to obtain a separation of 2 mm at the screen. Give the dimensions of the pole faces of an iron core electromagnet that produces such a field (approximately). The temperature of the oven is 1200°C and the sample consists of silver atoms. Note that $\langle mv^2/2 \rangle = \frac{3}{2}kT$, $v_p = (2kT/m)^{1/2}$.

PROBLEM 3

The nucleus of lead can be assumed to have a radius $R = 6 \times 10^{-15}$ m $= 6$ F. Electrons of energy 1000 MeV are incident on a lead target. Assume that the scattering is similar to that produced by *a slit* of width $d = 2R$.

(*a*) Make a plot of the relative number of electrons scattered at an angle θ.
(*b*) Discuss the resolution with which details of the structure of the lead nucleus can be observed in this experiment.

Note: $\hbar c \approx 200$ MeV-F; $m_e c^2 \approx 0.5$ MeV. Use the small-angle approximation $\sin\theta \approx \theta$. Make whatever other approximations are reasonable.

PROBLEM 4

A photon of energy $E = 100$ keV collides with a free electron at rest. It is scattered through 90°. Find the energy of the scattered photon.
Note: For the electron mass use $mc^2 = 500$ keV and recall that $E = [(pc)^2 + (mc^2)^2]^{1/2}$.

PROBLEM 5

The lattice constant of nickel is 2.15 Å. Calculate the radii of the diffraction rings at a distance $s = 50$ cm when cold neutrons of energy 10 eV are incident on nickel powder.

PROBLEMS 6

Professor X has applied to the NSF for funds to observe the quantum behavior of a small oscillator through a microscope using visible light. According to his proposal the oscillator consists of an object 10^{-6} m in diameter and of estimated mass 10^{-15} kg. It vibrates at the end of a thin fiber with maximum amplitude of 10^{-5} m and a frequency of 1000 Hz. Should the proposal be approved?

PROBLEM 7

Monochromatic light of wavelength $\lambda = 6000$ Å passes through a fast shutter. The shutter opens and closes periodically so that it is open for a time of 10^{-10} sec and closed for a time of 10^{-2} sec. Light passing through the shutter will then no longer be monochromatic but will show a spread in λ. Explain why this is so and estimate the spread in λ.

PROBLEM 8

The average energy of a particle in a gas (i.e., a system of noninteracting particles) is

$$E = \tfrac{3}{2}kT$$

Calculate the De Broglie wavelength for a nonrelativistic particle of mass m at temperature T. Suppose there are N such particles in a volume V and assume that they are in a *cubic* lattice. Find the mean interparticle spacing both in terms of N and V, and in terms of the mass of the particle and the density ρ of the gas. Quantum effects will be significant when $\lambda \gg d$. Show that this happens when

$$T \ll (\text{const})\left(\frac{N}{V}\right)^{2/3}$$

PROBLEM 9

From Problem 8 above find the temperature at which quantum effects should be exhibited by the following "gases."

(*a*) Liquid He^4; density 0.125×10^3 kg/m^3
(*b*) Free electrons in a metal $(N/V) \sim 10^{30}/\mathrm{m}^3$; free electrons in a semiconductor $(N/V) \sim 10^{21}/\mathrm{m}^3$.

PROBLEM 10

Compute the energy levels of the hydrogen atom by assuming that the electron moves in a circular orbit around the nucleus, and that the orbits are such that the circumference is an integral number of De Broglie wavelengths. This is an incorrect model, but it gives the correct answer.

PROBLEM 11

Use the uncertainty principle to compute how long an ordinary lead pencil can be balanced upside down on its point. This is an interesting problem.

PROBLEM 12

Calculate numerically the lowest energy of *two* protons confined in a potential well of width $2a = 2 \times 10^{-13}$ cm and of infinite height. Protons are Fermi particles since they have spin 1/2.

PROBLEM 13

Consider a one-dimensional wave packet $\psi(x, t)$ whose spectrum in momentum space is Gaussian with standard deviation σ_k

$$A(k) = \left(\frac{1}{\pi^{1/2}\sigma_K}\right)^{1/2} e^{-(k-k_0)^2/2\sigma_k^2}$$

Then

$$\psi(x, t) = \frac{1}{(2\pi)^{1/2}} \int_{-\infty}^{+\infty} A(k) e^{i(\omega t - kx)}\, dk$$

Perform the integration to find $\psi(x, t)$ and the standard deviation σ_x of the packet in coordinate space. See Appendix 1. Defining $\langle \Delta x^2 \rangle = \sigma_x^2/2$ and $\langle \Delta p_x^2 \rangle = \hbar^2 \sigma_k^2/2$, show that the exact expression for the minimum uncertainty is obtained

$$\langle \Delta x^2 \rangle \langle \Delta p_x^2 \rangle = \frac{\hbar^2}{4}$$

PROBLEM 14

Find expressions for the eigenfunctions and energy levels of a particle in a *two-dimensional* circular box that has perfectly rigid walls.

PROBLEM 15

Show that the following are representations of the delta function (see Appendix 2.)

(a) $\lim_{\varepsilon \to 0} \dfrac{\varepsilon}{\pi(x^2 + \varepsilon^2)}$

(b) $\lim_{\alpha \to 0} \left(\dfrac{\alpha}{\pi}\right)^{1/2} e^{-\alpha x^2}$

PROBLEM 16

Determine the value of the constant B such that

(a) $\lim_{\alpha \to \infty} Be^{-\alpha r} = \delta(\mathbf{r})$

(b) $\lim_{b \to 0} t_b(x) = \delta(x)$ with $t_b(x) = \begin{cases} 0, & \text{for } x^2 > b^2 \\ B\,|b - x|, & x^2 < b^2 \end{cases}$

PROBLEM 17

(a) Using the Gauss theorem

$$\int_{\Omega} (\operatorname{div} \mathbf{A})\, d^3r = \oint_S \mathbf{A} \cdot d\mathbf{S} \qquad (S \text{ is the surface of } \Omega)$$

show that

$$\nabla^2\left(\frac{1}{|\mathbf{r}|}\right) = -4\pi\delta^3(\mathbf{r})$$

(b) Using an integral representation for $1/|\mathbf{r}-\mathbf{r}'|$ show that

$$\nabla^2\left(\frac{1}{|\mathbf{r}-\mathbf{r}'|}\right) = -4\pi\delta^3(\mathbf{r}-\mathbf{r}')$$

Chapter 2

STATES, AMPLITUDES, AND OPERATORS

In this chapter we develop the *language* of quantum mechanics in which physical systems are described in terms of the quantum-mechanical *states* that the system can occupy. The state of the system can change under the influence of external forces, such external influences being represented by *operators* that act on the states. The overlap of two states of the system defines an *amplitude*, and the absolute square of the amplitude is interpreted as a *probability*.

We also introduce the concept of a *representation* for the states and operators. This is necessary in order to perform explicit calculations. States are represented by vectors in a linear vector space of appropriate dimension, admitting complex expansion coefficients. Operators are represented by matrices multiplying the vectors. The probability amplitude for finding a physical system in another state is given by the inner product of the two vectors. We use these concepts in Chapter 3 to discuss the process of measurement in quantum mechanics and to relate the state of a system to physically measurable observables. In more mathematical terms, we discuss the elements of *linear algebra* that are necessary for the description of quantum systems.

2.1. STATES OF A QUANTUM-MECHANICAL SYSTEM

In Chapter 1 we noted that the passage of an electron beam through two slits produces an interference pattern as shown in Fig. 1.9. We also showed that this phenomenon can be explained if we associate with the electron a properly normalized amplitude $\phi(x, t)$, the absolute square of which gives the probability that the electron would be at x, t (in an interval dx)

$$P(x, t)\, dx = |\phi(x, t)|^2\, dx \tag{2.1}$$

The function $\phi(x, t)$ is the probability amplitude for finding an electron at the *final* position x, given that it was emitted from the *initial* position s where the source is located. Thus, we see that a probability amplitude *connects* an initial and a final state of the electron. We use the notation

$$|s\rangle \quad \text{or} \quad |x\rangle$$

to represent particular states of the electron. In this case

$|s\rangle \Rightarrow$ state of the electrons emitted from the source s

$|x\rangle \Rightarrow$ state of the electrons reaching the screen at x

We will designate initial states by *kets*

$$|s\rangle \quad \textit{initial state}$$

and final states by *bras*

$$\langle x| \quad \textit{final state}$$

A probability amplitude is then designated by the *bra-ket*

$$\phi(x, t) = \langle x \mid s\rangle \qquad \textit{amplitude}$$

A bracket is read from right to left, that is, it is interpreted as the amplitude for a particle (an electron in this case) initially in the state $|s\rangle$ to be found in the final state $\langle x|$. This notation was first introduced by P. A. M. Dirac in 1928.

In general, the state of a physical system contains all the information about the system. However, depending on the particular problem at hand, we may only be interested in a limited number of properties of the system. For instance, in the example of electron interference we were only concerned with the position of the electron at any given time. Consequently, we talked only about states that described the position, ignoring for the moment all the other attributes of the electron. Electrons, on the other hand, also have intrinsic angular momentum, called *spin*, and associated with it a *magnetic moment.* In the Stern–Gerlach experiment the electrons emerge in two distinct beams, one "up" and one "down" along the

Z-axis. Electrons in the upper beam have their spin pointing *up* and in the lower beam the spin points *down*. Thus, one can describe the outcome of this experiment by introducing the spin states $|\text{up}\rangle$ and $|\text{down}\rangle$ for the electron and ignoring all other properties. For example, if the state of the electron before entering the apparatus is $|i\rangle$, then the amplitude for deflection up or down is given by

$$\phi(\text{up}) = \langle \text{up} \mid i\rangle, \qquad \phi(\text{down}) = \langle \text{down} \mid i\rangle$$

The probability of observing the electrons in the corresponding beam is the absolute square of the above amplitudes.

The expressions we have given so far are symbolic. To obtain quantitative results we must choose a *representation* for the states of the system. Any convenient representation can be chosen, just as we can use any suitable system of coordinates to represent a vector by its components. However, the amplitides must be *independent* of the choice of representation. This is evident from the fact that their absolute square is a physically observable probability that cannot depend on the system of coordinates. The probability amplitudes, therefore, can be complex functions of dynamical variables or simply a complex number.

To describe quantitatively the state of a physical system we must choose a set of *basis states* in terms of which we can express any arbitrary state of the system. As an analogy consider a two-dimensional vector. If we pick two linearly independent unit vectors, $\mathbf{u}_x$ and $\mathbf{u}_y$, we can express

$$\mathbf{v} = a_x\mathbf{u}_x + a_y\mathbf{u}_y$$

and write $\mathbf{v} = \{a_x, a_y\}$. The particular values of $\{a_x, a_y\}$ depend on the choice of the unit vectors. Furthermore, if the vector was three-dimensional, we would have to use a coordinate system with three unit vectors.

The same considerations hold in the choice of the basis states of a representation. For instance, consider an electron beam, and suppose we wish to describe only the orientations of its spin. As we have already discussed, when the beam passes through a Stern–Gerlach apparatus it will be deflected either up or down along the axis of the apparatus we take as the Z-axis. This is a consequence of the alignment of the magnetic moment, and we conclude that the spin angular momentum of the electron can be found in only two possible orientations. We choose these orientations as the basis states and designate them by

$$\begin{aligned} &|\text{up}\rangle \quad \text{or} \quad |+z\rangle \\ &|\text{down}\rangle \quad \text{or} \quad |-z\rangle \end{aligned} \tag{2.2a}$$

An arbitrary state of the electron's spin can be described in terms of the basis states of Eq. (2.2a) through

$$|i\rangle = a\,|\text{up}\rangle + b\,|\text{down}\rangle \tag{2.2b}$$

The doublet of numbers $\{a, b\}$ are the coordinates that describe the state $|i\rangle$ in the representation defined by the basis states of Eq. (2.2a).

Let us now perform an experiment by setting up two Stern–Gerlach apparatuses, one behind the other, oriented in the same way (see Fig. 2.1). If only the upper beam enters the second apparatus we will observe that all of the secondary beam is deflected up and none down. We designate by $|i'\rangle$ the state of the beam after emerging from the first apparatus. Our observations then imply

$$\begin{aligned} \langle \text{up} \mid i'\rangle &= \langle \text{up} \mid \text{up}\rangle = 1 \\ \langle \text{down} \mid i'\rangle &= \langle \text{down} \mid \text{up}\rangle = 0 \end{aligned} \tag{2.3}$$

In other words, the basis states defined by our apparatus are *normalized* and mutually *orthogonal.*

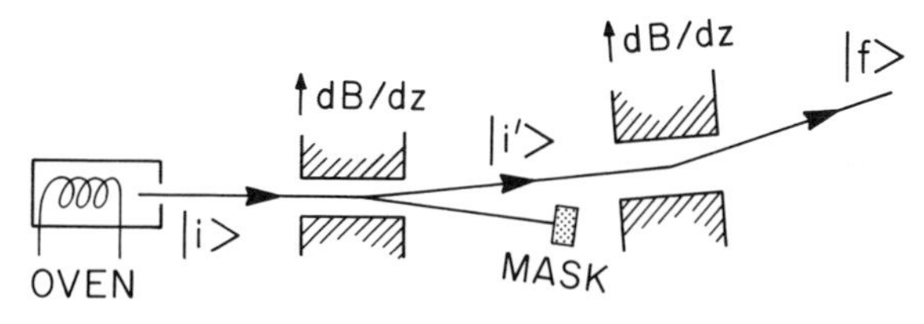

FIGURE 2.1. Schematic of the Stern–Gerlach experiment with two analyzing magnets placed one after the other. The gradient dB/dz is oriented along the same axis for both magnets.

We can examine how the state $|i'\rangle$ is related to the initial state $|i\rangle$, which has the general form of Eq. (2.2b). We know that $|i'\rangle = |\text{up}\rangle$ so that using Eqs. (2.3)

$$\langle i' \mid i\rangle = a\langle \text{up} \mid \text{up}\rangle + b\langle \text{up} \mid \text{down}\rangle = a$$

Thus, the coefficients $\{a, b\}$ used to represent the arbitrary state $|i\rangle$ can be found by forming the amplitude of the state $|i\rangle$ with each of the basis states. We further observe that the probability that the initial beam is deflected up is given by

$$P(\text{up}) = |\phi(\text{up})|^2 = |\langle \text{up} \mid i\rangle|^2 = |a|^2$$

and correspondingly for $P(\text{down})$. Since the beam cannot disappear the probability for deflections up and down must add to unity, i.e.,

$$P(\text{up}) + P(\text{down}) = |a|^2 + |b|^2 = 1 \tag{2.4}$$

This is a *normalization* condition on the expansion coefficients (amplitudes) $\{a, b\}$.

We can look at the normalization of a state from the point of view of forming the amplitude of the state with itself $\langle i \mid i\rangle$. By definition this must equal one. This will be so if we define the bra $\langle i|$ corresponding to the state of Eq. (2.2b) by

$$\langle i| = a^*\langle \text{up}| + b^*\langle \text{down}| \tag{2.5a}$$

Using the results of Eq. (2.3) we then find

$$\langle i \mid i\rangle = |a|^2 + |b|^2 \tag{2.5b}$$

which equals one by Eq. (2.4). We can generalize these results to the case where the representation is *spanned* by more than two states. For instance, the spin angular momentum of the deuteron is one and therefore its orientation is described by three basis states $|+1\rangle$, $|0\rangle$, and $|-1\rangle$; the position of a particle along the X-axis is described by a *continuous* and *non-denumerable* set of basis states. We will see how to deal with these cases in the following sections.

Example

A beam of electrons passing through a Stern–Gerlach apparatus is found to have equal probability of being deflected up or down. Express the state of these electrons in the basis provided by the apparatus.

We can use

$$|i\rangle = a\ |\text{up}\rangle + b\ |\text{down}\rangle \tag{2.2b}$$

and are given

$$P(\text{up}) = |a|^2 = P(\text{down}) = |b|^2$$

From the normalization condition

$$|a|^2 + |b|^2 = 1$$

Thus

$$|a|^2 = |b|^2 = \tfrac{1}{2} \tag{2.6a}$$

which defines the magnitude but not the phase of the expansion coefficients. We can choose the simple solution

$$|i\rangle = \frac{1}{\sqrt{2}}|\text{up}\rangle + \frac{1}{\sqrt{2}}|\text{down}\rangle \tag{2.6b}$$

but the state

$$|i'\rangle = \frac{1}{\sqrt{2}}|\text{up}\rangle + e^{i\phi}\frac{1}{\sqrt{2}}|\text{down}\rangle \tag{2.6c}$$

with ϕ real, is also a valid solution. The state $|i'\rangle$ is different from the state $|i\rangle$ above. A frequently occurring linear combination for the state $|i'\rangle$ is when the phase ϕ is equal to π.

2.2. REPRESENTATION OF QUANTUM-MECHANICAL STATES

In Section 2.1 we indicated how a quantum-mechanical state can be represented as a linear combination of basis states. This is in exact analogy to the description of an arbitrary vector $\mathbf{a}$ (in three-dimensional space) in terms of the unit vectors $\mathbf{u}_x$, $\mathbf{u}_y$, and $\mathbf{u}_z$

$$\mathbf{a} = a_x\mathbf{u}_x + a_y\mathbf{u}_y + a_z\mathbf{u}_z \tag{2.7}$$

Because of our familiarity with vectors, certain properties of the description indicated by Eq. (2.7) seem obvious. For instance, unless we use all three unit vectors we cannot describe an arbitrary vector; it is always possible to find the coefficients (a_x, a_y, a_z) needed to describe an arbitrary vector; we can use any system of coordinates to describe a vector, and the coefficients (a_x, a_y, a_z) are related to the coefficients (a'_x, a'_y, a'_z) in the new system; and we can always choose an orthonormal set of unit vectors.

All these properties carry over into the description of a quantum-mechanical state, and that is why we often say that a state is described by a *state vector.* The differences between ordinary vectors and quantum-mechanical state vectors are that: (i) The number of basis states depends on the representation and may be finite or infinite; the basis states can be discrete or continuous; (ii) the expansion coefficients $(a_1, a_2, \ldots)$ can be complex. Mathematically, these properties define a *Hilbert space,* and a quantum-mechanical state is a vector (or ray) in that space. This brings up the question of which is the appropriate representation for the description of a quantum-mechanical state. For the complete description of a quantum-mechanical state we may require a very extensive Hilbert space. For instance, if we wish to describe the position and the spin orientation of an electron, the basis states must span all possible combinations of position and spin orientation. However, in many physical problems we may be interested only in a limited number of properties of the system, and we can *choose the representation* in which to describe the system accordingly.

finite

A representation is always associated with a physical observable of the system and consists of a *complete* set of basis states. A complete set is a set of linearly independent vectors such that any arbitrary vector in that space can be expanded in terms of the vectors of the set. Thus, any quantum mechanical state of the system can be expressed as a linear combination of the basis state.

A familiar analogy to the description of a physical system in terms of discrete basis states can be found in classical physics. Consider polarized light propagating in the z-direction. The electric field vector can be written as a superposition of the fields in the x- and y-directions. For a fixed position z we write

how?

$$\mathbf{E} = E_x \mathbf{u}_x + E_y \mathbf{u}_y = \mathrm{Re}(Ae^{-i\omega t}\mathbf{u}_x + Be^{-i\omega t}\mathbf{u}_y)$$

where Re( ) means "take the real part of the expression in the parentheses." If $B = A$ the polarization is linear at 45° as shown in Fig. 2.2(a); if $B = 0$ it is linear at 0°; if $B = -A$ it is linear at 135°, and so on. If $B = iA = e^{i\pi/2}A$ the polarization vector rotates clockwise and for $B=-iA=e^{-i\pi/2}A$ counterclockwise. We speak of right-hand and left-hand circular (RHC and LHC, respectively) polarized light, as shown in Fig. 2.2(b, c). In the most general case $|B| \neq |A|$ and the polarization is elliptical.

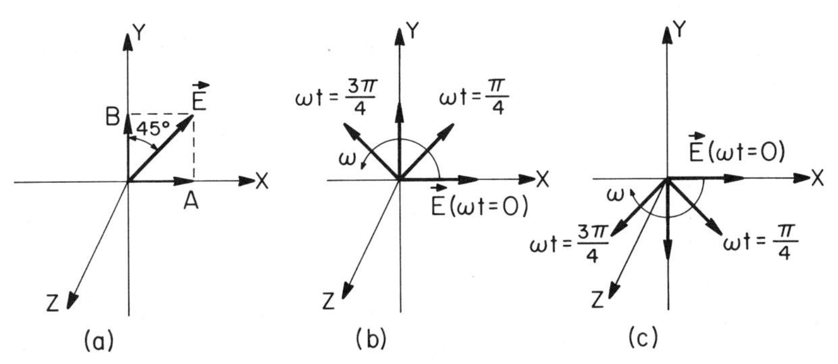

FIGURE 2.2. Polarized light propagating along the Z-axis. (a) Linearly polarized; (b) right-hand-circular polarization; (c) left-hand-circular polarization.

We can choose the unit vectors $\mathbf{u}_x$ and $\mathbf{u}_y$ as a complete set of basis states

$$|x\rangle = \mathbf{u}_x, \qquad |y\rangle = \mathbf{u}_y \tag{2.8a}$$

and express the electric field as a linear combination of these basis states

$$\mathbf{E} = \mathrm{Re}(|P\rangle) \quad \text{where} \quad |P\rangle = a\,|x\rangle + b\,|y\rangle \tag{2.8b}$$

The complex coefficients $a=Ae^{-i\omega t}$ and $b=Be^{-i\omega t}$ carry the time dependence of the field and must satisfy the normalization condition

$$|a|^2+|b|^2=|A|^2+|B|^2=|\mathbf{E}|^2 \tag{2.8c}$$

The set of basis states $|x\rangle, |y\rangle$ given by Eq. (2.8a) is *not unique.* We can use a different representation where the basis states are those of circular polarization (or helicity) defined as

$$\begin{aligned} |+\rangle &= \frac{1}{\sqrt{2}}(\mathbf{u}_x + i\mathbf{u}_y) = \frac{1}{\sqrt{2}}(|x\rangle + i\,|y\rangle) \qquad \text{right-hand-circular} \\ |-\rangle &= \frac{1}{\sqrt{2}}(\mathbf{u}_x - i\mathbf{u}_y) = \frac{1}{\sqrt{2}}(|x\rangle - i\,|y\rangle) \qquad \text{left-hand-circular} \end{aligned} \tag{2.9a}$$

In terms of these states the electric field vector is

$$\mathbf{E} = \mathrm{Re}(|P\rangle) \quad \text{where} \quad |P\rangle = c|+\rangle + d|-\rangle \tag{2.9b}$$

The complex coefficients c and d are related to a and b through

$$c = \frac{1}{\sqrt{2}}(a - ib) = \frac{1}{\sqrt{2}} e^{-i\omega t}(A - iB)$$

$$d = \frac{1}{\sqrt{2}}(a + ib) = \frac{1}{\sqrt{2}} e^{-i\omega t}(A + iB)$$

They satisfy the normalization condition

$$|c|^2 + |d|^2 = |\mathbf{E}|^2 \tag{2.9c}$$

as can be easily checked using Eq. (2.9a). In Fig. 2.2(b) we show the polarization vector $\mathbf{E}_+$ for the RHC state. In this case $|c|^2 = 1$, $|d|^2 = 0$, and thus $c = e^{-i\omega t}$ and

$$\mathbf{E}_+ = \cos \omega t \hat{u}_x + \sin \omega t \hat{u}_y$$

The polarization vector $\mathbf{E}_-$ of the LHC state, is shown in Fig. 2.2(c).

It is instructive to carry this analogy further. For instance, if we introduce a polaroid into the beam we can select the state of linear polarization along the x-axis. The state $|P'\rangle$ is then by necessity

$$|P'\rangle = a\,|x\rangle, \qquad a \neq 0 \tag{2.10a}$$

We now introduce a second polaroid into the beam, as shown in Fig. 2.3(a), oriented along the y-axis. No light will pass through the second polaroid. The polaroid functions as an analyzer in the sense that it measures the component of the light beam that has its polarization vector along y, namely, in the state $|y\rangle$. We conclude that the amplitude for finding the state $|P'\rangle$ in the (basis) state $|y\rangle$ is zero, i.e.,

$$\langle y \mid P'\rangle = 0$$

and in view of Eq. (2.10a)

$$\langle y \mid x\rangle = 0$$

Conversely, if the second polaroid was along the x-axis, as in Fig.

2.3(b), all the light would have been transmitted that is,

$$\langle x \mid P' \rangle = a \quad \text{or} \quad \langle x \mid x \rangle = 1$$

Thus, the basis states that are experimentally defined by using polaroids with their axes along the x and y directions are orthonormal

$$\langle x \mid y \rangle = \delta_{xy} \tag{2.10b}$$

and form a complete set.

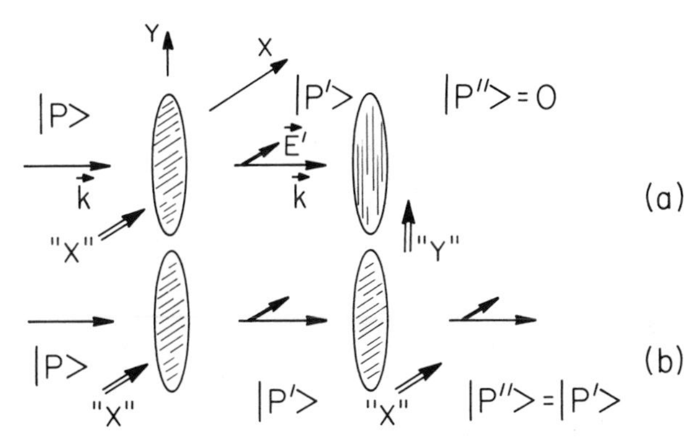

FIGURE 2.3. A light beam traversing two polaroids. The double arrows indicate the direction of polarization. (a) The polaroids are crossed and no light is transmitted. (b) The polaroids are aligned so that all light traversing the first polaroid traverses the second one as well.

To measure the polarization of the beam we have to introduce polaroids. It is important to realize that the polaroids not only measure the state of the beam but in so doing *modify* the state of the beam as well. Consider the situation shown in Fig. 2.4(a). Polaroid 1 establishes a light beam in the state $|x\rangle$

$$|P'\rangle = a \mid x\rangle$$

If this is followed by polaroid 3, which is along the y-direction, *no* light is transmitted (see Fig. 2.3). We can, however, introduce between 1 and 3 a polaroid 2 which is inclined at 45°; now light *will* be transmitted through the system. The state $|P''\rangle$ emerging after polaroid 2 is polarized along 45°, and the magnitude of the electric vector is

$$|\mathbf{E}''| = \frac{1}{\sqrt{2}} |\mathbf{E}'|$$

as shown in Fig. 2.4(b), where we are looking *into* the beam. If we write for the state $|P'\rangle$

$$|P'\rangle = a\,|x\rangle \qquad |a|^2 = 1$$

then

$$|P''\rangle = \frac{a}{\sqrt{2}}\,|x''\rangle = \frac{a}{2}\,|x\rangle + \frac{a}{2}\,|y\rangle$$

Since polaroid 3 has its axis along y the trnasmitted light will be in a state

$$|P'''\rangle = \frac{a}{2}\,|y\rangle$$

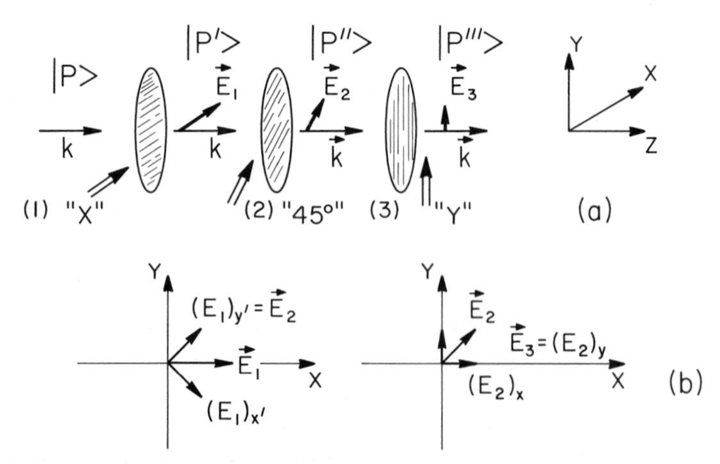

FIGURE 2.4. A light beam can be transmitted through two crossed polaroids if a third polaroid inclined at some other direction is inserted between them. (a) Polaroid 2 is at 45°. (b) The electric field vector after each polaroid.

The intensity of the transmitted light is proportional to the square of the electric field, which is given by the norm of the state $|P'''\rangle$

$$I_3 \propto |\mathbf{E}_3|^2 = |\langle P''' \, | P''' \rangle| = \frac{|a|^2}{4}$$

as compared to the intensity $I_1 \propto |a|^2$ after the first polaroid. The remaining three-fourths of the beam are absorbed in polaroids 2 and 3.

What we have demonstrated here is nontrivial. By introducing a device that *absorbs* light (in a preferential direction) we *increased* the light output of the system (as compared to the transmission when only two polaroids were present). Furthermore, we have seen that *the process of measurement alters the state of the system.* The analogy between the states of polarization of light and two-state quantum systems is a close one, except that for light the observable electric field $\mathbf{E}$ is given by the real part of the state vector. Quantum-mechanical state vectors are *not directly observable*; *only their absolute square can be measured* (which is equivalent to the intensity of the light beam).

Let us now return to an arbitrary quantum-mechanical state $|a\rangle$. We can represent it by an expansion in a *complete* set of orthonormal basis states $|u_i\rangle$ of a particular representation that is assumed to have n discrete basis states

$$\begin{gathered} |a\rangle = a_1 \, |u_1\rangle + a_2 \, |u_2\rangle + \cdots + a_n \, |u_n\rangle \\ \langle u_i \mid u_j \rangle = \delta_{ij} \end{gathered} \tag{2.11a}$$

By our definitions the probability amplitude of finding the state $|a\rangle$ in the basis state $|u_i\rangle$ is

$$\phi_i = \langle u_i \mid a \rangle$$

Because of the orthonormality of $|u_i\rangle$, Eq. (2.11a) immediately yields

$$\phi_i = \langle u_i \mid a \rangle = a_i \tag{2.12}$$

We have thus obtained the important result that the expansion coefficients defining a state $|a\rangle$ in a particular representation are simply the amplitudes for finding the arbitrary state in the corresponding basis state. Thus, we can rewrite Eq. (2.11) as

$$|a\rangle = \langle u_1 \mid a\rangle\, |u_1\rangle + \langle u_2 \mid a\rangle\, |u_2\rangle + \cdots + \langle u_n \mid a\rangle\, |u_n\rangle$$

or compactly as

$$|a\rangle = \sum_i \langle u_i \mid a\rangle\, |u_i\rangle$$

and even more compactly by using $|i\rangle$ to indicate the basis state $|u_i\rangle$

$$\boxed{|a\rangle = \sum_i |i\rangle\langle i \mid a\rangle} \qquad (2.11b)$$

Equation (2.11b) always implies summation over the *complete set* of basis states.

As an application of Eqs. (2.11) and (2.12) we consider the wave packet discussed in Section 1.6. To describe the state $|W\rangle$ of the particle we can use the *position* representation. In this representation the basis states $|x\rangle$ describe a particle found at x and are *continuous and nondenumerable.* The most general state is

$$|W\rangle = a_1\,|x_1\rangle + a_2\,|x_2\rangle + a_3\,|x_3\rangle + \cdots \qquad (2.13)$$

which is in analogy to Eq. (2.11a). Since the states $|x\rangle$ are continuous we must replace Eq. (2.11b) by an integral

$$|W\rangle = \int a(x)\,|x\rangle dx$$

For continuous basis states the orthonormality condition equivalent to Eq. (2.10b) is†

$$\langle x' \mid x\rangle = \delta(x - x')$$

Therefore, the amplitude of finding the particle at the position x' is

$$\phi(x') = \langle x' \mid W\rangle = \int a(x)\langle x' \mid x\rangle\, dx$$
$$= \int a(x)\delta(x - x')\, dx = a(x') \qquad (2.14)$$

We see that the probability wave $G(x, t)$ [see Eq. (1.57a)] corresponds to the expansion coefficient (function) $a(x)$ that describes the state $|W\rangle$ in the *position* representation. In that case $a(x)$ depended on time and so did the amplitude $\phi(x')$. Thus, we should have

† See Appendix 2 for a definition and properties of the δ-function.

written $a(x, t)$ and $\phi(x', t)$. We discuss the time dependence of amplitudes in Section 3.2.

We could have equally as well expressed the state $|W\rangle$ in the *momentum* representation where the basis states $|k\rangle$ describe a particle of sharp momentum $p = \hbar k$. Since the momentum basis states are continuous we must again use an integral

$$|W\rangle = \int a(k)\,|k\rangle\, dk$$

We leave it to the reader to show that the expansion coefficient (function) $a(k)$ is given by the probability wave $\tilde{\phi}(k, t)$ of Eq. (1.59).

2.3. PROPERTIES OF QUANTUM-MECHANICAL AMPLITUDES

We have seen that quantum-mechanical amplitudes play a role equivalent to the coordinates of a vector. To represent a state $|a\rangle$ we must know the amplitudes $\langle i \mid a\rangle$ between $|a\rangle$ and *all* the basis states $|i\rangle$ of the representation. This statement has a physical interpretation: by repeated measurements on (similarly prepared) states $|a\rangle$, we can determine the probability P_i that $|a\rangle$ is found in the basis state $|i\rangle$. By definition

$$P_i = |\langle i \mid a\rangle|^2 \tag{2.15}$$

so that we can, in principle, determine the *magnitude* of all amplitudes. The *relative phase* between amplitudes can be determined from interference experiments.

We will now establish some of the general properties obeyed by quantum-mechanical amplitudes. These follow from the fact that an arbitrary state $|a\rangle$ can be represented as a *linear superposition* of the basis states as in Eq. (2.11*b*)

$$|a\rangle = \sum_i |i\rangle\langle i \mid a\rangle \tag{2.11b}$$

To obtain the amplitude between the state $|a\rangle$ and some arbitrary state $|b\rangle$ we write

$$\boxed{\langle b \mid a\rangle = \sum_{\text{all } i} \langle b \mid i\rangle\langle i \mid a\rangle} \tag{2.16}$$

We will refer to the above result as the completeness or *closure relation.* In physical terms it means that the amplitude for going from the initial state $|a\rangle$ to the final state $|b\rangle$ is given by the sum of the products of the amplitudes for going from $|a\rangle$ to some intermediate state $\langle i|$ and then from $|i\rangle$ to the final state $\langle b|$. However, the summation must be over the *complete set* of the basis states $|i\rangle$ of any one particular representation.

The *normalization* condition on the amplitudes is obtained by demanding that the sum of the probabilities P_i of finding the state $|a\rangle$ in any one of the basis states $|i\rangle$ equal unity

$$\boxed{\sum_{\text{all } i} P_i = \sum_{\text{all } i} |\langle i \mid a\rangle|^2 = 1} \tag{2.17}$$

Similarly, the probability of finding the state $|a\rangle$ in the state $|a\rangle$ should obviously equal unity, or

$$P_{a,a} = |\langle a \mid a\rangle|^2 = 1 \tag{2.18a}$$

We choose the phase by *defining* the *amplitude* for finding the state $|a\rangle$ in the state $|a\rangle$ to be unity also

$$\langle a \mid a\rangle = 1 \tag{2.18b}$$

Equations (2.17) and (2.18*a*, *b*) can be combined if we express the amplitude $\langle a \mid a\rangle$ in terms of the closure relation of Eq. (2.16)

$$\langle a \mid a\rangle = \sum_{\text{all } i} \langle a \mid i\rangle\langle i \mid a\rangle = 1$$

Comparing this result to Eq. (2.17) we see that the following must be true

$$\boxed{\langle a \mid i\rangle = \langle i \mid a\rangle^*} \tag{2.19}$$

Equation (2.19) is a general statement on the behavior of amplitudes under *time reversal*: The amplitude for going from any state $|a\rangle$ to any state $|b\rangle$ is the complex conjugate of going from the state $|b\rangle$ to the state $|a\rangle$. It also follows that the expansion of a bra equivalent to Eq. (2.11*b*) is

$$\langle b| = \sum_i \langle b \mid i\rangle\langle i| = \sum_i \langle i \mid b\rangle^*\langle i| \tag{2.20}$$

Equations (2.16), (2.17), and (2.19) play the role of defining equations for quantum-mechanical amplitudes. For instance, by scratching $\langle b|$ from both sides of Eq. (2.16) we obtain Eq. (2.11b); by scratching $|a\rangle$ we obtain Eq. (2.20).

Example

Show that the circular polarization states $|+\rangle$ and $|-\rangle$ introduced in Eq. (2.9) are orthogonal namely, that the amplitude $\langle + \mid - \rangle$ is zero. We are given

$$|+\rangle = \frac{1}{2^{1/2}}(|x\rangle + i\,|y\rangle)$$

$$|-\rangle = \frac{1}{2^{1/2}}(|x\rangle - i\,|y\rangle)$$

with

$$\langle x \mid y\rangle = \delta_{xy}$$

Thus, from Eq. (2.20)

$$\langle +| = \frac{1}{2^{1/2}}\{\langle x| - i\langle y|\}$$

so that

$$\langle + \mid - \rangle = \tfrac{1}{2}\{\langle x \mid x\rangle - \langle y \mid y\rangle\} = 0$$

Example

A quantum system can be described in terms of a complete set of three basis states $|\alpha\rangle, |\beta\rangle$, and $|\gamma\rangle$. The state $|p\rangle$ has the following amplitudes in this representation

$$\langle \alpha \mid p\rangle = \frac{1}{3^{1/2}} \qquad \langle \beta \mid p\rangle = 0 \qquad \langle \gamma \mid p\rangle = i\left(\frac{2}{3}\right)^{1/2}$$

and a state $|q\rangle$ has amplitudes

$$\langle \alpha \mid q\rangle = \frac{1+i}{3^{1/2}} \qquad \langle \beta \mid q\rangle = \frac{1}{6^{1/2}} \qquad \langle \gamma \mid q\rangle = \frac{1}{6^{1/2}}$$

Find the amplitude $\langle p \mid q\rangle$ and the probability of finding the system in the state $\langle q|$ when it is initially in the state $|p\rangle$.

We use Eq. (2.16)

$$\langle p \mid q\rangle = \sum_{i=\alpha,\beta,\gamma} \langle p| i\rangle\langle i \mid q\rangle = \frac{1+i}{3} - \frac{i}{3} = \frac{1}{3}$$

and the probability is

$$P_{qp} = |\langle q \mid p\rangle|^2 = |\langle p \mid q\rangle|^2 = \tfrac{1}{9}$$

When we know the amplitudes describing the state $|a\rangle$ in a representation with basis states $|i\rangle$ it is possible to find the amplitudes describing the same state in a different representation, say with basis states $|\mu\rangle$. This statement is analogous to the transformation of the coordinates of a vector when we rotate the axes. Note, however, that the *amplitude* between two states $|a\rangle$ and $|b\rangle$ is invariant under such transformations of the representation (of the basis states).

To illustrate this we start by expanding the state $|a\rangle$ in both representations

$$\begin{aligned} |a\rangle &= \sum_{i=1}^{n} |i\rangle\langle i \mid a\rangle \\ |a\rangle &= \sum_{\mu=1}^{n} |\mu\rangle\langle \mu \mid a\rangle \end{aligned} \tag{2.21a}$$

Closing the first equation with $\langle\mu|$ and the second with $\langle i|$ we obtain

$$\begin{aligned} \langle \mu \mid a\rangle &= \sum_{i=1}^{n} \langle \mu \mid i\rangle\langle i \mid a\rangle \\ \langle i \mid a\rangle &= \sum_{\mu=1}^{n} \langle i \mid \mu\rangle\langle \mu \mid a\rangle \end{aligned} \tag{2.21b}$$

We see that the amplitudes $\langle \mu \mid a\rangle$ in the $|\mu\rangle$ basis are related to $\langle i \mid a\rangle$—the amplitudes in the $|i\rangle$ basis—*linearly* through the n^2 coefficients $\langle \mu \mid i\rangle$. The inverse relation is determined by the coefficients $\langle i \mid \mu\rangle = \langle \mu \mid i\rangle^*$.

It is convenient to arrange the transformation coefficients $\langle \mu \mid i\rangle$ into a square $(n \times n)$ matrix $S_{\mu i}$. This is shown below for $n = 3$. If we arrange the amplitudes $\langle i \mid a\rangle$ into a *column vector* **a**, then the vector **a** containing the amplitudes $\langle \mu \mid a\rangle$ for the new representation can be obtained by the rules of matrix multiplication.† Setting

$$S_{\mu i} = \begin{array}{c|ccc} & i & j & k \\ \hline \mu & \langle \mu \mid i \rangle & \langle \mu \mid j \rangle & \langle \mu \mid k \rangle \\ \nu & \langle \nu \mid i \rangle & \langle \nu \mid j \rangle & \langle \nu \mid k \rangle \\ \xi & \langle \xi \mid i \rangle & \langle \xi \mid j \rangle & \langle \xi \mid k \rangle \end{array}$$

$$\mathbf{a}_i = \begin{pmatrix} \langle i \mid a \rangle \\ \langle j \mid a \rangle \\ \langle k \mid a \rangle \end{pmatrix} \qquad \mathbf{a}'_\mu = \begin{pmatrix} \langle \mu \mid a \rangle \\ \langle \nu \mid a \rangle \\ \langle \xi \mid a \rangle \end{pmatrix} \tag{2.22a}$$

it holds that

$$a'_\mu = \sum_i S_{\mu i} a_i \tag{2.22b}$$

We also know how to express the inverse transformation in matrix language. Setting

$$S'_{i\mu} = \begin{array}{c|ccc} & \mu & \nu & \xi \\ \hline i & \langle i \mid \mu \rangle & \langle i \mid \nu \rangle & \langle i \mid \xi \rangle \\ j & \langle j \mid \mu \rangle & \langle j \mid \nu \rangle & \langle j \mid \xi \rangle \\ k & \langle k \mid \mu \rangle & \langle k \mid \nu \rangle & \langle k \mid \xi \rangle \end{array} = \begin{array}{|ccc} \mu & \nu & \xi \\ \hline \langle \mu \mid i \rangle^* & \langle \nu \mid i \rangle^* & \langle \xi \mid i \rangle^* \\ \langle \mu \mid j \rangle^* & \langle \nu \mid j \rangle^* & \langle \xi \mid j \rangle^* \\ \langle \mu \mid k \rangle^* & \langle \nu \mid k \rangle^* & \langle \xi \mid k \rangle^* \end{array} \tag{2.23a}$$

it follows from the second equation of Eqs. (2.21*b*) that

$$a_i = \sum_\mu S'_{i\mu} a'_\mu \tag{2.23b}$$

Because of the relation $\langle i \mid \mu \rangle = \langle \mu \mid i \rangle^*$, the matrix S' is the *Hermitian conjugate* of S, which we designate by $S^\dagger$ (pronounced S-dagger). Hermitian conjugation means to take the *transpose* of the original matrix and then *complex conjugate*

$$S^\dagger \equiv (S^T)^* \quad \text{or} \quad (S^\dagger)_{\mu i} = S^*_{i\mu} \tag{2.24}$$

† A review of the properties of matrices and of operations with matrices is given in Appendix 4.

It is evident from Eq. (2.23*a*) that

$$S' = S^{\dagger} \tag{2.25a}$$

If we now transform from the $\langle i \mid a\rangle$ representation to the $\langle \mu \mid a\rangle$ representation and then back to $\langle i \mid a\rangle$ we should regain the vector **a**. This is done by introducing Eq. (2.22*b*) into Eq. (2.23*b*)

$$\mathbf{a} = S'\mathbf{a}' = S'S\mathbf{a} = S^{\dagger}S\mathbf{a}$$

or, equivalently,

$$S^{\dagger}S = \mathbb{1} \tag{2.25b}$$

where the symbol $\mathbb{1}$ stands for the unit matrix in general. A matrix obeying Eq. (2.25*b*) is said to be *unitary*. In terms of matrix elements

$$\sum_{\beta} (S^{\dagger})_{\alpha\beta} S_{\beta\gamma} = \delta_{\alpha\gamma} \quad \text{or} \quad \sum_{\beta} S^{*}_{\beta\alpha} S_{\beta\gamma} = \delta_{\alpha\gamma}$$

A unitary matrix is a matrix whose inverse is its Hermitian conjugate. A unitary matrix that is real is an *orthogonal* matrix. Unitary transformations have the important properties of maintaining the normalization of state vectors and of leaving amplitudes invariant.

For instance, consider a state $|a\rangle$ that is normalized. In terms of the amplitudes in the $|i\rangle$ representation the normalization is expressed as in Eq. (2.17)

$$1 = \langle a \mid a\rangle = \sum_i |\langle i \mid a\rangle|^2 = \sum_i \langle a \mid i\rangle\langle i \mid a\rangle$$

We can use Eq. (2.21*b*) to express $\langle i \mid a\rangle$ in terms of the new representation $|\mu\rangle$. Then

$$\begin{aligned}
\langle a \mid a\rangle &= \sum_i \left(\sum_{\mu} \langle a \mid \mu\rangle\langle \mu \mid i\rangle\right)\left(\sum_{\nu} \langle i \mid \nu\rangle\langle \nu \mid a\rangle\right) \\
&= \sum_{\mu,\nu} \langle a \mid \mu\rangle\langle \mu \mid \nu\rangle\langle \nu \mid a\rangle \\
&= \sum_{\mu,\nu} \langle a \mid \mu\rangle\delta_{\mu\nu}\langle \nu \mid a\rangle \\
&= \sum_{\mu} \langle a \mid \mu\rangle\langle \mu \mid a\rangle = \sum_{\mu} |\langle \mu \mid a\rangle|^2
\end{aligned} \tag{2.26}$$

This is as expected because for a complete set $|\mu\rangle$, the amplitude $\langle a \mid a\rangle$ can always be expanded as in the last expression of Eq. (2.26).

The same result can be obtained using matrix notation. Just as we arranged the amplitudes $\langle i \mid a\rangle$ into a column vector $\mathbf{a}$, we can arrange the amplitudes $\langle a \mid i\rangle = \langle i \mid a\rangle^*$ into a *row vector* $\mathbf{a}^\dagger$ where

$$\begin{aligned}\mathbf{a}^\dagger &= [\langle a \mid i\rangle, \langle a \mid j\rangle, \langle a \mid k\rangle] \\ &= [\langle i \mid a\rangle^*, \langle j \mid a\rangle^*, \langle k \mid a\rangle^*] = [a_i^*]\end{aligned} \tag{2.27}$$

Since the column vector $\mathbf{a}$ contains the coordinates of the ket $|a\rangle$ in a particular representation, the row vector $\mathbf{a}^\dagger$ contains the coordinates of the bra $\langle a|$ in the same representation. The row vector $(\mathbf{a}')^\dagger$ containing the transformed amplitudes is given by

$$(\mathbf{a}')^\dagger = (S\mathbf{a})^\dagger = \mathbf{a}^\dagger S^\dagger \tag{2.28}$$

where the row vector $\mathbf{a}^\dagger$ multiplies the transformation matrix *from the left.*

In this notation the amplitude $\langle p \mid q\rangle$ can be expressed as the *scalar* product of the row vector $\mathbf{p}^\dagger$ and the column vector $\mathbf{q}$ corresponding to the states $\langle p|$ and $|q\rangle$

$$\langle p \mid q\rangle = \mathbf{p}^\dagger \mathbf{q} = \sum_i p_i^* q_i \tag{2.29}$$

In the new representation we use the vectors $(\mathbf{p}')^\dagger$ and $\mathbf{q}'$

$$\langle p \mid q\rangle = (\mathbf{p}')^\dagger \mathbf{q}' = (\mathbf{p}^\dagger S^\dagger)(S\mathbf{q}) = \mathbf{p}^\dagger (S^\dagger S)\mathbf{q} = \mathbf{p}^\dagger \mathbf{q}$$

where we used the unitarity of S, $S^\dagger S = \mathbb{1}$.

Example

Consider two states $|p\rangle$ and $|q\rangle$ given in the basis $|\alpha\rangle, |\beta\rangle, |\gamma\rangle$ by

$$\mathbf{p} = \begin{pmatrix} \langle \alpha \mid p\rangle \\ \langle \beta \mid p\rangle \\ \langle \gamma \mid p\rangle \end{pmatrix} = \frac{1}{\sqrt{2}} \begin{pmatrix} 1 \\ 0 \\ i \end{pmatrix} \qquad \mathbf{q} = \frac{1}{\sqrt{2}} \begin{pmatrix} 1 \\ 0 \\ -1 \end{pmatrix}$$

Let the transformation matrix S from the $|\alpha\rangle, |\beta\rangle, |\gamma\rangle$ basis to a basis

$|k\rangle, |l\rangle, |m\rangle$ be given by

$$S = \begin{array}{c|ccc} & |\alpha\rangle & |\beta\rangle & |\gamma\rangle \\ \hline |k\rangle & \frac{1}{2}(1+\cos\theta) & (1/2^{1/2})\sin\theta & \frac{1}{2}(1-\cos\theta) \\ |l\rangle & -(1/2^{1/2})\sin\theta & \cos\theta & (1/2^{1/2})\sin\theta \\ |m\rangle & \frac{1}{2}(1-\cos\theta) & -(1/2^{1/2})\sin\theta & \frac{1}{2}(1+\cos\theta) \end{array}$$

and consider the special case $\theta = \pi/2$. Find the amplitudes of the states $|p\rangle$ and $|q\rangle$ in the new basis and use them to calculate the amplitude $\langle p \mid q\rangle$.

For $\theta = \pi/2$ we have

$$S = \tfrac{1}{2}\begin{pmatrix} 1 & 2^{1/2} & 1 \\ -2^{1/2} & 0 & 2^{1/2} \\ 1 & -2^{1/2} & 1 \end{pmatrix} \qquad S^\dagger = \tfrac{1}{2}\begin{pmatrix} 1 & -2^{1/2} & 1 \\ 2^{1/2} & 0 & -2^{1/2} \\ 1 & 2^{1/2} & 1 \end{pmatrix}$$

Direct multiplication of these two matrices verifies that

$$S^\dagger S = \mathbb{1}$$

For the new amplitudes we obtain

$$\mathbf{p}' = S\mathbf{p} = \frac{1}{2}\begin{pmatrix} 1 & 2^{1/2} & 1 \\ -2^{1/2} & 0 & 2^{1/2} \\ 1 & -2^{1/2} & 1 \end{pmatrix} \frac{1}{2^{1/2}}\begin{pmatrix} 1 \\ 0 \\ i \end{pmatrix} = \frac{1}{2(2)^{1/2}}\begin{pmatrix} 1+i \\ 2^{1/2}(-1+i) \\ 1+i \end{pmatrix}$$

$$\mathbf{q}' = S\mathbf{q} = \frac{1}{2}\begin{pmatrix} 1 & 2^{1/2} & 1 \\ -2^{1/2} & 0 & 2^{1/2} \\ 1 & -2^{1/2} & 1 \end{pmatrix} \frac{1}{2^{1/2}}\begin{pmatrix} 1 \\ 0 \\ -1 \end{pmatrix} = \frac{1}{2(2)^{1/2}}\begin{pmatrix} 0 \\ -2(2)^{1/2} \\ 0 \end{pmatrix}$$

Thus

$$\langle p \mid q\rangle = (\mathbf{p}')^\dagger \mathbf{q}' = \tfrac{4}{8}(1+i)$$

which is the same result as obtained by using the amplitudes in the original representation

$$\mathbf{p}^\dagger \mathbf{q} = \tfrac{1}{2}(1+i)$$

2.4. OPERATORS AND CHANGE OF STATE

We know experimentally that the polarization vector of a light beam is changed when the beam passes through a polaroid. Similarly, if we look at only one of the beams emerging from a Stern–Gerlach apparatus, all the atoms would have their spins oriented in a specific direction. Thus, an apparatus acting on a quantum-mechanical system may change the state of the system.

We describe the effect of the apparatus by an *operator* $\hat{A}$ operating on the state $|\phi\rangle$ to produce a state $|\chi\rangle$. We will designate operators by using a caret, so that in our bra–ket notation

$$\boxed{|\chi\rangle = \hat{A}\,|\phi\rangle} \tag{2.30}$$

We say that an operator maps a state vector in Hilbert space onto another vector in the *same* Hilbert space. An operator is *linear* if

$$\hat{A}(a_1\,|\phi_1\rangle + a_2\,|\phi_2\rangle) = a_1\hat{A}\,|\phi_1\rangle + a_2\hat{A}\,|\phi_2\rangle \tag{2.31}$$

Since the superposition principle is strictly valid in quantum mechanics, all quantum-mechanical operators must be linear.

Just as we defined a state vector $|\phi\rangle$ by its coordinates—the amplitudes $\langle i \mid \phi\rangle$ in some complete basis $|i\rangle$—an operator is defined by giving its *matrix elements* in a particular representation. Consider the state $|\phi\rangle$ expressed in the basis $|i\rangle$, where

$$|\phi\rangle = \sum_i |i\rangle\langle i \mid \phi\rangle$$

When the operator $\hat{A}$ as given in Eq. (2.30) acts on this state we obtain

$$|\chi\rangle = \hat{A}\,|\phi\rangle = \sum_i \hat{A}\,|i\rangle\langle i \mid \phi\rangle \tag{2.32a}$$

The state $|\chi\rangle$ that we can express in the same representation through

$$|\chi\rangle = \sum_j |j\rangle\langle j \mid \chi\rangle \tag{2.32b}$$

will be completely defined if we know the amplitudes $\langle j \mid \chi\rangle$. These can be found from Eq. (2.32*a*) by closing it from the left with the basis states $\langle j|$

$$\langle j \mid \chi\rangle = \sum_i \langle j|\,\hat{A}\,|i\rangle\langle i \mid \phi\rangle$$

and introducing this result into Eq. (2.32*b*)

$$|\chi\rangle = \sum_j \sum_i |j\rangle\langle j|\,\hat{A}\,|i\rangle\langle i \mid \phi\rangle \tag{2.32c}$$

It is clear from the above result that if we know the amplitudes

$$\boxed{\langle j|\,\hat{A}\,|i\rangle = A_{ji}} \tag{2.33}$$

between *all* states $|i\rangle$ of this basis, we can determine the effect of the operator $\hat{A}$ on *any arbitrary* state $|\phi\rangle$: If the basis has n states, then the $(n \times n)$ amplitudes of Eq. (2.33) completely define the operator $\hat{A}$. In fact, it is sufficient to know the amplitudes of Eq. (2.33) in only one particular representation, because if we wish to work in a different basis $|\mu\rangle$ we can write Eq. (2.32*c*) in the following form

$$|\chi\rangle = \sum_{j,i} \sum_{\mu,\nu} |\nu\rangle\langle \nu \mid j\rangle\langle j|\,\hat{A}\,|i\rangle\langle i \mid \mu\rangle\langle \mu \mid \phi\rangle \tag{2.34a}$$

where we have used the closure relation [Eq. (2.16)] twice. Since $\langle \nu \mid j\rangle$ and $\langle i \mid \mu\rangle$ are the elements of the transformation from the basis $|j\rangle$ to the basis $|\nu\rangle$, and vice versa, we can always evaluate Eq. (2.34*a*) if the amplitudes $\langle j|\,\hat{A}\,|i\rangle$ are given in one basis.

We see that operators are defined by a *matrix* of $n \times n$ amplitudes given in one particular representation. This should be compared to a state vector which is defined by a *column* of n amplitudes, and to an amplitude which is defined by a single complex number (or function). The elements A_{ji} of the matrix are defined by Eq. (2.33). The first index (in this case j) labels the row of the matrix, while the second index (in this case i) labels the column. The amplitudes in the matrix defining $\hat{A}$ depend on the representation, just as the expansion coefficients of a state vector depend on the representation. Equation (2.34*a*) indicates how the *matrix elements* A_{ji} transform when the basis is changed

$$\langle \nu|\,\hat{A}\,|\mu\rangle = \sum_{i,j} \langle \nu \mid j\rangle\langle j|\,\hat{A}\,|i\rangle\langle i \mid \mu\rangle \tag{2.34b}$$

Eigenstates and Eigenvalues of an Operator

A matrix $A_{ji} = \langle j | \hat{A} | i \rangle$ is *Hermitian* if

$$A = A^{\dagger} = (A^{T})^{*} \tag{2.35a}$$

or in terms of the matrix elements

$$A_{ji} = A^{*}_{ij} \tag{2.35b}$$

If a matrix is Hermitian an important theorem states that it is always possible to find a basis $|\mu\rangle$ in which the matrix $A_{\nu\mu} = \langle \nu | A | \mu \rangle$ is *diagonal*. By diagonal we mean that

$$A_{\mu\nu} = 0, \qquad \text{if } \nu \neq \mu \tag{2.36a}$$

and only the elements $A_{\nu\nu}, A_{\mu\mu}, \ldots,$ etc., may be different from zero. Furthermore, the diagonal elements of a Hermitian matrix are *real*, since by the definition of Eqs. (2.35*a*) and (2.35*b*)

$$A_{\nu\nu} = (A^{\dagger})_{\nu\nu} = (A^{T}_{\nu\nu})^{*} = A^{*}_{\nu\nu} \tag{2.36b}$$

The elements $\hat{A}_{\nu\nu}$ of the matrix A when it is in diagonal form are called its *eigenvalues*. The n eigenvalues of a Hermitian matrix are real and can all be distinct or some of them may be identical. An eigenvalue can be positive, negative, or zero.

An operator $\hat{A}$ is said to be Hermitian if its matrix elements A_{ij} form a Hermitian matrix. Consider then a Hermitian operator and let the particular basis $|\mu\rangle$ be such that the corresponding matrix $A_{\nu\mu}$ is diagonal, or

$$\langle \nu | \hat{A} | \mu \rangle = 0, \qquad \text{if } \nu \neq \mu \tag{2.37a}$$

The basis states are orthonormal

$$\langle \nu \mid \mu \rangle = \delta_{\mu\nu}$$

We now use Eq. (2.32*c*) to find the effect of the operator $\hat{A}$ on one of the basis states, say $|\mu\rangle$

$$\begin{aligned} \hat{A} |\mu\rangle &= \sum_{\xi,\nu} |\xi\rangle\langle\xi| \hat{A} |\nu\rangle\langle \nu \mid \mu\rangle \\ &= \sum_{\xi,\nu} |\xi\rangle\langle\xi| \hat{A} |\nu\rangle\delta_{\nu\mu} = \sum_{\xi} |\xi\rangle\langle\xi| \hat{A} |\mu\rangle \end{aligned}$$

and in view of Eq. (2.37a)

$$\hat{A}\,|\mu\rangle = \sum_{\xi} |\xi\rangle\langle\xi|\,\hat{A}\,|\mu\rangle = |\mu\rangle\langle\mu|\,\hat{A}\,|\mu\rangle = |\mu\rangle A_{\mu\mu}$$

The diagonal matrix element $A_{\mu\mu}$ is the μth eigenvalue of $\hat{A}$, and we designate it by a_μ. Thus, when $\hat{A}$ operates on the state $|\mu\rangle$ it produces a new state $|\chi\rangle$, which is a *multiple* of the state $|\mu\rangle$

$$\hat{A}\,|\mu\rangle = a_\mu\,|\mu\rangle \qquad (2.37b)$$

We say that the state $|\mu\rangle$ is an eigenstate of operator $\hat{A}$ with the eigenvalue a_μ.

The great importance of the above results is that when Eq. (2.37b) is valid in one representation it remains valid in all representations, even in those where $\hat{A}$ is not any more diagonal. Of course, the matrix elements of $\hat{A}$ and the amplitudes describing the state $|\mu\rangle$ are different in the new basis, but the eigenvalues a_μ remain unchanged. In general, if an operator $\hat{A}$ acting on the state $|\phi\rangle$ changes the state only by a number a_ϕ (which may be complex), the state $|\phi\rangle$ is an eigenstate of $\hat{A}$ with the eigenvalue a_ϕ.

If $\hat{A}\,|\phi\rangle = a_\phi\,|\phi\rangle$, $|\phi\rangle$ is an eigenstate of $\hat{A}$ with eigenvalue a_ϕ

(2.37c)

In order to use this formalism we must also have a prescription for the action of an operator on a bra state. Given a state $|\phi\rangle$ and an operator $\hat{A}$ such that $\hat{A}\,|\phi\rangle = |\chi\rangle$, it holds that $\langle\chi| = \langle\phi|\,\hat{A}^\dagger$. The proof is straightforward, as given below, but the result is often used, and therefore we display it in Eq. (2.38)

$$\boxed{\text{if } \hat{A}\,|\phi\rangle = |\chi\rangle, \text{ then } \langle\phi|\,\hat{A}^\dagger = \langle\chi|} \qquad (2.38)$$

Proof of Eq. (2.38): From the definition of Hermiticity for a matrix [Eq. (2.24)] it follows that

$$\langle\phi|\,\hat{A}\,|\phi\rangle = A_{\phi\phi} = (A^\dagger)^*_{\phi\phi} = \langle\phi|\,\hat{A}^\dagger\,|\phi\rangle^* \qquad (2.39)$$

Since $\hat{A}\,|\phi\rangle = |\chi\rangle$ we have

$$\langle\phi|\,(\hat{A}\mid\phi\rangle) = \langle\phi\mid\chi\rangle = \langle\chi\mid\phi\rangle^*$$

Therefore

$$\langle \chi \mid \phi \rangle = \langle \phi | \hat{A} | \phi \rangle^* = (\langle \phi | \hat{A}^\dagger) | \phi \rangle$$

which implies $\langle \phi | \hat{A}^\dagger = \langle \chi |$ as stated.

It follows from Eq. (2.39) that if $|\phi\rangle$ is an eigenstate of $\hat{A}$ with eigenvalue a_ϕ, then $|\phi\rangle$ is also an eigenstate of $\hat{A}^\dagger$ with eigenvalue a_ϕ^*.

$$\text{If } \hat{A} |\phi\rangle = a_\phi |\phi\rangle, \text{ then } \hat{A}^\dagger |\phi\rangle = a_\phi^* |\phi\rangle \tag{2.40}$$

and also

$$\langle \phi | \hat{A} = a_\phi \langle \phi | \quad \text{and} \quad \langle \phi | \hat{A}^\dagger = a_\phi^* \langle \phi |$$

Of course, if $\hat{A}$ is Hermitian, $\hat{A}^\dagger = \hat{A}$, and all eigenvalues a_ϕ are real.

Example

Consider the discussion of polarized light, and let us depict the state of polarization of the electric field by the column vectors

$$|x\rangle = \begin{pmatrix} 1 \\ 0 \end{pmatrix} = \begin{matrix} \text{polarization} \\ \text{along } x \end{matrix} \qquad |y\rangle = \begin{pmatrix} 0 \\ 1 \end{pmatrix} = \begin{matrix} \text{polarization} \\ \text{along } y \end{matrix}$$

For instance, light polarized at 45° in the first quadrant is represented by

$$|\phi\rangle = \frac{1}{2^{1/2}} (|x\rangle + |y\rangle) = \frac{1}{2^{1/2}} \begin{pmatrix} 1 \\ 1 \end{pmatrix} \tag{2.41a}$$

[Compare to Eq. (2.6b).] We now introduce an active element that affects the two states of polarization differently. This is equivalent to an *operator* which we can express in matrix form as

$$\hat{A} = \begin{pmatrix} e^{i\theta} & 0 \\ 0 & e^{-i\theta} \end{pmatrix} \tag{2.41b}$$

with θ any arbitrary angle.

We note that $\hat{A}$ is already diagonal in this representation. The eigenvalues are $a_x = e^{i\theta}$ and $a_y = e^{-i\theta}$. The eigenstates are the basis vectors $|x\rangle$ and $|y\rangle$. The state $|\phi\rangle$ is *not* an eigenstate of $\hat{A}$ since

$$\hat{A} |\phi\rangle = |\chi\rangle = \frac{1}{2^{1/2}} \begin{pmatrix} e^{i\theta} \\ e^{-i\theta} \end{pmatrix} \tag{2.41c}$$

In fact, $|\chi\rangle$ represents elliptically polarized light, as can be seen from Eqs. (2.8), if we recall the time dependence of the electric field vector.

We note that the intensity of the light in the state $|\chi\rangle$ is the same as before: $\langle\chi \mid \chi\rangle = 1 = \langle\phi \mid \phi\rangle$. Note also that $\hat{A}$ as given in Eq. (2.41*b*) is *not* Hermitian (except for the trivial value $\theta = 0, \pi$).

In this example we used an active element rather than a polaroid, which would have been represented by a matrix of the form

$$\hat{M} = \begin{pmatrix} 1 & 0 \\ 0 & 0 \end{pmatrix} \tag{2.41d}$$

This is because $\hat{M}$, even though Hermitian, does not preserve the normalization. Physically this is evident from the fact that a polaroid absorbs one of the polarization components. As we will learn later, operators such as $\hat{M}$ are called projection operators.

A Complete Set of Basis States

We have seen that Hermitian operators have real eigenvalues. It is also true that the eigenstates of a Hermitian operator define a *complete set* of mutually orthogonal basis states. Because Hermitian operators satisfy these two conditions, they can be used to describe physical observables, as is discussed in detail in the next chapter.

We now prove the orthogonality of the eigenstates of the Hermitian operator $\hat{A}$. Let $|\phi\rangle$ and $|\chi\rangle$ be eigenstates corresponding to eigenvalues a_ϕ and a_χ, where $a_\chi \neq a_\phi$; $\hat{A}\,|\phi\rangle = a_\phi\,|\phi\rangle$ and $\hat{A}\,|\chi\rangle = a_\chi\,|\chi\rangle$ so that $\langle\chi|\,\hat{A}\,|\phi\rangle = a_\phi\langle\chi \mid \phi\rangle$. Since $\hat{A}$ is Hermitian we use Eq. (2.40) to write $\langle\chi|\,\hat{A}^\dagger = \langle\chi|\,\hat{A} = a_\chi\langle\chi|$ so that $\langle\chi|\,\hat{A}\,|\phi\rangle = a_\chi\langle\chi \mid \phi\rangle$. Subtracting the above two results

$$(a_\phi - a_\chi)\langle\chi \mid \phi\rangle = 0 \tag{2.42a}$$

and since $(\alpha_\phi - a_\chi) \neq 0$, it must hold that

$$\langle\chi \mid \phi\rangle = 0 \tag{2.42b}$$

In other words, the states $|\chi\rangle$ and $|\phi\rangle$ are orthogonal.

If $a_\phi = a_\chi$ we say that the two eigenstates $|\chi\rangle$ and $|\phi\rangle$ are *degenerate*. In that case Eq. (2.42*a*) cannot be used to establish the orthogonality of the states $|\chi\rangle$ and $|\phi\rangle$. Indeed, two degenerate eigenstates are not in general orthogonal. However, we can always form linear combinations of $|\chi\rangle$ and $|\phi\rangle$ that are mutually orthogonal. For instance, if

$$\hat{A}\,|\phi\rangle = a\,|\phi\rangle \quad \text{and} \quad \hat{A}\,|\chi\rangle = a\,|\chi\rangle$$

we normalize the state $\langle\phi \mid \phi\rangle = 1$ and choose

$$|\phi'\rangle = |\phi\rangle \quad \text{and} \quad |\chi'\rangle = |\chi\rangle - |\phi\rangle\langle\phi \mid \chi\rangle \tag{2.43}$$

Clearly the states $|\phi'\rangle$ and $|\chi'\rangle$ are eigenstates of $\hat{A}$ with eigenvalue a

$$\hat{A}\,|\phi'\rangle = \hat{A}\,|\phi\rangle = a\,|\phi\rangle = a\,|\phi'\rangle$$
$$\hat{A}\,|\chi'\rangle = \hat{A}\,|\chi\rangle - \hat{A}\,|\phi\rangle\langle\phi \mid \chi\rangle = a(|\chi\rangle - |\phi\rangle\langle\phi \mid \chi\rangle) = a\,|\chi'\rangle$$

and are mutually orthogonal since

$$\langle\phi' \mid \chi'\rangle = \langle\phi \mid \chi'\rangle = \langle\phi \mid \chi\rangle - \langle\phi \mid \phi\rangle\langle\phi \mid \chi\rangle = 0$$

Products of Operators

Two operators $\hat{A}$ and $\hat{B}$ may act one after the other on a given state. Let

$$\hat{A}\,|\phi\rangle = |\chi\rangle \quad \text{and} \quad \hat{B}\,|\chi\rangle = |\psi\rangle$$

We write

$$\hat{B}\hat{A}\,|\phi\rangle = |\psi\rangle \tag{2.44a}$$

and imply that $\hat{B}$ acts *after* $\hat{A}$. The effect of the combined operation $\hat{B}\hat{A}$ can be obtained when the matrices $\langle j|\,\hat{A}\,|i\rangle$ and $\langle j|\,\hat{B}\,|i\rangle$ defining the operators $\hat{A}$ and $\hat{B}$ are known. In this representation the state $|\psi\rangle$ is given by

$$|\psi\rangle = \sum_j |j\rangle\langle j \mid \psi\rangle$$

The amplitudes $\langle j \mid \psi\rangle$ that define the state $|\psi\rangle$ can be obtained from Eq. (2.44a)

$$\langle j \mid \psi\rangle = \langle j|\,\hat{B}\hat{A}\,|\phi\rangle = \sum_i \langle j|\,\hat{B}\hat{A}\,|i\rangle\langle i \mid \phi\rangle \tag{2.44b}$$

where the last step follows from closure properties. We can also use the closure property *between* the two operators to write

$$\langle j \mid \psi\rangle = \sum_{i,k} \langle j|\, \hat{B}\, |k\rangle\langle k|\, \hat{A}\, |i\rangle\langle i \mid \phi\rangle \tag{2.44c}$$

Comparison of Eqs. (2.44*b*) and (2.44*c*) shows that the matrix $\langle j|\, \hat{B}\hat{A}\, |i\rangle$ defining the combined operation $\hat{B}\hat{A}$ is the product of the matrices defining A and B separately. This is because

$$\langle j|\, \hat{B}\hat{A}\, |i\rangle = \sum_{k} \langle j|\, \hat{B}\, |k\rangle\langle k|\, \hat{A}\, |i\rangle \tag{2.45a}$$

can be written in matrix notation as

$$(BA)_{ji} = \sum_{k} B_{jk}A_{ki} \tag{2.45b}$$

Since matrix multiplication is noncommutative it is evident that in general the operation $\hat{B}\hat{A}$ may differ from the operation $\hat{A}\hat{B}$. If in a particular basis the matrix for $\hat{A}, \langle k|\, \hat{A}\, |i\rangle$, and the matrix for $\hat{B}, \langle j|\, \hat{B}\, |k\rangle$, are *both diagonal,* the two operations will *commute.*

Because quantum-mechanical operators can be defined in matrix notation many explicit calculations are facilitated by using a matrix representation. In fact, quantum mechanics was originally discovered by Heisenberg in the form of "matrix mechanics" and independently by Schrödinger in the form of "wave mechanics", which is discussed in the first chapter. However, these forms are only different representations of the same theory, that of quantum mechanics.

2.5. LANGUAGE OF QUANTUM MECHANICS

We now summarize what we have learned in this chapter about the description of quantum systems in terms of states defined in a linear vector space with complex coefficients—the Hilbert space. The first two statements are postulates of quantum mechanics; the next two are essential definitions of the theory. The remaining statements are consequent relations of the definitions and of the rules of linear algebra. They are, however, important enough to be included here.

A. *Postulates*

1. A physical system is completely described by its *quantum-mechanical state* $|\psi\rangle$. The state $|\psi\rangle$ contains all possible information that can be obtained about the system.

2. The occurrence of an event in Nature is specified by a complex number (or function), the *probability amplitude* ϕ, and we can predict only the *probability* P that any particular event will occur:

$$P = |\phi|^2$$

B. *Essential Definitions*

3. If the system is *initially* in the state $|\psi\rangle$ the amplitude for the occurrence of the *final* state $|\chi\rangle$ is given by

$$\phi = \langle \chi \mid \psi \rangle$$

4. Any arbitrary state can be represented by a *linear* superposition of a *complete* set of basis states $|i\rangle$

$$|\psi\rangle = \sum_i |i\rangle\langle i \mid \psi\rangle$$

The expansion coefficients are the amplitudes between the state $|\psi\rangle$ and the individual basis states $|i\rangle$. Thus, $|\psi\rangle$ is a vector in Hilbert space.

C. *Consequent Relations*

5. The basis states $|i\rangle$ of the Hilbert space play the role of a set of unit vectors and the amplitudes $\langle i \mid \psi\rangle$ play the role of a set of coordinates for the state $|\psi\rangle$. A Hilbert space can have a discrete or a continuous set of basis states. The number of states of a complete set may be denumerable or infinite.

6. It is always possible to choose the basis states within a complete set to be orthogonal to one another and normalized to unity

$$\langle i \mid j\rangle = \delta_{ij}$$

It is also customary to normalize the states $|\psi\rangle$ so that

$$\langle\psi \mid \psi\rangle = 1$$

7. The amplitude for the process where the initial and final states have been interchanged (time reversal) is the complex conjugate of the original amplitude

$$\langle\psi \mid \chi\rangle = \langle\chi \mid \psi\rangle^*$$

8. Two orthonormal basis states spanning the same Hilbert space are connected by a unitary transformation

$$|\mu\rangle = |i\rangle S^\dagger = \sum_i |i\rangle\langle i \mid \mu\rangle$$

The coordinates of an arbitrary state $|\psi\rangle$ in the two bases are related through

$$\langle\mu \mid \psi\rangle = S\langle i \mid \psi\rangle = \sum_i \langle\mu \mid i\rangle\langle i \mid \psi\rangle$$

where the transformation matrix S must be unitary

$$S^{-1} = S^\dagger \quad \text{or} \quad S^\dagger S = \mathbb{1}$$

A unitary transformation preserves the normalization of state vectors.

9. For a *complete* set of states the *closure* relation is always valid

$$\sum_i |i\rangle\langle i| = 1$$

It can be used to expand or contract relations between amplitudes and states.

10. *Operators* act on states to produce a different or the same state

$$\hat{A}\,|\psi\rangle = |\chi\rangle$$

An operator $\hat{A}$ is completely specified when its matrix elements between all the states of a complete set are known

$$|\chi\rangle = \hat{A}\,|\psi\rangle = \sum_{ji} |j\rangle\langle j|\,\hat{A}\,|i\rangle\langle i \mid \psi\rangle$$

11. When an operator reproduces the same state on which it acts

modulo a (complex) number a

$$\hat{A}\,|\psi\rangle = a\,|\psi\rangle$$

the state $|\psi\rangle$ is an *eigenstate* of the operator $\hat{A}$ with the *eigenvalue* a.

12. Hermitian operators are defined through

$$\hat{H}^\dagger = \hat{H}$$

Therefore,

$$\langle j|\,\hat{H}\,|i\rangle = \langle i|\,\hat{H}^\dagger\,|j\rangle^* = \langle i|\,\hat{H}\,|j\rangle^*$$

The totality of the eigenstates of a Hermitian operator forms a complete set and can be chosen mutually orthogonal and normalized. Furthermore, their eigenvalues are real.

13. The following conjugation relations hold

$$\text{if} \quad |\chi\rangle = \hat{A}\,|\psi\rangle, \text{ then } \langle\chi| = \langle\psi|\,\hat{A}^\dagger$$
$$\text{if } \hat{A}\,|\psi\rangle = a_\psi\,|\psi\rangle, \text{ then } \hat{A}^\dagger\,|\psi\rangle = a_\psi^*\,|\psi\rangle$$

14. The matrix elements of the product of two operators $\hat{C} = \hat{A}\hat{B}$ are given by

$$\langle i|\,\hat{C}\,|j\rangle = \sum_k \langle i|\,\hat{A}\,|k\rangle\langle k|\,\hat{B}\,|j\rangle$$

Two operators $\hat{A}$ and $\hat{B}$ are said to commute when $\hat{A}\hat{B} = \hat{B}\hat{A}$. In general, however,

$$\hat{A}\hat{B} \neq \hat{B}\hat{A}$$

Furthermore,

$$(\hat{A}\hat{B})^\dagger = \hat{B}^\dagger\hat{A}^\dagger$$

15. Under a unitary transformation S of the basis states the matrix elements of an operator transform according to

$$\hat{A}' = S\hat{A}S^{-1}$$

or

$$\langle \nu|\,\hat{A}\,|\mu\rangle = \sum_{ij} \langle \nu \mid i\rangle\langle i|\,\hat{A}\,|j\rangle\langle j \mid \mu\rangle$$

Problems

PROBLEM 1

Consider the matrix

$$\hat{A} = \begin{vmatrix} 0 & -i & 0 \\ i & 0 & -i \\ 0 & i & 0 \end{vmatrix}$$

(*a*) Bring it into diagonal form $\hat{A}'$.
(*b*) Find the corresponding eigenvectors and normalize them to 1.
(*c*) Find the matrix S that makes $\hat{A}$ diagonal, namely,

$$\hat{A}' = S\hat{A}S^{-1}$$

Consult Appendix 4.

PROBLEM 2

Consider the transformation matrix S given in the last example of Section 2.3 (p. 76). Introduce the new basis states

$$|\xi_x\rangle = -\frac{1}{\sqrt{2}}(|\alpha\rangle - |\gamma\rangle) \qquad |p_x\rangle = -\frac{1}{\sqrt{2}}(|k\rangle - |m\rangle)$$

$$|\xi_y\rangle = -\frac{i}{\sqrt{2}}(|\alpha\rangle + |\gamma\rangle) \qquad |p_y\rangle = \frac{i}{\sqrt{2}}(|k\rangle + |m\rangle)$$

$$|\xi_z\rangle = |\beta\rangle \qquad |p_z\rangle = |l\rangle$$

(*a*) Form the transformation matrix corresponding to the nine amplitudes

$$\langle \xi \mid p \rangle_{ij}$$

(*b*) Compare it to the transformation matrix for a vector rotated around the Y-axis.

PROBLEM 3

Which of the following operators are linear?

(*a*) $\Omega\phi(x) = \phi(-x)$
(*b*) $\Omega\phi(x) = \phi^2(x)$
(*c*) $\Omega\phi(x) = \phi(x) + k$ (k const)
(*d*) $\Omega\phi(x) = \phi(x + k)$ (k const)
(*e*) $\Omega\phi(x) = \phi(x/2)$
(*f*) $\Omega\phi(x) = \int_{-\infty}^{\infty} K(x, x')\phi(x')\, dx'$, with $K(x, x') = K^*(x', x)$
(*g*) $\Omega\phi(x) = \int_{-\infty}^{\infty} K(x, x')\phi(x')\, dx'$, with $K(x, x') = -K(x', x)$

PROBLEM 4

Which of the following operators are Hermitian?

(*a*) $\Omega\phi(x) = \phi(x + a)$
(*b*) $\Omega\phi(x) = \phi^*(x)$
(*c*) $\Omega\phi(x) = \phi(-x)$
(*d*) $\Omega\phi(x) = \int_{-\infty}^{\infty} K(x, x')\phi(x')\, dx'$, where $K(x, x')$ is real and $K(x, x') = -K(x', x)$

PROBLEM 5

If $\hat{A}$, $\hat{B}$, and $\hat{C}$ are Hermitian operators, determine if the following combinations are Hermitian.

(*a*) $\hat{A}+\hat{B}$
(*b*) $\frac{1}{2i}[\hat{A},\hat{B}]=\frac{1}{2i}(\hat{A}\hat{B}-\hat{B}\hat{A})$
(*c*) $(\hat{A}\hat{B}\hat{C}-\hat{C}\hat{B}\hat{A})$
(*d*) $\hat{A}^2+\hat{B}^2+\hat{C}^2$
(*e*) $(\hat{A}+i\hat{B})$

PROBLEM 6

(*a*) $\hat{A}$ and $\hat{B}$ are Hermitian operators. Show that $\hat{A}\hat{B}$ is Hermitian if and only if $\hat{A}$ commutes with $\hat{B}$, that is $\hat{A}\hat{B}=\hat{B}\hat{A}$.
(*b*) If $\hat{A}$ is a Hermitian operator that obeys

$$\hat{A}^4=1$$

find its eigenvalues. Find the eigenvalues when $\hat{A}$ is not Hermitian.

PROBLEM 7

Let $\hat{U}$ be a unitary operator

$$\hat{U}^{\dagger}\hat{U}=1$$

(*a*) Show that if $\langle\psi\mid\psi\rangle=1$, then $\langle\hat{U}\psi\mid\hat{U}\psi\rangle=1$.
(*b*) If $|u_i\rangle$ is a complete orthonormal set

$$\langle u_i\mid u_j\rangle=\delta_{ij}$$

show that $|v_i\rangle=\hat{U}\,|u_i\rangle$ is also an orthonormal set.

PROBLEM 8

(*a*) Discuss how a quarter-wave plate transforms linearly polarized light into circularly polarized light.

(*b*) Two quarter-wave plates are placed one behind the other (i) so that their principal axes coincide and (ii) so that their principal axes are at 90° with respect to each other. Assume that the incident light is linearly polarized at 45° with respect to the principal axis of the first quarter-wave plate.

Express the incident light in terms of appropriate base states and the effect of quarter-wave plates by appropriate 2×2 matrices in the same base states. Find the state of the outgoing light and obtain the transmitted intensity and polarization of the light.

PROBLEM 9

(*a*) Unpolarized light is incident on an imperfect polaroid that transmits 90% of the light polarized along its axis and 10% of the light polarized along the direction perpendicular to its axis. What is the *intensity* of the transmitted light?

(*b*) You can express unpolarized light by a density matrix

$$\rho_{\text{incident}} = \frac{1}{2}\begin{pmatrix} 1 & 0 \\ 0 & 1 \end{pmatrix}$$

and the effect of the quarter-wave plates by 2×2 matrices, as in Problem 8. Show that the intensity of the transmitted light is given by

$$I = (\rho_{\text{out}})_{11} + (\rho_{\text{out}})_{22}$$

Chapter 3

OBSERVABLES AND THE DESCRIPTION OF QUANTUM SYSTEMS

In this chapter we show through simple examples how quantum mechanics is applied to the description of physical systems. We start by discussing the process of measurement and show that we can predict only the *probability* that a measurement will yield a particular result. We then study the dependence of quantum-mechanical amplitudes on time and position. This allows us to treat simple examples and to exhibit the correspondence between the classical and quantum results. In discussing the position dependence of amplitudes we introduce the concept of the wavefunction with which some readers may already be familiar.

We also show how probabilities are combined, i.e. for indistinguishable processes we must add amplitudes, whereas for distinguishable processes we must add probabilities. Finally, we consider briefly systems with identical particles and indicate why the amplitude must be either symmetric or antisymmetric under the exchange of any two identical particles.

The concepts introduced in this chapter are fundamental and absolutely central to the understanding of quantum mechanics; they are summarized in Section 3.7. We have purposely kept the mathematics as simple as possible so as not to obscure the basic ideas. Further along in the book, as these concepts are applied to more complex physical systems, we develop the necessary mathematical techniques which enable us to quantitatively solve realistic problems.

3.1. PROCESS OF MEASUREMENT

Measurement implies that we can assign a numerical value to a physical property of a system. To achieve this we must use some apparatus that performs the measurement. In simple cases the human senses are adequate for this task, as when measuring the position of a macroscopic body. If the body is microscopic a more sensitive apparatus may be needed, such as a fluorescent screen or a bubble chamber. In all cases, however, the apparatus must *interact* with the physical system, the properties of which are being measured. We shall identify as *physical observables* those properties of a system that can be determined by direct observation.

As an example, consider the measurement of the state of linear polarization of a beam of light: it can be achieved by using a polaroid as discussed in the previous chapter. The process of measurement, which involves the insertion of the polaroid into the beam, forces the beam to emerge in a state of linear polarization along the principal axis of the polaroid. A similar effect takes place when a Stern–Gerlach apparatus is used to measure the spin projection of the atoms in a beam. By an appropriate arrangement of apertures only atoms with a particular projection of their spin are transmitted, and one can measure the fraction of the total beam that is in that particular state. For example spin-1 atoms emerging from a Stern–Gerlach experiment can be selected so that they all are in one of the three possible basis states defined by the apparatus. One concludes that the process of measurement affects the system that is being measured.

In quantum mechanics we describe a physical system by a *state*, and we have already remarked that the process of measurement may change the state of the system. In this respect, a measuring device can be represented by an operator. After the measurement the system will be in one of the eigenstates of the operator that describes the measuring apparatus. This was the case in the two examples of measurement that we considered, and it is true in general. We will use the operator $\hat{Q}$ to designate the measuring device and let the complete set of eigenstates of $\hat{Q}$ be $|q_1\rangle, |q_2\rangle, \ldots, |q_n\rangle$, corresponding to the eigenvalues $\lambda_1, \lambda_2, \ldots, \lambda_n$. The state that is to be measured is designated by $|\phi\rangle$, and we can expand it in the eigenstates of $\hat{Q}$ as

$$|\phi\rangle = \sum_j |q_j\rangle\langle q_j \mid \phi\rangle \tag{3.1}$$

A complete measurement of the state $|\phi\rangle$ would consist in determining all the amplitudes $\langle q_j \mid \phi\rangle$. To achieve this we can let the device $\hat{Q}$ act on the state $|\phi\rangle$ and observe the probability of finding $\hat{Q}\,|\phi\rangle$ in one of the eigenstates $|q_i\rangle$. In our formalism we operate with $\hat{Q}$ on $|\phi\rangle$ as given by Eq. (3.1) and then close from the left (take the inner product) with the bra $\langle q_i|$. Remembering that the $|q_j\rangle$ are eigenstates of $\hat{Q}$ we obtain

$$\begin{aligned}\langle q_i|\,\hat{Q}\,|\phi\rangle &= \sum_j \langle q_i|\,\hat{Q}\,|q_j\rangle\langle q_j \mid \phi\rangle \\ &= \sum_j \lambda_i\delta_{ij}\langle q_j \mid \phi\rangle = \lambda_i\langle q_i \mid \phi\rangle\end{aligned} \tag{3.2}$$

Equation (3.2) describes the measurement process in quantum mechanics. *The result of a measurement with an apparatus $\hat{Q}$ is always one of its eigenvalues λ_i*. Immediately after the measurement the system will be in the corresponding eigenstate $|q_i\rangle$. Since the eigenstates $|q_i\rangle$ are orthogonal, it is not possible that a *single* measurement can produce two different eigenvalues λ_i and λ_j. We cannot predict which of the eigenvalues λ will result from the measurement, but the *probability* of obtaining the particular eigenvalue λ_i is given by

$$P(\lambda_i) = |\langle q_i \mid \phi\rangle|^2 \tag{3.3}$$

If the state $|\phi\rangle$ is already an eigenstate of $\hat{Q}$, say $|\phi\rangle = |q_k\rangle$, then we can predict with *certainty* that the results of the measurement will be the eigenvalue λ_k. This can be easily seen from Eq. (3.2), where, if $|\phi\rangle = |q_k\rangle$, then $\langle q_i \mid \phi\rangle = \langle q_i \mid q_k\rangle = \delta_{ik}$, and Eq. (3.3) gives $P(\lambda_k) = 1$. We note that the interpretation we have given for Eq. (3.2) agrees with the conclusions drawn from the Stern–Gerlach experiment. In that device a single atom may emerge only from *one* of the slits, depending on whether it is in the (+), the (0) or the (−) state. A *single* atom never emerges from two slits, implying that it is always in one of the eigenstates of the measuring apparatus.

Suppose now that we make a measurement on the arbitrary state $|\phi\rangle$ with the apparatus $\hat{Q}$ and immediately repeat the measurement with an identical apparatus $\hat{Q}$. If the first measurement yields the eigenvalue λ_i we can predict with certainty that the second measurement will result in the same eigenvalue λ_i. This follows because if the first measurement resulted in λ_i, then the system is by necessity in the state $|q_i\rangle$. But a measurement with $\hat{Q}$ on the state $|q_i\rangle$ always results in λ_i. As one would expect, we see that measurements made in *short* intervals of time are *repeatable.* We stress the condition that the time interval between the measurements must be short because any quantum-mechanical state can change in time. Thus, the interval must be short with respect to the time evolution of the system, as will be discussed in more detail later.

A complete measurement of an arbitrary state $|\phi\rangle$ implies the determination of all the amplitudes $\langle q_i \mid \phi\rangle$. Therefore, we must perform *repeated* measurements on the same state $|\phi\rangle$. Since the state $|\phi\rangle$ is changed by the measurement, we must make the measurement on *similarly prepared* states $|\phi\rangle$. If the eigenvalue λ_i is obtained in a fraction P_i of all measurements, then from Eq. (3.3) we can determine the magnitude of $\langle q_i \mid \phi\rangle$. (More sophisticated measurements can reveal the relative phase of the amplitudes as well.) We see that even if we are given the state $|\phi\rangle$ and the apparatus $\hat{Q}$, we can only predict the *probability* of the outcome of a measurement unless $|\phi\rangle$ is an eigenstate of $\hat{Q}$. This is in agreement with the uncertainty principle and the interpretation of probability waves discussed in Chapter 1.

As an illustration of the measurement process, the oven of a Stern–Gerlach apparatus can be considered to produce atoms similarly prepared in the state $|\phi\rangle$. As we count the atoms reaching the detectors, we are performing repeated measurements (one measurement for each atom) on the state $|\phi\rangle$. If N_+, N_0, and N_- atoms reach each detector we conclude that

$$\frac{N_+}{N_+ + N_0 + N_-} = P(+) = |\langle + \mid \phi\rangle|^2, \qquad \text{etc.}$$

We are in a position then to predict the probability that an atom will arrive at the (+), (0), or (−) detector; however, we cannot predict with certainty that the next atom will reach a particular detector.

Expectation Values

We defined a physical observable as a property of our system that can be determined by direct observation. Consequently, the measurement of a physical observable is expressed by *real* numbers. Since a measurement of the observable yields one of the eigenvalues λ of the corresponding operator, these eigenvalues must all be real. We conclude that physical observables can be represented only by *Hermitian operators.*

As an example of a physical observable, we may consider the total energy $E = T + V$ of, say, a hydrogen atom. The eigenstates of the operator $\hat{E}$ can be taken as discrete, and we label them by

$$|\psi_1\rangle, |\psi_2\rangle, \ldots, |\psi_n\rangle$$

with eigenvalues $\varepsilon_1, \varepsilon_2, \ldots, \varepsilon_n$. A measurement of the energy of the atom will always yield one of the eigenvalues ε_i. Therefore, the observed energies may form a discrete (quantized) spectrum as observed experimentally. If the atom is in an eigenstate $|\psi_n\rangle$ of the energy operator, then the measurement will yield the eigenvalue ε_n according to

$$\hat{E}\,|\psi_n\rangle = \varepsilon_n\,|\psi_n\rangle \tag{3.4}$$

On the other hand, the atom need not be in an eigenstate of $\hat{E}$, but may be in an arbitrary state $|\phi\rangle$. This state can be expanded into the eigenstates of the energy operator

$$|\phi\rangle = \sum_n |\psi_n\rangle\langle\psi_n \mid \phi\rangle = \sum_n |\psi_n\rangle C_n \tag{3.5a}$$

where we designate the amplitudes $\langle\psi_n \mid \phi\rangle$ by C_n. We assume that the state $|\phi\rangle$ is properly normalized, and this implies that

$$\langle\phi \mid \phi\rangle = \sum_n |\langle\psi_n \mid \phi\rangle|^2 = \sum_n |C_n|^2 = 1 \tag{3.5b}$$

provided that the eigenstates $|\psi_n\rangle$ were chosen to be orthonormal.

A measurement of the energy of an atom in the state $|\phi\rangle$ can yield any one of the eigenvalues ε_n. If we make repeated measurements on similarly prepared atoms we can find the *average value* $\langle\varepsilon\rangle$ of the energy for the state $|\phi\rangle$; this is given by

$$\langle \varepsilon \rangle = \frac{\sum_n \varepsilon_n P_n}{\sum_n P_n} = \frac{\sum_n \varepsilon_n |\langle \psi_n \mid \phi \rangle|^2}{\sum_n |\langle \psi_n \mid \phi \rangle|^2}$$

$$= \sum_n \varepsilon_n |\langle \psi_n \mid \phi \rangle|^2 = \sum_n \varepsilon_n |C_n|^2 \tag{3.6a}$$

where P_n is the probability of obtaining the eigenvalue ε_n in any one measurement. In the third step we made use of Eq. (3.5*b*). We now show that the average value of the energy in the state $|\phi\rangle$ is given directly by the matrix element of the energy operator

$$\langle \varepsilon \rangle = \langle \phi | \hat{E} | \phi \rangle \tag{3.6b}$$

Proof: We first operate with $\hat{E}$ on Eq. (3.5*a*)

$$\hat{E} |\phi\rangle = \sum_n \hat{E} |\psi_n\rangle\langle \psi_n \mid \phi \rangle = \sum_n \varepsilon_n |\psi_n\rangle\langle \psi_n \mid \phi \rangle$$

and then we close from the left with the bra $\langle \phi |$ to obtain

$$\langle \phi | \hat{E} |\phi\rangle = \sum_k \langle \phi \mid \psi_k\rangle\langle \psi_k | \hat{E} |\phi\rangle = \sum_{k,n} \langle \phi \mid \psi_k\rangle \varepsilon_n \langle \psi_k \mid \psi_n \rangle\langle \psi_n \mid \phi \rangle$$

$$= \sum_{k,n} \varepsilon_n \langle \phi \mid \psi_n \rangle \delta_{kn} \langle \psi_n \mid \phi \rangle = \sum_n \varepsilon_n \langle \phi \mid \psi_n \rangle\langle \psi_n \mid \phi \rangle$$

$$= \sum_n \varepsilon_n |\langle \psi_n \mid \phi \rangle|^2 = \langle \varepsilon \rangle$$

where in the last step the result of Eq. (3.6*a*) was used.

The expression obtained in Eq. (3.6*b*) is obviously not restricted to the energy operator. For any operator we define its *expectation value* in the state $|\phi\rangle$ as the diagonal matrix element

$$\langle \phi | \hat{Q} |\phi\rangle \qquad \textit{expectation value}$$

For Hermitian operators the expectation value equals the average value of repeated measurements of $\hat{Q}$ on similarly prepared states $|\phi\rangle$

$$\boxed{\langle \phi | \hat{Q} |\phi\rangle = \langle q \rangle} \tag{3.7}$$

When $|\phi\rangle$ is an eigenstate $|q_n\rangle$ of $\hat{Q}$, then the expectation value is the corresponding eigenvalue q_n.

Interpretation

That the measurement of a physical observable $\hat{Q}$ on similarly prepared states $|\phi\rangle$ yields *different* values q_n, is contrary to the intuition gained in classical physics. In classical physics the state of a system completely determines the outcome of a measurement, whereas the quantum-mechanical interpretation of the measurement process asserts that we can predict only the probability of the occurrence of an event. The probabilistic interpretation of quantum mechanics led to much debate in the first half of this century, often in the form of paradoxes that appeared to lead to contradictions. One such famous paradox discussed by Einstein† can be illustrated by the decay of the π^0-meson.

The π^0-meson is an elementary particle that decays into two photons with lifetime $\tau = 10^{-16}$ sec

$$\pi^0 \to \gamma\gamma$$

When the π^0 is at rest the two photons are emitted back to back with equal energies as shown in Fig. 3.1. It is known that the π^0 has spin 0, whereas the photon has spin 1. Along the direction of motion of the two photons there can be no orbital angular momentum (since $\mathbf{l} = \mathbf{r} \times \mathbf{p} = 0$ when $\mathbf{r}$ is parallel to $\mathbf{p}$) and, therefore, the projection of the angular momentum on that axis must be zero. To satisfy this requirement the photons must both be either right-hand-circular (RHC) polarized or left-hand-circular (LHC) polarized, as shown in (a) and (b) of the figure. Thus, the state of the two-photon system is described as

$$|\phi_a\rangle = |R_1 R_2\rangle \tag{3.8a}$$

or

$$|\phi_b\rangle = |L_1 L_2\rangle \tag{3.8b}$$

for the two cases. $R_{1,2}$ refers to RHC polarization for photons 1, 2 and, similarly, $L_{1,2}$ refers to LHC polarization. Since either configuration is equally probable, we must take a linear combination of the two

$$|\phi\rangle = \frac{1}{2^{1/2}}(|R_1 R_2\rangle \pm |L_1 L_2\rangle) \tag{3.9a}$$

† A. Einstein, B. Podolsky, and N. Rosen, *Phys. Rev.* **47,** 777 (1935).

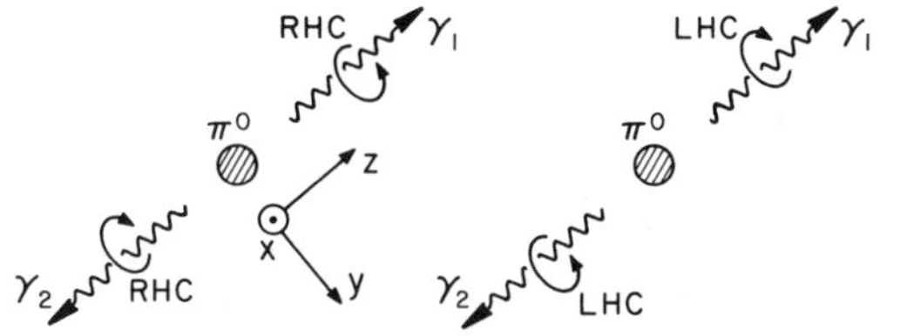

FIGURE 3.1. Decay of a π^0-meson into two photons. Both photons must have the same sense of circular polarization in order to conserve angular momentum.

It is found experimentally that the minus sign is appropriate for the π^0-meson (the π^0 is pseudoscalar under inversion of the coordinates; see Example 2), and therefore

$$|\phi\rangle = \frac{1}{2^{1/2}}(|R_1R_2\rangle - |L_1L_2\rangle) \tag{3.9b}$$

The states $R_{1,2}$ are orthogonal to $L_{1,2}$ and can be decomposed into linear polarization states according to Eq. (2.9a).

From Eq. (3.9b) we can obtain the amplitude for finding the system with both photons RHC polarized

$$\begin{aligned}\langle R_1R_2 \mid \phi\rangle &= \frac{1}{2^{1/2}}\langle R_1R_2 \mid R_1R_2\rangle - \frac{1}{2^{1/2}}\langle R_1R_2 \mid L_1L_2\rangle \\ &= \frac{1}{2^{1/2}}\langle R_1R_2 \mid R_1R_2\rangle = \frac{1}{2^{1/2}}\end{aligned}$$

and thus the probability

$$P(R_1R_2) = |\langle R_1R_2 \mid \phi\rangle|^2 = \tfrac{1}{2}$$

Similarly, $P(L_1L_2) = \frac{1}{2}$, whereas $P(R_1L_2) = P(L_1R_2) = 0$. Consequently, if we measure photon 1 to be RHC polarized, we can predict with certainty that photon 2 also will be RHC; the same holds true for LHC polarization. So far, our argument is not different from the conclusions we would derive from classical physics.

Alternately, we can express the state of the two photons in terms of their linear polarization along the X and Y axes shown in Fig. 3.1. One finds

$$|\phi\rangle = \frac{i}{2^{1/2}}(|X_1Y_2\rangle + |Y_1X_2\rangle) \tag{3.9c}$$

and in view of the orthogonality $\langle X_1 | Y_1 \rangle = 0$, etc.,

$$\langle X_1 Y_2 | \phi \rangle = \frac{i}{2^{1/2}} \qquad \langle Y_1 X_2 | \phi \rangle = \frac{i}{2^{1/2}}$$

$$\langle X_1 X_2 | \phi \rangle = \langle Y_1 Y_2 | \phi \rangle = 0$$

In other words, if photon 1 is polarized along the X-axis, we can predict with certainty that photon 2 is polarized along the Y-axis, and likewise for the opposite combination.

We see that if we use a circular polarization analyzer for photon 1, we are certain to find photon 2 circularly polarized as well. But if we use a linear polarization analyzer for photon 1, we are certain to find photon 2 to be linearly polarized in an orthogonal orientation. What bothered Einstein was the following: how can photon 2 know what type of analyzer photon 1 will encounter and then adjust its own polarization accordingly? After all, the π^0 decays, and the two photons have separated long before photon 1 reaches the polarizer. If the photons are circularly polarized, then either linear polarization is *equally* probable, and what we do to photon 1 should not affect photon 2.

In quantum mechanics, on the other hand, there is no difficulty in interpreting the experimental observations. The two photons form a single system described by the state $|\phi\rangle$. Any measurement of the state $|\phi\rangle$ will yield one of the eigenvalues of the measuring device. If we use a circular polarization analyzer the possible eigenstates are only

$$|R_1 R_2\rangle \quad \text{or} \quad |L_1 L_2\rangle$$

If we use a linear polarization analyzer the possible eigenstates are only

$$|X_1 Y_2\rangle \quad \text{or} \quad |Y_1 X_2\rangle$$

and any measurement will yield one of these two possibilities. Such behavior may appear to be peculiar, but it is confirmed by *all* experiments in which it is tested.

Example 1

Let us show here how one obtains Eq. (3.9*c*) from Eq. (3.9*b*). We have the following connection between states of linear and circular polarization [see Eqs. (2.9)]

$$|R_1\rangle = \frac{1}{2^{1/2}}(|X_1\rangle + i\,|Y_1\rangle)$$

$$|L_1\rangle = \frac{1}{2^{1/2}}(|X_1\rangle - i\,|Y_1\rangle)$$

Thus

$$|R_1R_2\rangle = \frac{1}{2}(|X_1X_2\rangle - |Y_1Y_2\rangle) + \frac{i}{2}(|X_1Y_2\rangle + |Y_1X_2\rangle)$$

$$|L_1L_2\rangle = \frac{1}{2}(|X_1X_2\rangle - |Y_1Y_2\rangle) - \frac{i}{2}(|X_1Y_1\rangle + |Y_1X_2\rangle)$$

and, therefore,

$$|\phi\rangle = \frac{1}{2^{1/2}}(|R_1R_2\rangle - |L_1L_2\rangle) = \frac{i}{2^{1/2}}(|X_1Y_2\rangle + |Y_1X_2\rangle)$$

Example 2

Let us show here that the state $|\phi\rangle$ defined by Eq. (3.9*b*) changes sign under an inversion of the coordinates:

We refer to Fig. 3.1(a) and note that for photon 1 the sense of rotation is from $X \to Y$, whereas the momentum is along the Z-axis. If we invert the coordinate axes $X' = -X$, $Y' = -Y$, $Z' = -Z$. With respect to the new coordinate system, the sense of rotation is still from $X' \to Y'$, but the momentum is along $-Z'$. Thus, the helicity of the photons has reversed sign. This is the familiar effect whereby inversion changes a right-handed coordinate system into a left-handed one, and vice-versa. Thus, using R', L' to denote the helicity in the new system, we have

$$|\phi\rangle = \frac{1}{2^{1/2}}(|L_1'L_2'\rangle - |R_1'R_2'\rangle)$$

$$= -\frac{1}{2^{1/2}}(|R_1'R_2'\rangle - |L_1'L_2'\rangle) = -|\phi'\rangle$$

where $|\phi'\rangle$ stands for the same functional relationship as $|\phi\rangle$ but is expressed in the new coordinate system.

3.2. TIME DEPENDENCE OF QUANTUM-MECHANICAL AMPLITUDES

We are familiar with the notion that observables describing a physical system may be changing in time. We speak of time dependence, a very simple example being the position of a moving particle. Something analogous must hold in quantum mechanics, and we must be prepared to allow the state describing a quantum-mechanical system to evolve in time. We will adopt the convention that the basis vectors that span the Hilbert space are fixed in time. Thus, the coordinates of the state vector, namely, the amplitudes between the state and the basis vector, will carry the time dependence.†

The time dependence of a state vector, and thus of the corresponding amplitudes, is governed by the equation of motion of quantum mechanics, which we will introduce in the next chapter. Here we would like to use a heuristic argument to obtain the time dependence of probability amplitudes for some simple cases. Consider a particle of mass m that is at rest, and let its state be designated by $|\phi_0\rangle$. If $|x\rangle$ represents the eigenstates of the position operator, then $\langle x \mid \phi_0\rangle$ is the amplitude for finding this particle at the position x.

Since the particle is at rest, its momentum is well defined and equal to zero, or $\Delta p_x = 0$. From the uncertainty principle, $\Delta x \to \infty$, namely, the probability is the same for finding the particle at any point on the X-axis. This means that

$$P(x) = |\langle x \mid \phi_0\rangle|^2 = \text{independent of } x \tag{3.10}$$

† This approach is known as the *Schrödinger picture* of quantum mechanics. One can also adopt the opposite convention where the basis vectors change continuously in time and the amplitudes for any particular state remain fixed—this is known as the *Heisenberg picture*.

Furthermore, $P(x)$ cannot depend on time since a particle that is at rest remains at rest in the absence of external forces. The probability $P(x)$ for this case must be independent of time; therefore the amplitude $\langle x \mid \phi_0 \rangle$ can contain at most a time-dependent phase

$$\langle x \mid \phi_0 \rangle = Ae^{-i\alpha t} \tag{3.11a}$$

where α is a *real* constant.

The constant α must be related to the properties of the particle. If we ignore internal structure, then the only such property for a particle at rest is its mass m, and we postulate that

$$\langle x \mid \phi_0 \rangle = e^{-i(mc^2/\hbar)t} \tag{3.11b}$$

This result is consistent with the De Broglie relations for matter waves introduced in Section 1.3. For a particle at rest, the energy E_0 is

$$E_0 = mc^2$$

so that Eq. (3.11*b*) can be written as

$$\langle x \mid \phi_0 \rangle = Ae^{-i(E_0/\hbar)t} = Ae^{-i\omega_0 t} \tag{3.11c}$$

Equation (3.11*c*) implies that even though the particle is at rest and the position probability $P(x)$ is time- and space-independent, the position amplitude $\langle x \mid \phi_0 \rangle$ oscillates rapidly with angular frequency $\omega_0 = E_0/\hbar$. This is an important result which we will generalize to amplitudes involving states of well-defined energy. If $|j\rangle$ is the complete set of basis states of well-defined energy E_j for our system, and $|\psi\rangle$ is any arbitrary state of the system, then the amplitudes $\langle j \mid \psi \rangle$ depend on time as $\exp(-iE_jt/\hbar)$. We write

$$\boxed{\langle j \mid \psi \rangle = C_{j\psi}(\lambda)e^{-i(E_j/\hbar)t}} \tag{3.12}$$

The functions $C_{j\psi}(\lambda)$ may depend on various physical observables that we designate by λ, but they may *not* contain the time. Of course, they depend on the state $|\psi\rangle$ and on the corresponding basis vector $|j\rangle$.

Equation (3.12) can be used to express the time dependence of any arbitrary state $|\psi\rangle$ in terms of the simple oscillatory time dependence of states with well-defined energy, namely,

$$|\psi\rangle = \sum_j |j\rangle C_{j\psi}(\lambda) e^{-i(E_j/\hbar)t} \tag{3.13a}$$

In view of the orthonormality of the basis states $\langle k \mid j\rangle = \delta_{kj}$ it follows from Eq. (3.13*a*) that if the state $|\psi\rangle$ is normalized, then

$$\sum_j |C_{j\psi}(\lambda)|^2 = 1 \tag{3.13b}$$

The expansion of an arbitrary state $|\psi\rangle$ into states of well-defined energy is most useful when the coefficients $C_{j\psi}(\lambda)$ are known or can be easily found. In the special case that *only one* of the coefficients $C_{j\psi}(\lambda)$ is different from zero, the state $|\psi\rangle$ has a very simple time dependence. If

$$\begin{aligned} C_{j\psi}(\lambda) &= 1, \qquad j = k \\ &= 0, \qquad \text{otherwise} \end{aligned}$$

then

$$|\psi\rangle = |k\rangle e^{-i(E_k/\hbar)t} \tag{3.14a}$$

States that have the form of Eq. (3.14*a*) are called *stationary* because the probability of obtaining a particular value for any physical observable does not change in time. For instance, let an observable be represented by the operator $\hat{Q}$, and let $|q_i\rangle$ be the corresponding basis states. Then the probability that a measurement of $\hat{Q}$ on the system $|\psi\rangle$ will yield the eigenvalue q_i is

$$P(q_i) = |\langle q_i \mid \psi\rangle|^2 = |\langle q_i \mid k\rangle|^2 \tag{3.14b}$$

where we have used Eq. (3.14*a*) to express $|\psi\rangle$. Since both $|k\rangle$ and $|q_i\rangle$ represent basis states, they are (in our convention) fixed in time, thus $\langle q_i \mid k\rangle$ does *not* depend on time and $P(q_i)$ is constant for all physical observables.

When more than one of the coefficients $C_{j\psi}(\lambda)$ in Eq. (3.13*a*) are different from zero (and they correspond to basis states with different energy), the state $|\psi\rangle$ is not any more stationary. In this case the probability of finding a specific value of a physical observable may change in time. This can be seen in the example given below, as well as in the following section.

Example 3

Consider a state

$$|\psi\rangle = \frac{1}{2^{1/2}}(|j\rangle e^{-i(E_j/\hbar)t} + |k\rangle e^{-i(E_k/\hbar)t}) \tag{3.15a}$$

with $E_j \neq E_k$. The probability that a measurement of the observable $\hat{Q}$ will yield the value q_i is obtained from

$$\langle q_i \mid \psi\rangle = \frac{1}{2^{1/2}}(\langle q_i \mid j\rangle e^{-i\omega_j t} + \langle q_i \mid k\rangle e^{-i\omega_k t})$$

For simplicity we set

$$\langle q_i \mid j\rangle = A \qquad \langle q_i \mid k\rangle = B \qquad \omega_j - \omega_k = \Delta\omega$$

so that

$$\begin{aligned} P(q_i) &= |\langle q_i \mid \psi\rangle|^2 \\ &= \tfrac{1}{2}(|A|^2 + |B|^2) + [\mathrm{Re}(AB^*) \cos \Delta\omega t + \mathrm{Im}(AB^*) \sin \Delta\omega t] \end{aligned} \tag{3.15b}$$

The above example shows that for nonstationary states the time dependence of probabilities is governed by the *difference* in the frequencies ω_j of the states of well-defined energy needed to describe the system. Consequently, if we change the energy of *all* the states by the same fixed amount W, the physically observable results remain unchanged. This is analogous to the case of the potential energy in classical physics where the absolute value can be freely chosen without affecting the dynamics.

Because of this, there is a certain freedom in expressing the time dependence of quantum-mechanical amplitudes such as those given in Eq. (3.12). When no particles are created or destroyed, then the total mass of the system remains fixed and E_j can be set equal to the sum of the kinetic and potential energy of the state $|j\rangle$. As a further illustration of this argument we note that in principle we should include in E_j the potential energy due to the gravitational attraction of the earth, but when—as is almost always the case—this is the same for all states $|j\rangle$, we can omit it. Thus, the value of E_j that governs the time dependence of amplitudes has no absolute meaning but, of course, must be chosen consistently for all the states $|j\rangle$ used to describe the system under consideration.

3.3. OBSERVABLES WITH NO CLASSICAL ANALOGUE: SPIN

We are familiar from classical physics with such physical observables as position, velocity, angular momentum, energy, etc. In describing a quantum system it is found that one needs to specify new observables that have *no* counterpart in a classical description of physical systems. The most familiar of these observables is the *spin* of microscopic particles, e.g., atoms, nuclei, or electrons.

That electrons have spin was first discovered in the fine structure of atomic spectra and in the interpretation of the so-called "anomalous" Zeeman effect. Spin has many similarities to angular momentum and is measured in units of $\hbar = h/2\pi$. A molecule or a nucleus, or an elementary particle, is said to have spin s when the *maximum* value of the projection of the spin on some reference axis is $S_z = s\hbar$. The spin quantum number† s can take only half-integer or integer values, and the only possible projections on the reference axis are $S_z = m_s\hbar$, where $m_s = s, s-1, \ldots, (-s+1), -s$. The number of possible m_s values is $(2s+1)$; for example, a spin 1/2 particle can have only two projections, $\pm\hbar/2$.

The spin of quantum systems appears to be a very fundamental property because it determines the behavior of aggregates of particles. Particles with half-integer spin obey Fermi–Dirac statistics, which leads to the Pauli exclusion principle, whereas particles with zero or integer spin obey Bose–Einstein statistics. Associated with the spin $\mathbf{S}$ there is a magnetic dipole moment $\boldsymbol{\mu}$ related to it through

$$\boldsymbol{\mu} = g\frac{e}{2m}\mathbf{S} \tag{3.16}$$

where we have introduced the dimensionless factor g. The quantity $\mu_0 = e\hbar/2m_e$ is known as the Bohr magneton,‡ and in terms of this the electron has a g-factor that is very close to $g \sim 2.0$. Thus, the projections of the electron's magnetic moment on the reference axis

† As indicated in Section 1.1 we can think of the magnitude of the spin as being $\hbar[s(s+1)]^{1/2}$, but this is not very relevant since in all observable interactions we are concerned with the projection of the spin on a particular axis.

‡ In Chapter 1 we used the symbol μ_B.

are $\pm\mu_0$. Often we refer to these two projections simply as spin "up" or "down." The magnetic moment of nuclei are measured in terms of the nuclear magneton, which is given by Eq. (3.16) with the electron mass replaced by the proton mass $\mu_N = e\hbar/2m_p$.

The magnitude of the magnetic moment can be determined from its interaction energy in a magnetic field that modifies the overall energy of any particular state of the system. It can also be measured directly by the deflection of particles in a Stern–Gerlach apparatus or by the precession of the magnetic moment when the particle is placed in a uniform magnetic field. We will describe this latter process because it provides a very clear example of how to handle the time dependence of quantum-mechanical amplitudes discussed in Section 3.2.

Consider a uniform magnetic field B along the Z-axis, as shown in Fig. 3.2, and an electron at rest with its spin along the X-axis. By this we mean that the state of the electron is described by having spin projection on the X-axis equal to $\hbar/2$. We will designate this state by $|+x\rangle$, and for this choice of axes the complete set of states consists of the two possible projections, $|+x\rangle$ or $|-x\rangle$. We could have chosen a different basis, for instance, along the Z-axis, in which case the complete set of states would have been $|+z\rangle$ or $|-z\rangle$. These different bases are connected by a unitary transformation of the general form given by Eq. (2.22).

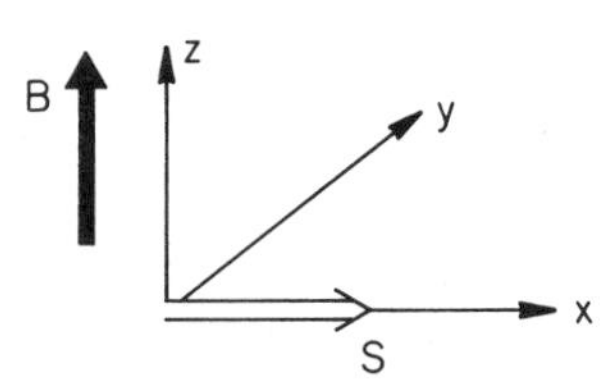

FIGURE 3.2. Coordinate system indicating a constant magnetic field **B** along the Z-axis, while the projection of the electron's spin onto the X-axis is $+\frac{1}{2}$.

The physical connection between the basis states is a rotation as shown in Fig. 3.3. Here the spin is along the X-axis of the original coordinate system, or along the Z'-axis of the primed coordinates. The latter can be brought into coincidence with the original system by rotating 90% counterclockwise around the Y'-axis. We will not derive the general form of the rotation matrix that connects the two systems of basis states for spin 1/2 systems, but give the result for

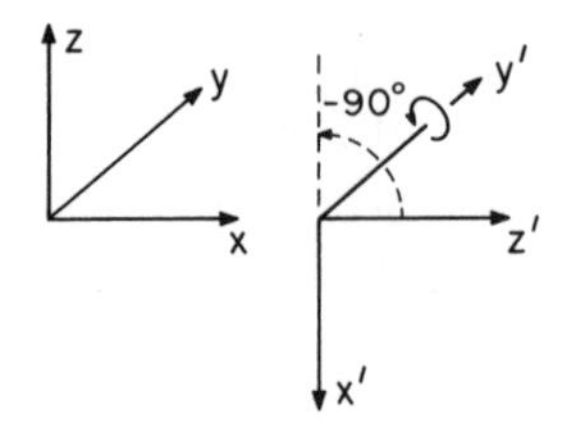

FIGURE 3.3. Rotation of the coordinates around the Y' axis by 90° counter-clockwise.

arbitrary θ and for $\theta = -\pi/2$

$$R_y(\theta) = \begin{array}{c|cc} & |+z'\rangle & |-z'\rangle \\ \hline |+z\rangle & \cos\theta/2 & \sin\theta/2 \\ |-z\rangle & -\sin\theta/2 & \cos\theta/2 \end{array}$$

$$R_y\left(-\frac{\pi}{2}\right) = \begin{array}{c|cc} & |+x\rangle & |-x\rangle \\ \hline |+z\rangle & 1/2^{1/2} & -1/2^{1/2} \\ |-z\rangle & 1/2^{1/2} & 1/2^{1/2} \end{array} \tag{3.17}$$

For the particular case $\theta = -\pi/2$, $|+z'\rangle$ is equivalent to $|+x\rangle$, and we have so indicated in the second matrix of Eq. (3.17).

The state $|\phi\rangle$ of the electron at $t = 0$ is such that its spin projection is along the $+X$-axis, namely,

$$|\phi(t=0)\rangle = |+x\rangle \tag{3.18a}$$

The energy of the electron is given by

$$E = m_e c^2 + U$$

where U is the interaction energy between the magnetic moment and the uniform magnetic field. Classically,

$$U = -\boldsymbol{\mu} \cdot \mathbf{B} \tag{3.19a}$$

In quantum mechanics the projection of the magnetic moment of a spin 1/2 particle onto the B axis can take only two values, and thus $U = \pm\mu B$ when the electron is in the states $|+z\rangle$ or $|-z\rangle$ (the Z-axis is along $\mathbf{B}$). We must express the state $|\phi\rangle$ in terms of its projections along the Z-axis

$$|\phi(t=0)\rangle = |+x\rangle = |+z\rangle\langle +z \mid +x\rangle + |-z\rangle\langle -z \mid +x\rangle \quad (3.18b)$$

The states $|+z\rangle$ and $|-z\rangle$ are eigenstates of energy with eigenvalues

$$\begin{aligned} |+z\rangle \qquad & E = \mu_0 B \\ |-z\rangle \qquad & E = -\mu_0 B \end{aligned} \quad (3.19b)$$

where we have ignored the rest energy term since it is the same for both states. The signs in Eq. (3.19b) are reversed [with respect to Eq. (3.19a)] because for an electron the magnetic moment is opposite to the spin direction.

We can now find the time dependence of the state $|\phi\rangle$ by using Eq. (3.13a) and the amplitudes $\langle +z \mid +x\rangle$ and $\langle -z \mid +x\rangle$ as given by the second matrix of Eq. (3.17).

$$|\phi(t)\rangle = \frac{1}{2^{1/2}}\left(|+z\rangle e^{i\omega t} + |-z\rangle e^{-i\omega t}\right) \quad (3.20)$$

with

$$\omega = \frac{\mu_0 B}{\hbar}$$

From Eq. (3.20) we can form the amplitudes $\langle +z \mid \phi(t)\rangle$ and $\langle -z \mid \phi(t)\rangle$, and see that the probability of finding the spin projection along the $+Z$ or $-Z$ axis is 1/2. However, if we ask for the probabilities of the projection along the X-axis we obtain

$$\langle +x \mid \phi(t)\rangle = \frac{1}{2^{1/2}}\left(\langle +x \mid +z\rangle e^{i\omega t} + \langle +x \mid -z\rangle e^{-i\omega t}\right)$$

and similarly for $\langle -x \mid \phi(t)\rangle$. We can use the second matrix of Eq. (3.17) to evaluate $\langle +x \mid +z\rangle = \langle +z \mid +x\rangle^*$ and $\langle +x \mid -z\rangle = \langle -z \mid +x\rangle^*$. Hence

$$P(+x, t) = |\langle +x \mid \phi(t)\rangle|^2 = |\tfrac{1}{2}(e^{i\omega t} + e^{-i\omega t})|^2 = \cos^2 \omega t \quad (3.21a)$$

$$P(-x, t) = |\langle -x \mid \phi(t)\rangle|^2 = |\tfrac{1}{2}(-e^{i\omega t} + e^{-i\omega t})|^2 = \sin^2 \omega t \quad (3.21b)$$

As a check we note that

$$P(+x, t) + P(-x, t) = 1$$

which is what we would expect since the two states $|+x\rangle$ and $|-x\rangle$ provide a complete basis.

The result of Eq. (3.20) is plotted in Fig. 3.4(a) and is equivalent to a continuous rotation (a precession) of the spin in the X–Y plane with angular frequency

$$\Omega = 2\omega = \frac{2\mu_0 B}{\hbar} \tag{3.22a}$$

as shown in part (b) of the figure. The direction of precession is counterclockwise because of the negative charge of the electron. This can be established by calculating the probability $P(+y, t)$, which becomes equal to one at time $t = \pi/4\omega$.

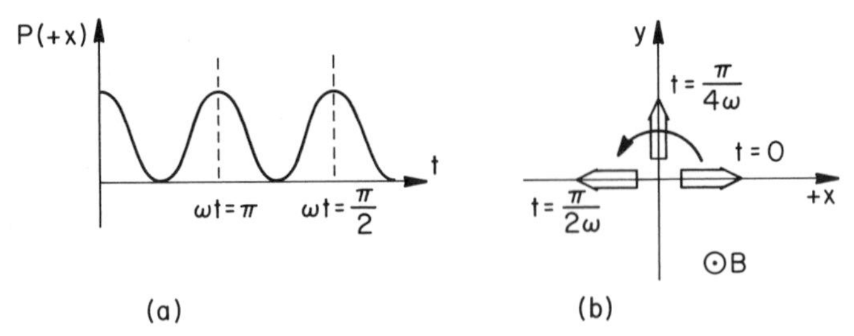

FIGURE 3.4. Spin precession in a magnetic field. (a) The probability of finding the spin projection in the state $|+x\rangle$. (b) Sketch indicating the classical interpretation of spin precession.

The calculated motion of the (projection of the) spin is exactly the same as that obtained in a classical calculation for a particle that has an angular momentum $\mathbf{L}$ projected onto the X–Y plane with the value $\hbar/2$ and has a magnetic moment $\boldsymbol{\mu} = (2\mu_0/\hbar)\mathbf{L}$. Due to the external field $\mathbf{B}$ a torque $\boldsymbol{\tau}$ acts on the particle, and it precesses with angular frequency $\boldsymbol{\Omega}$ such that

$$\boldsymbol{\tau} = \boldsymbol{\mu} \times \mathbf{B} = \boldsymbol{\Omega} \times \mathbf{L}$$

Using the projected values of $\boldsymbol{\mu}$ and $\mathbf{L}$, which are in a plane perpendicular to $\mathbf{B}$, we can solve for $|\boldsymbol{\Omega}|$, which is along the $\mathbf{B}$ direction

$$|\boldsymbol{\Omega}| = \frac{|\boldsymbol{\mu}|\,|\mathbf{B}|}{|\mathbf{L}|} = \frac{2\mu_0 L}{\hbar}\frac{B}{L} = \frac{2\mu_0 B}{\hbar} \tag{3.22b}$$

In this case the classical and quantum-mechanical results agree because we are comparing the classical motion with the *most probable* orientation of the spin axis as calculated by quantum mechanics; this is a manifestation of the correspondence principle.

Spin-precession experiments such as the one we have discussed can be performed with great accuracy using μ-mesons, which like electrons have spin 1/2. The μ-mesons are produced from the decay of π-mesons, $\pi^+ \to \mu^+ \nu_\mu$. As shown in Fig. 3.5(a) the μ^+-mesons have their spin direction opposite to their momentum vector. The μ^+-meson beam is then brought to rest in a target that is located in a magnetic field. The μ-mesons decay† with a mean lifetime of 2.2 μsec into a positron, a neutrino, and an antineutrino

$$\mu^+ \to e^+ \nu_e \bar{\nu}_\mu$$

In this decay process the positrons are emitted preferentially along the direction of the spin of the μ^+-meson [Fig. 3.5(c)]. Therefore, by measuring the direction in which the positrons are emitted, we can infer the orientation of the spin vector of the parent μ^+-meson.

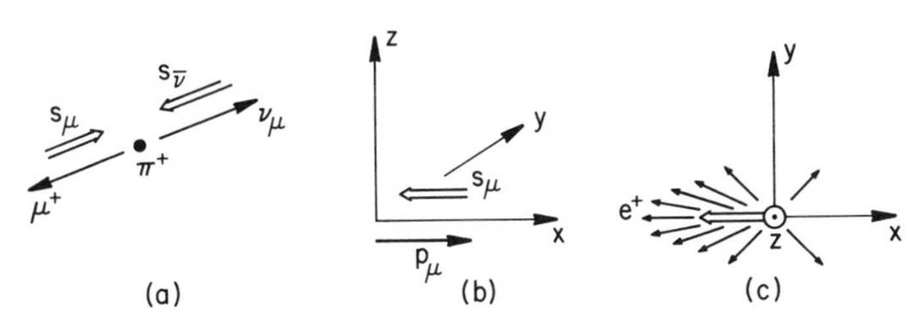

FIGURE 3.5. Measurement of the magnetic moment of the μ-meson. (a) μ^+-mesons produced in π^+-meson decay. (b) The spin projection is opposite to the momentum vector. (c) Positrons from the decay of the μ-meson are preferentially emitted along the spin direction.

If we place a detector along some fixed direction in the X–Y plane, say the X-axis, we see that the yield of positrons varies sinusoidally with the angular frequency Ω given by Eq. (3.22). Data from an actual experiment [J. Sandweiss *et al.*, *Phys. Rev. Lett.* **30,** 1002 (1973)] are shown in Fig. 3.6. The precession frequency is

† The decay is due to the weak interaction which does not conserve parity.

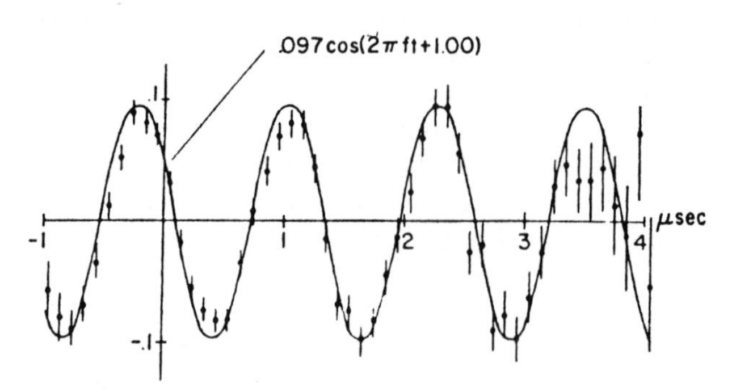

FIGURE 3.6. Precession of the muon spin in a magnetic field of $6\times 10^{-3}\ T$ (60 G). [From J. Sandweiss *et al.*, *Phys. Rev. Lett.* **30,** 1002 (1973).]

found to be $\nu_p = 807.5$ kHz in a static magnetic field of 60 G. Thus

$$\mu = \frac{\Omega\hbar}{2B} = \nu_p \frac{\pi\hbar}{B}$$

for the above value of ν_p and $B = 6\times 10^{-3} T$ we find in MKS units

$$\mu = 4.46\times 10^{-26}\ \text{A-m}^2 \tag{3.23a}$$

In these units the equivalent of the Bohr magneton for the μ-meson is

$$\mu_\mu = \frac{e\hbar}{2m_\mu} = \frac{e\hbar c^2}{2m_\mu c^2} = 4.50\times 10^{-26}\ \text{A-m}^2 \tag{3.23b}$$

where we use $m_\mu c^2 = 105.4$ MeV for the rest mass of the μ-meson. Comparing the results of Eqs. (3.23*a*) and (3.23*b*), and recalling that the μ-meson has spin 1/2, these data imply a g-factor

$$\mu = g_\mu \frac{e}{2m_\mu} S_\mu = g_\mu \frac{1}{2}\frac{e\hbar}{2m_\mu}$$

or

$$g_\mu \simeq 1.98$$

The correct value of the g-factor of the μ-meson is $g_\mu = 2.0033\ldots$, and has been measured to the remarkable accuracy of 1 part per billion, which is in agreement with the theoretical prediction of quantum electrodynamics. This agreement represents one of the great triumphs of modern physics.

3.4. DEPENDENCE OF QUANTUM-MECHANICAL AMPLITUDES ON POSITION: THE WAVEFUNCTION

In addition to their time dependence quantum-mechanical amplitudes may also depend on physical observables that are properties of the system, on position in particular. We cannot derive the position dependence, but we will follow an argument similar to that used to obtain the time dependence in Section 3.2. We postulated there that for a particle at rest the amplitude for finding it at a position x was

$$\langle x \mid \phi \rangle = A e^{-i(mc^2/\hbar)t} \tag{3.24a}$$

which leads to a time- and position-independent probability.

For an observer moving with uniform velocity $v = \beta c$ along the negative X-axis, the particle will appear to be in motion with velocity v along the positive X'-axis. All the laws of physics will be the same as viewed from the moving and stationary observers so that the amplitude $\langle x' \mid \phi \rangle$ in the moving system can be obtained by a Lorentz transformation of Eq. (3.24a). In Eq. (3.24a) the phase factor is given by

$$\exp[-(i/h)(E_0 t)]$$

For any Lorentz transformation, expressions such as

$$Et - \mathbf{p} \cdot \mathbf{x}$$

are *invariant.* Since the particle is at rest in the unprimed coordinate frame, $E = E_0$ and $\mathbf{p} = 0$, we can write

$$E_0 t = Et - px = E't' - p'x' \tag{3.24b}$$

where E' is the energy and p is the momentum (along the X'-axis) of the particle in the moving frame of reference.

Example 4

Let us now prove Eq. (3.24*b*). We recall the properties of the Lorentz transformation

$$ct = \frac{ct' + \mathbf{x} \cdot (\mathbf{v}/c)}{[1-(v/c)^2]^{1/2}} = \gamma(ct' - x'\beta)$$
$$E' = \frac{E_0 - \mathbf{v} \cdot \mathbf{p}_0}{[1-(v/c)^2]^{1/2}} = \gamma[E_0 + (cp_0)\beta] \tag{3.24c}$$

with the notation $\gamma = 1/[1-(v/c)^2]^{1/2}$. Because the primed coordinate frame is moving toward the negative axis we set $\beta = -v/c$. Since $p_0 = 0$, cross multiplication of Eqs. (3.24*c*) yields

$$(ct)(\gamma E_0) = [\gamma(ct' - x'\beta)]E'$$

or

$$E_0 t = E't' - \frac{E'\beta}{c}x' = E't' - p'x'$$

where the last step follows because

$$E'\frac{\beta}{c} = \gamma' m_0 c^2 \frac{v'}{c^2} = \gamma' m_0 v' = p'$$

We now use the phase expressed in the primed coordinates [as in Eq. (3.24*a*)] to find the amplitude for a particle moving with uniform velocity along the positive X' axis

$$\langle x' \mid \phi \rangle = Ae^{-i/\hbar(E't' - p'x')}$$

We can relabel $x' \rightarrow x$, etc., without any loss of generality and use the notation $\omega = E/\hbar$, $k = p/\hbar$ to write

$$\boxed{\langle x \mid \phi \rangle = Ae^{-i(\omega t - kx)}} \tag{3.25}$$

This is the amplitude for finding a particle of *well-defined* velocity (and, thus, of momentum $p = \hbar k$) along the X-axis at the position x. This is identical to the probability waves introduced in Section 1.3 and gives a time- and position-independent probability as required by the uncertainty principle. Amplitudes of the form of Eq. (3.25) are referred to as plane waves and as we saw in Section 1.6 can be used to build up wave packets.

In Section 3.2 we expressed an arbitrary state $|\psi\rangle$ as an expansion in states of definite energy. We can perform a similar expansion in a basis whose states correspond to definite position along the X-axis. We label these states $|x'\rangle$ and remember that they form a continuous spectrum

$$|\psi\rangle = \sum_{\text{all } x'} |x'\rangle\langle x' \mid \psi\rangle = \int |x'\rangle A_{x'} e^{-i(\omega t - kx')}\, dx'$$

where we use Eq. (3.25) to express $\langle x' \mid \psi\rangle$. We replace the summation by an integral because the base states are continuous. We also add an index to the constants $A_{x'}$, since they may be different for each position (basis state). If we write $\psi(x', t) = A_{x'} e^{-i(\omega t - kx')}$ we obtain

$$|\psi\rangle = \int |x'\rangle \psi(x', t)\, dx' \tag{3.26a}$$

as the complete description of an arbitrary state in the position representation.

The amplitude for finding the state $|\psi\rangle$ at the position x (that is, in the basis state $|x\rangle$) is

$$\langle x \mid \psi\rangle = \int \langle x \mid x'\rangle \psi(x', t)\, dx'$$

The continuum basis states $|x\rangle$ are normalized through $\langle x \mid x'\rangle = \delta(x - x')$, so that

$$\langle x \mid \psi\rangle = \int \delta(x - x')\psi(x', t)\, dx' = \psi(x, t) \tag{3.26b}$$

The function $\psi(x, t)$ in Eq. (3.26b) is called the *wavefunction* of the particle. It is the amplitude for finding the state $|\psi\rangle$ at the position x at the time t

$$\boxed{\langle x \mid \psi\rangle = \psi(x, t)} \tag{3.27}$$

To see the usefulness of the wavefunction, consider the state $|\psi\rangle$ and suppose we seek the amplitude for the overlap of $|\psi\rangle$ with the arbitrary state $|\phi\rangle$. By definition, this is $\langle \phi \mid \psi\rangle$. We represent $|\psi\rangle$ and $\langle \phi|$ through Eq. (3.26a)

$$|\psi\rangle = \int |x\rangle \psi(x, t)\, dx$$

$$\langle \phi| = \int \langle x'| \phi^*(x', t)\, dx'$$

and thus

$$\begin{aligned}\langle \phi \mid \psi \rangle &= \iint \langle x' \mid x \rangle \phi^*(x', t)\psi(x, t)\, dx'\, dx \\ &= \int \delta(x' - x)\phi^*(x', t)\psi(x, t)\, dx'\, dx \\ &= \int \phi^*(x, t)\psi(x, t)\, dx \qquad (3.28a)\end{aligned}$$

In the same way we can obtain the matrix elements of an operator $\hat{Q}$ between two arbitrary states

$$\langle \phi | \hat{Q} | \psi \rangle = \int \phi^*(x, t) Q(x) \psi(x, t)\, dx \qquad (3.28b)$$

However, $\hat{Q}$ must be expressed in the continuous position representation, i.e., as a differential operator acting on the position coordinates.

The plane-wave amplitude of Eq. (3.25) describes a freely moving particle. In the presence of external forces the amplitude will be modified. This can be taken into account by considering the change in the potential energy U of the particle. Since the total energy must include the potential energy, the angular frequency ω in Eq. (3.25) is changed to

$$\omega = \frac{E}{\hbar} = \frac{1}{\hbar}(mc^2 + T + U)$$

In the presence of a constant potential energy U the amplitude acquires an additional phase δ such that

$$e^{i\delta} = e^{-i(U/\hbar)t} \qquad (3.29a)$$

If the potential energy depends on position, then the phase change between positions 1 and 2 is

$$\delta_{2,1} = -\frac{1}{\hbar}\int_1^2 U(x)\, dt \qquad (3.29b)$$

where the integration is along the trajectory of the particle.

As an example, a particle of charge e moving in an electrostatic (scalar) potential $\phi(x)$ acquires a phase

$$\delta_{2,1} = -\frac{e}{\hbar}\int_1^2 \phi(x)\, dt \qquad (3.29c)$$

An analogous expression can be found for the motion of a particle in a magnetic field, and we will make use of it in the following section. In quantum mechanics the concept of force, and therefore of electric or magnetic fields, plays a consistently smaller role in favor of the potentials. This is because quantum-mechanical amplitudes propagate with wavelike properties [see Eqs. (3.12) and (3.25)], so that the physically observable phenomena depend sensitively on the phase of the amplitudes.

3.5. SUPERPOSITION OF AMPLITUDES: INTERFERENCE

An essential step in the use of quantum-mechanical amplitudes is the ability to combine several amplitudes into one. This can be illustrated by considering again the two-slit interference experiment discussed in Section 1.2. Let electrons be emitted from a source A that is equidistant from the two slits s_1 and s_2 as shown in Fig. 3.7. The slits are separated by a distance d, and the electrons are observed on a screen as a function of the position x. The electrons can arrive at x by passing through either s_1 or s_2.

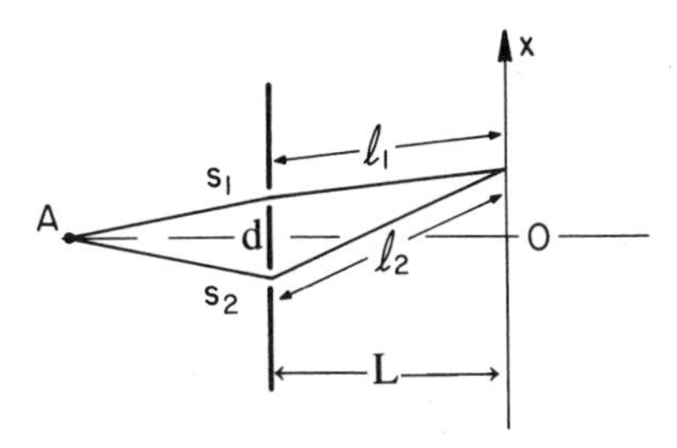

FIGURE 3.7. Interference of amplitudes when an electron emitted from the source A can reach the point x of the screen by following two different paths, either through slit s_1 or through slit s_2.

The amplitude for the arrival at x of an electron that passed through s_1 can be written as the product of two amplitudes

$$\phi_1 = \langle x \mid A\rangle_1 = \langle x \mid s_1\rangle\langle s_1 \mid A\rangle$$

Similarly, if the electron passed through s_2 the amplitude is

$$\phi_2 = \langle x \mid A\rangle_2 = \langle x \mid s_2\rangle\langle s_2 \mid A\rangle$$

We *postulate* that the amplitude ϕ for the arrival at the point x of the screen of an electron emitted from the source A is the *linear superposition* of ϕ_1 and ϕ_2

$$\begin{aligned}\langle x \mid A\rangle = \phi = \phi_1 + \phi_2 &= \langle x \mid s_1\rangle\langle s_1 \mid A\rangle + \langle x \mid s_2\rangle\langle s_2 \mid A\rangle \\ &= \sum_{s_i} \langle x \mid s_i\rangle\langle s_i \mid A\rangle \end{aligned} \tag{3.30}$$

Equation (3.30) is a *closure* relation over all intermediate positions of the electron exactly analogous to Eq. (2.16); the latter was based on purely formal considerations in Hilbert space.

As can be easily seen the linear superposition of amplitudes leads to interference: Electrons arriving at slit 1 or slit 2 have the same phase; however, when they reach the screen their phase has evolved according to Eq. (3.25) and therefore depends on the path length from the corresponding slit to the observation point

$$\langle x \mid s_1\rangle = Be^{-i\omega t}e^{ikl_1} \qquad \langle x \mid s_2\rangle = Be^{-i\omega t}e^{ikl_2}$$

From the geometry of Fig. 3.7 we find $l_2 = l_1 + \Delta l = l_1 + d \sin\theta$, where $\theta \sim x/L$, and therefore Eq. (3.30) reads

$$\langle x \mid A\rangle = Be^{-i(\omega t - kl_1)}(1 + e^{ik\, d \sin\theta})$$

The probability of detecting an electron at x is the absolute square of this amplitude

$$\begin{aligned} P(x) = |\langle x \mid A\rangle|^2 &= 4\,|B|^2 \cos^2\!\left(\frac{k\, d \sin\theta}{2}\right) \\ &= 4\,|B|^2 \cos^2\!\left(\frac{\pi\, d \sin\theta}{\lambda}\right) \end{aligned}$$

which exhibits a typical interference pattern, as shown in Fig. 1.9(b).

In Section 1.4 we discussed in some detail that if we place a counter near the slits s_1 and s_2 so as to be able to distinguish through which slit the photon passes, the interference pattern disappears. We conclude therefore that the resultant amplitude is the linear superposition of the amplitudes for all *indistinguishable* processes that lead to the same final state. If the processes are distinguishable, then we must *add the probabilities* and not the amplitudes. This is true as long as the processes involved are distinguishable *irrespective* of whether we distinguish them or not. Thus, the probability for the occurrence of a particular event that can proceed through intermediate states is

$$P(x) = \left| \sum_i \langle x \mid s_i \rangle \langle s_i \mid A \rangle \right|^2 \quad \text{intermediate state } i \text{ indistinguishable}$$

$$P(x) = \sum_i |\langle x \mid s_i \rangle \langle s_i \mid A \rangle|^2 \quad \text{intermediate state } i \text{ distinguishable}$$

(3.31)

A beautiful verification of the law of addition of amplitudes is found in the diffraction of neutrons from crystals. A neutron emerging at an angle θ may have scattered from any of the N nuclei in the crystal. The amplitude $\phi(\theta)$ is

$$\phi(\theta) = \sum_{i=1}^{N} \langle \theta \mid i \rangle a(\theta) \langle i \mid s \rangle$$

where $\langle i \mid s \rangle$ is the amplitude for the propagation of the neutron from the source to the ith nucleus, $a(\theta)$ is the amplitude for scattering from a single nucleus through the angle θ, and $\langle \theta \mid i \rangle$ is the amplitude for propagation to a detector located at the angle θ. Because the nuclei are regularly spaced, the amplitudes $\langle \theta \mid i \rangle$, $\langle \theta \mid i+1 \rangle$, $\langle \theta \mid i+2 \rangle$, etc., differ by a fixed phase. Since N is very large, $\phi(\theta)$ will be very small, except in the case where the phase angle is 2π, as shown schematically in Fig. 3.8(a). In that case, sharp peaks appear at distinct angles as shown in Fig. 3.9.

If the nuclei have spin 1/2 (as does the neutron) it is possible that the neutron and the nucleus will *flip* their spins in the scattering process, as indicated symbolically in Fig. 3.8(b). Suppose that the kth nucleus flipped its spin. In principle then we could tell that the scattering occurred on the kth nucleus so that the spin-flip amplitudes for scattering from any one of the N nuclei would be distinguishable. Therefore, the probability of observing a particle at the angle θ (the cross section) is different for the nonflip and spin-flip cases and is given, respectively, by

$$P(\theta) = |a(\theta)|^2 \left| \sum_{i=1}^{N} \langle \theta \mid i \rangle \langle i \mid s \rangle \right|^2 \qquad \text{nonflip}$$

$$P(\theta) = |b(\theta)|^2 \sum_{i=1}^{N} |\langle \theta \mid i \rangle \langle i \mid s \rangle|^2 \qquad \text{spin-flip}$$

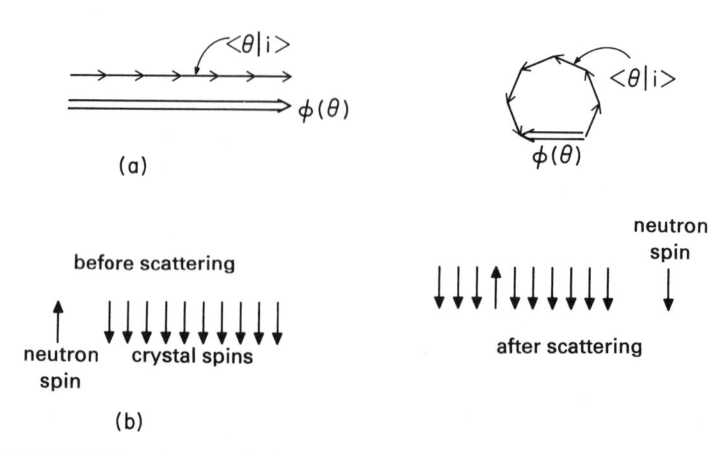

FIGURE 3.8. Interference of a large number of amplitudes. (a) When the individual amplitudes have the same phase, the resulting amplitude is much larger. (b) Schematic of spin-flip scattering of a neutron from a crystal.

where $b(\theta)$ represents the spin-flip scattering amplitude. The interference effects are greatly reduced in the spin-flip case, and the peaks are "washed out." Even though we have no practical way of finding out from which nucleus the neutron scattered, nature has given this information away by flipping the spin: thus, the "game is lost," and there can be no interference.

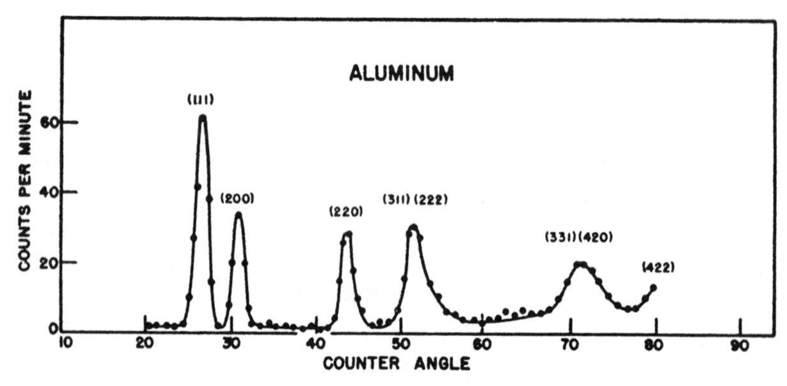

FIGURE 3.9. Neutron diffraction pattern from aluminum powder. The wavelength is $\lambda = 1.08$ Å and the Miller indices of the diffraction plane are indicated. From E. O. Wollan and C. G. Shull, *Physical Review* **73,** 830 (1948).

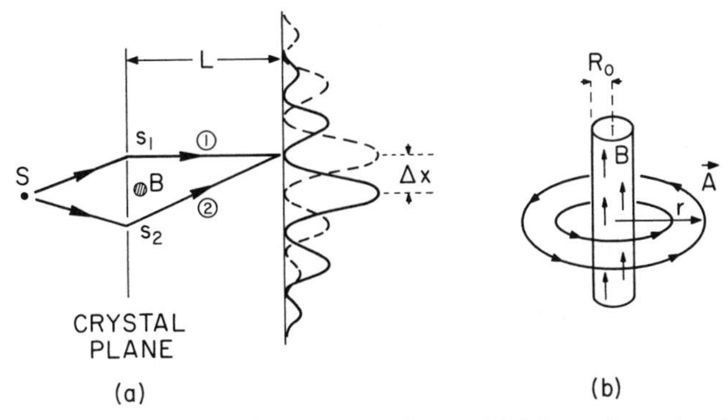

FIGURE 3.10. The Bohm–Aharonov experiment. (a) The path of the electrons through the crystal does not traverse the magnetic field **B**. (b) Vector potential describing a confined axial magnetic field.

Evidence for the interference of quantum-mechanical amplitudes abounds. An interesting experiment is the Bohm–Aharonov effect first observed in 1956. A beam of electrons was diffracted from a crystal indicated schematically by the two slits in Fig. 3.10(a). A magnetic field perpendicular to the diffraction plane was established by introducing a magnetized filament behind the crystal. The magnetic field B was *confined* to the filament, and no force acted on the electrons. However, a vector potential **A** was directed along concentric circles such that†

$$\nabla \times \mathbf{A} = \mathbf{B}$$

surrounding the filament, as shown in (b) of the figure. Therefore, the electrons traversed a region of space where $\mathbf{A} \neq 0$, and according to our discussion in the previous section we could expect their amplitude to acquire an additional phase given by

$$\delta = \frac{e}{\hbar} \int_s \mathbf{A} \cdot d\mathbf{l} \tag{3.32a}$$

This is analogous to Eq. (3.29c), except that we must take the line integral of the vector potential along the path followed by the electron.

† For a constant field **B** in the region $r < R_0$, the vector potential **A** for $r > R_0$ is given by $\mathbf{A} = \frac{1}{2}(\mathbf{B} \times \mathbf{r})(R_0^2/r^2)$.

For paths such as ① and ② [in Fig. 3.10(a)] which go around the filament, the phase difference is simply

$$\delta_1 - \delta_2 = \frac{e}{\hbar}\left[\int_{①} \mathbf{A}\cdot d\mathbf{l} - \int_{②} \mathbf{A}\cdot d\mathbf{l}\right] = \frac{e}{\hbar}\oint \mathbf{A}\cdot d\mathbf{l} \qquad (3.32b)$$

where the combination of path ① and the inverse of path ② make up a closed loop. From the Stokes theorem

$$\oint \mathbf{A}\cdot d\mathbf{l} = \iint (\boldsymbol{\nabla}\times\mathbf{A})\cdot d\mathbf{S} = \iint \mathbf{B}\cdot d\mathbf{S} = \Phi$$

where Φ is the magnetic flux contained in the closed loop. If the radius of the magnetic filament is R, then the shift in phase as derived from Eq. (3.32b) is

$$\delta_1 - \delta_2 = \frac{e}{\hbar}(B\pi R^2)$$

assuming that the direction of the magnetic field is pointing out of the paper. As a result, the entire interference pattern is displaced on the screen (located at a distance L) by

$$\Delta x = L\theta = L\delta\frac{\lambda}{2\pi d} = \frac{1}{d}\frac{\lambda}{2\pi}\frac{e}{\hbar}(B\pi R^2) \qquad (3.32c)$$

and in the direction indicated in the figure, d is the spacing between the slits, and $\lambda = 2\pi/k$, the electron wavelength.

The great significance of the Bohm–Aharonov experiment is in demonstrating that even though no magnetic field (i.e., no force) acts on the electrons, the phase of the quantum-mechanical amplitude is shifted because of the presence of the potential. This general topic is an important point in the interpretation of quantum mechanics that is still being debated today.

3.6. IDENTICAL PARTICLES

So far we have been concerned with the description of states that contain only one particle. For instance, in the two-slit experiment the electrons or photons arrive one at a time, and the interference is due to the superposition of the amplitudes for different indistinguishable paths. There are, however, many quantum-mechanical systems that contain several particles. A multielectron atom, a nucleus when viewed in terms of its constituent protons and neutrons, any many-body system such as a crystal—all are obvious examples. To describe a many-body system the basis states must span a Hilbert space of very high dimensionality which can accommodate the variables of all the particles. In many cases, when the interaction between the particles is weak or absent, it is possible to construct the many-body state $|\phi\rangle$ from linear combinations of the *direct product* of single-particle states. To illustrate this point consider the hydrogen atom that is a bound state of a proton and an electron. We can express any state $|\phi\rangle$ of the hydrogen atom as

$$|\phi\rangle = \sum_{p,e} |p\rangle\,|e\rangle$$

where, symbolically, $|p\rangle$ are the states of the proton, for instance its spin states, and $|e\rangle$ are the states of the electron.

A particularly important case is when the particles in a many-body system are *identical.* Because of the absence of determinism in quantum mechanics it is impossible to label particles, and, consequently, identical particles are *indistinguishable.* Thus, all physically observable results must be invariant under the *interchange* of any two identical particles. This is a fundamental principle of quantum mechanics and imposes the condition that amplitudes for identical particles be either even or odd under particle exchange. Before we derive this general result we can demonstrate it by a familiar example.

We consider the helium atom which consists of two electrons bound to the helium nucleus He^4. Even though the two electrons interact with each other, for the present discussion we can neglect this effect as well as the motion of the nucleus. Under these conditions the ground state of the atom will be the state where both electrons are in their lowest single-particle state, which we designate by $|s\rangle$. The amplitudes for finding electron 1 or 2 in this state are written as

$$\langle 1 \mid s\rangle \quad \text{and} \quad \langle 2 \mid s\rangle$$

Electrons have spin 1/2, and thus for the state $|s\rangle$ there are two possible spin orientations, up and down, which we designate by $|s+\rangle$ and $|s-\rangle$. The corresponding ground-state single-particle amplitudes for each electron are then

$$\langle 1 \mid s+\rangle, \langle 1 \mid s-\rangle \quad \text{and} \quad \langle 2 \mid s+\rangle, \langle 2 \mid s-\rangle \tag{3.33a}$$

We can now construct the amplitude for the ground state of the helium atom by using the above single-particle amplitudes. There are four possible two-particle amplitudes

$$\begin{aligned} \psi_1 &= \langle 1 \mid s+\rangle\langle 2 \mid s+\rangle \\ \psi_2 &= \langle 1 \mid s+\rangle\langle 2 \mid s-\rangle \\ \psi_3 &= \langle 1 \mid s-\rangle\langle 2 \mid s+\rangle \\ \psi_4 &= \langle 1 \mid s-\rangle\langle 2 \mid s-\rangle \end{aligned} \tag{3.33b}$$

and they, or their linear combinations, should be suitable for describing the ground state of the helium atom.

It is a remarkable experimental fact that of the four possible linear combinations only one appears in nature. It is the combination

$$\psi_a = \langle 1 \mid s+\rangle\langle 2 \mid s-\rangle - \langle 1 \mid s-\rangle\langle 2 \mid s+\rangle \tag{3.33c}$$

which has the property that when we exchange electrons 1 and 2, the amplitude changes sign. This observation is a direct result of the *exclusion principle*, also called the Pauli principle after its discoverer, which states that *the amplitude for a system containing several electrons must be antisymmetric under the exchange of any two electrons*. The reader should verify that it is impossible to construct another linear combination of Eqs. (3.33*b*) that obeys the exclusion principle. An equivalent way of stating this principle is that in a system consisting of several electrons, no two electrons can be in the same single-particle state. This follows directly from the fact that if any two electrons are in the same state, the totally antisymmetric amplitude always vanishes.

To treat the general case we consider a system with two identical particles 1 and 2 and the single particle states $|a\rangle$ and $|b\rangle$. Physically the amplitudes

$$\begin{aligned} \phi_{\mathrm{I}} &= \langle 1 \mid a\rangle\langle 2 \mid b\rangle \\ \phi_{\mathrm{II}} &= \langle 1 \mid b\rangle\langle 2 \mid a\rangle \end{aligned} \tag{3.34a}$$

are *indistinguishable* because we have no way of telling which is particle 1 and which is particle 2. Therefore, the amplitude for finding one particle in $|a\rangle$ and the other in $|b\rangle$ must be a linear superposition of $\phi_{\rm I}$ and $\phi_{\rm II}$

$$\phi = \phi_{\rm I} \pm \phi_{\rm II} = \langle 1 \mid a\rangle\langle 2 \mid b\rangle \pm \langle 1 \mid b\rangle\langle 2 \mid a\rangle \tag{3.34b}$$

We have used a $\pm$ sign in Eq. (3.34*b*) because when particles 1 and 2 are *identical* we have no way of knowing whether we must add or subtract the two interfering amplitudes. In the previous section we postulated that the amplitudes for all indistinguishable paths of a single particle are additive. In the present case the indistinguishability arises from the presence of two identical particles: a different situation!

Let us denote by ϕ' the amplitude after the interchange of the particles. The probability $|\phi'|^2$ after the interchange must equal the probability $|\phi|^2$ before the interchange since the two amplitudes correspond to indistinguishable physical conditions. From

$$|\phi'|^2 = |\phi|^2$$

we conclude that

$$\phi' = e^{i\delta}\phi, \qquad \delta \text{ real} \tag{3.34c}$$

The interchange modifies the phase of the overall amplitude. If we interchange once more we return to the original situation so that

$$\phi'' = \phi$$

From Eq. (3.34*c*)

$$\phi'' = e^{i\delta}\phi' = e^{2i\delta}\phi$$

and therefore

$$e^{2i\delta} = 1 \quad \text{or} \quad e^{i\delta} = \pm 1$$

We thus reach the important conclusion that a quantum-mechanical amplitude for two identical particles must either *retain* or *reverse* its sign under interchange of the particles. If we choose the plus sign in Eq. (3.34*b*)

$$\phi' = \phi \qquad \text{symmetric amplitude}$$

while if we choose the minus sign

$$\phi' = -\phi \qquad \text{antisymmetric amplitude}$$

We have already mentioned that according to the Pauli principle the amplitude for a state containing two or more electrons must be antisymmetric. In fact, this is true whenever the identical particles have half-integer spin; we call these *Fermi-particles* or *fermions.* On the other hand, when the identical particles have an integer spin such as the photon, the π-meson, etc., the amplitude must be symmetric; we call these *Bose-particles* or *bosons.* As a consequence, the number of identical bosons that can be in the same state is unrestricted.

This generalization of the exclusion principle is known as the spin-statistics theorem. It has far-reaching consequences well beyond the structure of the Periodic Table, and is found to be strictly valid in all physical processes observed up to today. We will return to a quantitative discussion of systems with identical particles in Chapter 10.

3.7. CENTRAL CONCEPTS OF QUANTUM MECHANICS

In Section 2.5 we summarized postulates and definitions of quantum mechanics under the heading of "Language of Quantum Mechanics." In this section we give the rules governing the evolution of quantum-mechanical amplitudes and the connection of the formalism with physically observable quantities as introduced in this chapter. We hope that this section in conjunction with Section 2.5 will provide the reader with a compact reference to the principles of quantum mechanics. In the following chapters of the text we will be concerned mainly with the application of these concepts and of the corresponding formalism to the solution of physical problems.

A. *Process of Measurement*

1. Physical observables are represented by hermitian operators. The result of a measurement of a physical observable always results

in one of the eigenvalues of the corresponding operator. If the system that is measured is an eigenstate $|i\rangle$ of the operator, the outcome of the measurement is predictable and will yield the eigenvalue λ_i of the particular eigenstate. If the system is in a state $|\psi\rangle$, which is not an eigenstate $|j\rangle$ of the operator, one can only predict the probability $P_i = |\langle i \mid \psi\rangle|^2$ of obtaining the eigenvalue λ_i as a result of the measurement. Immediately after the measurement the system will be in the state $|i\rangle$.

2. The average value of *repeated* measurements of an observable $\hat{A}$ on identically prepared systems in the state $|\psi\rangle$ is

$$\langle A\rangle = \langle\psi|\, \hat{A}\, |\psi\rangle$$

and is called the *expectation value* of $\hat{A}$ in the state $|\psi\rangle$.

B. General Properties of Quantum-Mechanical Amplitudes

We will work in the Schrödinger picture where the basis states defined by any operator are fixed in time. Therefore, all the time dependence is contained in the state vector $|\psi\rangle$.

3. Amplitudes between an arbitrary state $|\psi\rangle$ and an eigenstate of *energy* $|j\rangle$ have the explicit time dependence

$$\langle j \mid \psi\rangle = C_{j\psi}(\lambda)e^{-(i/\hbar)E_j t}$$

where E_j is the eigenvalue of (the energy of the system in) the state $|j\rangle$, and $C_{j\psi}(\lambda)$ is a complex function not depending on time. Therefore the evolution in time of the state $|\psi\rangle$ can be represented by an expansion in energy eigenstates

$$|\psi\rangle = \sum_j |j\rangle C_{j\psi}(\lambda)e^{-(i/\hbar)E_j t}$$

4. When $|\psi\rangle$ is an eigenstate of energy it is called a *stationary state.* The probability of finding a stationary state $|\psi_s\rangle$ in any one basis state $|\mu\rangle$ is independent of time

$$P_{\psi_s}(\mu) = |\langle\mu \mid \psi_s\rangle|^2 = |\langle\mu \mid s\rangle|^2\, |C_{s\psi}(\lambda)|^2$$

5. The amplitude for finding the state $|\psi\rangle$ in the basis state of position $|x\rangle$ is called the *wavefunction* (in coordinate or configura-

tion space) of the system

$$\psi(\mathbf{x}, t) = \langle \mathbf{x} \mid \psi \rangle$$

For a free particle of well-defined momentum $\mathbf{p}$

$$\psi_{\mathbf{p}}(\mathbf{x}) = Ae^{-(i/\hbar)(Et-\mathbf{p}\cdot\mathbf{x})} = Ae^{-i(\omega t-\mathbf{k}\cdot\mathbf{x})}$$

with

$$\omega = \frac{E}{\hbar} \qquad \mathbf{k} = \frac{\mathbf{p}}{\hbar}$$

6. In the coordinate representation the overlap of two states is given by

$$\langle \chi \mid \psi \rangle = \int \chi^*(\mathbf{x}, t)\psi(\mathbf{x}, t)\, d^3x$$

and the matrix element of an operator $\hat{A}$ is

$$\langle \chi | \hat{A} | \psi \rangle = \int \chi^*(\mathbf{x}, t) A(\mathbf{x}) \psi(\mathbf{x}, t)\, d^3x$$

where $A(\mathbf{x})$ is a differential operator acting on $\psi(\mathbf{x}, t)$.

7. When a final state $|b\rangle$ can be reached from an initial state $|a\rangle$ through two or more *indistinguishable* intermediate states $|i\rangle$, the amplitude $\phi = \langle a \mid b \rangle$ is the sum of all amplitudes $\phi_i = \langle a \mid i \rangle \langle i \mid b \rangle$

$$\phi = \phi_1 + \phi_2 + \cdots = \sum_i \langle a \mid i \rangle \langle i \mid b \rangle$$

and the probability

$$P = |\phi|^2 = |\phi_1 + \phi_2 + \cdots|^2 = \left| \sum_i \langle a \mid i \rangle \langle i \mid b \rangle \right|^2$$

which may lead to *interference* between the amplitudes. When the intermediate states are *distinguishable* the probability P is given by

$$P = |\phi_1|^2 + |\phi_2|^2 + \cdots = \sum_i |\langle a \mid i \rangle \langle i \mid b \rangle|^2$$

8. Amplitudes involving two or more *identical* particles have definite properties under the exchange of any two particles. If the particles have integral spin, i.e., they are bosons, the amplitude must be *symmetric* under exchange. If the particles have half-integral spin, i.e., they are fermions, the amplitude must be *antisymmetric* under exchange.

Problems

PROBLEM 1

Consider the four Pauli spin matrices defined by Eqs. (4.68c).

(a) Calculate the expressions (commutators)

$$\hat{\sigma}_x\hat{\sigma}_y - \hat{\sigma}_y\hat{\sigma}_x$$
$$\hat{\sigma}_y\hat{\sigma}_z - \hat{\sigma}_z\hat{\sigma}_y$$
$$\hat{\sigma}_z\hat{\sigma}_x - \hat{\sigma}_x\hat{\sigma}_z$$

and discuss your result.

(b) Define $\hat{S}_x = (\hbar/2)\hat{\sigma}_x$, $\hat{S}_y = \cdots$, etc., and find the eigenvalues of the matrix

$$\hat{S}^2 = \hat{S}_x^2 + \hat{S}_y^2 + \hat{S}_z^2$$

PROBLEM 2

A quantum mechanical system is in the state

$$|\psi\rangle = c_1\,|1\rangle + c_2\,|2\rangle + \cdots + c_n\,|n\rangle$$

where the orthogonal normalized states $|1\rangle, |2\rangle, \ldots, |n\rangle$ are the *complete* set of states in which the Hamiltonian matrix H is *diagonal* (energy eigenstates). Find the average value of repeated measurements of the energy on similarly prepared states $|\psi\rangle$. Express your result in terms of the energy of the stationary states and the amplitudes c_i.

PROBLEM 3

Consider Problem 2 above and let $|i\rangle$ represent a *complete* set of base states and $|\mu\rangle$ represent a different complete set of basis states.

Show that if $\hat{A}$ is a hermitian operator, then

$$D^2 = \sum_i \sum_\mu |\langle\mu| \hat{A} |i\rangle|^2$$

(the summations are over the *entire* sets) is independent of the choice of basis and equals the sum of the squares of all the eigenvalues of $\hat{A}$. This result is called a *sum rule.*

PROBLEM 4

Let $\hat{Q}$ be an operator that does *not* contain time explicitly. If $|\psi\rangle$ is an eigenstate of energy, show that

$$\langle\psi| \hat{Q} |\psi\rangle$$

is independent of time. (You can consult Chapter 6.)

PROBLEM 5

The quantum-mechanical equation of motion for a quasi-free electron moving in a one-dimensional lattice (crystal) is of the form

$$i\hbar \frac{dC_n}{dt} = E_0 C_n - AC_{n-1} - AC_{n+1}$$

where $C_n = \langle n \mid \phi\rangle$ are the amplitudes for finding the electron at the nth site (atom) and A is a constant. For stationary states $|\phi\rangle$

$$C_n(x, t) = a(x_n)e^{-i(E/\hbar)t}$$

(*a*) Find the *range* of energies E of the stationary states by setting

$$a(x_n) = e^{ikx_n}$$
$$a(x_{n+1}) = e^{ik(x_n+b)}$$
$$a(x_{n+2}) = e^{ik(x_n+2b)}$$

where b is the lattice spacing, and k is a parameter. Plot E versus k.

(*b*) Give the *amplitude* C_n and the *probability* P_n for finding the electron at the nth, $(n+1)$th, and $(n+2\pi/b)$th sites. Interpret your result. What is the physical significance of the parameter k?

Note: The spectrum of energies E of the stationary states is continuous but restricted to an *energy band*, as you will derive in part (*a*).

PROBLEM 6

The data in the figure indicate the distribution of decay electrons from polarized muons which are moving in a circular orbit under the influence of a uniform magnetic field $B = 1.47$ Tesla. As time advances a peak appears when the mean direction of the decay electrons coincides with the momentum vector of the muon and a valley when the mean direction of the decay electrons is opposite the momentum vector of the muon. The muon momentum vector

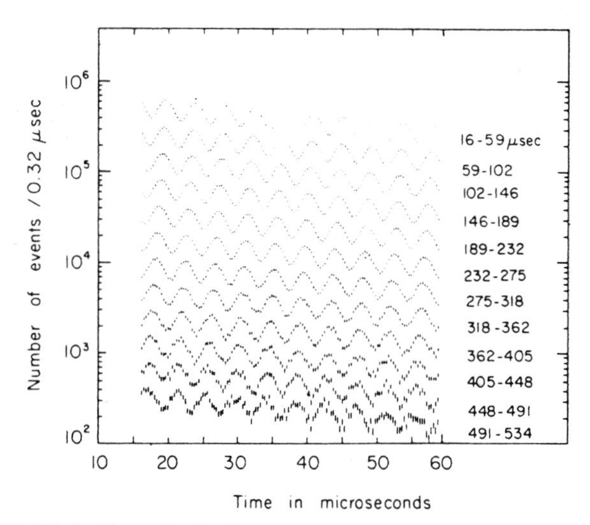

FIGURE P3.6. Time distribution of decay electrons as a function of time. The electrons arise from the decay of polarized muons precessing in a magnetic field of 1.47 *T*.

rotates in the laboratory with the Larmor frequency

$$\omega_L = \frac{eB}{m_\mu}$$

From the data find the magnetic moment of the muon in units of $\mu = e\hbar/2m_\mu$. The spin precession frequency is given by Eq. (3.22b). The reference to the data is J. Bailey *et al.*, *Phys. Lett.* **68B,** 191 (1977).

Note: To simplify the calculation the above relations refer to nonrelativistic muons. This is not true in the actual experiment.

PROBLEM 7

A neutron beam is split by Bragg scattering in a crystal and recombined as shown in the figure. The path S_1 is 30 cm *higher* than path S_2, and the length along the horizontal direction is 1 m. The wavelength of the neutrons is $\lambda = h/p = 1$ Å. Due to the gravitational interaction a phase difference is induced between the two paths. Calculate by how many fringes the interference pattern shifts when the apparatus is rotated so that path S_1 is 30 cm *lower* than S_2.

Note: This is a beautiful experiment performed by Colella, Overhauser, and Werner demonstrating the influence of the gravitational force on the time evolution of quantum-mechanical amplitudes. [*Phys. Rev. Lett.* **34,** 1472 (1975).]

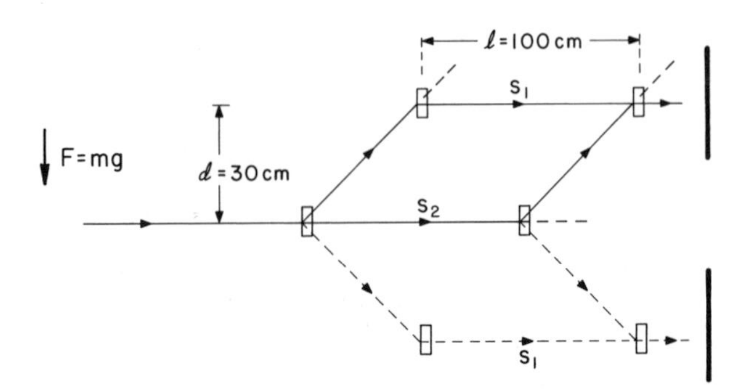

FIGURE P3.7. Interferometer for the measurement of the phase shift of the neutron wavefunction as a function of the gravitational potential.

Chapter 4

STATIONARY STATES OF A QUANTUM SYSTEM

We have already seen that the quantum-mechanical state of a system can be described as a linear superposition of basis states in a particular representation. One such set of convenient basis states is given by the energy eigenstates. We know that if a system is in a definite energy eigenstate, then it is stationary, i.e., the probability of finding such a system in any other eigenstate of energy vanishes at all times. It is clear, therefore, that if we are considering an isolated system it is sufficient to determine the energy eigenstates completely. We discuss in this chapter various ways of determining the energy eigenvalues and eigenstates from the knowledge of the Hamiltonian or energy operator. In any given representation the Hamiltonian, of course, can be represented as a matrix whose form is completely fixed once we know the forces (and hence the potentials) acting on the system.

We derive the quantum-mechanical equation of motion, otherwise known as the Schrödinger equation. The time-evolution operator which causes temporal changes in a quantum-mechanical state is introduced, and it is shown that it contains all the dynamical information about the system. It is clear from the form of the time-evolution operator that the study of the Hamiltonian matrix is all that is needed to determine how states change with time.

In nature, however, there are only a handful of physical systems whose Hamiltonian matrices can be solved to yield the exact energy eigenvalues. Fortunately, most physical systems have Hamiltonians that deviate only slightly from systems that can be solved exactly. We develop the theory of time-independent perturbation in this chapter and discuss how we can use this to obtain approximate solutions that are very close to the true energy eigenvalues of complicated systems.

As an application of the methods of perturbation theory we

consider the two-level system in complete detail. The ammonia molecule, the $K^0-\bar{K}^0$ system, or a spin 1/2 particle in a magnetic field, are only a few of the physical examples that can be discussed as two-level systems. We examine how such systems behave under an external perturbation. A two-level system is simple enough so that exact energy eigenvalues can be obtained. We then compare the perturbative results to the exact values in order to bring out the powerful nature of perturbation theory.

4.1. THE HAMILTONIAN MATRIX AND THE TIME EVOLUTION OF QUANTUM-MECHANICAL STATES

In Chapter 3 we discussed the Schrödinger picture or description of quantum mechanics wherein the basis states are fixed in time, but the states describing a physical system depend on or change with time. The wavefunction that represents the overlap of the state of the system with a basis state is therefore time dependent. The operators, on the other hand, which represent physical observables, are not.

Let us consider a general quantum-mechanical state $|\phi(t)\rangle$. Let $|\phi(t_1)\rangle$ and $|\phi(t_2)\rangle$ denote the states of the system at times t_1 and t_2, respectively. We can relate the two states as

$$|\phi(t_2)\rangle = \hat{U}(t_2, t_1)\,|\phi(t_1)\rangle \tag{4.1}$$

where $\hat{U}(t_2, t_1)$ is a quantum-mechanical operator since it causes a change in a state, taking it from a state at time t_1 to one at time t_2. Consequently, it is known as the time-evolution operator or the time-translation operator.

By definition we note that

$$\hat{U}(t, t) = 1 \tag{4.2}$$

which simply reflects the fact that if the initial and the final time are the same, then the state remains unchanged.

Furthermore,

$$|\phi(t_1)\rangle = \hat{U}(t_1, t_2)\,|\phi(t_2)\rangle \tag{4.3a}$$

But Eq. (4.1) gives

$$|\phi(t_1)\rangle = \hat{U}^{-1}(t_2, t_1)\,|\phi(t_2)\rangle \tag{4.3b}$$

which leads to the relation

$$\hat{U}^{-1}(t_2, t_1) = \hat{U}(t_1, t_2) \tag{4.4}$$

Time evolution does not change the normalization of states since that would amount to nonconservation of probability. Therefore, if a state is initially normalized to unity it must remain so normalized at the end of any finite interval of time. Thus

$$\langle\phi(t_2)\mid\phi(t_2)\rangle = 1 = \langle\phi(t_1)|\,\hat{U}^{\dagger}(t_2, t_1)\hat{U}(t_2, t_1)\,|\phi(t_1)\rangle \tag{4.5a}$$

This implies that

$$\hat{U}^{\dagger}(t_2, t_1)\hat{U}(t_2, t_1) = 1 \tag{4.5b}$$

Consequently,

$$\hat{U}^{\dagger}(t_2, t_1) = \hat{U}^{-1}(t_2, t_1) \tag{4.5c}$$

which shows that the time-evolution operator is unitary. The operator $\hat{U}$ also satisfies

$$\hat{U}(t_3, t_1) = \hat{U}(t_3, t_2)\hat{U}(t_2, t_1) \tag{4.6}$$

This physically represents the fact that a state evolving from t_1 to t_3 is equivalent to the evolution of the state through an intermediate time t_2. We can thus summarize the properties of the time evolution operator as

$$\begin{aligned} &\hat{U}^{\dagger}(t_2, t_1) = \hat{U}^{-1}(t_2, t_1) = \hat{U}(t_1, t_2) \\ &\hat{U}(t_3, t_1) = \hat{U}(t_3, t_2)\hat{U}(t_2, t_1) \\ &\hat{U}(t, t) = 1 \end{aligned} \tag{4.7}$$

The operator $\hat{U}$, and hence the time evolution of the system, is completely determined when we know all the matrix elements of this operator in a particular basis. We now show how this can be achieved in practice and give a simple derivation of the differential equation that governs the time evolution of quantum-mechanical states.

We denote a set of basis states by $|i\rangle$ and consider two states $|\phi(t)\rangle$ and $|\phi(t+\Delta t)\rangle$ separated by an infinitesimal time interval. By definition

$$|\phi(t+\Delta t)\rangle = \hat{U}(t+\Delta t, t)\,|\phi(t)\rangle$$

or

$$\langle i \mid \phi(t+\Delta t)\rangle = \langle i|\,\hat{U}\,|\phi(t)\rangle = \sum_j \langle i|\,\hat{U}\,|j\rangle\langle j \mid \phi(t)\rangle \tag{4.8}$$

We also define

$$\begin{aligned} \langle i \mid \phi(t)\rangle &\equiv C_i(t) \\ \langle i|\,\hat{U}(t+\Delta t, t)\,|j\rangle &= U_{ij} = \delta_{ij} + \Delta t K_{ij} \end{aligned} \tag{4.9a}$$

Furthermore, from the observation that the time-evolution operator is unitary and from the fact that, classically, time translation has the form e^{-iHt}, we write

$$U_{ij} = \delta_{ij} - \Delta t \frac{i}{\hbar} H_{ij} \tag{4.9b}$$

and identify H_{ij} as the Hamiltonian matrix. Thus Eq. (4.8) becomes

$$\begin{aligned} C_i(t+\Delta t) &= \sum_j \left(\delta_{ij} - \Delta t \frac{i}{\hbar} H_{ij}\right) C_j(t) \\ &= C_i(t) - \frac{i}{\hbar}\Delta t \sum_j H_{ij} C_j(t) \end{aligned}$$

or

$$\frac{C_i(t+\Delta t) - C_i(t)}{\Delta t} = -\frac{i}{\hbar}\sum_j H_{ij} C_j(t) \tag{4.10}$$

In the limit $\Delta t \to 0$, this reduces to the differential equation

$$\frac{dC_i}{dt} = -\frac{i}{\hbar}\sum_j H_{ij} C_j \tag{4.11}$$

This is the basic equation that governs the time development of probability amplitudes in quantum mechanics. It is also known as the Schrödinger equation. The operator form of this equation is obtained by simply expressing the $C_i(t)$ by their definition as amplitudes:

$$\begin{aligned} i\hbar \frac{d}{dt}\langle i \mid \phi(t)\rangle &= \sum_j \langle i|\,\hat{H}\,|j\rangle\langle j \mid \phi(t)\rangle \\ &= \langle i|\,\hat{H}\,|\phi(t)\rangle \end{aligned}$$

or

$$\boxed{i\hbar \frac{d}{dt}|\phi(t)\rangle = \hat{H}\,|\phi(t)\rangle} \tag{4.12}$$

If the Hamiltonian operator does not depend on time explicitly, then Eq. (4.12) can be integrated to give

$$|\phi(t)\rangle = e^{-(i/\hbar)\hat{H}t}\,|\phi(0)\rangle \tag{4.13a}$$

which shows that the time-evolution operator has the form

$$\hat{U}(t_2, t_1) = e^{-(i/\hbar)\hat{H}(t_2-t_1)} \tag{4.13b}$$

It is clear from these results that the time development of a quantum-mechanical system is completely determined if the Hamiltonian matrix H_{ij} is known. We will now apply these ideas to a few simple examples and find the explicit time-dependence of the quantum states.

Consider first a system that has only one state. Then Eq. (4.11) reduces to

$$i\hbar\frac{dC_1}{dt} = H_{11}C_1 \tag{4.14a}$$

Furthermore,

$$H_{11} = \langle 1|\,\hat{H}\,|1\rangle = E_1 \tag{4.14b}$$

is the energy of that state. Hence the solution of Eq. (4.14a) is

$$C_1(t) = \text{const}\cdot \mathrm{e}^{-(i/\hbar)E_1t} \tag{4.15}$$

The constant is easily obtained from the normalization of the amplitude

$$|C_1(t)|^2 = 1$$

If the system has two basis states and if the Hamiltonian happens to be diagonal in the particular representation $|1\rangle$ and $|2\rangle$, then Eq. (4.11) reduces to two decoupled differential equations

$$\begin{aligned} i\hbar\frac{dC_1}{dt} &= H_{11}C_1 \\ i\hbar\frac{dC_2}{dt} &= H_{22}C_2 \end{aligned} \tag{4.16a}$$

Furthermore, if E_1 and E_2 are the energy eigenvalues associated with the basis states $|1\rangle$ and $|2\rangle$, respectively, then the solution to Eq. (4.16a) is given by

$$\begin{aligned} \langle 1\,|\,\phi(t)\rangle &= C_1(t) = Ae^{-(i/\hbar)E_1t} \\ \langle 2\,|\,\phi(t)\rangle &= C_2(t) = Be^{-(i/\hbar)E_2t} \end{aligned} \tag{4.16b}$$

Thus, the state $|\phi(t)\rangle$ can be written as

$$|\phi(t)\rangle = Ae^{-(i/\hbar)E_1t}|1\rangle + Be^{-(i/\hbar)E_2t}|2\rangle \tag{4.17}$$

where the constants A and B can be uniquely fixed from the initial condition as well as from normalization. Such a state was discussed in Section 3.2.

The solutions of Eq. (4.16b) correspond to amplitudes for stationary states, and we conclude that stationary states are given by the basis states in the representation that makes the Hamiltonian matrix H_{ij} diagonal. The eigenvalues of H_{ij} give the energy of the stationary states. If the Hamiltonian is not diagonal in a particular representation, we can always find an orthonormal set of basis states in which it will be diagonal. This follows from the fact that the Hamiltonian matrix is hermitian, and hence one can always find a unitary matrix that would diagonalize it, i.e.,

$$\hat{H}' = S\hat{H}S^{-1} = S\hat{H}S^{\dagger} \tag{4.18a}$$

as indicated in Eq. (2.34a). The convention for the transformation matrix $S_{\mu i}$ is that of Eq. (2.22a). The basis states in the representation that diagonalizes the Hamiltonian are labeled $|\mu\rangle$, whereas in the original representation they were labeled $|i\rangle$. Then

$$\langle\mu|\,\hat{H}\,|\nu\rangle = \sum_{i,j} \langle\mu \mid i\rangle\langle i|\,\hat{H}\,|j\rangle\langle j \mid \nu\rangle$$

or

$$H_{\mu\nu} = \sum_{i,j} S_{\mu i}H_{ij}(S^{\dagger})_{j\nu} \tag{4.18b}$$

The elements of the unitary matrix S are given by the amplitudes

$$\begin{aligned} S_{\mu i} &= \langle\mu \mid i\rangle \\ (S^{\dagger})_{j\nu} &= \langle j \mid \nu\rangle = \langle\nu \mid j\rangle^* \end{aligned} \tag{4.18c}$$

as defined in Eq. (2.22a), with S a unitary matrix. The new basis states are obtained through

$$\begin{aligned} |\mu\rangle &= \sum_i |i\rangle\langle i \mid \mu\rangle \\ &\equiv \sum_i |i\rangle\,(S^{\dagger})_{i\mu} \end{aligned} \tag{4.19a}$$

Finally, the amplitudes transform as in Eq. (2.21b), which for an arbitrary state $|\phi\rangle$ gives

$$\begin{aligned}\langle\mu \mid \phi\rangle &= \sum_i \langle\mu \mid i\rangle\langle i \mid \phi\rangle \\ &= \sum_i S_{\mu i} \langle i \mid \phi\rangle\end{aligned} \tag{4.19b}$$

Note that the amplitudes transform inversely from the basis states. While Eqs. (4.18b) and (4.19b) can be interpreted directly as matrix multiplications, this is not the case for Eq. (4.19a).

Hermiticity of the Hamiltonian Matrix

It is clear from Eq. (4.13b) that the time-evolution operator can be unitary only if the Hamiltonian matrix is hermitian. We will now show that the Hamiltonian matrix must be hermitian if probability is to be conserved. We know that the sum of the probabilities over all basis states of a complete set must equal unity, i.e.,

$$\sum_i |C_i(t)|^2 = \sum_i C_i^*(t)C_i(t) = 1 \tag{4.20}$$

Taking the derivative with respect to time, we have

$$0 = \sum_i \left[\frac{dC_i^*}{dt} C_i(t) + C_i^*(t) \frac{dC_i}{dt}\right] \tag{4.21}$$

The quantum-mechanical equation of motion [Eq. (4.11)] tells us that

$$\frac{dC_i}{dt} = -\frac{i}{\hbar} \sum_j H_{ij} C_j$$

and

$$\frac{dC_i^*}{dt} = \frac{i}{\hbar} \sum H_{ij}^* C_j^* \tag{4.22}$$

Substituting into Eq. (4.21) we obtain

$$\sum_{i,j} \left(\frac{i}{\hbar} H_{ij}^* C_j^* C_i - \frac{i}{\hbar} C_i^* H_{ij} C_j\right) = 0$$

and since we sum over the indices i, j we can interchange them in the first term

$$\sum_{i,j} \frac{i}{\hbar} (C_i^* H_{ji}^* C_j - C_i^* H_{ij} C_j) = 0$$

or

$$\sum_{i,j} \frac{i}{\hbar} C_i^* (H_{ji}^* - H_{ij}) C_j = 0 \tag{4.23a}$$

For arbitrary amplitudes this relation can be true only if

$$H_{ji}^* = H_{ij}$$

or

$$(H^{\dagger})_{ij} = H_{ij} \tag{4.23b}$$

That is, the Hamiltonian matrix must be hermitian for probability to remain conserved. Furthermore, its eigenvalues, which represent the energy of the stationary states of the system, will therefore be real.

4.2. TIME-INDEPENDENT PERTURBATION OF AN ARBITRARY SYSTEM

As we have seen, the knowledge of the eigenvalues and eigenvectors of the Hamiltonian matrix is all that is needed to determine the time development of a quantum-mechanical system. However, for most physical systems an exact solution of the Hamiltonian is impossible to obtain. Surprisingly, approximate solutions arbitrarily close to the exact results can be obtained almost without effort if the Hamiltonian can be split into a dominant part which can be diagonalized and another part which is complicated but can be treated as a perturbation on the first part.

It is worthwhile to digress for a moment, to illustrate the power of perturbation theory through the example of an algebraic equation. Suppose we want to find the roots of the cubic equation

$$x^3 - 4.001x + 0.003 = 0 \tag{4.24a}$$

This equation is extremely difficult to solve exactly. However, let us introduce a small parameter

$$\varepsilon = 0.001$$

so that the equation is written

$$x^3-(4+\varepsilon)x+3\varepsilon=0 \tag{4.24b}$$

It is clear that the solution to this equation would be a function of the parameter ε, and hence one can expand the solution in the form

$$x(\varepsilon)=\sum_{n=0}^{\infty} a_n\varepsilon^n \tag{4.25}$$

We substitute this expansion into Eq. (4.24b) and solve for x consistently order-by-order in ε. For example, to order ε^0 the equation becomes

$$x^3-4x=0$$

with solutions

$$x=0, \pm 2 \tag{4.26}$$

To find the roots consistently up to order ε, we keep the first order terms in Eq. (4.25) for the three roots

$$x_1=0+a\varepsilon \tag{4.27a}$$

$$x_2=2+b\varepsilon \tag{4.27b}$$

$$x_3=-2+c\varepsilon \tag{4.27c}$$

Substituting for example x_1, from Eq. (4.27a) into Eq. (4.24b) we obtain

$$(a\varepsilon)^3-(4+\varepsilon)a\varepsilon+3\varepsilon=0 \tag{4.28a}$$

If we keep only terms linear in ε, Eq. (4.28a) reduces to

$$-4a\varepsilon+3\varepsilon=0$$

with the solution

$$a=\tfrac{3}{4}$$

Thus

$$x_1=0+\tfrac{3}{4}\varepsilon=\tfrac{3}{4}\varepsilon \qquad \text{up to order } \varepsilon \tag{4.28b}$$

If we substitute x_2 from Eq. (4.27b) into Eq. (4.24b) we obtain

$$(2+b\varepsilon)^3-(4+\varepsilon)(2+b\varepsilon)+3\varepsilon=0$$

or

$$8+12b\varepsilon+6b^2\varepsilon^2+b^3\varepsilon^3-(8+2\varepsilon+4b\varepsilon+b\varepsilon^2)+3\varepsilon=0 \tag{4.29a}$$

Therefore, to order ε the equation becomes

$$12b\varepsilon - 4b\varepsilon - 2\varepsilon + 3\varepsilon = 0$$

with solution

$$b = -\tfrac{1}{8} \tag{4.29b}$$

Thus

$$x_2 = 2 - \tfrac{1}{8}\varepsilon \qquad \text{up to order } \varepsilon$$

Similarly, it is easy to show that up to order ε the third root becomes

$$x_3 = -2 - \tfrac{5}{8}\varepsilon \tag{4.29c}$$

Thus, to this order the roots become

$$\begin{aligned} x_1 &= \tfrac{3}{4}(0.001) \\ x_2 &= 2 - \tfrac{1}{8}(0.001) \\ x_3 &= -2 - \tfrac{5}{8}(0.001) \end{aligned} \tag{4.30}$$

One can similarly obtain the roots up to order ε^2. And it is clear that the roots in order ε^2 would differ from the exact roots only by 1 part in 10^{-6}. We do not have to stress further the significance of this approximation scheme.

In quantum mechanics, however, we deal with matrices, and we develop below a perturbation scheme to handle them. We assume that we are looking for the solutions of the equation

$$\hat{H}\,|n\rangle = E_n\,|n\rangle \tag{4.31a}$$

We also assume that the Hamiltonian can be separated into two parts

$$\hat{H} = \hat{H}_0 + \varepsilon\hat{H}' \tag{4.31b}$$

As before, it is evident all entities will be functions of the parameter ε. For instance, the eigenvalue for the nth state is

$$E_n(\varepsilon) = E_n^{(0)} + \varepsilon E_n^{(1)} + \varepsilon^2 E_n^{(2)} + \cdots \tag{4.32a}$$

and the corresponding state vector is

$$|n(\varepsilon)\rangle = |n^{(0)}\rangle + \varepsilon\,|n^{(1)}\rangle + \varepsilon^2\,|n^{(2)}\rangle + \cdots \tag{4.32b}$$

Here $|n^{(0)}\rangle$ corresponds to the stationary state of the unperturbed Hamiltonian $\hat{H}_0$, and for simplicity we consider the case where the perturbing Hamiltonian $\hat{H}'$ is independent of time. If we substitute the expansion given in Eqs. (4.32) into Eq. (4.31a), we can solve consistently for various quantities order-by-order in ε. Thus

$$(\hat{H}_0+\varepsilon\hat{H}')(|n^{(0)}\rangle+\varepsilon\,|n^{(1)}\rangle+\varepsilon^2\,|n^{(2)}\rangle+\cdots$$
$$=(E_n^{(0)}+\varepsilon E_n^{(1)}+\varepsilon^2 E_n^{(2)}+\cdots)(|n^{(0)}\rangle+\varepsilon\,|n^{(1)}\rangle+\varepsilon^2\,|n^{(2)}\rangle+\cdots) \tag{4.33}$$

Thus up to

$$\text{order } \varepsilon^0\text{:}\quad \hat{H}_0\,|n^{(0)}\rangle=E_n^{(0)}\,|n^{(0)}\rangle \tag{4.34a}$$

$$\text{order } \varepsilon^1\text{:}\quad \hat{H}_0\,|n^{(1)}\rangle+\hat{H}'\,|n^{(0)}\rangle=E_n^{(0)}\,|n^{(1)}\rangle+E_n^{(1)}\,|n^{(0)}\rangle \tag{4.34b}$$

$$\text{order } \varepsilon^2\text{:}\quad \hat{H}_0\,|n^{(2)}\rangle+\hat{H}'\,|n^{(1)}\rangle$$
$$=E_n^{(0)}\,|n^{(2)}\rangle+E_n^{(1)}\,|n^{(1)}\rangle+E_n^{(2)}\,|n^{(0)}\rangle \tag{4.32c}$$

Equation (4.34a) simply reflects our choice of notation where $|n^{(0)}\rangle$ are the stationary states of the Hamiltonian $\hat{H}_0$ with eigenvalues $E_n^{(0)}$. Taking the inner product with $\langle n^{(0)}|$ in Eq. (4.34b) we obtain

$$E_n^{(1)}=\langle n^{(0)}|\,\hat{H}'\,|n^{(0)}\rangle \tag{4.35a}$$

Similarly, taking the inner product with the state $\langle m^{(0)}|$, where $m\neq n$, we obtain

$$\langle m^{(0)}\,|\hat{H}_0\,|n^{(1)}\rangle+\langle m^{(0)}\,|\hat{H}'\,|n^{(0)}\rangle=E_n^{(0)}\,\langle m^{(0)}\mid n^{(1)}\rangle+E_n^{(1)}\,\langle m^{(0)}\mid n^{(0)}\rangle$$

or

$$\langle m^{(0)}\mid n^{(1)}\rangle=\frac{\langle m^{(0)}|\,H'\,|n^{(0)}\rangle}{E_n^{(0)}-E_m^{(0)}},\qquad m\neq n \tag{4.35b}$$

We can thus find $|n^{(1)}\rangle$ in the usual way

$$|n^{(1)}\rangle=\sum_m |m^{(0)}\rangle\langle m^{(0)}\,|n^{(1)}\rangle$$
$$=a\,|n^{(0)}\rangle+\sum_{m\neq n}\frac{\langle m^{(0)}|\,\hat{H}'\,|n^{(0)}\rangle}{E_n^{(0)}-E_m^{(0)}}\,|m^{(0)}\rangle \tag{4.36a}$$

where a is undetermined in view of Eq. (4.35b). Thus, up to order ε,

$$|n\rangle=|n^{(0)}\rangle+\varepsilon\,|n^{(1)}\rangle$$
$$=(1+a\varepsilon)\,|n^{(0)}\rangle+\varepsilon\sum_{m\neq n}\frac{\langle m^{(0)}|\,H'\,|n^{(0)}\rangle}{E_n^{(0)}-E_m^{(0)}}\,|m^{(0)}\rangle \tag{4.36b}$$

Normalization of the state $|n\rangle$ up to order ε further requires that

$$a=0$$

Thus, to order ε we obtain

$$\begin{aligned} E_n &= E_n^{(0)} + \varepsilon \langle n^{(0)}| \hat{H}' |n^{(0)}\rangle \\ |n\rangle &= |n^{(0)}\rangle + \varepsilon \sum_{m \neq n} \frac{\langle m^{(0)}| \hat{H}' |n^{(0)}\rangle}{E_n^{(0)} - E_m^{(0)}} |m^{(0)}\rangle \end{aligned} \tag{4.36c}$$

Similarly, one can carry out the calculation up to order ε^2 and show that

$$E_n = E_n^{(0)} + \varepsilon \langle n^{(0)}| \hat{H}' |n^{(0)}\rangle + \varepsilon^2 \langle n^{(0)}| \hat{H}' |n^{(1)}\rangle$$

$$\boxed{E_n = E_n^{(0)} + \varepsilon \langle n^{(0)}| \hat{H}' |n^{(0)}\rangle + \varepsilon^2 \sum_{m \neq n} \frac{\langle n^{(0)}| \hat{H}' |m^{(0)}\rangle \langle m^{(0)}| \hat{H}' |n^{(0)}\rangle}{E_n^{(0)} - E_m^{(0)}}} \tag{4.37}$$

In a quantum-mechanical system it is not easy to identify a parameter ε, unlike the algebraic example we considered above. Thus, ε has been introduced for bookkeeping purposes and can be set equal to unity. However, looking at the expressions in Eqs. (4.36c) and (4.37) it is clear that the true expansion parameter is

$$\xi = \frac{\langle m^{(0)}| \hat{H}' |n^{(0)}\rangle}{E_n^{(0)} - E_m^{(0)}} = \frac{H'_{mn}}{E_n^{(0)} - E_m^{(0)}}, \qquad m \neq n$$

and for the perturbation scheme to be useful we must have

$$|\xi| \ll 1$$

In other words, the perturbation expansion works only if the off-diagonal elements of the Hamiltonian are much smaller than the difference in any two energy levels of the unperturbed Hamiltonian.

Clearly the perturbation expansion that we introduced breaks down if any two levels $|m^{(0)}\rangle$ and $|n^{(0)}\rangle$ are very close to each other or if they are degenerate in energy, unless the corresponding matrix elements H'_{mn} vanish. In such a case, if the first-order correction does not remove the degeneracy or increase the separation between the two levels, then a partial diagonalization of the Hamiltonian in the space of degenerate states is necessary. Physically, one can understand the difficulty in the following way. When degenerate states are present any linear combination of such states is a good starting point for the perturbative calculation. An ideal starting point is the linear combination to which the perburbative solution reduces in the limit of vanishing perturbation. We, of course, have no prior knowledge of this combination and hence, in the presence of degenerate states, we must first select the right zeroth-order linear combination of the degenerate states. This is accomplished by diagonalizing the Hamiltonian in the subspace of degenerate states and then applying the perturbation theory developed in this section. This will be illustrated in Section 4.6 in connection with the ammonia molecule.

The perturbation method that we described for obtaining approximate energy eigenvalues is due to Rayleigh and Schrödinger. Note that although we assumed in our previous discussion that the energy levels were discrete, the method works as well for continuous states also. In this case the matrix elements H'_{mn} can be written as

$$\begin{aligned} H'_{mn} &= \langle m| H' |n\rangle \\ &= \int d^3x\, d^3x' \langle m \mid \mathbf{x}\rangle\langle \mathbf{x}| \hat{H}' |\mathbf{x}'\rangle\langle \mathbf{x}' \mid n\rangle \\ &= \int d^3x\, d^3x' \psi^*_m(\mathbf{x})\, \langle \mathbf{x}| \hat{H}' |\mathbf{x}'\rangle\, \psi_n(\mathbf{x}') \end{aligned} \tag{4.38a}$$

Thus, to obtain the perturbative results we need to know the matrix elements of the perturbing Hamiltonian $\hat{H}'$ in the coordinate representation. In this case the perturbation Hamiltonian is given by a differential operator $\hat{H}'(x)$ such that

$$\int d^3x' \langle \mathbf{x}| \hat{H}' |\mathbf{x}'\rangle\, \psi_n(\mathbf{x}') \equiv \hat{H}'(\mathbf{x})\psi_n(\mathbf{x})$$

and therefore

$$H'_{mn} = \int d^3x \psi^*_m(\mathbf{x}) \hat{H}'(\mathbf{x}) \psi_n(\mathbf{x}) \tag{4.38b}$$

We see that given the wave functions of a system the problem is reduced to finding the differential operator $\hat{H}'(\mathbf{x})$ that describes the perturbation. This can often be achieved (but not always) by correspondence with classical physics. For example, when an electric field $\vec{\mathscr{E}}$ is applied the perturbation energy acquired by an electron of charge e is described in the coordinate representation by the differential operator

$$\hat{H}'(\mathbf{x}) = e\vec{\mathscr{E}} \cdot \hat{\mathbf{r}} \tag{4.39a}$$

If the electric field is along the Z-axis, then

$$\hat{H}'(\mathbf{x}) = e\mathscr{E}\hat{z} \tag{4.39b}$$

where $\hat{z}$ is the differential operator (in this case, it is simply the coordinate z). Similarly, if a magnetic moment $\boldsymbol{\mu} = (\mu/\hbar)\mathbf{L}$, where $\mathbf{L}$ is the angular momentum, is subjected to a magnetic field $\mathbf{B}$, then the perturbation energy is described in the coordinate representation by the differential operator

$$\hat{H}'(\mathbf{x}) = -\hat{\boldsymbol{\mu}} \cdot \mathbf{B} = -\frac{\mu}{\hbar}\hat{\mathbf{L}} \cdot \mathbf{B} \tag{4.40a}$$

where $\hat{\mathbf{L}}$ is the angular momentum operator. If the magnetic field is along the z-direction

$$\hat{H}'(\mathbf{x}) = -\frac{\mu B}{\hbar}\hat{L}_z \tag{4.40b}$$

We have already discussed this interaction in Section 3.3 for the case of an electron. The construction of general differential operators in the coordinate representation is taken up in Chapter 6. In the next section we give simple examples of perturbation theory applied to matrix mechanics.

4.3. SIMPLE MATRIX EXAMPLES OF TIME-INDEPENDENT PERTURBATION

As a numerical example consider the 3×3 matrix given by

$$H = \begin{pmatrix} 1 & \varepsilon & \varepsilon \\ \varepsilon & 2 & \varepsilon \\ \varepsilon & \varepsilon & 3 \end{pmatrix} \tag{4.41a}$$

Here ε is a small parameter and we wish to determine the eigenvalues and eigenvectors of this operator. Note that this appears to be a highly complicated problem, and the degree of complexity increases as we go to matrices of increasing dimensionality. We will therefore use the perturbation techniques developed in the previous section and calculate the eigenvalues only up to order ε^2 and the eigenvectors only up to order ε. If ε is sufficiently small, then this would give a reasonably close approximation to the real answer. To use perturbation theory we rewrite the Hamiltonian as

$$\begin{aligned}\hat{H} &= H_0 + \varepsilon H' \\ &= \begin{pmatrix} 1 & 0 & 0 \\ 0 & 2 & 0 \\ 0 & 0 & 3 \end{pmatrix} + \varepsilon \begin{pmatrix} 0 & 1 & 1 \\ 1 & 0 & 1 \\ 1 & 1 & 0 \end{pmatrix}\end{aligned} \tag{4.41b}$$

The matrix H_0 is diagonal. By inspection we see that its eigenvalues and eigenvectors are

$$\begin{aligned} E_1^{(0)} &= 1 \qquad \psi_1^{(0)} = \begin{pmatrix} 1 \\ 0 \\ 0 \end{pmatrix} \\ E_2^{(0)} &= 2 \qquad \psi_2^{(0)} = \begin{pmatrix} 0 \\ 1 \\ 0 \end{pmatrix} \\ E_3^{(0)} &= 3 \qquad \psi_3^{(0)} = \begin{pmatrix} 0 \\ 0 \\ 1 \end{pmatrix} \end{aligned} \tag{4.42}$$

The matrix elements of the perturbing Hamiltonian in this basis are given by

$$\left.\begin{aligned} H'_{mn} &= 1, \qquad \text{for } m \neq n \\ &= 0, \qquad \text{for } m = n \end{aligned}\right\} \qquad m, n = 1, 2, 3 \tag{4.43}$$

Thus the first-order correction to the eigenvalues are all zero

$$\begin{aligned} E_1^{(1)} &= \varepsilon H'_{11} = 0 \\ E_2^{(1)} &= \varepsilon H'_{22} = 0 \\ E_3^{(1)} &= \varepsilon H'_{33} = 0 \end{aligned} \tag{4.44a}$$

and the eigenvalues are unchanged through this order. The second-order change in the eigenvalues is given by

$$
\begin{aligned}
E_1^{(2)} &= \varepsilon^2 \sum_{m\neq 1} \frac{H'_{1m}H'_{m1}}{E_1^{(0)}-E_m^{(0)}} \\
&= \varepsilon^2\left(\frac{H'_{12}H'_{21}}{E_1^{(0)}-E_2^{(0)}}+\frac{H'_{13}H'_{31}}{E_1^{(0)}-E_3^{(0)}}\right) \\
&= \varepsilon^2\left(\frac{1\cdot 1}{1-2}+\frac{1\cdot 1}{1-3}\right) = -\tfrac{3}{2}\varepsilon^2 \\
E_2^{(2)} &= \varepsilon^2 \sum_{m\neq 2} \frac{H'_{2m}H'_{m2}}{E_2^{(0)}-E_m^{(0)}} \\
&= \varepsilon^2\left(\frac{H'_{21}H'_{12}}{E_2^{(0)}-E_1^{(0)}}+\frac{H'_{23}H'_{32}}{E_2^{(0)}-E_3^{(0)}}\right) \\
&= \varepsilon^2\left(\frac{1\cdot 1}{2-1}+\frac{1\cdot 1}{2-3}\right) = 0 \\
E_3^{(2)} &= \varepsilon^2 \sum_{m\neq 3} \frac{H'_{3m}H'_{m3}}{E_3^{(0)}-E_m^{(0)}} \\
&= \varepsilon^2\left(\frac{H'_{31}H'_{13}}{E_3^{(0)}-E_1^{(0)}}+\frac{H'_{32}H'_{23}}{E_3^{(0)}-E_2^{(0)}}\right) \\
&= \varepsilon^2\left(\frac{1\cdot 1}{3-1}+\frac{1\cdot 1}{3-2}\right) = \tfrac{3}{2}\varepsilon^2
\end{aligned}
\tag{4.44b}
$$

and the eigenvalues up to order ε^2 are given by

$$
\begin{aligned}
E_1 &= E_1^{(0)}+E_1^{(1)}+E_1^{(2)} = 1-\tfrac{3}{2}\varepsilon^2 \\
E_2 &= E_2^{(0)}+E_2^{(1)}+E_2^{(2)} = 2 \\
E_3 &= E_3^{(0)}+E_3^{(1)}+E_3^{(2)} = 3+\tfrac{3}{2}\varepsilon^2
\end{aligned}
\tag{4.44c}
$$

The first-order correction to the eigenvectors are obtained according to Eq. (4.36*c*)

$$
\begin{aligned}
\psi_1^{(1)} &= \sum_{m\neq 1} \frac{H'_{m1}\psi_m^{(0)}}{E_1^{(0)}-E_m^{(0)}} \\
&= \frac{H'_{21}\psi_2^{(0)}}{E_1^{(0)}-E_2^{(0)}}+\frac{H'_{31}\psi_3^{(0)}}{E_1^{(0)}-E_3^{(0)}} \\
&= \frac{1}{1-2}\begin{pmatrix}0\\1\\0\end{pmatrix}+\frac{1}{1-3}\begin{pmatrix}0\\0\\1\end{pmatrix}=\begin{pmatrix}0\\-1\\-1/2\end{pmatrix}
\end{aligned}
$$

$$\begin{aligned}\psi_2^{(1)} &= \sum_{m\neq 2} \frac{H'_{m2}\psi_m^{(0)}}{E_2^{(0)}-E_m^{(0)}} \\ &= \frac{H'_{12}\psi_1^{(0)}}{E_2^{(0)}-E_1^{(0)}} + \frac{H'_{32}\psi_3^{(0)}}{E_2^{(0)}-E_3^{(0)}} \\ &= \frac{1}{2-1}\begin{pmatrix}1\\0\\0\end{pmatrix} + \frac{1}{2-3}\begin{pmatrix}0\\0\\1\end{pmatrix} = \begin{pmatrix}1\\0\\-1\end{pmatrix}\end{aligned} \tag{4.45a}$$

$$\begin{aligned}\psi_3^{(1)} &= \sum_{m\neq 3} \frac{H'_{m3}\psi_m^{(0)}}{E_3^{(0)}-E_m^{(0)}} \\ &= \frac{H'_{13}}{E_3^{(0)}-E_1^{(0)}}\psi_1^{(0)} + \frac{H'_{23}}{E_3^{(0)}-E_2^{(0)}}\psi_2^{(0)} \\ &= \frac{1}{3-1}\begin{pmatrix}1\\0\\0\end{pmatrix} + \frac{1}{3-2}\begin{pmatrix}0\\1\\0\end{pmatrix} = \begin{pmatrix}1/2\\1\\0\end{pmatrix}\end{aligned}$$

Thus, the eigenvectors up to order ε are given by

$$\begin{aligned}\psi_1 &= \psi_1^{(0)} + \varepsilon\psi_1^{(1)} = \begin{pmatrix}1\\0\\0\end{pmatrix} + \varepsilon\begin{pmatrix}0\\-1\\-1/2\end{pmatrix} = \begin{pmatrix}1\\-\varepsilon\\-\varepsilon/2\end{pmatrix} \\ \psi_2 &= \psi_2^{(0)} + \varepsilon\psi_2^{(1)} = \begin{pmatrix}0\\1\\0\end{pmatrix} + \varepsilon\begin{pmatrix}1\\0\\-1\end{pmatrix} = \begin{pmatrix}\varepsilon\\1\\-\varepsilon\end{pmatrix} \\ \psi_3 &= \psi_3^{(0)} + \varepsilon\psi_3^{(1)} = \begin{pmatrix}0\\0\\1\end{pmatrix} + \varepsilon\begin{pmatrix}1/2\\1\\0\end{pmatrix} = \begin{pmatrix}\varepsilon/2\\\varepsilon\\1\end{pmatrix}\end{aligned} \tag{4.45b}$$

We have thus obtained the desired information about the eigenvalues and eigenvectors of the matrix of Eq. (4.41a).

As a second example, let us consider a simple physical system of angular momentum (or spin) 1 in a uniform magnetic field of magnitude B along the z-direction. We have seen before that the perturbing Hamiltonian in this case is

$$\hat{H}' = -\frac{\mu B}{\hbar}\hat{L}_z \tag{4.46a}$$

Here $\hat{L}_z$ is the angular momentum operator that measures the projection of angular momentum onto the Z-axis. We know that the eigenvalues of $\hat{L}_z$ can take only the values $+1$, 0, or -1 (in units of $\hbar$) so that the basis states are defined to be

$$\begin{aligned}\hat{L}_z\,|+\rangle &= \hbar\,|+\rangle \\ \hat{L}_z\,|0\rangle &= 0 \\ \hat{L}_z\,|-\rangle &= -\hbar\,|-\rangle\end{aligned} \tag{4.46b}$$

The basis states form a complete orthonormal set. Thus, the perturbing Hamiltonian can be written in matrix form as

$$\hat{H}' = \begin{array}{c|ccc} & |+\rangle & |0\rangle & |-\rangle \\ \hline |+\rangle & -\mu B & 0 & 0 \\ |0\rangle & 0 & 0 & 0 \\ |-\rangle & 0 & 0 & \mu B \end{array} \tag{4.47}$$

The above perturbation is diagonal in the $\{|+\rangle, |0\rangle, |-\rangle\}$ representation. If these states are stationary in the absence of the magnetic field (namely, if they are eigenstates of the unperturbed Hamiltonian $\hat{H}_0$), they will remain stationary even in the presence of the magnetic field that induces only first-order corrections. Let the three states be degenerate (rotational symmetry requires this) and have the energy E_0 in the absence of the field. Then, in the presence of the magnetic field, the degeneracy is lifted, and the energy values for the three levels are given by

$$\begin{aligned}E_+ &= E_0 - \mu B \\ E_0 &= E_0 \\ E_- &= E_0 + \mu B\end{aligned} \tag{4.48}$$

The change in the energy levels is shown in Fig. 4.1. This phenomenon of the splitting of otherwise degenerate energy levels due to the presence of a magnetic field was first observed in atomic spectra and is called the Zeeman effect after its discoverer. In that case since the electron has negative electric charge, μ is negative and the energy of the $|+\rangle$ state increases, while that of the $|-\rangle$ state decreases due to the perturbation. The number of degenerate states depends, of course, on the value of the angular momentum and is equal to $(2l+1)$.

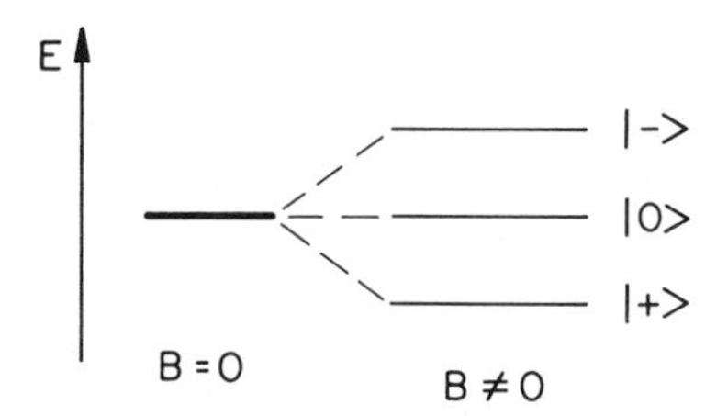

FIGURE 4.1. Splitting of degenerate energy levels in a magnetic field. The three levels correspond to states with $l = 1$, and have different projections of angular momentum onto the Z-axis.

4.4. ENERGY EIGENSTATES OF A TWO-STATE SYSTEM

We have so far developed a formalism (a mathematical machinery) for describing a quantum-mechanical state and its evolution. In order to apply this formalism to a physical system it is natural to begin by considering the basis states in which to write our equations. This is a formidable task because, in principle, any real system has a very large number of degrees of freedom. In fact, we do not know all the possible degrees of freedom of, say, an electron. We assume that it is a point particle with spin 1/2, and thus for a free electron its momentum and spin projection form the complete set of basis states. However, the electron may have internal structure (this has not as yet been discovered) in which case we would need additional basis states to describe the state of the internal structure of the electron. For instance, in the case of the proton we could specify the states of its constituent quarks.

Such lack of knowledge does not prevent us from obtaining physical predictions using quantum mechanics in all cases of interest. For calculational purposes we assume that the electron is a point and use the appropriate subset of basis states as a complete set. We can ignore the basis states describing a particular degree of freedom as long as they do not affect the problem under consideration. For instance, when we studied the precession of the electron's spin projection in Section 3.3 we ignored the momentum basis states because the electron was at rest. Often, even if the additional degrees of freedom affect the problem, we can ignore them as long as their influence is adequately small. This approach is the one used in all practical calculations and agrees with the spirit of perturbation theory that we discussed.

As an example of a system that has only two basis states we will consider the ammonia molecule in its lowest energy state. Ammonia (NH_3) consists of three hydrogen atoms and one nitrogen atom arranged as shown in Fig. 4.2(a). The three hydrogen atoms define a plane and are symmetrically positioned; the nitrogen lies out of the plane. The nitrogen may lie either above or below the symmetry plane as shown in Fig. 4.2(b). These two configurations represent two different states of the molecule and we label them as $|1\rangle$ and $|2\rangle$. Of course, the molecule can be in many other states. For instance, the hydrogen molecules may vibrate about their equilibrium position, the rotational motion may involve more units of angular momentum, the electron in one or more of the hydrogen atoms may be in an excited state, and so on. In addition, the molecule as a whole may be in motion. For our present considerations we can ignore these degrees of freedom and treat the two basis states $|1\rangle$ and $|2\rangle$ as a complete set.

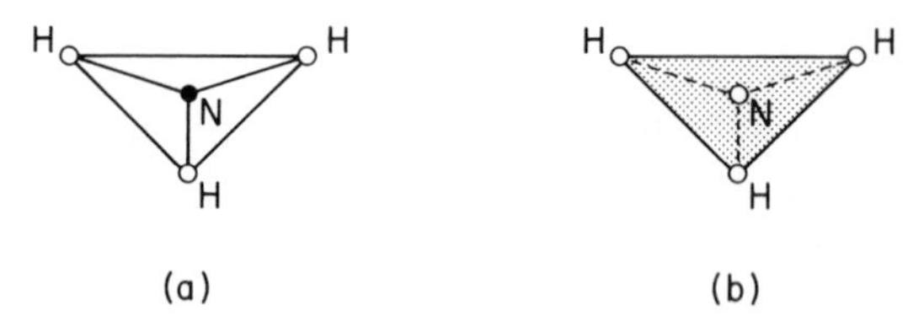

FIGURE 4.2. Two possible configurations of the ammonia molecule. The nitrogen atom can be found (a) above or (b) below the plane defined by the three hydrogen atoms.

Let us construct the energy matrix for the two states $|1\rangle$ and $|2\rangle$. When in the state $|1\rangle$, let the energy of the molecule be E_0. From the symmetry between the two configurations the energy of the system when in the state $|2\rangle$ must also be E_0. Thus, we have established two of the four matrix elements

$$\langle 1|\hat{H}|1\rangle = E_0 \qquad \langle 2|\hat{H}|2\rangle = E_0$$

What about the matrix elements $\langle 2|\hat{H}|1\rangle$ and $\langle 1|\hat{H}|2\rangle$? If these matrix elements are different from zero there will be a finite probability for a transition from the state $|1\rangle$ to state $|2\rangle$, and vice versa. In fact, this is the case for the ammonia molecule which is very "flat." When the nitrogen atom is in the symmetry plane it is strongly repulsed by the hydrogens; in other words, the potential energy is very high. There is, however, a small probability for the nitrogen atom to *tunnel* through this potential barrier so that starting from the configuration of Fig. 4.2(a) it could reach the configuration of Fig. 4.2(b). Therefore, the amplitude $\langle 2|\hat{H}|1\rangle$ is different from zero, and we let it have the value $-A$. Again, from the symmetry of the two configurations we expect that the amplitude $\langle 1|\hat{H}|2\rangle$ has the same value. Therefore, in the $|1\rangle, |2\rangle$ representation, the energy matrix is given by

$$\hat{H} = \begin{array}{c|cc} & |1\rangle & |2\rangle \\ \hline |1\rangle & E_0 & -A \\ |2\rangle & -A & E_0 \end{array} \qquad E_0, A \text{ real} \tag{4.49a}$$

and $\hat{H}$ is hermitian as expected.

Having obtained the energy matrix we can write the equation of motion for any arbitrary state $|\phi(t)\rangle$ of the ammonia molecule. As in Section 4.1 we introduce coefficients $C(t)$ as shorthand for the amplitudes

$$C_1(t) = \langle 1 \mid \phi\rangle \quad \text{and} \quad C_2(t) = \langle 2 \mid \phi\rangle$$

which obey the equation of motion, Eq. (4.11)

$$\begin{aligned} i\hbar \frac{d}{dt} C_1(t) &= E_0 C_1(t) - A C_2(t) \\ i\hbar \frac{d}{dt} C_2(t) &= -A C_1(t) + E_0 C_2(t) \end{aligned} \tag{4.49b}$$

We can readily see that the basis states $|1\rangle$ and $|2\rangle$ do not correspond to stationary states. From Eq. (4.49a) we note that the energy matrix is not diagonal in the $|1\rangle, |2\rangle$ representation. From Eq. (4.49b) we see that even if $C_2(t) = 0$, [and therefore $C_1(t) = 1$] $dC_2/dt \neq 0$, so that at a later time t', $C_2(t')$ will be nonzero.

Equation (4.49b) can be solved exactly. If we add and subtract the two equations we get

$$i\hbar \frac{d}{dt}(C_1+C_2)=E_0(C_1+C_2)-A(C_1+C_2)$$
$$i\hbar \frac{d}{dt}(C_1-C_2)=E_0(C_1-C_2)+A(C_1-C_2) \tag{4.50a}$$

We introduce the notation

$$C_{\mathrm{I}}=(C_1+C_2) \qquad C_{\mathrm{II}}=(C_1-C_2) \tag{4.50b}$$

and integrate Eqs. (4.50a) to obtain

$$C_{\mathrm{I}}=C_1+C_2=ae^{-(i/\hbar)(E_0-A)t}$$
$$C_{\mathrm{II}}=C_1-C_2=be^{-(i/\hbar)(E_0+A)t} \tag{4.51a}$$

Here a and b are arbitrary constants to be determined from the initial conditions. Adding and subtracting Eqs. (4.51a) we solve for the original amplitudes

$$C_1(t)=\frac{a}{2}e^{-(i/\hbar)(E_0-A)t}+\frac{b}{2}e^{-(i/\hbar)(E_0+A)t}$$
$$C_2(t)=\frac{a}{2}e^{-(i/\hbar)(E_0-A)t}-\frac{b}{2}e^{-(i/\hbar)(E_0+A)t} \tag{4.51b}$$

If at time $t=0$ the molecule is in the basis state $|1\rangle$, then $C_1(t=0)=\langle 1\,|\,\phi(t=0)\rangle=1$ and $C_2(t=0)=\langle 2\,|\,\phi(t=0)\rangle=0$. For these initial conditions the constants a and b in Eqs. (4.51b) must satisfy $a=b=1$, and therefore

$$C_1(t)=e^{-(i/\hbar)E_0t}\left(\frac{e^{(i/\hbar)At}+e^{-(i/\hbar)At}}{2}\right)=e^{-(i/\hbar)E_0t}\cos\left(\frac{At}{\hbar}\right)$$
$$C_2(t)=e^{-(i/\hbar)E_0t}\left(\frac{e^{(i/\hbar)At}-e^{-(i/\hbar)At}}{2}\right)=ie^{-(i/\hbar)E_0t}\sin\left(\frac{At}{\hbar}\right)$$
$$\tag{4.52a}$$

At any time t the state $|\phi(t)\rangle$ is then given by

$$|\phi(t)\rangle=|1\rangle\, C_1(t)+|2\rangle\, C_2(t)$$
$$=e^{-(i/\hbar)E_0t}\left[|1\rangle\cos\left(\frac{At}{\hbar}\right)+i\,|2\rangle\sin\left(\frac{At}{\hbar}\right)\right] \tag{4.52b}$$

The state described by Eq. (4.52b) remains properly normalized at all times, since $\langle\phi(t)\mid\phi(t)\rangle=1$.

Clearly, the probability of finding the molecule in the basis state $|1\rangle$ or the basis state $|2\rangle$ changes in time as

$$\begin{aligned} P(1) &= |\langle 1\mid\phi(t)\rangle|^2 = |C_1|^2 = \cos^2\left(\frac{At}{\hbar}\right) \\ P(2) &= |\langle 2\mid\phi(t)\rangle|^2 = |C_2|^2 = \sin^2\left(\frac{At}{\hbar}\right) \end{aligned} \tag{4.52c}$$

This time dependence is exactly the same as that found in Eqs. (3.21) for a spin-1/2 particle precessing in a magnetic field. This should not be surprising since that system also consisted of only two states, the spin projections onto the $+X$ and $-X$ axes. These states were not stationary.

Finding the Stationary States by Diagonalizing the Energy Matrix

We have just seen that the basis states $|1\rangle$ and $|2\rangle$ are not stationary. The stationary states are therefore given by some linear combination of the states $|1\rangle$ and $|2\rangle$. We can find this linear combination either by the explicit solution of the equation of motion as obtained in Eqs. (4.51a) or equivalently by finding the representation in which the energy matrix is diagonal. In the first approach we note that the amplitudes

$$C_{\mathrm{I}}=\langle \mathrm{I}\mid\phi\rangle \quad \text{and} \quad C_{\mathrm{II}}=\langle \mathrm{II}\mid\phi\rangle \tag{4.53a}$$

given by Eqs. (4.51a), have the time dependence corresponding to a well-defined energy; thus the basis states $|\mathrm{I}\rangle$ and $|\mathrm{II}\rangle$ are stationary. Making use of Eq. (4.50b) we can express these states in terms of the original basis of $|1\rangle$ and $|2\rangle$

$$\begin{aligned} |\mathrm{I}\rangle &= \frac{1}{2^{1/2}}[|1\rangle+|2\rangle] \\ |\mathrm{II}\rangle &= \frac{1}{2^{1/2}}[|1\rangle-|2\rangle] \end{aligned} \tag{4.53b}$$

The factor of $1/2^{1/2}$ is needed to assure the normalization of the states $|\mathrm{I}\rangle$ and $|\mathrm{II}\rangle$. Note also that $|\mathrm{I}\rangle$ and $|\mathrm{II}\rangle$ are orthogonal, i.e., $\langle \mathrm{I} | \mathrm{II} \rangle = 0$; thus Eqs. (4.53*b*) provide a *complete orthonormal* set of basis states. They correspond to stationary states of the system (the ammonia molecule). The energy of the molecule in these states is

$$E_{\mathrm{I}} = E_0 - A \quad \text{and} \quad E_{\mathrm{II}} = E_0 + A \tag{4.53c}$$

as can be seen directly from Eqs. (4.51*a*).

We could have obtained these results without explicitly solving the equation of motion [Eqs. (4.50*a*)], because in the representation in which the basis states correspond to the stationary states of the system the energy matrix must be *diagonal*. Therefore, let us diagonalize the matrix of Eq. (4.49*a*). This is achieved by setting the secular determinant of the matrix equal to zero

$$\det \begin{vmatrix} E_0 - \lambda & -A \\ -A & E_0 - \lambda \end{vmatrix} = 0 \tag{4.54a}$$

Namely,

$$(E_0 - \lambda)^2 - A^2 = 0$$

with the two solutions $(E_0 - \lambda) = \pm A$ or

$$\lambda_1 = E_0 - A \qquad \lambda_2 = E_0 + A \tag{4.54b}$$

λ_1 and λ_2 are the *eigenvalues* of $\hat{H}$, and in the new representation $\hat{H}$ takes the form

$$\hat{H}' = S\hat{H}S^{-1} = \begin{vmatrix} E_0 - A & 0 \\ 0 & E_0 + A \end{vmatrix} \tag{4.55a}$$

To find the new basis states we must first find the unitary matrix S that induces the transformation indicated by Eq. (4.55*a*), namely, between the basis states $|1\rangle$ and $|2\rangle$ and the basis states that make $\hat{H}$ diagonal. Using the convention of Eqs. (4.18) and designating the new basis states by $|\mathrm{I}\rangle$ and $|\mathrm{II}\rangle$ we find

$$S = \begin{array}{c|cc} & |1\rangle & |2\rangle \\ \hline |\mathrm{I}\rangle & \langle \mathrm{I} | 1 \rangle & \langle \mathrm{I} | 2 \rangle \\ |\mathrm{II}\rangle & \langle \mathrm{II} | 1 \rangle & \langle \mathrm{II} | 2 \rangle \end{array} = \frac{1}{2^{1/2}} \begin{pmatrix} 1 & 1 \\ 1 & -1 \end{pmatrix} \tag{4.55b}$$

It is easy to verify that $S^{\dagger}S = 1$ and that Eq. (4.55*a*) is satisfied. To find the new basis states we have from Eq. (4.19*a*)

$$|\mu\rangle = \sum_i |i\rangle\,(S^\dagger)_{i\mu}$$

and with S given by Eq. (4.55*b*)—(note that for this special case $S = S^\dagger = S^{-1}$)—we obtain

$$\begin{aligned}|\mathrm{I}\rangle &= \frac{1}{2^{1/2}}[|1\rangle + |2\rangle]\\ |\mathrm{II}\rangle &= \frac{1}{2^{1/2}}[|1\rangle - |2\rangle]\end{aligned} \tag{4.56a}$$

which is in agreement with Eqs. (4.53*b*). The energy eigenvalues for these eigenstates can be obtained directly from Eq. (4.55*a*) as

$$E_{\mathrm{I}} = E_0 - A \quad \text{and} \quad E_{\mathrm{II}} = E_0 + A \tag{4.56}$$

which is in agreement with Eq. (4.53*c*). We see that *diagonalization* of the energy matrix is *completely equivalent* to solving the equation of motion.

Discussion

That the nitrogen atom can tunnel through the symmetry plane is due to the structure of the ammonia molecule which is extremely flat. That is, the equilibrium position of the nitrogen atom is very close to the symmetry plane. If the nitrogen atom could not interchange its position, there would still be two possible states for the molecule, the states $|1\rangle$ and $|2\rangle$, but they would be stationary states and have the same energy E_0. The energy matrix would be diagonal ($A \to 0$) in the $|1\rangle, |2\rangle$ representation and would be *degenerate.*

In reality, there exists a finite probability for the tunneling of the nitrogen atom. As a consequence, the degeneracy is removed and the two energy eigenstates $|\mathrm{I}\rangle$ and $|\mathrm{II}\rangle$ acquire a small difference in energy. From a study of the microwave absorption spectrum of ammonia one finds that the splitting between these two states is of order

$$\Delta E = 2A = 1 \times 10^{-4}\ \mathrm{eV}$$

This is much smaller than the spacing to the next rotational state of ammonia, which differs from the ground state by $\sim 10^{-1}$ eV, or to the vibrational states, which differ by 1 eV. Clearly, the other possible basis states of the system are far removed on an energy scale compared to the energy difference between the states $|\mathrm{I}\rangle$ and $|\mathrm{II}\rangle$. It is for this reason that we can ignore these other basis states and treat the states $|1\rangle$, $|2\rangle$ or $|\mathrm{I}\rangle$, $|\mathrm{II}\rangle$ as a complete set.

It is worth noting that we were able to describe the ground state of the molecule without knowing the exact values of E_0 and A. In fact, it is hardly possible to calculate A from first principles, its value being obtained from the experimental observations.

The relation between the states $|1\rangle$, $|2\rangle$, and $|\mathrm{I}\rangle$, $|\mathrm{II}\rangle$ is analogous to the behavior of a mechanical system with two coupled modes. For instance, consider two identical pendula connected by a weak spring as shown in Fig. 4.3. If we set pendulum A in oscillatory motion while B is at rest [Fig. 4.3(a)], after some time pendulum A will come to rest and B will be in motion [Fig. 4.3(b)]. These modes of oscillation are not stationary in time in the sense that if we start with mode (a) at a later time the system will be in mode (b), and as time advances it will revert back to mode (a), and so on. Modes (a) and (b) correspond to the description of the ammonia molecule by the states $|1\rangle$ and $|2\rangle$. If, however, we set both pendula in motion simultaneously, as in Fig. 4.3(c), they will remain in this mode (neglecting friction) forever. The same is true for the mode of Fig. 4.3(d) where the two pendula oscillate with phases differing by 180°. Modes (c) and (d) are the *normal modes*, i.e., the eigenstates, of the coupled pendula system and correspond to the description of the ammonia molecule by the basis states $|\mathrm{I}\rangle$ and $|\mathrm{II}\rangle$. This equivalence holds because the equations of motion have the same mathematical form for both systems.

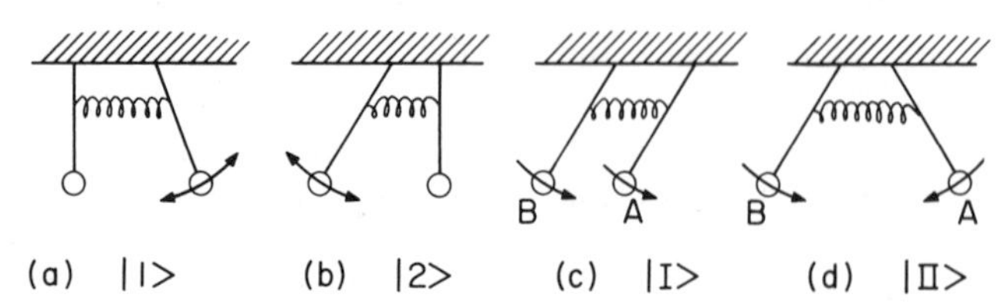

FIGURE 4.3. The motion of two coupled pendula. The normal modes of the system labeled $|\mathrm{I}\rangle$ and $|\mathrm{II}\rangle$ are the stationary states.

4.5. TIME-INDEPENDENT PERTURBATION OF A TWO-STATE SYSTEM

We have seen that when weak forces, not previously considered, act on a system, it is best to treat them as perturbations. For the ammonia molecule, as for all atomic and molecular systems, these forces arise from the presence of electric or magnetic fields. The ammonia molecule possesses an *electric dipole* moment because the valence electrons are clustered closer to the nitrogen nucleus than to the protons. As a result, even though the molecule as a whole is electrically neutral, the mean position of the negative charges is displaced with respect to the positive charges as shown in Fig. 4.4(a). Thus, the electric dipole moment is directed opposite to the position of the nitrogen atom with respect to the symmetry plane of the molecule, as shown in Figs. 4.4(b) and 4.4(c).

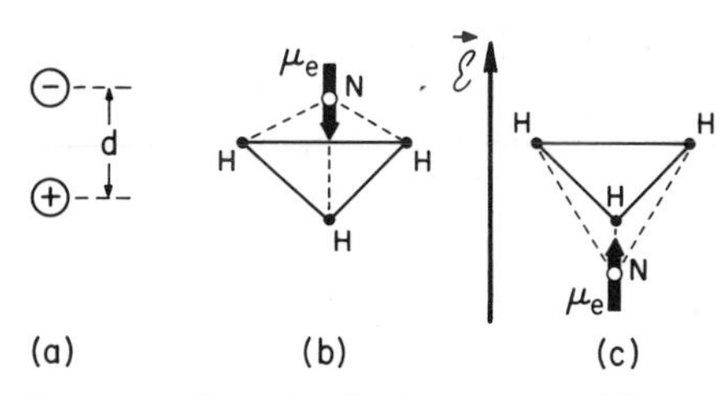

FIGURE 4.4. In the ammonia molecule the mean positions of the positive and negative charges are displaced from one another, giving rise to an electric dipole moment $\boldsymbol{\mu}_e$. In the presence of an electric field $\vec{\mathscr{E}}$ the two orientations of the dipole moment make different contributions to the total energy of the molecule.

We shall designate the electric dipole moment by $\boldsymbol{\mu}_e$ and we can roughly estimate its magnitude. Assuming that the average charge is of the order of the electron's charge e and that the displacement of positive from negative charges is of the order of molecular dimensions, that is, 1 Å, we find

$$\mu = |\boldsymbol{\mu}_e| \sim e \times 10^{-10} \text{ coul-m} \tag{4.57a}$$

Suppose an external electric field $\vec{\mathscr{E}}$ is applied and the ammonia molecule is positioned as in Fig. 4.4 so that the symmetry plane is perpendicular to the direction of $\vec{\mathscr{E}}$. If the molecule is in state $|1\rangle$

[Fig. 4.4(b)] it will acquire an additional potential energy

$$U = -\boldsymbol{\mu}_e \cdot \vec{\mathscr{E}} = \mu\mathscr{E} \qquad \text{state } |1\rangle \tag{4.57b}$$

and if in state $|2\rangle$ [Fig. 4.4(c)] the additional potential energy is

$$U = -\boldsymbol{\mu}_e \cdot \vec{\mathscr{E}} = -\mu\mathscr{E} \qquad \text{state } |2\rangle \tag{4.57c}$$

For experimentally realizable fields, $\mathscr{E} \simeq 10^4$ V/cm, Eq. (4.57*a*) gives $U \sim 10^{-4}$ eV; that is, U is of the same order of magnitude as the energy required for tunneling from the state $|1\rangle$ to the state $|2\rangle$.

In the presence of the external electric field the energy in the state $|1\rangle$ is $H_{11} = E_0 + \mu\mathscr{E}$ and in the state $|2\rangle$, $H_{22} = E_0 - \mu\mathscr{E}$. Thus the energy matrix of Eq. (4.49*a*) takes the form

$$\hat{H} = \begin{array}{c|cc} & |1\rangle & |2\rangle \\ \hline |1\rangle & E_0 + \mu\mathscr{E} & -A \\ |2\rangle & -A & E_0 - \mu\mathscr{E} \end{array} \tag{4.58a}$$

The states $|1\rangle$ and $|2\rangle$ are not stationary since $\hat{H}$ is not diagonal. The energies of the stationary states are obtained by diagonalizing $\hat{H}$ and are given by the roots of the secular equation

$$\det \begin{vmatrix} (E_0 + \mu\mathscr{E} - \lambda) & -A \\ -A & (E_0 - \mu\mathscr{E} - \lambda) \end{vmatrix} = 0$$

or

$$[(E_0 - \lambda) + \mu\mathscr{E}][(E_0 - \lambda) - \mu\mathscr{E}] - A^2 = 0$$

with solutions

$$\begin{aligned} \lambda_\alpha &= E_0 + (A^2 + \mu^2\mathscr{E}^2)^{1/2} \\ \lambda_\beta &= E_0 - (A^2 + \mu^2\mathscr{E}^2)^{1/2} \end{aligned} \tag{4.58b}$$

For weak electric fields such that $(\mu\mathscr{E}/A)^2 \ll 1$ Eqs. (4.58*b*) are expanded to yield the approximate expressions

$$\left.\begin{aligned} \lambda_\alpha &\approx (E_0 + A) + \frac{\mu^2\mathscr{E}^2}{2A} \\ \lambda_\beta &\approx (E_0 - A) - \frac{\mu^2\mathscr{E}^2}{2A} \end{aligned}\right\} \text{weak field} \tag{4.59a}$$

Correspondingly, for strong electric fields the expansion is carried out in the small parameter $(A/\mu\mathscr{E})^2 \ll 1$, yielding the approximate expressions

$$\left.\begin{aligned}\lambda_\alpha &\simeq E_0+\mu\mathscr{E}+\frac{A^2}{2\mu\mathscr{E}}\\ \lambda_\beta &\simeq E_0-\mu\mathscr{E}-\frac{A^2}{2\mu\mathscr{E}}\end{aligned}\right\}\text{ strong field} \tag{4.59b}$$

These results can be visualized by plotting the energy eigenvalues λ_α and λ_β as a function of field strength, as shown in Fig. 4.5. The dashed lines in the figure correspond to the functions $(E_0+\mu\mathscr{E})$ and $(E_0-\mu\mathscr{E})$. As $\mathscr{E}\to 0$ the eigenvalues reduce to (E_0+A) and (E_0-A), as also found in Eq. (4.56*b*). The corresponding eigenstates are the states $|\mathrm{I}\rangle$ and $|\mathrm{II}\rangle$. The presence of the weak electric field is not sufficient to prevent the occurrence of tunneling from state $|1\rangle$ to state $|2\rangle$. It perturbs the ammonia molecule quadratically in the small quantity $(\mu\mathscr{E}/A)$ as indicated by Eqs. (4.59*a*).

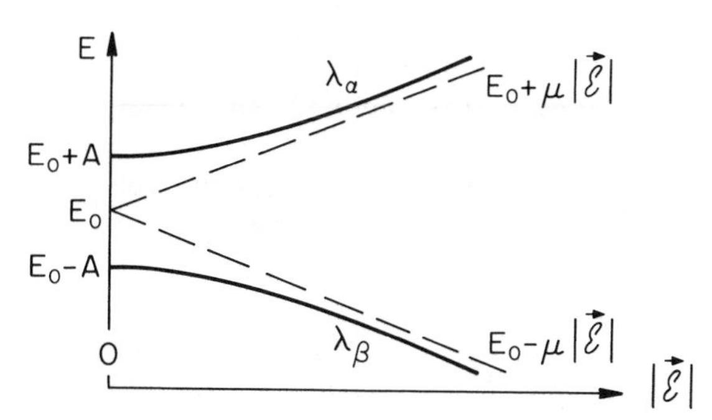

FIGURE 4.5. The energy of the two states of the ammonia molecule as a function of the strength of a perturbing electric field $\vec{\mathscr{E}}$.

For a strong electric field the energy eigenvalues tend to $(E_0+\mu\mathscr{E})$, and $(E_0-\mu\mathscr{E})$ and the corresponding eigenstates approximate the states $|1\rangle$ and $|2\rangle$. The interaction of the electric dipole moment with the electric field is now the dominant factor and *prevents* the nitrogen atom from tunneling through the symmetry plane. The energy separation between the states $|1\rangle$ and $|2\rangle$ increases linearly with the strength of the electric field. The tunneling effect is a perturbation which contributes only quadratically in the small quantity $(A/\mu\mathscr{E})$ as given by Eqs. (4.59*b*). This contribution is indicated by the difference between the dashed and solid lines in the figure.

For intermediate values of the electric field, that is, when $\mu\mathscr{E}$ is comparable to A, the stationary states approximate neither the basis states $|1\rangle$ or $|2\rangle$ nor the basis states $|\mathrm{I}\rangle$ or $|\mathrm{II}\rangle$. This is because in the absence of tunneling and an electric field the two states of the molecule are degenerate, as mentioned previously. There are large off-diagonal elements in the energy matrix in either the $|1\rangle$, $|2\rangle$ or the $|\mathrm{I}\rangle$, $|\mathrm{II}\rangle$ representation, and they mix the states that they connect.†

The Perturbative Solution

The techniques of perturbation theory enable us to obtain approximate eigenvalues such as Eqs. (4.59*a*) or (4.59*b*) without having to perform the exact diagonalization of the energy matrix. This is very important when the energy matrix is of high dimensionality. We illustrate the application of the method using a simple example and begin by considering the energy matrix in the absence of the tunneling as well as electric field perturbations. As we know, it has the form

$$\hat{H}_0 = \begin{array}{c|cc} & |1\rangle & |2\rangle \\ \hline |1\rangle & E_0 & 0 \\ |2\rangle & 0 & E_0 \end{array} \tag{4.60a}$$

Next we introduce both the electric field and the tunneling effect so that the total Hamiltonian, still expressed in the $|1\rangle$, $|2\rangle$ representation, becomes

$$\hat{H} = \begin{array}{c|cc} & |1\rangle & |2\rangle \\ \hline |1\rangle & E_0 + \mu\mathscr{E} & -A \\ |2\rangle & -A & E_0 - \mu\mathscr{E} \end{array} \tag{4.60b}$$

† We can think of an off-diagonal element H_{ij} as a *connection* between the states $|i\rangle$ and $|j\rangle$.

Clearly two distinct cases arise:

Case I: Weak-Field Case ($\mu\mathscr{E} \ll A$)

Here we can treat the electric field as a perturbation and write

$$\hat{H} = \hat{H}_0 + \hat{H}_p$$

where

$$\hat{H}_0 = \begin{array}{c|cc} & |1\rangle & |2\rangle \\ \hline |1\rangle & E_0 & -A \\ |2\rangle & -A & E_0 \end{array} \tag{4.61a}$$

and

$$\hat{H}_p = \begin{array}{c|cc} & |1\rangle & |2\rangle \\ \hline |1\rangle & \mu\mathscr{E} & 0 \\ |2\rangle & 0 & -\mu\mathscr{E} \end{array} \tag{4.61b}$$

We have to diagonalize $\hat{H}_0$, which has already been done in Eq. (4.55a). In the basis in which $\hat{H}_0$ is diagonal, it has the form

$$\hat{H}'_0 = \begin{array}{c|cc} & |\mathrm{I}\rangle & |\mathrm{II}\rangle \\ \hline |\mathrm{I}\rangle & E_0 - A & 0 \\ |\mathrm{II}\rangle & 0 & E_0 + A \end{array} \tag{4.62a}$$

The matrix elements of $\hat{H}_p$ in the new representation are found in the usual way, $\hat{H}'_p = S\hat{H}_p S^{-1}$, where the transformation matrix S is given by Eq. (4.55b). We can use matrix multiplication or calculate the elements directly by writing

$$\begin{aligned}\langle \mathrm{I}|\,\hat{H}_p\,|\mathrm{I}\rangle &= \langle \mathrm{I} \mid 1\rangle\langle 1|\,\hat{H}_p\,|1\rangle\langle 1 \mid \mathrm{I}\rangle + \langle \mathrm{I} \mid 1\rangle\langle 1 \mid H_p\,|2\rangle\langle 2 \mid \mathrm{I}\rangle \\ &\quad + \langle \mathrm{I} \mid 2\rangle\langle 2|\,\hat{H}_p\,|1\rangle\langle 1 \mid \mathrm{I}\rangle + \langle \mathrm{I} \mid 2\rangle\langle 2|\,\hat{H}_p\,|2\rangle\langle 2 \mid \mathrm{I}\rangle \\ &= \frac{1}{2^{1/2}}\,\mu\mathscr{E}\,\frac{1}{2^{1/2}} + 0 + 0 + \frac{1}{2^{1/2}}(-\mu\mathscr{E})\,\frac{1}{2^{1/2}} = 0\end{aligned}$$

and similarly

$$\begin{aligned}\langle \mathrm{I}|\,\hat{H}_p\,|\mathrm{II}\rangle &= \mu\mathscr{E} \\ \langle \mathrm{II}|\,\hat{H}_p\,|\mathrm{I}\rangle &= \mu\mathscr{E} \\ \langle \mathrm{II}|\,\hat{H}_p\,|\mathrm{II}\rangle &= 0\end{aligned}$$

so that

$$\hat{H}'_{\mathrm{p}} = \begin{array}{c|cc} & |\mathrm{I}\rangle & |\mathrm{II}\rangle \\ \hline |\mathrm{I}\rangle & 0 & \mu\mathscr{E} \\ |\mathrm{II}\rangle & \mu\mathscr{E} & 0 \end{array} \tag{4.62b}$$

This shows that

$$\begin{aligned} (H'_{\mathrm{p}})_{mn} \equiv H'_{mn} &= \mu\mathscr{E} \qquad \text{for } m \neq n \ (m, n = \mathrm{I}, \mathrm{II}) \\ &= 0 \qquad \text{for } m = n \end{aligned}$$

Clearly, the first-order change to the energy eigenvalues vanishes. The second-order change is given by

$$\begin{aligned} E_{\mathrm{I}}^{(2)} &= \sum_{m \neq \mathrm{I}} \frac{H'_{m\mathrm{I}} H'_{\mathrm{I}m}}{E_{\mathrm{I}}^{(0)} - E_m^{(0)}} \\ &= \frac{H'_{\mathrm{II}\,\mathrm{I}} H'_{\mathrm{I}\,\mathrm{II}}}{E_{\mathrm{I}}^{(0)} - E_{\mathrm{II}}^{(0)}} = \frac{\mu\mathscr{E} \cdot \mu\mathscr{E}}{E_0 - A - E_0 - A} = -\frac{\mu^2\mathscr{E}^2}{2A} \end{aligned}$$

Similarly

$$\begin{aligned} E_{\mathrm{II}}^{(2)} &= \sum_{m \neq \mathrm{II}} \frac{H'_{m\mathrm{II}} H'_{\mathrm{II}m}}{E_{\mathrm{II}}^{(0)} - E_m^{(0)}} \\ &= \frac{H'_{\mathrm{I}\,\mathrm{II}} H'_{\mathrm{II}\,\mathrm{I}}}{E_{\mathrm{II}}^{(0)} - E_{\mathrm{I}}^{(0)}} = \frac{\mu\mathscr{E} \cdot \mu\mathscr{E}}{E_0 + A - E_0 + A} = \frac{\mu^2\mathscr{E}^2}{2A} \end{aligned} \tag{4.63a}$$

Thus, up to second-order in the electric fields, the energy eigenvalues become

$$\begin{aligned} E_{\mathrm{II}} &= E_0 + A + \frac{\mu^2\mathscr{E}^2}{2A} \\ E_{\mathrm{I}} &= E_0 - A - \frac{\mu^2\mathscr{E}^2}{2A} \end{aligned} \tag{4.63b}$$

which is in agreement with the expansion of the exact solution as indicated by Eqs. (4.59*a*).

The change in the stationary states up to linear order is given by

$$\begin{aligned} |\mathrm{I}^{(1)}\rangle &= \sum_{m \neq \mathrm{I}} \frac{\langle m^{(0)}| H'_{\mathrm{p}} |\mathrm{I}^{(0)}\rangle}{E_{\mathrm{I}}^{(0)} - E_m^{(0)}} |m^{(0)}\rangle \\ &= \frac{H'_{\mathrm{II}\,\mathrm{I}}}{E_{\mathrm{I}}^{(0)} - E_{\mathrm{II}}^{(0)}} |\mathrm{II}^{(0)}\rangle = \frac{\mu\mathscr{E}}{E_0 - A - E_0 - A} |\mathrm{II}^{(0)}\rangle = -\frac{\mu\mathscr{E}}{2A} |\mathrm{II}^{(0)}\rangle \end{aligned}$$

and

$$|\mathrm{II}^{(1)}\rangle = \sum_{m\neq \mathrm{II}} \frac{\langle m^{(0)}|\, H'_{\mathrm{p}}\, |\mathrm{II}^{(0)}\rangle}{E_{\mathrm{II}}^{(0)} - E_m^{(0)}} |m\rangle$$

$$= \frac{H'_{\mathrm{I\,II}}}{E_{\mathrm{II}}^{(0)} - E_{\mathrm{I}}^{(0)}} |\mathrm{I}^{(0)}\rangle = \frac{\mu\mathscr{E}}{E_0 + A - E_0 + A} |\mathrm{I}^{(0)}\rangle = \frac{\mu\mathscr{E}}{2A} |\mathrm{I}^{(0)}\rangle \quad (4.64a)$$

Thus, the eigenstates up to first-order in the perturbation are given by

$$\begin{aligned} |\mathrm{I}\rangle &= |\mathrm{I}^{(0)}\rangle - \frac{\mu\mathscr{E}}{2A} |\mathrm{II}^{(0)}\rangle \\ |\mathrm{II}\rangle &= |\mathrm{II}^{(0)}\rangle + \frac{\mu\mathscr{E}}{2A} |\mathrm{I}^{(0)}\rangle \end{aligned} \quad (4.64b)$$

Case II: Strong-Field Case ($\mu\mathscr{E} \gg A$)

In this case we can treat the tunneling as a perturbation and separate the Hamiltonian in the following way.

$$\hat{H} = \hat{H}_0 + \hat{H}_{\mathrm{p}}$$

where

$$\hat{H}_0 = \begin{array}{c|cc} & |1\rangle & |2\rangle \\ \hline |1\rangle & E_0 + \mu\mathscr{E} & 0 \\ |2\rangle & 0 & E_0 - \mu\mathscr{E} \end{array}$$

$$\hat{H}_{\mathrm{p}} = \begin{array}{c|cc} & |1\rangle & |2\rangle \\ \hline |1\rangle & 0 & -A \\ |2\rangle & -A & 0 \end{array} \quad (4.65a)$$

The unperturbed Hamiltonian is already diagonal with eigenvalues $E_0 + \mu\mathscr{E}$ and $E_0 - \mu\mathscr{E}$. The perturbing Hamiltonian has only the off-diagonal elements nonzero. Hence

$$\begin{aligned} (\hat{H}_{\mathrm{p}})_{mn} \equiv H''_{mn} &= -A \qquad \text{for } m \neq n \ (m, n = 1, 2) \\ &= 0 \qquad \text{for } m = n \end{aligned} \quad (4.65b)$$

Furthermore, the off-diagonal elements are small in magnitude compared to the difference in the energy levels of the unperturbed Hamiltonian. Therefore, the theory of nondegenerate perturbation is applicable. Since the perturbing Hamiltonian has no diagonal elements it is also clear that the first-order change to the energy eigenvalues vanishes. The second-order change is given by

$$E_1^{(2)} = \sum_{m\neq 1} \frac{H''_{m1}H''_{1m}}{E_1^{(0)} - E_m^{(0)}}$$
$$= \frac{H''_{21}H''_{12}}{E_1^{(0)} - E_2^{(0)}} = \frac{(-A)(-A)}{E_0 + \mu\mathscr{E} - E_0 + \mu\mathscr{E}} = \frac{A^2}{2\mu\mathscr{E}}$$

and

$$E_2^{(2)} = \sum_{m\neq 2} \frac{H''_{m2}H''_{2m}}{E_2^{(0)} - E_m^{(0)}}$$
$$= \frac{H''_{12}H''_{21}}{E_2^{(0)} - E_1^{(0)}} = \frac{(-A)(-A)}{E_0 - \mu\mathscr{E} - E_0 - \mu\mathscr{E}} = -\frac{A^2}{2\mu\mathscr{E}} \tag{4.66a}$$

Thus, the energy eigenvalues up to second-order in the perturbation are given by

$$\begin{aligned} E_1 &= E_0 + \mu\mathscr{E} + \frac{A^2}{2\mu\mathscr{E}} \\ E_2 &= E_0 - \mu\mathscr{E} - \frac{A^2}{2\mu\mathscr{E}} \end{aligned} \tag{4.66b}$$

which are in agreement with the expansion of the exact solution as indicated by Eqs. (4.59*b*).

The first-order change to the eigenstates is given by

$$|1^{(1)}\rangle = \sum_{m\neq 1} \frac{\langle m^{(0)}|\, H''_p\, |1\rangle}{E_1^{(0)} - E_m^{(0)}} |m^{(0)}\rangle$$
$$= \frac{H''_{21}}{E_1^{(0)} - E_2^{(0)}} |2^{(0)}\rangle = \frac{-A}{E_0 + \mu\mathscr{E} - E_0 + \mu\mathscr{E}} |2^{(0)}\rangle$$
$$= -\frac{A}{2\mu\mathscr{E}} |2^{(0)}\rangle$$

and

$$\begin{aligned}
|2^{(1)}\rangle &= \sum_{m\neq 2} \frac{\langle m^{(0)}| H_p'' |2\rangle}{E_2^{(0)} - E_m^{(0)}} |m^{(0)}\rangle \\
&= \frac{H_{12}'}{E_2^{(0)} - E_1^{(0)}} |1^0\rangle \\
&= \frac{-A}{E_0 - \mu\mathscr{E} - E_0 - \mu\mathscr{E}} |1^0\rangle \\
&= \frac{A}{2\mu\mathscr{E}} |1^0\rangle
\end{aligned} \tag{4.67a}$$

Thus, the eigenstates up to first-order in perturbation are given by

$$\begin{aligned}
|1\rangle &= |1^{(0)}\rangle - \frac{A}{2\mu\mathscr{E}} |2^{(0)}\rangle \\
|2\rangle &= |2^{(0)}\rangle + \frac{A}{2\mu\mathscr{E}} |1^{(0)}\rangle
\end{aligned} \tag{4.67b}$$

This simple example should illustrate to the reader the power and consistency of perturbation theory.

4.6. GENERAL DESCRIPTION OF TWO-STATE SYSTEMS: PAULI MATRICES

We have seen that the evolution of any two-state system is completely defined by its corresponding 2×2 energy matrix $\hat{H}$; this matrix is hermitian. Therefore, if we develop a formalism describing the most general hermitian 2×2 matrix it will be applicable to any and all two-state systems. Such a description is provided by the Pauli matrices, first introduced by W. Pauli in order to describe the effects of the electron's spin.

Consider the most general 2×2 matrix with real elements. It is obvious that any such matrix can be expressed as a linear combination, with real coefficients, of the four matrices

$$\begin{pmatrix} 1 & 0 \\ 0 & 0 \end{pmatrix} \quad \begin{pmatrix} 0 & 0 \\ 0 & 1 \end{pmatrix} \quad \begin{pmatrix} 0 & 1 \\ 0 & 0 \end{pmatrix} \quad \begin{pmatrix} 0 & 0 \\ 1 & 0 \end{pmatrix} \tag{4.68a}$$

It is also evident that the four matrices of Eq. (4.68*a*) are linearly independent.† Of course, we can also use any other set of linear combinations of these four matrices. We can form the sum and difference of the first two and last two to obtain the set

$$\begin{pmatrix} 1 & 0 \\ 0 & 1 \end{pmatrix} \quad \begin{pmatrix} 1 & 0 \\ 0 & -1 \end{pmatrix} \quad \begin{pmatrix} 0 & 1 \\ 1 & 0 \end{pmatrix} \quad \begin{pmatrix} 0 & 1 \\ -1 & 0 \end{pmatrix} \qquad (4.68b)$$

The matrices of Eq. (4.68*b*) can be used to expand any arbitrary 2×2 real matrix. If we allow the expansion coefficients to be complex we can construct all 2×2 matrices with complex elements.

At present we are only interested in hermitian matrices and will demand that the expansion coefficients be *real*. The first three matrices are already hermitian, but not the fourth. Multiplying it by $-i$ we obtain the set of four hermitian matrices

$$\hat{\sigma}_0 = \mathbb{1} = \begin{pmatrix} 1 & 0 \\ 0 & 1 \end{pmatrix} \quad \hat{\sigma}_1 = \hat{\sigma}_x = \begin{pmatrix} 0 & 1 \\ 1 & 0 \end{pmatrix} \quad \hat{\sigma}_2 = \hat{\sigma}_y = \begin{pmatrix} 0 & -i \\ i & 0 \end{pmatrix}$$

$$\hat{\sigma}_3 = \hat{\sigma}_z = \begin{pmatrix} 1 & 0 \\ 0 & -1 \end{pmatrix} \qquad (4.68c)$$

Any hermitian 2×2 matrix with real or complex elements can be expressed as a linear combination with *real* coefficients of the four matrices in the above set. We write

$$\hat{H} = a_0\mathbb{1} + A_x\hat{\sigma}_x + A_y\hat{\sigma}_y + A_z\hat{\sigma}_z, \qquad a_0, A_x, A_y, A_z \text{ real} \qquad (4.69)$$

The four matrices of Eq. (4.68*c*) are the Pauli spin matrices. Of course, Eq. (4.68*c*) is a particular representation of the Pauli matrices since we can obtain other equivalent representations by performing similarity transformations

$$\hat{\sigma}'_x = S\hat{\sigma}_x S^{-1}, \qquad \text{etc.}$$

with S a unitary matrix ($S^\dagger S = 1$). The representation of Eq. (4.68*c*) is used most frequently, and we will adopt it unless otherwise indicated.

The particular set of Eq. (4.68*c*) has important properties [which are absent in the choice of Eq. (4.68*a*)] because the four matrices

† It is not possible to construct any one of these matrices as a linear combination of the other three.

provide the representation of a *group*.† This is so because: (1) the matrix operation

$$\hat{\sigma}_\alpha \hat{\sigma}_\beta, \qquad \alpha, \beta, \ldots = 0, x, y, z$$

reproduces one of the four matrices; (2) the set contains the identity element; (3) the inverse of all elements is contained in the set; and (4) matrix multiplication is associative. It can be easily shown that the matrices in the set of Eq. (4.68*c*) satisfy

$$\hat{\sigma}_0^2 = \hat{\sigma}_x^2 = \hat{\sigma}_y^2 = \hat{\sigma}_z^2 = 1 \tag{4.70a}$$

$$\left.\begin{aligned} \hat{\sigma}_x\hat{\sigma}_y - \hat{\sigma}_y\hat{\sigma}_x &= 2i\hat{\sigma}_z \\ \hat{\sigma}_y\hat{\sigma}_z - \hat{\sigma}_z\hat{\sigma}_y &= 2i\hat{\sigma}_x \\ \hat{\sigma}_z\hat{\sigma}_x - \hat{\sigma}_x\hat{\sigma}_z &= 2i\hat{\sigma}_y \end{aligned}\right\} \tag{4.70b}$$

This last property can be written in compact form as

$$[\hat{\sigma}_i, \hat{\sigma}_j] = 2i\hat{\sigma}_k, \qquad i, j, k \text{ cyclic} \tag{4.70c}$$

where i, j, k stands for the indices $1, 2, 3$ or x, y, z as given in Eq. (4.68*c*).

The four matrices can be separated into two classes according to the following properties

$$\begin{aligned} &\det(\hat{\sigma}_0) = +1 \qquad \mathrm{Tr}(\hat{\sigma}_0) = 2 \\ &\det(\hat{\sigma}_i) = -1 \qquad \mathrm{Tr}(\hat{\sigma}_i) = 0, \qquad i = 1, 2, 3 \end{aligned}$$

The three matrices $\hat{\sigma}_i$ can be thought of as the components‡ of a vector $\boldsymbol{\sigma}$, which, of course, is a matrix-vector, a fact we have anticipated by writing $\sigma_1 = \sigma_x, \ldots$, etc. We can then use the shorthand notation

$$\hat{H} = a_0 \mathbb{1} + \mathbf{A} \cdot \boldsymbol{\sigma} \tag{4.71}$$

to express the most general 2×2 hermitian matrix as given by Eq. (4.69). Here $\mathbf{A}$ is a real vector (A_x, A_y, A_z), and a_0 is a real number. Often we will drop the symbol $\mathbb{1}$, implying that in matrix equations a real number is always understood to be multiplied by the unit matrix $\mathbb{1}$.

† This is the $SU(2)$ group, which can be defined through the algebra indicated by Eqs. (4.70).

‡ The justification for this identification is discussed in Section 5.6.

To see the usefulness of the Pauli matrices let us examine the special case where: $a_0 = 0$, $A_x = A_y = 0$, and $A_z = 1$. Then $\hat{H}$ takes the form

$$\hat{H} = \hat{\sigma}_z = \begin{pmatrix} 1 & 0 \\ 0 & -1 \end{pmatrix} \tag{4.72a}$$

and the system has two eigenstates $|1\rangle$ and $|2\rangle$ with eigenvalues $\lambda_1 = +1$ and $\lambda_2 = -1$. Since it must hold that

$$\hat{H}\,|1\rangle = \lambda_1\,|1\rangle = |1\rangle \quad \text{and} \quad \hat{H}\,|2\rangle = \lambda_2\,|2\rangle = -|2\rangle \tag{4.72b}$$

the two states $|1\rangle$ and $|2\rangle$ can be represented as column vectors of the form

$$|1\rangle = \begin{pmatrix} 1 \\ 0 \end{pmatrix} \qquad |2\rangle = \begin{pmatrix} 0 \\ 1 \end{pmatrix} \tag{4.73a}$$

They are orthogonal since their inner product

$$\langle 1 \,|\, 2\rangle = (1 \quad 0)\begin{pmatrix} 0 \\ 1 \end{pmatrix} = 0 \tag{4.73b}$$

vanishes; they are also normalized. The most general state of the system is described by

$$|\phi\rangle = b_1\,|1\rangle + b_2\,|2\rangle = \begin{pmatrix} b_1 \\ b_2 \end{pmatrix} \tag{4.73c}$$

where b_1, b_2 can be complex but must obey the normalization condition

$$|b_1|^2 + |b_2|^2 = 1$$

The hermitian $\hat{\sigma}$-matrices represent operators, and we can use the representation of Eqs. (4.68*c*) and (4.73*c*) to find their effect on an arbitrary state $|\phi\rangle$. Of particular interest are the combinations

$$\hat{\sigma}_+ = \frac{\hat{\sigma}_x + i\hat{\sigma}_y}{2} = \begin{pmatrix} 0 & 1 \\ 0 & 0 \end{pmatrix} \quad \text{and} \quad \sigma_- = \frac{\hat{\sigma}_x - i\hat{\sigma}_y}{2} = \begin{pmatrix} 0 & 0 \\ 1 & 0 \end{pmatrix} \tag{4.74a}$$

which have the properties

$$\hat{\sigma}_+\,|1\rangle = 0, \quad \hat{\sigma}_+\,|2\rangle = |1\rangle \quad \text{and} \quad \hat{\sigma}_-\,|1\rangle = |2\rangle, \quad \hat{\sigma}_-\,|2\rangle = 0$$

When acting on the arbitrary state $|\phi\rangle$, the operators $\hat{\sigma}_+, \hat{\sigma}_-$ yield

$$\hat{\sigma}_+ |\phi\rangle = \begin{pmatrix} 0 & 1 \\ 0 & 0 \end{pmatrix} \begin{pmatrix} b_1 \\ b_2 \end{pmatrix} = \begin{pmatrix} b_2 \\ 0 \end{pmatrix}$$

and

$$\hat{\sigma}_- |\phi\rangle = \begin{pmatrix} 0 & 0 \\ 1 & 0 \end{pmatrix} \begin{pmatrix} b_1 \\ b_2 \end{pmatrix} = \begin{pmatrix} 0 \\ b_1 \end{pmatrix} \tag{4.74b}$$

Therefore, $\hat{\sigma}_+$ and $\hat{\sigma}_-$ act as operators for *raising* and *lowering*, respectively, the entries in a column vector.

Examples of Two-State Systems

Consider light propagating along the Z-axis. We represent linear polarization along the X-axis by the state $|+z\rangle$ and linear polarization along the Y-axis by the state $|-z\rangle$. Then

$$|P_x\rangle = \begin{pmatrix} 1 \\ 0 \end{pmatrix} \qquad |P_y\rangle = \begin{pmatrix} 0 \\ 1 \end{pmatrix} \tag{4.79a}$$

These are eigenstates of the σ_z operator with eigenvalues ± 1. According to Eqs. (2.9a) RHC and LHC polarized light will then be represented as

$$|P_R\rangle = \frac{1}{2^{1/2}} \begin{pmatrix} 1 \\ i \end{pmatrix} \quad \text{and} \quad |P_L\rangle = \frac{1}{2^{1/2}} \begin{pmatrix} 1 \\ -i \end{pmatrix} \tag{4.79b}$$

These are eigenstates of the operator $\hat{\sigma}_y$ with eigenvalues ± 1, as can be easily checked. A linear polarizer along the X-(or Y-)axes is represented by the operator

$$\hat{M} = \frac{\mathbb{1} \pm \hat{\sigma}_z}{2} \tag{4.80a}$$

as given also in Eq. (2.41d). A circular polarizer will be represented by

$$\hat{N} = \frac{\mathbb{1} \pm \hat{\sigma}_y}{2} \tag{4.80b}$$

As another example we examine the ammonia molecule that we analyzed in the previous sections. For instance, in the $|1\rangle, |2\rangle$ representation the energy matrix of Eq. (4.49a) can be written as

$$\hat{H} = E_0 \mathbb{1} - A\hat{\sigma}_x \tag{4.81a}$$

In the presence of an external electric field the Hamiltonian takes the form of Eq. (4.58a), which again can be compactly expressed in terms of the Pauli matrices as

$$\hat{H} = E_0 \mathbb{1} - A\hat{\sigma}_x + \mu\mathscr{E}\hat{\sigma}_z \tag{4.81b}$$

Another well-studied two-state quantum system is that of the neutral K-mesons. This consists of two states, $|K^0\rangle$ and $|\overline{K^0}\rangle$, which are distinguished from one another by having opposite *strangeness* quantum numbers. The $|K^0\rangle$ state is assigned, by convention, strangeness -1, whereas the $|\overline{K^0}\rangle$ is assigned strangeness $+1$. Otherwise the two states are identical. They can be thought of as the $|1\rangle, |2\rangle$ states of the ammonia molecule.

Weak interactions do not conserve the strangeness quantum number and thus provide a connection between the two states, which is in exact analogy to the connection resulting from the tunneling of the nitrogen in the ammonia molecule. Therefore, the energy matrix is of the form

$$\hat{M} = \begin{array}{c|cc} & |K^0\rangle & |\overline{K^0}\rangle \\ \hline |K^0\rangle & m & \dfrac{\Delta m}{2} \\ |\overline{K^0}\rangle & \dfrac{\Delta m}{2} & m \end{array} \quad \text{or} \quad \hat{M} = m\mathbb{1} + \frac{\Delta m}{2}\hat{\sigma}_x \tag{4.82a}$$

which is in exact analogy to Eq. (4.81a).

If we also take into account the possibility that the K^0-mesons may decay we must introduce a hermitian decay matrix

$$\hat{\Gamma} = \begin{pmatrix} \gamma & \gamma_{12} \\ \gamma_{12} & \gamma \end{pmatrix} \quad \text{or} \quad \hat{\Gamma} = \gamma\mathbb{1} + \gamma_{12}\hat{\sigma}_x \tag{4.82b}$$

with γ, γ_{12} real. The overall energy matrix is not hermitian anymore because the $K^0, \overline{K^0}$ decay. We can, however, represent it by

$$\hat{H} = \hat{M} + i\hat{\Gamma} = (m + i\gamma)\mathbb{1} + \left(\frac{\Delta m}{2} + i\gamma_{12}\right)\hat{\sigma}_x \tag{4.83a}$$

When this matrix is brought to diagonal form, the basis states in the new representation are known as the $|K_1^0\rangle$ and $|K_2^0\rangle$ states, and the energy matrix in this basis has the form

$$\hat{H}' = \begin{pmatrix} m + \dfrac{\Delta m}{2} & 0 \\ 0 & m - \dfrac{\Delta m}{2} \end{pmatrix} + i \begin{pmatrix} \gamma + \gamma_{12} & 0 \\ 0 & \gamma - \gamma_{12} \end{pmatrix}$$

$$= (m + i\gamma)\mathbb{1} + \left(\frac{\Delta m}{2} + i\gamma_{12}\right)\hat{\sigma}_z \quad (4.83b)$$

The mass (energy) of the $|K_1^0\rangle$, $|K_2^0\rangle$ states is given by the real part of the diagonal elements of the matrix

$$m(K_1^0) = m + \frac{\Delta m}{2} \qquad m(K_2^0) = m - \frac{\Delta m}{2}$$

The imaginary parts of the diagonal energy matrix give the probability for the decay of the two states, the so-called decay width,† which we indicate by Γ

$$\gamma(K_1^0) = \gamma + \gamma_{12} \qquad \gamma(K_2^0) = \gamma - \gamma_{12}$$

Note that $\hat{\sigma}_x$, $\hat{\sigma}_y$, $\hat{\sigma}_z$ do not refer to any particular direction in space but are defined in terms of their action on the states of the neutral K^0 meson system.

As a final example, we consider the proton and the neutron as two states of the same system—the *nucleon.* We represent the proton as the $|+z\rangle$ eigenstate of the nucleon and the neutron as the $|-z\rangle$ eigenstate. Again, such a representation has nothing to do with directions in space but can be thought of as defining the Z-axis in the *isotopic spin* Hilbert space. If $|\psi\rangle$ represents an arbitrary state of an assembly of nucleons, then the electric charge Q of the system can be obtained from the expectation value

$$Q = \left\langle \psi \middle| \frac{\mathbb{1} + \hat{\sigma}_z}{2} \middle| \psi \right\rangle \quad (4.84)$$

This result can be easily checked by using the representation of Eqs. (4.68c) and (4.73a).

† See the discussion in Section 5.7. In the presence of CP violation the off-diagonal elements of the mass and decay matrices can be complex, an effect we have ignored here. CP is the product of the charge conjugation (C) and parity (P) symmetries.

4.7. SUMMARY

We have tried to emphasize that the time evolution of a quantum-mechanical state is completely determined once the time-evolution operator is known. This, in turn, depends on the Hamiltonian matrix. Therefore, knowledge of the eigenvalues of the Hamiltonian matrix determines the time development of the quantum-mechanical states. However, physical systems possess complicated Hamiltonians which do not lend themselves readily to exact solutions. In such situations we take recourse to approximation methods. In this chapter we have developed the theory of time-independent perturbation. We have emphasized that this method is used to determine the approximate stationary states of the theory. Transitions between different states involve time-dependent perturbations and are dealt with in the next chapter. We have discussed simple examples to indicate the power of perturbation theory. We have solved exactly for the energy levels of the ammonia molecule which represents a two-level system. We have also used perturbation theory to obtain approximate solutions for this system and have compared these with the exact solutions. Finally, we have introduced the Pauli matrices which lead to a general description of all two-level systems.

Problems

PROBLEM 1

An atom is in a stationary state of energy E_0 where the eigenvalue of the total angular momentum is $j = 3/2$. A magnetic field B_z is applied along the Z-axis. The perturbation Hamiltonian is

$$\hat{H}' = -g\frac{\mu_0}{\hbar}(\mathbf{B} \cdot \hat{\mathbf{J}})$$

where $\mu_0 = (e\hbar/2m_e)$ is the Bohr magneton. Find the energy of all the stationary states having $j = 3/2$ in the presence of the perturbation.

Note: The magnetic field is weak so that the perturbation energy is small as compared to the energy spacing between the unperturbed states of different j-values.

PROBLEM 2

Consider a Hamiltonian matrix of the form

	$F_{1/2}$ $m=-1/2$	$F_{1/2}$ $m=+1/2$	$F_{3/2}$ $m=-3/2$	$F_{3/2}$ $m=-1/2$	$F_{3/2}$ $m=+1/2$	$F_{3/2}$ $m=+3/2$
$F_{1/2}, m=-1/2$	$-A-\frac{2}{3}B$	0	0	$-\frac{2^{1/2}}{3}B$	0	0
$F_{1/2}, m=+1/2$	0	$-A+\frac{2}{3}B$	0	0	$\frac{2^{1/2}}{3}B$	0
$F_{3/2}, m=-3/2$	0	0	$\frac{A}{2}-B$	0	0	0
$F_{3/2}, m=-1/2$	$-\frac{2^{1/2}}{3}B$	0	0	$\frac{A}{2}-\frac{1}{3}B$	0	0
$F_{3/2}, m=+1/2$	0	$\frac{2^{1/2}}{3}B$	0	0	$\frac{A}{2}+\frac{1}{3}B$	0
$F_{3/2}, m=+3/2$	0	0	0	0	0	$\frac{A}{2}+B$

(a) Diagonalize the above matrix and plot the energy of the six stationary states as a function of H for the range $B \ll A$ to $B \sim 100$ A. *Hint*: Rearrange the ordering of the states so that the matrix becomes *block diagonal.*

(b) Check your results by using perturbation theory in the limits $B \ll A$ and $A \gg B$ using the Hamiltonian in the (nondiagonal) form given above.

PROBLEM 3

A perturbation due to an electric field

$$\hat{H}' = e\mathscr{E}\hat{z} = e\mathscr{E}r\cos\theta$$

is applied to the hydrogen atom.

(a) Construct the perturbation matrix for the *four* $n=2$ states Use the eigenfunctions given by Eqs. (8.32).

(b) Find the energy of the eigenstates in the presence of the electric field. Note that the energy will depend linearly on the electric field. Why?

PROBLEM 4

Consider Problem 3 above, but include also the $n=1$, $l=0$, and $m=0$ state in the energy matrix. The wave function for this state is [see Eqs. (8.32)]

$$\psi_{n=1,l=0,m=0}(\mathbf{r}) = \frac{1}{2(\pi a^3)^{1/2}} e^{-r/a_0}$$

Find the energy of the eigenstates in the presence of the electric field. Note that you will now also obtain a contribution *quadratic* in the electric field. You may use perturbations theory through second order where applicable.

PROBLEM 5

(*a*) Show that any matrix that commutes with $\hat{\boldsymbol{\sigma}}$ is a multiple of the unit matrix.
(*b*) Show that we cannot find a matrix that anticommutes with all three Pauli matrices.

PROBLEM 6

Let $\hat{\mathbf{A}}$ and $\hat{\mathbf{B}}$ be two vector operators that commute with the Pauli matrices but do not commute between themselves. Prove the Dirac identity

$$(\hat{\boldsymbol{\sigma}} \cdot \hat{\mathbf{A}})(\hat{\boldsymbol{\sigma}} \cdot \hat{\mathbf{B}}) = (\hat{\mathbf{A}} \cdot \hat{\mathbf{B}}) + i(\hat{\mathbf{A}} \times \hat{\mathbf{B}}) \cdot \hat{\boldsymbol{\sigma}}$$

PROBLEM 7

Show that any 2×2 complex matrix $\hat{M}$ can be written as

$$\hat{M} = \sum_{\alpha=0}^{3} m_\alpha \hat{\sigma}_\alpha$$

where $m_\alpha = \frac{1}{2}\mathrm{Tr}(\hat{\sigma}_\alpha \hat{M})$ and $\sigma_0 = \mathbb{1}$, σ_α for $\alpha = 1, 2, 3$, are the Pauli matrices.

PROBLEM 8

The neutral K-mesons can be found in the $|K^0\rangle$ or $\overline{|K^0\rangle}$ states which have different strong interactions. Another representation of the neutral K-mesons is in the two states $|K_1\rangle$ and $|K_2\rangle$, which have different masses and lifetimes. The time evolution of the projection of an arbitrary state $\psi(t)$ on the basis states $|K_1\rangle$ and $|K_2\rangle$ is given by

$$\langle K_1|\psi(t)\rangle = \langle K_1|\psi(0)\rangle e^{-i\alpha_1 t-\beta t} \qquad \text{decays in time}$$

$$\langle K_2|\psi(t)\rangle = \langle K_2|\psi(0)\rangle e^{-i\alpha_2 t} \qquad \text{stable in time}$$

Here $\alpha_1 \neq \alpha_2$ are real; β real. The *basis* states $|K_1\rangle$, $|K_2\rangle$ and $|K^0\rangle$, $|\overline{K^0}\rangle$ are mutually orthogonal and normalized.

In terms of the $|K_1\rangle$ and $|K_2\rangle$ states the $|K^0\rangle$ and $|\overline{K^0}\rangle$ are given by

$$|K^0\rangle = \frac{1}{2^{1/2}}(|K_1\rangle + |K_2\rangle)$$

$$|\overline{K^0}\rangle = \frac{1}{2^{1/2}}(|K_1\rangle - |K_2\rangle)$$

Consider a neutral K-meson *at rest* in a state $|\psi\rangle$ such that at time $t=0$ it is a pure K^0, i.e.,

$$\langle K^0 \mid \psi(t=0)\rangle = 1 \qquad \langle \overline{K^0} \mid \psi(t=0)\rangle = 0$$

(*a*) Find *as a function of time* the *probability* that the state $|\psi\rangle$ remains a K^0.

(*b*) Plot your result for the case

$$\alpha \equiv \alpha_1 - \alpha_2 = 2\pi(2\beta) \quad \text{and} \quad 2\beta = 10^{10}\,\text{sec}$$

(*c*) Give a physical interpretation of the constants α_1, α_2, and β, and of the result obtained in part (*a*).

PROBLEM 9

Consider the K^0 system introduced in Problem 8 above. Plot the probability for finding a $\overline{K^0}$ in a beam of particles which at time $t=0$ is pure K^0. Use the exact values for the parameters of the system

$$\frac{\Delta mc^2}{\hbar} = \alpha = \alpha_1 - \alpha_2 = 0.535 \times 10^{10}\,\text{sec}^{-1}$$

$$\tau_s \equiv \frac{1}{2\beta} = 0.893 \times 10^{-10}\,\text{sec}$$

Extend the plot for at least two complete oscillations.

Chapter 5

TRANSITIONS BETWEEN STATIONARY STATES

In Chapter 4 we indicated that for any quantum system the eigenstates of the energy operator are stationary. Finding the spectrum of these stationary states is an important task, and we showed how to calculate the shifts in the energy of the stationary states due to small perturbations of the system: we were concerned with the *static* properties of quantum systems. In this chapter we will discuss the *dynamics* of the system, namely, the changes of state. These are treated in quantum mechanics by considering the stationary states of the system and evaluating the probability that a transition will occur from one to some other, stationary state. For anyone familiar with quantum mechanics such a concept seems natural, but it is far removed from the classical view of the change in the state of motion of physical systems. In quantum mechanics the system can undergo changes only between specified states, which may be discrete. We calculate the probability that the change will take place without knowledge of the details of the intermediate steps: this is why we speak of *transitions.*

In general, the external forces that are responsible for the transition are presumed weak as compared to the energy of the stationary states. Otherwise, as we saw in Chapter 4, it does not make sense to use the spectrum of the stationary states as calculated in the absence of the external forces. For weak forces we can resort to methods of perturbation theory similar to those introduced for the static case. Brief reflection convinces us that in order to induce transitions between stationary states the perturbation must be time dependent.

In this chapter we begin by calculating the transition probability for a two-level system. This can be done with only few approximations. As an example, we consider the ammonia molecule discussed in Section 4.4. Next, we develop the formalism for time-dependent perturbations of an arbitrary system and show how it leads to

Fermi's "golden rule" of quantum mechanics. This expression was coined by Fermi because of the vast number of physical applications of time-dependent perturbation theory in all areas of physics.

We use the golden rule to discuss the emission and absorption of radiation from quantum systems in Section 5.3. Scattering from a fixed potential calculated in the lowest-order approximation is presented in the following section, again using the golden rule. As a further example of transitions between stationary states we discuss nuclear magnetic resonance and the conditions for observing such transitions. Since this is a two-state system we use the Pauli spin matrices introduced in Chapter 4.

A system can sometimes undergo transitions spontaneously. When this happens its states are not truly stationary. If the mean time for the occurrence of a transition is τ, then, according to the uncertainty principle, measurements of the energy of the state will have a dispersion $\Delta E = \hbar/\tau$ around the mean value of the energy. This subject is discussed in the final section and illustrated by specific examples.

5.1. TRANSITIONS IN A TWO-STATE SYSTEM

We begin the discussion of quantum-mechanical transitions by considering a two-state system. In this case the problem is simplified since the only possible transitions are from state 1 to state 2 ($1 \to 2$), and vice versa ($2 \to 1$). In particular, when the perturbation causing the transition varies harmonically in time, it is possible to obtain an exact solution of the quantum-mechanical equation of motion of the system. If the angular frequency of the perturbation is ω, and ΔE is the energy difference between the two states, the transition probability is significant only when $\hbar\omega = \Delta E$. Two-state systems are often encountered in practice, as when a proton is placed in a magnetic field where the transitions between the two states lead to the phenomenon of nuclear magnetic resonance.

We will develop the general results for transitions in two-state systems by using the ammonia molecule as an example. We have already analyzed this system in Section 4.4 and identified the states

$|1\rangle$ and $|2\rangle$ as corresponding to two different configurations of the nitrogen atom with respect to the plane of the hydrogens. Because of the possibility of tunneling from one configuration to the other, the stationary states are given by linear combinations of $|1\rangle$ and $|2\rangle$ which we labeled $|\mathrm{I}\rangle, |\mathrm{II}\rangle$

$$|\mathrm{I}\rangle = \frac{1}{2^{1/2}}(|1\rangle + |2\rangle \qquad |\mathrm{II}\rangle = \frac{1}{2^{1/2}}(|1\rangle - |2\rangle) \tag{5.1a}$$

with energies

$$E_{\mathrm{I}} = E_0 - A = \hbar\omega_{\mathrm{I}} \quad \text{and} \quad E_{\mathrm{II}} = E_0 + A = \hbar\omega_{\mathrm{II}} \tag{5.1b}$$

If the system is initially in the state $|\mathrm{I}\rangle$ (or in $|\mathrm{II}\rangle$) it will remain in this state. If a static perturbation is applied the stationary states are modified, but we can always find the corresponding two stationary states.

If, however, the perturbation is time-dependent, we cannot strictly speak of stationary states because the energy of the system is no longer constant. When the perturbation is weak it is still useful to treat the eigenstates of the unperturbed system as stationary. As a result of the perturbation the system may undergo a *transition* from one to the other stationary state. By "weak" we mean that the matrix elements of the perturbation matrix are small compared to the energy difference between the unperturbed states.

In Section 4.5 we examined the changes in the energy spectrum when an external electric field $\vec{\mathscr{E}}$ is applied, as in Fig. 4.4. We now assume that the electric field is time-dependent

$$\mathscr{E}(t) = \mathscr{E}_0 \cos \omega t = \tfrac{1}{2}\mathscr{E}_0(e^{i\omega t} + e^{-i\omega t}) \tag{5.2}$$

If $\mu\mathscr{E}_0 \ll A$, we can treat the states $|\mathrm{I}\rangle$ and $|\mathrm{II}\rangle$ as stationary, and in this representation the energy matrix $H = H_0 + H'$ is given by Eqs. (4.62). The equation of motion [Eq. (4.11)] for the two-state system takes the form

$$\begin{aligned} i\hbar \frac{d}{dt} C_{\mathrm{I}}(t) &= (E_0 - A)C_{\mathrm{I}}(t) + \mu\mathscr{E}(t)C_{\mathrm{II}}(t) \\ i\hbar \frac{d}{dt} C_{\mathrm{II}}(t) &= \mu\mathscr{E}(t)C_{\mathrm{I}}(t) + (E_0 + A)C_{\mathrm{II}}(t) \end{aligned} \tag{5.3}$$

Our task is to solve the coupled Eqs. (5.3) for the two coefficients $C_{\mathrm{I}}(t)$ and $C_{\mathrm{II}}(t)$.

In the absence of the perturbation ($\mu\mathscr{E}=0$), Eqs. (5.3) are uncoupled and have the simple solutions

$$C_{\mathrm{I}}(t)=ae^{-i\omega_{\mathrm{I}}t} \quad \text{and} \quad C_{\mathrm{II}}(t)=be^{-i\omega_{\mathrm{II}}t}$$

with $\omega_{\mathrm{I}}, \omega_{\mathrm{II}}$ defined by Eqs. (5.1*b*). Since $\mu\mathscr{E}_0 \ll A$ the coefficients $C_{\mathrm{I}}(t), C_{\mathrm{II}}(t)$ will not be very different from their values in the absence of the perturbation, and we therefore write them as a product of a slowly varying function of time $c_{\mathrm{I}}(t), c_{\mathrm{II}}(t)$ and of the rapid exponential behavior of the unperturbed solution

$$C_{\mathrm{I}}(t)=c_{\mathrm{I}}(t)e^{-i\omega_{\mathrm{I}}t} \qquad C_{\mathrm{II}}(t)=c_{\mathrm{II}}(t)e^{-i\omega_{\mathrm{II}}t} \tag{5.4a}$$

We introduce these expressions into Eq. (5.3), multiply the first equation by $e^{i\omega_{\mathrm{I}}t}$ (the second by $e^{i\omega_{\mathrm{II}}t}$) and define

$$\omega_0=\omega_{\mathrm{II}}-\omega_{\mathrm{I}}=\frac{2A}{\hbar} \tag{5.4b}$$

to find the differential equations for the slowly varying coefficients

$$\begin{aligned} \frac{d}{dt}c_{\mathrm{I}}(t) &= -\frac{i}{\hbar}\mu\mathscr{E}(t)c_{\mathrm{II}}(t)e^{-i\omega_0 t} \\ \frac{d}{dt}c_{\mathrm{II}}(t) &= -\frac{i}{\hbar}\mu\mathscr{E}(t)c_{\mathrm{I}}(t)e^{i\omega_0 t} \end{aligned} \tag{5.4c}$$

Given the explicit form of $\mathscr{E}(t)$ we can solve Eq. (5.4*c*) to obtain the coefficients $c_{\mathrm{I}}(t), c_{\mathrm{II}}(t)$ and using the definition of Eqs. (5.4*a*) the coefficients $C_{\mathrm{I}}(t), C_{\mathrm{II}}(t)$ for any initial condition of the system. We then know the probabilities (as a function of time) $|C_{\mathrm{I}}(t)|^2$ and $|C_{\mathrm{II}}(t)|^2$ of finding the system in the state $|\mathrm{I}\rangle$ or $|\mathrm{II}\rangle$ in the presence of the time-dependent perturbation.

As an illustration of the case when the perturbation has a harmonic time dependence we use the form of $\mathscr{E}(t)$ given by Eq. (5.2) and introduce it into Eq. (5.4*c*). We obtain

$$\begin{aligned} \frac{d}{dt}c_{\mathrm{I}}(t) &= -\frac{i}{2\hbar}\mu\mathscr{E}_0(e^{i(\omega-\omega_0)t}+e^{-i(\omega+\omega_0)t})c_{\mathrm{II}}(t) \\ \frac{d}{dt}c_{\mathrm{II}}(t) &= -\frac{i}{2\hbar}\mu\mathscr{E}_0(e^{i(\omega+\omega_0)t}+e^{-i(\omega-\omega_0)t})c_{\mathrm{I}}(t) \end{aligned} \tag{5.5a}$$

which is a fairly complicated set of coupled equations with the formal solution

$$c_{\mathrm{I}}(t) = -\frac{i}{2\hbar}\mu\mathscr{E}_0\left[\int_0^t e^{i(\omega-\omega_0)t'}c_{\mathrm{II}}(t')\,dt' + \int_0^t e^{-i(\omega+\omega_0)t'}c_{\mathrm{II}}(t')\,dt'\right] \tag{5.5b}$$

Since we have assumed that $c_{\mathrm{I}}(t)$ and $c_{\mathrm{II}}(t)$ vary in time much slower than $e^{i\omega_0 t}$ the integrals in Eq. (5.5*b*) will average out to zero unless $(\omega-\omega_0)$ is very small. We can therefore ignore the terms $e^{\pm i(\omega+\omega_0)t}$ and remember that $c_{\mathrm{I}}(t)$ will be different from zero only for values of ω very close to ω_0. Under this assumption Eqs. (5.5*a*) simplify to

$$\begin{aligned} \frac{d}{dt}c_{\mathrm{I}}(t) &= -\frac{i}{2\hbar}\mu\mathscr{E}_0 e^{i(\omega-\omega_0)t}c_{\mathrm{II}}(t) \\ \frac{d^2}{dt^2}c_{\mathrm{II}}(t) &= -\frac{i}{2\hbar}\mu\mathscr{E}_0 e^{-i(\omega-\omega_0)t}c_{\mathrm{I}}(t) \end{aligned} \tag{5.5c}$$

which can be solved exactly.† We will consider only the special case of exact resonance $\omega=\omega_0$. Differentiating Eqs. (5.5*c*) and substituting one in the other we obtain the uncoupled equations

$$\begin{aligned} \frac{d^2}{dt^2}c_{\mathrm{I}}(t) &= -\left(\frac{\mu\mathscr{E}_0}{2\hbar}\right)^2 c_{\mathrm{I}}(t) \\ \frac{d^2}{dt^2}c_{\mathrm{II}}(t) &= -\left(\frac{\mu\mathscr{E}_0}{2\hbar}\right)^2 c_{\mathrm{II}}(t) \end{aligned} \tag{5.6a}$$

with the solutions

$$\begin{aligned} c_{\mathrm{I}}(t) &= a_0\cos\left(\frac{\mu\mathscr{E}_0}{2\hbar}t\right) + b_0\sin\left(\frac{\mu\mathscr{E}_0}{2\hbar}t\right) \\ c_{\mathrm{II}}(t) &= -ia_0\sin\left(\frac{\mu\mathscr{E}_0}{2\hbar}t\right) + ib_0\cos\left(\frac{\mu\mathscr{E}_0}{2\hbar}t\right) \end{aligned} \tag{5.6b}$$

The constants a_0 and b_0 depend on the initial conditions. For instance, if at $t=0$ the system is in state $|\mathrm{I}\rangle$, $a_0=1$ and $b_0=0$. The

---

† See, for instance, L. D. Landau and E. M. Lifshitz, *Quantum Mechanics*, Second Edition, Addison-Wesley, Reading, MA (1965), p. 139.

probability of finding the system in the state $|\mathrm{II}\rangle$ is then

$$P_{\mathrm{II}}(t)=|C_{\mathrm{II}}(t)|^2=|c_{\mathrm{II}}(t)|^2=\sin^2\left(\frac{\mu\mathscr{E}_0}{2\hbar}t\right)$$

and for finding it in the state $|\mathrm{I}\rangle$ (5.7a)

$$P_{\mathrm{I}}(t)=|C_{\mathrm{I}}(t)|^2=|c_{\mathrm{I}}(t)^2=\cos^2\left(\frac{\mu\mathscr{E}_0}{2\hbar}t\right)$$

Probability conservation is satisfied since the condition

$$P_{\mathrm{I}}(t)+P_{\mathrm{II}}(t)=1$$

holds at all times. The system must be in one of its two states.

Equations (5.7a) are plotted in Fig. 5.1 and we see that after a time $t=(\hbar\pi/\mu\mathscr{E}_0)$ the system has changed from the stationary state $|\mathrm{I}\rangle$ to the stationary state $|\mathrm{II}\rangle$. The time-dependent perturbation has caused a *transition*. When $(\mu\mathscr{E}_0/2\hbar)t\ll 1$

$$P_{\mathrm{II}}(t)\simeq\left(\frac{\mu\mathscr{E}_0}{2\hbar}\right)^2 t^2 \tag{5.7b}$$

which shows that the transition probability is proportional to the square of the electric field. The average intensity of an electromagnetic wave is given by

$$\langle I\rangle=\left(\frac{\varepsilon_0}{\mu_0}\right)^{1/2}\frac{\mathscr{E}_0^2}{2} \tag{5.8}$$

so that when an electromagnetic wave of resonant frequency ω_0 is incident on an ammonia molecule the transition probability is proportional to the intensity of the radiation.

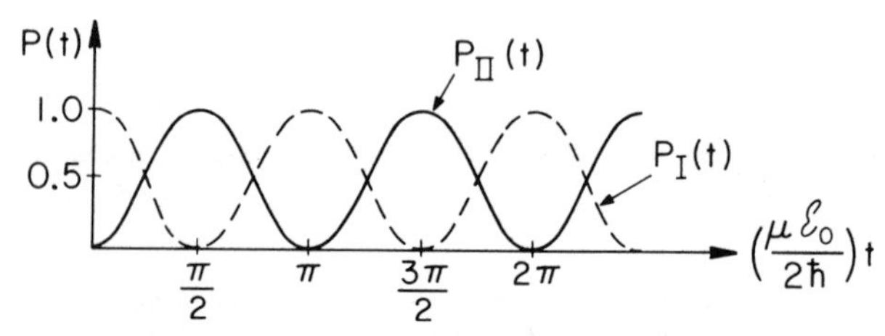

FIGURE 5.1. The probability of finding the two-level system in state $|\mathrm{I}\rangle$ (dashed curve) and for finding it in state $|\mathrm{II}\rangle$ (solid curve), plotted as a function of time.

The energy of the system when it is in state $|\mathrm{II}\rangle$ is larger by the amount $(E_{\mathrm{II}}-E_{\mathrm{I}})=\hbar\omega_0$ than when it is in state $|\mathrm{I}\rangle$. The external time-dependent electric field has supplied energy $\hbar\omega=\hbar\omega_0$ to the system. Note that $\hbar\omega$ is the energy carried by one photon of the electromagnetic field at the frequency ω. Thus, the system *absorbed* one photon in undergoing the transition $|\mathrm{I}\rangle\rightarrow|\mathrm{II}\rangle$. Conversely, when the system is in state $|\mathrm{II}\rangle$ (at the time $t=\hbar\pi/\mu\mathscr{E}_0$), there exists a finite probability that it will make a transition back to the state $|\mathrm{I}\rangle$ at some later time. In this case the system loses energy, since $E_{\mathrm{I}}<E_{\mathrm{II}}$, and in undergoing the transition $|\mathrm{II}\rangle\rightarrow|\mathrm{I}\rangle$, *emits* a photon frequency ω. The two states of the system are indicated in Fig. 5.2 by their corresponding energy, and the arrows indicate the possible transitions.

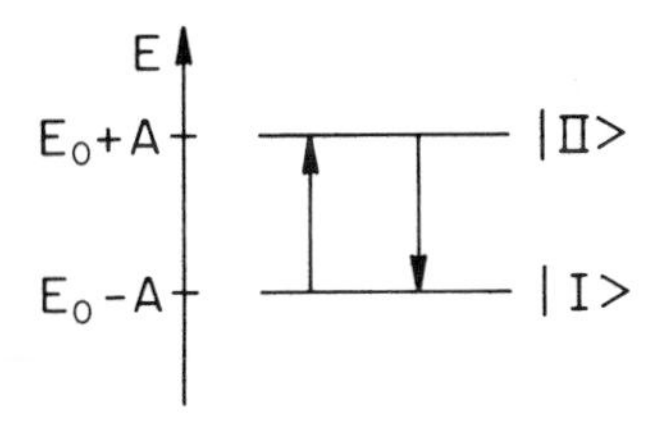

FIGURE 5.2. Representation of the transitions between the states of a two-level system.

The results obtained in this section are idealized because we assume that the electric field has the *exact* frequency $\omega=\omega_0$. In practice, the frequency of the field will be centered at ω_0 but will have some spread around the central value. Similarly, whenever we consider an aggregate of ammonia molecules, as is the case in real experiments, not all molecules have their stationary states at exactly the same energy E_{I} and E_{II} because of collisions with the walls, random motion, etc. We discuss how to take account of these effects in the following sections.

5.2. TIME-DEPENDENT PERTURBATIONS: THE "GOLDEN RULE" OF QUANTUM MECHANICS

We have considered a perturbation that depends explicitly on time. A static perturbation, that is present for only a finite amount of time (say from $t=0$ to $t=T$) depends implicitly on time and, therefore, can also induce transitions between the stationary states of the unperturbed system. In this case, transitions occur only between states of the same energy. We will now develop a general formalism for calculating the probability per unit time for a transition between the stationary states of an arbitrary system in the presence of a time-dependent perturbation.

We designate the unperturbed energy matrix by $\hat{H}_0$ and the perturbation by $\hat{H}'$. With the basis states taken as the eigenstates of $\hat{H}_0$ the equation of motion is of the form

$$i\hbar\frac{d}{dt}C_s(t)=\sum_n H_{sn}C_n(t) \tag{5.9a}$$

where the matrix element H_{sn} is given by

$$H_{sn}=(H_0)_{sn}+(H')_{sn}=E_s^{(0)}\delta_{sn}+H'_{sn} \tag{5.9b}$$

Here $E_s^{(0)}$ is the energy eignevalue of the sth state of $\hat{H}_0$. As in the previous section we assume that $H'_{sn}\ll(E_s^{(0)}-E_n^{(0)})$ and therefore write

$$C_s(t)=c_s(t)e^{-(i/\hbar)E_s^{(0)}t} \tag{5.10a}$$

where $c_s(t)$ is a slowly varying function of time. Substituting this expression into Eq. (5.9a) and making use of Eq. (5.9b)

$$i\hbar\left[\frac{d}{dt}c_s(t)\right]e^{-(i/\hbar)E_s^{(0)}t}+E_s^{(0)}C_s(t)=E_s^{(0)}C_s(t)+\sum_n H'_{sn}c_n(t)e^{-(i/\hbar)E_n^{(0)}t}$$

and multiplying by $e^{(i/\hbar)E_s^{(0)}t}$ we obtain the equation of motion for the slowly varying coefficients

$$\boxed{i\hbar\frac{d}{dt}c_s(t)=\sum_n H'_{sn}c_n(t)e^{(i/\hbar)(E_s^{(0)}-E_n^{(0)})t}} \tag{5.10b}$$

This result represents a set of N coupled equations (where N is the number of basis states), the solution of which determines the N coefficients $c_s(t)$. It is an *exact* equation and reduces to Eq. (5.4c) for a two-state system.

We must solve Eq. (5.10b) by approximation methods based on our assumption that the matrix elements H'_{sn} are weak as specified above. Then the time derivatives are small

$$\frac{d}{dt}c_n(t) \ll 1 \tag{5.11a}$$

Given the initial conditions

$$c_i(t=0)=1 \quad \text{and} \quad c_s(t=0)=0 \qquad \text{for } s \neq i \tag{5.11b}$$

and the assumption of Eq. (5.11*a*), we can set all the coefficients $c_n(t)$ in the right-hand side of Eq. (5.10*b*) equal to zero, except for $c_i(t)$, which we equate to its initial value of 1. In this approximation Eq. (5.10*b*) reduces to N uncoupled equations

$$\frac{d}{dt}c_s(t) = -\frac{i}{\hbar}(H')_{si}e^{(i/\hbar)(E_s^{(0)}-E_i^{(0)})t} \tag{5.12a}$$

which can be integrated directly to yield the coefficients $c_s(t)$

$$c_s(t) = -\frac{i}{\hbar}\int_0^t (H')_{si}e^{(i/\hbar)(E_s^{(0)}-E_i^{(0)})t'}\,dt' \tag{5.12b}$$

where all $c_s(t)$ are assumed much smaller than 1.

To perform the integration in Eq. (5.12*b*) we must know the time dependence of the perturbation. In particular, we may assume that $\hat{H}'$ depends harmonically on time as already given in Eq. (5.2)

$$\hat{H}' \equiv 2\hat{H}_1 \cos \omega t = \hat{H}_1(e^{i\omega t}+e^{-i\omega t}) \tag{5.13a}$$

where $\hat{H}_1$ is time independent. We then have

$$c_s(t) = -\frac{i}{\hbar}\int_0^t (H_1)_{si}[e^{(i/\hbar)(E_s^{(0)}-E_i^{(0)}+\hbar\omega)t'} + e^{(i/\hbar)(E_s^{(0)}-E_i^{(0)}-\hbar\omega)t'}]\,dt'$$

The integral is elementary and gives

$$c_s(t) = (H_1)_{si}\left[\frac{1-e^{(i/\hbar)(E_s^{(0)}-E_i^{(0)}+\hbar\omega)t}}{E_s^{(0)}-E_i^{(0)}+\hbar\omega} + \frac{1-e^{(i/\hbar)(E_s^{(0)}-E_i^{(0)}-\hbar\omega)t}}{E_s^{(0)}-E_i^{(0)}-\hbar\omega}\right] \tag{5.13b}$$

Since $(H_1)_{si} \ll E_s^{(0)} - E_i^{(0)}$ the only significant contribution to $c_s(t)$ will come from the term for which the denominator is close to zero. For instance, when $E_s^{(0)} > E_i^{(0)}$ the second term is relevant, and then, *only if*

$$\hbar\omega \simeq E_s^{(0)} - E_i^{(0)}$$

The probability of finding the system at time t in the state $|s\rangle$ (given that at $t=0$ it was in the state $|i\rangle$) is

$$P_{si}(t) = |C_s(t)|^2 = |c_s(t)|^2 = 4\,|(H_1)_{si}|^2 \frac{\sin^2[E_s^{(0)} - E_i^{(0)} - \hbar\omega)t/2\hbar]}{(E_s^{(0)} - E_i^{(0)} - \hbar\omega)^2} \tag{5.13c}$$

Equation (5.13c) is the general result for weak harmonic perturbations; it has the form of the function

$$\frac{\sin^2 ax}{x^2} \quad \text{with} \quad x = E_s^{(0)} - E_i^{(0)} - \hbar\omega \quad \text{and} \quad a = \frac{t}{2\hbar} \tag{5.13d}$$

This function peaks at $x=0$, that is, when $E_s^{(0)} - E_i^{(0)} = \hbar\omega$, as shown in Fig. 5.3 and becomes narrower as t increases. In the limit $t \to \infty$ it becomes a δ-function. We see that the transition probability is significant only when the angular frequency ω of the perturbation is such that $\hbar\omega$ almost coincides with the energy difference between the initial and the final states.

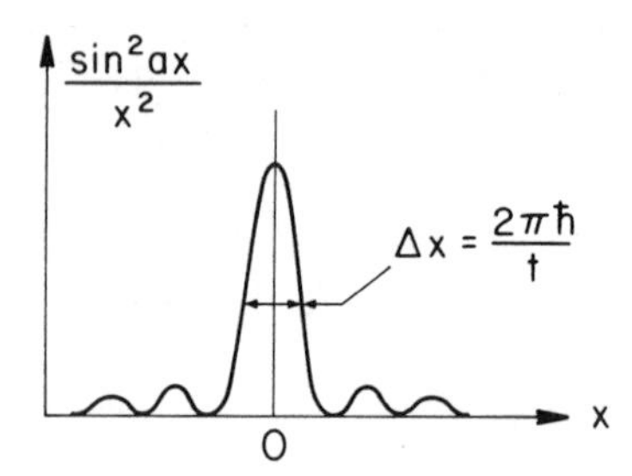

FIGURE 5.3. The function $(\sin^2 ax)/x^2$ plotted as a function of x. Here $x = (E_s^{(0)} - E_i^{(0)} - \hbar\omega)$ and $a = t/2\hbar$.

At this point we have obtained an expression for the transition probability $P_{si}(t)$, but have not taken into account the finite width of the frequency of the perturbation nor the finite width of the energy levels. The latter is shown in Fig. 5.4 where we assume that in the vicinity of the final state $|s\rangle$ there are *many* stationary states with energies differing very little from $E_s^{(0)}$. In that case the total probability for a transition from $|i\rangle$ to any one of the group of states $|s\rangle$ is

$$P_{\text{all}|s\rangle}(t) = \sum_s P_{si}(t) = \sum_s |c_s(t)|^2 \tag{5.14a}$$

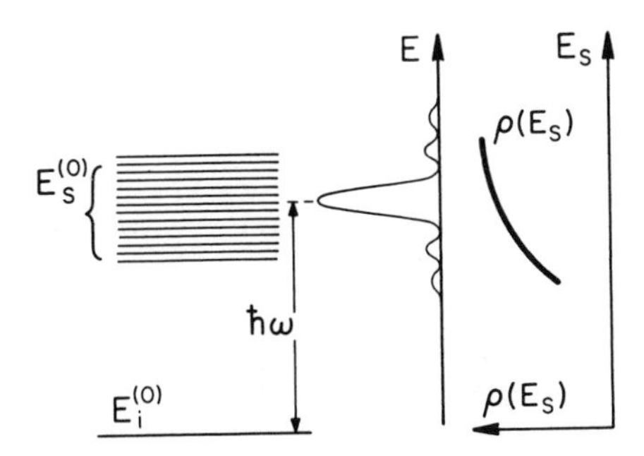

FIGURE 5.4. When many states are available near the final energy $E_s^{(0)}$ they are represented by a density function $\rho(E_s)$. This is indicated by the curve.

If the states $|s\rangle$ are continuously distributed we must replace the sum in the above equation by an integral, taking into account how many states $|s\rangle$ are found per unit energy interval in the vicinity of the energy $E_s^{(0)}$. The conversion factor $\rho(E_s)$ is called the *density of final states factor* and is defined as

$$\rho(E)=\frac{dN}{dE}$$

where N represents the number of stationary states with energy E. Therefore,

$$P_{\mathrm{all}|s\rangle}(t)=\int |c_s(t)|^2\,\rho(E_s)\,dE_s \tag{5.14b}$$

Since $|c_s(t)|^2$ is so sharply peaked at $E_s=E_i+\hbar\omega$ we can treat $(H_1)_{si}$ and $\rho(E_s)$ as constants for the purpose of the integral in Eq. (5.14*b*)

$$P_{\mathrm{all}|s\rangle}(t)=4\,\overline{|(H_1)_{si}|^2}\,\rho(E_s)\int\frac{\sin^2 ax}{x^2}\,dx$$

where we have used the notation of Eq. (5.13*d*), and $\overline{|(H_1)_{si}|^2}$ is the value of the absolute square of the perturbation matrix element averaged over the final states. The integral has the value

$$\int_{-\infty}^{+\infty}\frac{\sin^2 ax}{x^2}\,dx=\pi a=\frac{\pi t}{2\hbar}$$

and we see that the *transition rate*

$$R_{si}(t) = \frac{P_{\text{all}|s\rangle}(t)}{t} \tag{5.15}$$

is independent of time. Thus the *probability for a transition per unit time* is given by

$$\boxed{R_{si} = \frac{2\pi}{\hbar}\overline{|(H_1)_{si}|^2}\,\rho(E_s)} \tag{5.16}$$

Equation (5.16) has been named by Enrico Fermi the "*Golden rule*" of quantum mechanics because of its many important applications to practical problems. It is valid only when the perturbation is weak, so that the assumptions of Eqs. (5.11*a*) and (5.11*b*) remain justified. The frequency of the perturbation must satisfy the resonance condition

$$\hbar\omega = \pm(E_s^{(0)} - E_i^{(0)}) \tag{5.17}$$

From Fig. 5.3, or equivalently from Eq. (5.13*c*), it is evident that a reciprocal relation exists between the time interval over which the perturbation acts and the sharpness of the resonance condition. This is a manifestation of the uncertainty principle, which can be illustrated as follows: the half-width (at half-maximum) of the resonance curve in Fig. 5.3 is $\Delta\omega = \pi/\Delta t$, corresponding to an energy width

$$\Delta E = \hbar\,\Delta\omega = \hbar\pi/\Delta t \quad \text{or} \quad \Delta E\,\Delta t = h/2 \geqslant \hbar \tag{5.18}$$

Namely, if a transition occurs in a time interval Δt, it can lead to a final state with energy differing from $(E_i^{(0)} \pm \hbar\omega)$ by an amount $\Delta E = \hbar/\Delta t$. As a result, the transition energy is distributed around its central value with a width of order ΔE. This phenomenon is evident in all short-lived systems and is discussed in Section 5.7.

5.3. PHASE SPACE

To use the "golden rule" we must be able to evaluate the density of final states $\rho(E_s)$. In many cases $\rho(E_s)$ can be approximated by the density of final states for a free particle that can be calculated from the corresponding volume in phase space. Phase space is the space spanned by the position and momentum coordinates of the system.

Consider a particle of momentum $\mathbf{p}$ confined inside a cubic volume with sides of length $2L$. As discussed in Section 1.8, because the probability of finding the particle outside the cubic box vanishes, the allowed values of the momentum are given by

$$p_x = \hbar k_x = \pm\hbar n_x \frac{\pi}{2L} \tag{5.19a}$$

and similarly for p_y and p_z. The numbers n_x, n_y, and n_z are positive integers, and p_x can be along either direction on the X-axis. Every combination of the integers n_x, n_y, n_z represents a different state. The number of states dn_x included in an interval of momentum dp_x is given by

$$dp_x = 2\hbar \frac{\pi}{2L}\, dn_x \tag{5.19b}$$

where the factor of 2 is included because p_x can be either positive or negative.

The total number of states dN available to a particle of momentum $|\mathbf{p}|$ in the interval $d^3\mathbf{p}$ is

$$dN = dn_x\, dn_y\, dn_z = \left(\frac{2L}{2\pi\hbar}\right)^3 dp_x\, dp_y\, dp_z = \frac{V}{(2\pi\hbar)^3}\, d^3\mathbf{p} \tag{5.20}$$

where we have used $(2L)^3 = V$ for the confining volume. Since $d^3\mathbf{p}$ is the volume element in momentum space in which the particle can be found, Eq. (5.20) has the natural interpretation that

$$dN = \frac{\text{phase space volume}}{(h)^3}$$

In other words, the number of available states equals the phase space volume divided by $(h)^3$. Alternately, we can say that every state of a free particle occupies a volume $(h)^3$ in phase space.

The momentum space element $d^3\mathbf{p}$ can be conveniently expressed in spherical coordinates, as shown in Fig. 5.5. We have

$$\begin{aligned} d^3\mathbf{p} &= p^2\, d\phi\, d\cos\theta\, dp \\ &= p^2\, dp\, d\Omega \end{aligned}$$

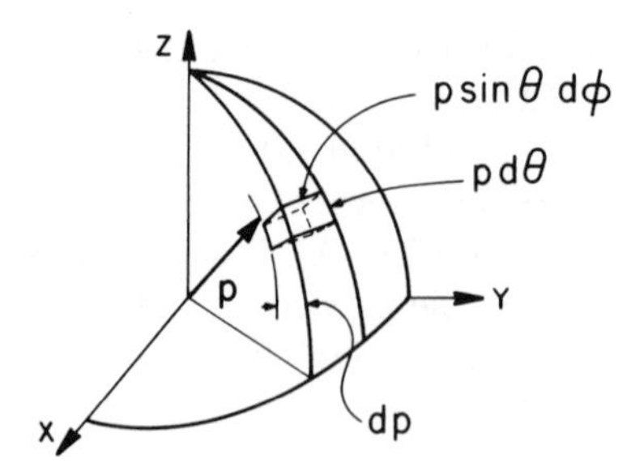

FIGURE 5.5. The differential volume element in spherical coordinates in momentum space.

with $d\Omega$ the solid angle. The density of states is then given by

$$\rho(E)=\frac{dN}{dE}=\frac{V}{(2\pi\hbar)^3}\frac{p^2\,dp\,d\Omega}{dE} \tag{5.21a}$$

For a free particle

$$\frac{dp}{dE}=\frac{1}{v}$$

whether the particle is moving slowly or near the velocity of light. Note that for a slowly moving particle we may write

$$E=mc^2+T=mc^2+\tfrac{1}{2}mv^2$$

$$\frac{dE}{dv}=mv \quad \text{or} \quad \frac{dp}{dE}=m\frac{dv}{dE}=\frac{1}{v}$$

whereas, relativistically,

$$E^2=(mc^2)^2+p^2c^2 \quad \text{or} \quad 2E\,dE=2c^2p\,dp$$

$$\frac{dp}{dE}=\frac{E}{c^2p}=\frac{\gamma mc^2}{c^2m\gamma\beta c}=\frac{1}{\beta c}=\frac{1}{v}$$

Introducing this result for dp/dE into Eq. (5.21a) we obtain

$$\rho(E)=\frac{V}{(2\pi\hbar)^3}\frac{p^2}{v}\,d\Omega \tag{5.21b}$$

This is a very useful expression to which we will refer often. As it stands, it contains the arbitrary volume V into which the particle has been confined. However, when $\rho(E)$ is used to calculate observable phenomena, as in the "golden rule", the matrix element contains a corresponding factor of $1/V$, so that the final result is always independent of the *quantization volume* V. This should be true since the use of a quantization volume is a calculational artifact on which physical observables should not depend.

5.4. EMISSION AND ABSORPTION OF RADIATION

Emission of visible light by atoms is one of the most spectacular but also key phenomena in our universe. It takes place when an atom undergoes a transition from one stationary state to another of lower energy. As mentioned in Chapter 1, the spectrum of atomic radiation is discrete, implying that the stationary states have discrete energies. Therefore, we must calculate the transition probability between states of discrete energy. This is precisely where the "golden rule" is applicable. We will use it to first find the transition probability under the influence of an external time-dependent electromagnetic field. Then, using an argument on the statistical properties of an ensemble of atoms in equilibrium with the radiation that they emit, we will be able to deduce the probability for a spontaneous transition, that is, a transition in the absence of external fields.

To orient ourselves we recall that the energy spacing between the two states of the ammonia molecule we considered, is

$$\Delta E \simeq 10^{-4}\,\text{eV}$$

and therefore the electric field must oscillate at an angular frequency

$$\omega = \frac{\Delta E}{\hbar} = \frac{10^{-4}\,\text{eV}}{6.6\times 10^{-16}\,\text{eV-sec}} \simeq 1.5\times 10^{11}\,\text{rad/sec}$$

which corresponds to microwave frequencies. For transitions between atomic levels the energy differences are typically $\Delta E \simeq 1$–2 eV so that $\omega \simeq 10^{15}$ rad/sec, which corresponds to visible light. The electric field associated with a monochromatic electromagnetic (em) wave of average intensity I is given by

$$\langle I\rangle = c\langle\rho\rangle = c\left(\varepsilon_0\frac{\langle\vec{\mathscr{E}}_0^2\rangle}{2}+\frac{1}{\mu_0}\frac{\langle\mathbf{B}_0^2\rangle}{2}\right) = \left(\frac{\varepsilon_0}{\mu_0}\right)^{1/2}\frac{\mathscr{E}_0^2}{2} = c\varepsilon_0\frac{\mathscr{E}_0^2}{2} \tag{5.23a}$$

where ρ is the energy density of the field, and $|\vec{\mathscr{E}}_0|$ and $|\mathbf{B}_0| = (1/c)\,|\vec{\mathscr{E}}_0|$ are the maximum amplitudes of the electric and magnetic field, respectively, in the wave. This is the same expression as in Eq. (5.8) and is given in MKS units. Since, in practice, the em wave contains a spread of frequencies, we must specify the intensity $dI/d\omega$ in a differential frequency interval $d\omega$,

$$\left(\frac{dI}{d\omega}\right)d\omega = cu(\omega)\,d\omega \tag{5.23b}$$

Where $u(\omega)$ is the energy density per unit frequency interval at the frequency ω.

As for the ammonia molecule the perturbation energy is

$$\boldsymbol{\mu}\cdot\vec{\mathscr{E}}(t) = (\boldsymbol{\mu}\cdot\boldsymbol{\varepsilon})\,|\vec{\mathscr{E}}(t)| = (\boldsymbol{\mu}\cdot\boldsymbol{\varepsilon})\frac{\mathscr{E}_0}{2}(e^{i\omega t}+e^{-i\omega t}) \tag{5.24a}$$

where $\boldsymbol{\varepsilon}$ is the vector along the direction of $\vec{\mathscr{E}}$, and $\boldsymbol{\varepsilon}$ is referred to as the polarization of the wave. Introducing this result in Eq. (5.13c) we find that the transition probability induced by a *monochromatic* wave of frequency ω is

$$P_{fi}(\omega, t) = 4\,|\langle f|\,\boldsymbol{\mu}\cdot\boldsymbol{\varepsilon}\,|i\rangle|^2\frac{\mathscr{E}_0^2}{4}\frac{\sin^2\left[(E_f-E_i-\hbar\omega)\dfrac{t}{2\hbar}\right]}{(E_f-E_i-\hbar\omega)^2}$$

where $E_f - E_i = \hbar\omega_0$. We can now take account of the spread of frequencies of the em wave around the resonant value ω_0, by integrating $P_{fi}(\omega, t)$ over $d\omega$. To do this we note that $\mathscr{E}_0^2 = 2\langle I\rangle/(c\varepsilon_0)$, and replace $\langle I\rangle$ by $(dI/d\omega)\,d\omega$

$$\begin{aligned} P_{fi}(t) &= \int_0^\infty P_{fi}(\omega, t)\,d\omega \\ &= \frac{2}{c\varepsilon_0}\left(\frac{dI}{d\omega}\right)_{\omega_0}\overline{|\langle f|\,\boldsymbol{\mu}\cdot\boldsymbol{\varepsilon}\,|i\rangle|^2}\int_0^\infty\frac{\sin^2\left[(\hbar\omega_0-\hbar\omega)\dfrac{t}{2\hbar}\right]}{(\hbar\omega_0-\hbar\omega)^2}\,d\omega \end{aligned} \tag{5.24b}$$

In the second step of Eq. (5.24b) we assumed that $dI/d\omega$ and the matrix element $|\langle f|\,\boldsymbol{\mu}\cdot\boldsymbol{\varepsilon}\,|i\rangle|^2$ varied slowly with frequency as compared to the $(\sin^2 \alpha x)/x^2$ resonance term, and therefore we have taken them outside the integral. This is true in most cases of interest. With ω_0 fixed the integral over $d\omega$ yields $\pi t/2\hbar^2$, and we obtain a form equivalent to the "golden rule"

$$R_{fi} = \frac{P_{fi}(t)}{t} = \frac{\pi}{c\varepsilon_0\hbar^2}|\langle f|\,\boldsymbol{\mu}\cdot\boldsymbol{\varepsilon}\,|i\rangle|^2\left(\frac{dI}{d\omega}\right)_{\omega_0} \tag{5.25}$$

The only difference between Eq. (5.16) and Eq. (5.25) is that instead of the density of final states we have introduced the density of the incoming radiation. Of course, the frequency of the radiation must satisfy the resonance condition $\hbar\omega_0 = E_f - E_i$.

Equation (5.25) gives the rate for transitions *induced* (the term *stimulated* is also used) by an em field acting on a molecular or atomic system that has an electric dipole moment. For instance, in the laser the electric field of the radiation trapped between the two mirrors induces transitions from a state of higher energy to a state of lower energy. The electric field [see Eq. (5.24a)] contains both an $e^{i\omega t}$ and an $e^{-i\omega t}$ term (unless the radiation is circularly polarized) and therefore the rate of induced transitions from the state $|i\rangle$ of energy E_i to a lower state $|f\rangle$ of energy $E_f < E_i$ equals the rate of induced transitions from $|f\rangle$ to $|i\rangle$

$$R_{if} = R_{fi} \tag{5.26}$$

In addition to induced transitions, atomic and molecular systems are observed to also undergo *spontaneous* transitions, which occur from a state of higher energy to one of lower energy in the absence of an external field. This is difficult to explain within our framework since the state $|i\rangle$ is assumed stationary and there is no perturbation present.† We will use a statistical equilibrium argument first introduced by Einstein to show that if the probability for induced transitions is finite, spontaneous transitions must take place, and we will calculate their rate.

Consider an assembly of identical atoms in a cavity containing em radiation, the system being in thermodynamic equilibrium. Let E_1

---

† In reality, the state $|i\rangle$ is quasi-stationary because it interacts with the fluctuations of the vacuum which are present even in the absence of an external em field.

and E_2 be the energies of two states of the atom with $E_2 > E_1$. When equilibrium is reached the number of atoms in the two states is determined by the Boltzmann distribution

$$\frac{N_2}{N_1} = \frac{Ne^{-E_2/kT}}{Ne^{-E_1/kT}} = e^{-[(E_2-E_1)/kT]} \tag{5.27a}$$

The number of atoms (per unit time) undergoing the transition $1 \to 2$ is proportional to the rate R_{21} induced by the radiation and to the number N_1 of atoms in the initial state

$$\frac{dN}{dt}(1 \to 2) = N_1 R_{21}$$

The number of atoms (per unit time) undergoing the transition $2 \to 1$ is proportional to N_2 and to $(R_{12} + A)$, where A is the *spontaneous* transition rate as shown in Fig. 5.6

$$\frac{dN}{dt}(2 \to 1) = N_2(R_{12} + A)$$

In equilibrium, these two rates must be equal. Using Eq. (5.27*a*) we find

$$\frac{R_{12} + A}{R_{21}} = \frac{N_1}{N_2} = e^{(E_2-E_1)/kT} = e^{(\hbar\omega/kT)} \tag{5.27b}$$

where as usual $\hbar\omega = E_2 - E_1$.

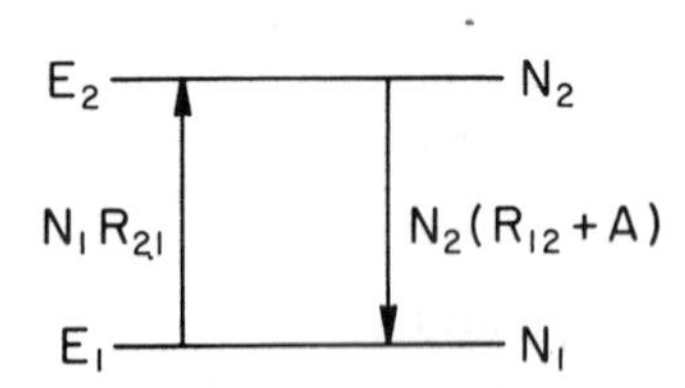

FIGURE 5.6. Induced and spontaneous transitions between two atomic levels.

We now refer to our result for the induced transition rate [Eq. (5.25)] and use Eq. (5.23*b*) to express it in terms of the energy density per unit frequency interval $u(\omega)$ in the cavity

$$R_{21} = \frac{\pi}{\varepsilon_0 \hbar^2} |\langle 2 | \boldsymbol{\mu} \cdot \boldsymbol{\varepsilon} | 1 \rangle|^2 u(\omega) = B_{21} u(\omega)$$

where we have used the notation $B_{21}=\pi/\varepsilon_0\hbar^2\,|\langle 2|\,\boldsymbol{\mu}\cdot\boldsymbol{\varepsilon}\,|1\rangle|^2$. For em radiation in *equilibrium* at a temperature T the energy density per unit frequency interval is given by Planck's law

$$u(\omega)=\frac{1}{\pi^2c^3}\frac{\hbar\omega^3}{e^{\hbar\omega/kT}-1} \tag{5.28}$$

Introducing these results in Eq. (5.27b) we obtain

$$\frac{B_{12}}{B_{21}}+\frac{A}{B_{21}}\frac{\pi^2c^3}{\hbar\omega^3}e^{\hbar\omega/kT}-\frac{A}{B_{21}}\frac{\pi^2c^3}{\hbar\omega^3}=e^{\hbar\omega/kT}$$

which must hold for all temperatures T. This is possible only if

$$\frac{A}{B_{21}}\frac{\pi^2c^3}{\hbar\omega^3}=1 \quad \text{and} \quad \frac{B_{12}}{B_{21}}=1$$

The second condition $B_{12}=B_{21}$ implies that the rate for induced transitions is equal for $1\rightarrow 2$ and $2\rightarrow 1$, as we already know from Eq. (5.26).† The first condition gives the rate for spontaneous transitions

$$A=\frac{\hbar\omega^3}{\pi^2c^3}B_{21}=\frac{\omega^3}{\varepsilon_0\pi\hbar c^3}|\langle 1|\,\boldsymbol{\mu}\cdot\boldsymbol{\varepsilon}\,|2\rangle|^2 \tag{5.29}$$

Equation (5.29) is a landmark result in that it determines the probability for the emission of light by atomic systems. It allows us to calculate the intensity of atomic spectral lines in terms of the matrix element of the electric dipole moment operator $(\boldsymbol{\mu}\cdot\boldsymbol{\varepsilon})$. To compare it with experiment we note that the lifetime τ of the upper state is the inverse of the spontaneous transition rate

$$\tau=\frac{1}{A} \tag{5.30a}$$

For the matrix element entering Eq. (5.29) we make a rough estimate as follows. The electric dipole moment is of the order $\langle\mu\rangle=\langle x\rangle e$, where e is the charge of the electron and $\langle x\rangle$ is typical of atomic dimensions. We must also include a factor of $\frac{1}{3}$ from averaging over all oreintations in $(\boldsymbol{\mu}\cdot\boldsymbol{\varepsilon})$, since the moments are not

† Note also that $B_{12}=B_{21}$ are the absolute squares of time-reversed matrix elements and therefore must be equal.

aligned†

$$\frac{1}{\tau} = A = \frac{4}{3} \frac{\omega^3}{\hbar c^3} \frac{e^2}{4\pi\varepsilon_0} |\langle x \rangle|^2 \qquad (5.30b)$$

Numerically $e^2/4\pi\varepsilon_0\hbar c = \alpha = 1/137$; $\omega/c = 2\pi/\lambda$, and setting $\langle x \rangle \simeq 1$ Å

$$\frac{1}{\tau} = \frac{4}{3} \frac{1}{137} c \left(\frac{2\pi}{\lambda}\right)^3 (1 \text{ Å})^2 \simeq \frac{6 \times 10^{18}}{[\lambda(\text{Å})]^3} \sec^{-1}$$

For a typical wavelength in the visible spectrum $\lambda = 4 \times 10^3$ Å, our estimate gives

$$\tau \simeq 10^{-8} \sec$$

which agrees well with the observed typical lifetimes of atomic states.

It is also interesting to compare Eq. (5.30b) with the result obtained from classical radiation theory. The power radiated by an accelerated particle of charge e is given by the Larmor formula‡

$$P = \frac{2}{3} \frac{e^2}{4\pi\varepsilon_0} \frac{(\dot{v})^2}{c^3}$$

where $\dot{v}$ is the acceleration. If we assume that the particle moves in a circular orbit of radius r, with uniform angular velocity ω, its acceleration is $\dot{v} = \omega^2 r$. We can argue that the time τ required for the classical system to radiate energy $\hbar\omega/2$ is equivalent to the lifetime τ. Thus

$$\frac{1}{\tau} \Rightarrow \frac{2P}{\hbar\omega} = \frac{1}{\hbar\omega} \frac{4}{3} \frac{e^2}{4\pi\varepsilon_0} \frac{\omega^4 r^2}{c^3} = \frac{4}{3} \frac{\omega^3}{\hbar c^3} \frac{e^2}{4\pi\varepsilon_0} r^2 \qquad \text{(classical)}$$

The qualitative agreement between the classical and quantum-mechanical results is a manifestation of the correspondence principle. However, the mechanism for the emission of the radiation is completely different in the two cases, and the classical argument can never produce the discrete spectrum of the radiation.

---

† $(\boldsymbol{\mu} \cdot \boldsymbol{\varepsilon})^2 = (ex)^2(\cos\theta_{x,\varepsilon})^2$ and $\langle \cos^2\theta \rangle = \frac{1}{3}$.

‡ See, for instance, J. D. Jackson, *Classical Electrodynamics*, Wiley, New York 1962.

5.5. SCATTERING OF A PARTICLE FROM A STATIC POTENTIAL

As another application of the "golden rule" we will consider the scattering of a fast-moving particle from a potential that is fixed in a region of space as shown in Fig. 5.7. In this case the perturbation does not depend explicitly on time. However, it acts only for a finite time interval, $-T$ to $+T$, when the particle is within the range of the potential. It can therefore cause transitions between different stationary states as long as they have the *same energy*. Since every direction of motion corresponds to a different state the particle can be deflected by the potential but will retain the same energy. We say that it is *scattered elastically*.

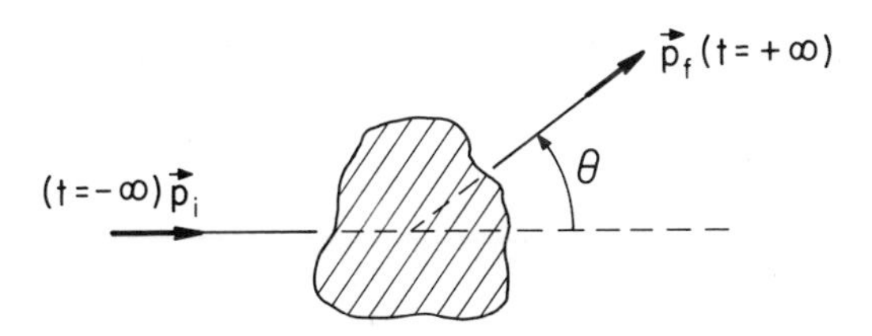

FIGURE 5.7. Scattering from a localized potential. The region where the potential is different from zero is indicated schematically by the shading.

In scattering experiments we observe the direction and not the position of the particles. Thus, we choose stationary states of well-defined momentum. These form a continuous set, and it is convenient to use the coordinate (or position) representation introduced in Section 3.4. In this representation a state of the system is specified by the amplitude for finding the particle at the position $\mathbf{x}$, which we call the wavefunction. Since the particle is free and its momentum $\mathbf{p} = \hbar\mathbf{k}$ is well defined, the wavefunctions are plane waves

$$\psi_i(\mathbf{x}) = \frac{1}{[(2L)^3]^{1/2}}\, e^{-i(\omega t - \mathbf{k}\cdot\mathbf{x})} \qquad \psi_f(x) = \frac{1}{[(2L)^3]^{1/2}}\, e^{-i(\omega t - \mathbf{k}'\cdot\mathbf{x})} \tag{5.31}$$

The wavefunctions are normalized to a density of one particle in a cubic volume $V=(2L)^3$, as can be seen from the fact that $\int_V |\psi|^2\, d^3x = 1$ when the integration is extended over the volume V. This is the same normalization condition used in the discussion of phase space [Eq. (5.20)]. As usual, $\omega = E/\hbar = p^2/2\hbar m$ and $\mathbf{k} = \mathbf{p}/\hbar$. The wave vector for the initial state is $\mathbf{k}$, and for the final state it is $\mathbf{k}' \neq \mathbf{k}$. The two wave vectors have the same magnitude, $|\mathbf{k}| = |\mathbf{k}'|$, because the scattering is assumed to be elastic.

We are interested in the probability that a particle incident along the direction of the wave vector $\mathbf{k}$ will emerge after the scattering along the direction of the wave vector $\mathbf{k}'$. That is, we must find the rate of transition from the state $\mathbf{k}$ to the state $\mathbf{k}'$. This is given by the golden rule [Eq. (5.16)]. We therefore need to find the matrix element of the perturbation Hamiltonian

$$\langle \mathbf{k}' |\, \hat{H}' \,| \mathbf{k} \rangle \tag{5.32}$$

between the initial and final states. In this case the perturbation is given by the interaction energy of the particle in the potential of the scattering center. In general, this will be a function of $\mathbf{x}$, and we write

$$\hat{H}' = U(\mathbf{x})$$

For instance, when a particle of charge $+e$ is scattered from a nucleus of charge $+Ze$ we have

$$U(\mathbf{x}) = \frac{1}{4\pi\varepsilon_0} \frac{e(Ze)}{|\mathbf{x}|} \tag{5.33}$$

provided the scattering center is placed at the origin of the coordinates. We also assume that the position of the scattering center remains fixed.

In the coordinate representation the matrix element of Eq. (5.32) is given by the overlap integral [see Eqs. (3.28)]

$$\langle \mathbf{k}' |\, \hat{H}' \,| \mathbf{k} \rangle = \int \psi_f^*(\mathbf{x}) U(\mathbf{x}) \psi_i(\mathbf{x})\, d^3x = \frac{1}{(2L)^3} \int e^{i\mathbf{k}'\cdot\mathbf{x}} U(\mathbf{x}) e^{-i\mathbf{k}\cdot\mathbf{x}}\, d^3x$$

where the time dependence has cancelled out because ω has the same value for the final as the initial state. Note that the matrix element depends only on the difference $(\mathbf{k}' - \mathbf{k})$ of the wave vectors since the space dependence is integrated out. We introduce the notation

$$\mathbf{q} = \hbar(\mathbf{k}' - \mathbf{k}) \tag{5.34a}$$

where the vector $\mathbf{q}$ represents the momentum transferred during the scattering. $\mathbf{q}$ is referred to as the *momentum transfer* vector. As can be seen from Fig. 5.8 the magnitude of $\mathbf{q}$ is

$$|\mathbf{q}| = 2\,|\mathbf{p}|\sin\frac{\theta}{2} \tag{5.34b}$$

where θ is the scattering angle.

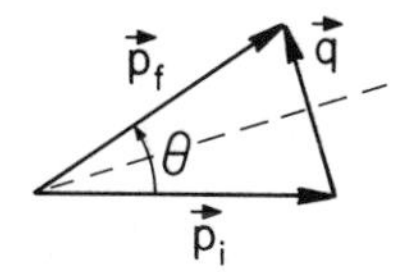

FIGURE 5.8. Definition of the momentum transfer vector $\mathbf{q}$.

The matrix element is now written in the form

$$\langle\mathbf{k}'|\,\hat{H}'\,|\mathbf{k}\rangle = \frac{1}{(2L)^3}\int U(\mathbf{x})e^{i(\mathbf{k}'-\mathbf{k})\cdot\mathbf{x}}\,d^3x = \frac{1}{(2L)^3}\int U(\mathbf{x})e^{(i/\hbar)\mathbf{q}\cdot\mathbf{x}}\,d^3x \tag{5.35a}$$

We recognize that this is the three-dimensional Fourier transform of the interaction energy with a momentum equal to the momentum transfer (see Appendix 1). A complementarity is apparent between the spatial dependence of the scattering potential and the angular distribution of the scattering.

We can evaluate the Fourier transform of the Coulomb potential of Eq. (5.33) by standard techniques. It is best to work in spherical coordinates as shown in Fig. 5.9 and so we choose the polar axis along the $\mathbf{q}$ direction. We let γ be the angle between the $\mathbf{q}$ vector and the arbitrary vector $\mathbf{x}$. Thus

$$\langle\mathbf{k}'|\,\hat{H}'\,|\mathbf{k}\rangle = \frac{Ze^2}{4\pi\varepsilon_0}\frac{1}{(2L)^3}\int\frac{e^{(i/\hbar)\mathbf{q}\cdot\mathbf{x}}}{|\mathbf{x}|}\,d^3x$$

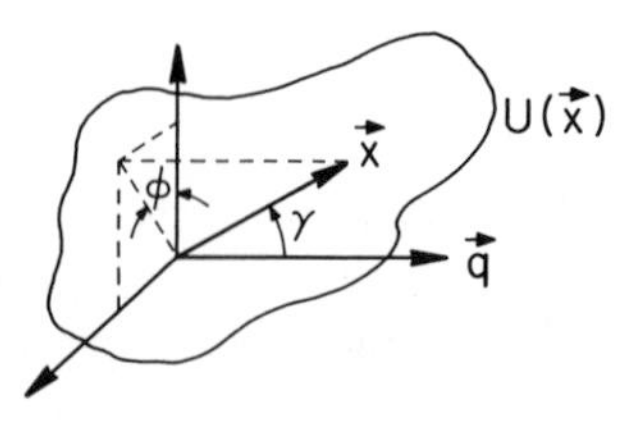

FIGURE 5.9. Coordinates used for performing the integration indicated by Eq. (5.35a). For fixed direction of $\mathbf{q}$ we integrate over the region where the potential is different from zero.

and

$$\int \frac{e^{(i/\hbar)\mathbf{q}\cdot\mathbf{x}}}{|\mathbf{x}|} d^3x = \int \frac{e^{(i/\hbar)qr\cos\gamma}}{r} r^2\, dr\, d\cos\gamma\, d\phi$$

$$= \frac{2\pi\hbar}{iq} d\left(\frac{i}{\hbar} qr \cos\gamma\right) \int_0^\infty dr \int_{\cos\gamma=-1}^{+1} e^{(i/\hbar)qr\cos\gamma}\, d\left(\frac{i}{\hbar} qr\cos\gamma\right)$$

$$= \frac{4\pi\hbar}{q} \int_0^\infty \sin\left(\frac{qr}{\hbar}\right) dr = \frac{4\pi\hbar^2}{q^2} \tag{5.35b}$$

In the last step of this derivation the integral at the upper limit is of the form $\cos(qr/\hbar)$ with $r \to \infty$, which is ambiguous. It can, however, safely be taken to equal zero.† Therefore, the matrix element entering into the "golden rule" is

$$|\langle\mathbf{k}'|\, \hat{H}'\, |\mathbf{k}\rangle|^2 = \left| \frac{Ze^2}{4\pi\varepsilon_0} \frac{1}{(2L)^3} \frac{4\pi\hbar^2}{q^2} \right|^2 \tag{5.35c}$$

We also need to evaluate the density of final states $\rho(E_f)$. Since the particle in the final state (after the scattering) is free, we can use the result of Eq. (5.21b)

$$\rho(E_f) = \frac{(2L)^3}{(2\pi\hbar)^3} \frac{p^2}{v} d\Omega \tag{5.21b}$$

Introducing the matrix element of Eq. (5.35c) and $\rho(E_f)$ into the "golden rule" [Eq. (5.16)], the transition rate is given as

$$R_{\mathbf{k}',\mathbf{k}} = \left(\frac{Ze^2}{4\pi\varepsilon_0}\right)^2 \frac{1}{(2L)^3} \frac{4p^2}{v} \frac{1}{q^4} d\Omega \tag{5.36}$$

† This result can also be found in most tables of finite integrals.

Even though we have obtained the transition rate this is not a directly observable physical quantity. In a scattering process we determine the differential cross section $d\sigma/d\Omega$, which is defined as

$$\frac{d\sigma(\theta,\phi)}{d\Omega}=\frac{\text{probability of observing one particle per unit time at angle } \theta,\phi \text{ per unit solid angle } d\Omega}{\text{when one particle is incident per unit area per unit time}} \tag{5.37a}$$

The numerator on the rhs of Eq. (5.37a) is the transition rate $R_{\mathbf{k}',\mathbf{k}}$ divided by $d\Omega$, and the denominator is the incident flux S. Therefore

$$S\frac{d\sigma(\theta,\phi)}{d\Omega}=R_{\mathbf{k}',\mathbf{k}}\frac{1}{d\Omega} \tag{5.37b}$$

Since we have normalized the wave functions to a density ρ of one particle in a volume $(2L)^3$, the flux is given by

$$S=\rho v=\frac{1}{(2L)^3}v$$

Introducing this value for S and the result for $R_{\mathbf{k}'\mathbf{k}}$ into Eq. (5.37b) we obtain the differential cross section for Coulomb scattering

$$\boxed{\frac{d\sigma}{d\Omega}=\left(\frac{Ze^2}{4\pi\varepsilon_0}\right)^2\frac{4p^2}{v^2}\frac{1}{q^4}=\left(\frac{Ze^2}{4\pi\varepsilon_0}\right)^2\frac{m^2}{4p^4}\frac{1}{\sin^4(\theta/2)}} \tag{5.38}$$

We have used the expression of Eq. (5.34b) for the magnitude of the momentum transfer $|\mathbf{q}|$ so that the differential cross section is also expressed in terms of the scattering angle θ.

As for the result obtained in the previous section, Eq. (5.38) is central to modern physics and contains several features that need to be discussed. First, it agrees exactly with the classical result for Coulomb scattering, the well-known Rutherford cross section. This is a coincidence due to a mathematical peculiarity of the $1/r$ potential that does not occur for other forms of the potential. In general, however, whenever a quantum-mechanical result does not contain $\hbar$, it is equal to the classical calculation. Second, the arbitrary normalization volume $(2L)^3$, even though required for the intermediate steps, does not appear in the expression for the physical observables. Furthermore, since Coulomb scattering is proportional to $1/q^4$, both $d\sigma/d\Omega$ and $\sigma = \int (d\sigma/d\Omega)\, d\Omega$ diverge as $\theta \to 0$. This last property is due to the infinite range of the $1/r$ potential and is not present for the short-range potential of the nuclear forces. Finally, the scattering is independent of the sign of the charge of the incident particle, because the cross section depends on the square of the matrix element.

The calculation of scattering by the "golden rule" is the leading term of a perturbation expansion. This is known as the Born series after Max Born who first used it in collision problems. It can be applied only when the effect of the scattering potential is weak and the incident particles have high momentum. If the range of the potential is a and its strength U, then the condition for the validity of the Born approximation is

$$\frac{a}{\hbar}[(p^2 - 2mU)^{1/2} - p] \ll 1$$

5.6. NUCLEAR MAGNETIC RESONANCE

The phenomenon of nuclear magnetic resonance (nmr), first observed in 1945, has become an important analytical tool in physics, chemistry, and other sciences. In nmr experiments one observes transitions between the two spin states of the proton when it is placed in a magnetic field. The proton has spin 1/2 and thus, in the presence of the magnetic field, there are two basis states which we choose to correspond to the projections of the spin onto the Z-axis. We label these states $|+z\rangle$ and $|-z\rangle$. The potential energy due to the interaction of the proton's magnetic moment $\boldsymbol{\mu}$ with the field $\mathbf{B}$ directed along the Z-axis is given by

$$\begin{aligned} U\ (\text{for } |+z\rangle) &= -\mu\ |\mathbf{B}| \\ U\ (\text{for } |-z\rangle) &= \mu\ |\mathbf{B}| \end{aligned} \tag{5.39a}$$

As in our previous discussions μ stands for the maximal value of the projection of the magnetic moment onto a reference axis. The magnetic moment of the proton is defined in complete analogy to Eq. (3.16), but in terms of the nuclear magneton μ_N

$$\boldsymbol{\mu} = g\mu_N \frac{\mathbf{S}}{\hbar} \qquad \mu_N = \frac{e\hbar}{2m_p} \tag{5.39b}$$

where $\mathbf{S}$ is the proton spin and g is the g-factor. For protons $s = 1/2$ and therefore m_s can take only the two values $m_s = \pm 1/2$. Consequently, the possible projections of the magnetic moment onto any reference axis are

$$(\boldsymbol{\mu} \cdot \mathbf{u}_z) \Rightarrow g\mu_N m_s = \pm\tfrac{1}{2} g\mu_N = \pm\mu$$

It is common usage to call μ the *magnetic moment of the proton*, and it has the value $\mu = 2.792846\mu_N$.

If we take into account the rest energy of the proton which is $E_0 = m_p c^2$ and the potential energy given by Eq. (5.39a), then the energy matrix in the $|\pm z\rangle$ representation is given by

$$\hat{H} = \begin{array}{c|cc} & |+z\rangle & |-z\rangle \\ \hline |+z\rangle & mc^2 - \mu\,|\mathbf{B}| & 0 \\ |-z\rangle & 0 & mc^2 + \mu\,|\mathbf{B}| \end{array} = \begin{pmatrix} mc^2 & 0 \\ 0 & mc^2 \end{pmatrix} - \begin{pmatrix} \mu\,|\mathbf{B}| & 0 \\ 0 & -\mu\,|\mathbf{B}| \end{pmatrix}$$

where we use m to designate the proton mass m_p. Since the proton in a magnetic field is a two-state system we can use the Pauli matrices introduced in Section 4.6 to describe the system in concise notation. The above energy matrix is then written as

$$\hat{H} = \hat{H}_0 + \hat{H}_z = mc^2\mathbb{1} - (\mu\,|\mathbf{B}|)\hat{\sigma}_z \tag{5.40a}$$

If the $\mathbf{B}$-field is oriented along the X-axis the potential energy matrix is diagonal in the $|\pm x\rangle$ representation

$$\hat{U}_x = \begin{array}{c|cc} & |+x\rangle & |-x\rangle \\ \hline |+x\rangle & -\mu\,|\mathbf{B}| & 0 \\ |-x\rangle & 0 & +\mu\,|\mathbf{B}| \end{array} \tag{5.40b}$$

We can, however, express the potential energy operator $\hat{U}_x$ in the $|\pm z\rangle$ representation by transforming the above matrix according to

Eq. (2.34b)

$$\hat{U}'_x = S\hat{U}_x S^{-1} \tag{5.41a}$$

where the unitary matrix S rotates the $+x$ axis into the $+z$ direction. This is the same transformation as shown in Fig. 3.3 and the matrix S is given by Eq. (3.17)

$$S = \begin{array}{c|cc} & |+x\rangle & |-x\rangle \\ \hline |+z\rangle & 1/\sqrt{2} & -1/\sqrt{2} \\ |-z\rangle & 1/\sqrt{2} & 1/\sqrt{2} \end{array} \qquad S^{-1} = S^{\dagger} = \begin{pmatrix} 1/\sqrt{2} & 1/\sqrt{2} \\ -1/\sqrt{2} & 1/\sqrt{2} \end{pmatrix}$$

Using these values in Eq. (5.41a) we obtain

$$\hat{U}'_x = \begin{array}{c|cc} & |+z\rangle & |-z\rangle \\ \hline |+z\rangle & 0 & -\mu\,|\mathbf{B}| \\ |-z\rangle & -\mu\,|\mathbf{B}| & 0 \end{array} = -\mu\,|\mathbf{B}|\,\hat{\sigma}_x$$

Thus, in terms of the Pauli matrices, the energy matrix in the $|\pm z\rangle$ representation when the **B** field is along the X-axis, is written as

$$\hat{H} = \hat{H}_0 + \hat{U}'_x = mc^2\mathbb{1} - \mu\,|\mathbf{B}|\,\hat{\sigma}_x \tag{5.41b}$$

Similarly, when the field is along the Y-axis the interaction energy in the $|\pm z\rangle$ representation is given by $\hat{U}'_y = -\mu\,|\mathbf{B}|\,\hat{\sigma}_y$.

Combining the above results, we see that for an arbitrarily oriented magnetic field **B** with components (B_x, B_y, B_z) the interaction energy matrix is

$$\hat{U} = -\mu(B_x\hat{\sigma}_x + B_y\hat{\sigma}_y + B_z\hat{\sigma}_z) = -\mu\mathbf{B}\cdot\boldsymbol{\sigma} \tag{5.42a}$$

If the Pauli matrices are chosen as in Eq. (4.68c) where $\hat{\sigma}_z$ is diagonal, then Eq. (5.42a) gives $\hat{U}$ in the $|\pm z\rangle$ representation. Of course, the magnetic field components must be given with reference to the same system of coordinates. On the other hand, the expression $\hat{U} = -\mu\mathbf{B}\cdot\hat{\boldsymbol{\sigma}}$ is independent of the choice of coordinate axes since it is the scalar product of two vectors.† If we identify the operator

$$\mu\hat{\boldsymbol{\sigma}} = \hat{\boldsymbol{\mu}} \tag{5.42b}$$

† More precisely, both **B** and $\hat{\boldsymbol{\sigma}}$ are pseudovectors, but $\mathbf{B}\cdot\hat{\boldsymbol{\sigma}}$ is a true scalar.

as the quantum-mechanical magnetic moment operator, then the interaction-energy operator is simply written as

$$\hat{U} = -\hat{\boldsymbol{\mu}} \cdot \mathbf{B} \tag{5.42c}$$

This expression is directly analogous to the classical potential energy of a magnetic dipole $\boldsymbol{\mu}_c$ in a field $\mathbf{B}$, $U_{\text{classical}} = -(\boldsymbol{\mu}_c \cdot \mathbf{B})$. The complete energy matrix for a proton in a field $\mathbf{B}$ is then

$$\hat{H} = mc^2\mathbb{1} - \mu(\mathbf{B} \cdot \hat{\boldsymbol{\sigma}})$$

We can now apply this formalism to the description of nmr. Let a sample of protons be placed in a constant and homogeneous magnetic field $\mathbf{B}_0$. We choose the orientation of the coordinates so that the Z-axis is along $\mathbf{B}_0$. Then, as we already know, the energy matrix is

$$\hat{H} = mc^2\mathbb{1} - \mu B_0\hat{\sigma}_z = \begin{pmatrix} mc^2 - \mu B_0 & 0 \\ 0 & mc^2 + \mu B_0 \end{pmatrix} \tag{5.43a}$$

The magnetic field lifts the degeneracy between the two states $|+z\rangle$ and $|-z\rangle$, which become spaced by an energy difference $\Delta E = 2\mu B_0$. This is indicated in Fig. 5.10(a). Next, a time-varying magnetic field $B_x = B' \cos \omega t$ is applied in the x-direction. In practice this is accomplished as shown in Fig. 5.10(b). An iron core electromagnet is used to provide the field B_0, and the sample of protons (say, water) is located between the pole faces surrounded by a solenoid coil coupled to a tunable radio frequency circuit. The coil is oriented so that the magnetic radiofrequency (rf) field lines are along the X-axis.

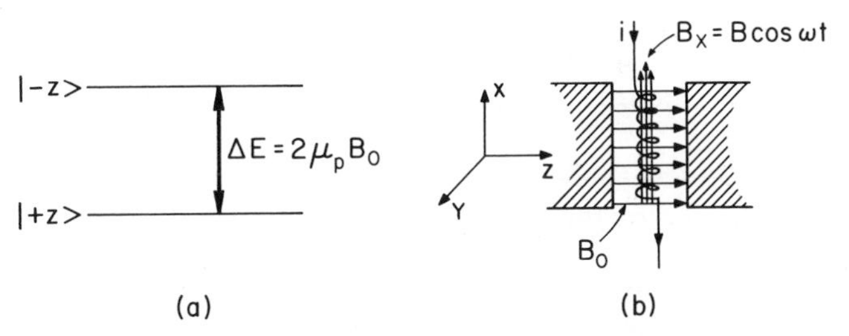

FIGURE 5.10. Nuclear magnetic resonance. (a) The two states of a proton placed in a homogeneous magnetic field B_0 have energy differing by $\Delta E = 2\mu_p B_0$. (b) Physical arrangement of an nmr experiment. The time-dependent magnetic field must be perpendicular to the constant field.

The potential energy due to the time varying field gives rise to an interaction energy

$$\hat{U} = -\mu B_x \hat{\sigma}_x = -\frac{\mu B'}{2}(e^{i\omega t} + e^{-i\omega t})\hat{\sigma}_x$$

where the X-axis is completely determined by our previous choice of placing the Z-axis along $\mathbf{B}_0$. Thus, the complete energy matrix is

$$H = mc^2 \mathbb{1} - \mu B_0 \hat{\sigma}_z - \mu B_x \hat{\sigma}_x$$

$$= mc^2 \mathbb{1} - \begin{pmatrix} -\mu B_0 & -\frac{\mu B'}{2}(e^{i\omega t} + e^{-i\omega t}) \\ -\frac{\mu B'}{2}(e^{i\omega t} + e^{-i\omega t}) & +\mu B_0 \end{pmatrix} \tag{5.43b}$$

This energy matrix is exactly the same as for the time-dependent perturbation of the ammonia molecule given by Eq. (5.3). Thus, if the radiofrequency is such that

$$\hbar\omega = 2\mu B_0 \tag{5.43c}$$

the off-diagonal elements will induce transitions† between the states $|+z\rangle$ and $|-z\rangle$. The rate for transitions "up" equals the rate for transitions "down" as found in Eqs. (5.7). If there are N_+ and N_- protons in the two states and we designate the transition rate by R, we can write for the number of protons undergoing transitions per unit time

$$\begin{aligned} \Delta N(+ \to -) &= RN_+ \\ \Delta N(- \to +) &= RN_- \end{aligned} \tag{5.44a}$$

Under equilibrium conditions we must have $\Delta N[(+) \to (-)] = \Delta N[(-) \to (+)]$, and therefore if the rf magnetic field is strong enough it will *equalize* the population of the two states, leading to $N_+ = N_-$.

Initially, however, the distribution of protons between the two states is governed by the Boltzmann law [also given in Eq. (5.27*a*)]

$$\frac{N_-}{N_+} = \frac{N_0 e^{-(E_-/kT)}}{N_0 e^{-(E_+/kT)}} = \frac{N_0 e^{-(mc^2 + \mu B_0)/kT}}{N_0 e^{-(mc^2 - \mu B_0)/kT}} = e^{-(2\mu B_0/kT)} \tag{5.44b}$$

† Note that if the time-dependent field is parallel to the direction of $\mathbf{B}_0$ no transitions occur since the off-diagonal elements will be zero.

Therefore $N_+ > N_-$, and according to Eq. (5.44*a*) the number of transitions 'up'' exceeds the number of transitions "down''

$$\Delta N[(+) \to (-)] > \Delta N[(-) \to (+)]$$

This implies that the proton sample absorbs more energy from the rf field than it delivers to it. With suitable apparatus we can detect this energy absorption which occurs only when the frequency is exactly $\omega = 2\mu B_0/\hbar$. But even in the presence of the rf magnetic field the upper level $|-z\rangle$ contains fewer protons because of the interaction of the protons with neighboring nuclei and collisions in the sample. These processes, which are called *relaxation mechanisms*, result in the transfer of protons from the state $|-z\rangle$ to the state $|+z\rangle$ without the emission of a photon of energy $\omega = 2\mu B_0/\hbar$. Thus, the population difference given by Eq. (5.44*b*) is maintained even in the presence of the rf field and the absorption of energy from the field can be observed under steady-state conditions.

In practical applications the constant magnetic field is modulated by a small field $\Delta B_0(t)$ at a low frequency (60 Hz) by using sweep coils [Fig. 5.11(a)]. When $[B_0 + \Delta B_0(t)] = \hbar\omega/2\mu$, an absorption signal is observed, as shown in Fig. 5.11(b). Since ΔB_0 is often of sinusoidal form, resonance is crossed twice in each cycle and a typical pattern displayed on an oscilloscope is as shown in (c) of the figure.

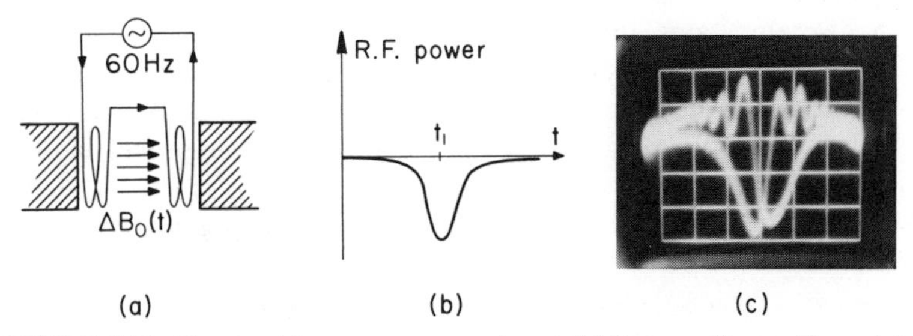

FIGURE 5.11. The signal from an nmr experiment. (a) Sweep coils provide modulation of the constant field. (b) As a result of the modulation, power is absorbed only during some fraction of the sweep. (c) The observed signal on an oscilloscope. Note that the field passes through resonance twice in every cycle.

The detailed structure of the observed signal depends on the homogeneity of the constant field B_0 and on the strength of the externally applied rf field. In addition, it depends critically on the interaction of the protons with the other nuclei and atoms in the sample. Thus, nmr can provide detailed information about the chemical structure of compounds, biological samples, and many different materials.

To estimate the frequency at which proton nmr can be observed under laboratory conditions, we note that the magnetic moment of the proton is

$$\mu = g\tfrac{1}{2}\mu_N = g\frac{1}{2}\frac{e\hbar}{2m_p} = 5.586(\tfrac{1}{2})(5.051\times 10^{-27}\ \text{A-m}^2)$$

Thus, for a magnetic field of 7 kG (0.7 T) we obtain

$$\nu = \frac{\omega}{2\pi} = \frac{2\mu B}{2\pi\hbar} = 29.807\ \text{MHz}$$

which is a convenient radiofrequency. Note that the g-factor of the proton $g = 5.586$ differs from the value $g \simeq 2$ that we had found for the electron. This gross deviation from the prediction of the Dirac equation for spin 1/2 point-particles is due to the extended structure of the proton. It indicates that the proton is a composite system. Indeed there is now overwhelming evidence that the proton is composed of elementary spin-1/2 particles, named *quarks*.

It is instructive to consider a semiclassical interpretation of nuclear magnetic resonance. Classically, there is no way to explain the two discrete states that the projection of the proton's spin can occupy. We must postulate that the spin projection onto the direction of the magnetic field is fixed and given by $\pm\hbar/2$. We can then use the classical picture of a spinning magnetic dipole precessing in a magnetic field B_0 as discussed in Section 3.3. This is shown in Fig. 5.12(a), the precession angular frequency Ω being given by Eq. (3.22*b*)

$$\Omega = \frac{2\mu B_0}{\hbar} \tag{5.45}$$

We also show in the figure the field B_x oriented along the X-axis. Such a field will exert a torque on the magnetic moment, tending to precess it around the B_x direction. As a result the spin projection would change from $+\hbar/2$ to $-\hbar/2$, and vice versa. However, for the field B_x to be effective it must always remain in the plane defined by the spin and the constant field B_0. This implies that B_x must rotate in the $X-Y$ plane with the precession angular frequency of Eq. (5.45), which is exactly equal to the resonance frequency at which quantum-mechanical transitions occur between the two states, as given by Eq. (5.43*c*). An oscillatory field in the x-direction is equivalent to two counterrotating fields in the $X-Y$ plane, as shown in Fig. 5.12(b). One sense of rotation follows the positive projection of the spin and the other follows the negative projection. Thus, transitions "down" are as probable as transitions "up," as we had already found from the quantum-mechanical calculation of the transition rate.

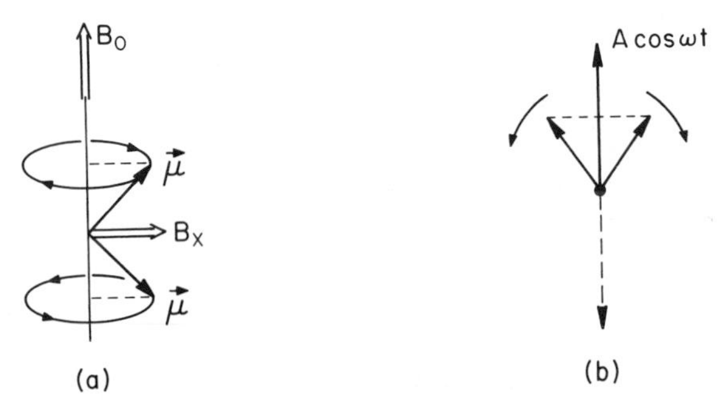

FIGURE 5.12. (a) Precession of a magnetic dipole in a homogeneous magnetic field B_0. The perturbing field B_x must rotate around the direction of B_0 at the precession frequency. (b) An oscillatory field is equivalent to two counterrotating fields.

The phenomenon of nmr gloriously and unambiguously demonstrates the existence of stationary states of discrete energy and the quantization of angular momentum. It further demonstates that transitions induced by an external em field can occur between these states; nmr is easy to observe experimentally and can be analyzed in terms of the most simple two-state quantum system. It should come as no surprise that it has so many practical applications.

5.7. ENERGY WIDTH OF QUASI-STATIONARY STATES

Throughout this chapter we have been considering stationary states and the transitions that take place between them. In Section 5.4, however, we showed that an atom in an excited state will make a spontaneous transition—we say it will *decay*—to its ground state. Since the transition is spontaneous, the excited state is not precisely stationary but characterized by a lifetime τ as given by Eqs. (5.30). Since the state can be observed on the average only for a time interval $\Delta t \simeq \tau$, any measurement of its energy will be correspondingly uncertain by $\Delta E = \hbar/\Delta t = \hbar/\tau$. Even though individual measurements fluctuate by ΔE from that central value, the central value of the energy of the state remains precisely defined and can be determined by repeated measurements. This is reflected in the width of the spectral lines that are emitted in the transitions from or between quasi-stationary states, or in direct measurements of the energy.

We can calculate the shape of the linewidth by a slight mathematical extension of our formalism for the description of stationary states. We know that such a state is an eigenstate of the energy operator with eigenvalue ε. Thus the time dependence of the wavefunction is given by

$$\psi(t) = e^{-(i/\hbar)\varepsilon t}$$

We now assume that the eigenvalue is complex

$$\varepsilon = E_0 - i\Gamma/2, \qquad \Gamma \text{ real, positive} \tag{5.46a}$$

This choice of ε implies that the energy operator is no longer hermitian. This is a necessary step if we wish to describe a decaying state, namely, a situation where probability does not appear to be conserved. In terms of Eq. (5.46*a*) the time dependence of the wavefunction becomes

$$\psi(t) = e^{-i(E_0/\hbar)t - (\Gamma/2\hbar)t} \tag{5.46b}$$

and its absolute square is

$$|\psi(t)|^2 = e^{-(\Gamma/\hbar)t} \tag{5.46c}$$

The probability of finding the system in the state ψ decreases in time exponentially, with a lifetime $\tau = \hbar/\Gamma$. Thus Γ represents the width or uncertainty in the energy of the state.

To find the distribution around the central energy E_0 of the state, we can perform a Fourier transform of Eq. (5.46*b*) from the time variable to the conjugate frequency variable. Any function of time

$\psi(t)$ can be represented by a Fourier integral

$$\psi(t) = \frac{1}{(2\pi)^{1/2}} \int_0^\infty A(\omega) e^{i\omega t}\, d\omega$$

where

$$A(\omega) = \frac{1}{(2\pi)^{1/2}} \int_0^\infty \psi(t) e^{i\omega t}\, dt \tag{5.47a}$$

Introducing Eq. (5.46c) for $\psi(t)$ and setting $\omega_0 = E_0/\hbar$ we obtain directly

$$\begin{aligned} A(\omega) &= \frac{1}{(2\pi)^{1/2}} \int_0^\infty e^{-i(\omega_0-\omega)t-(\Gamma/2\hbar)t}\, dt \\ &= \frac{1}{(2\pi)^{1/2}} \frac{1}{i(\omega_0-\omega)+\Gamma/2\hbar} \end{aligned} \tag{5.47b}$$

The absolute square of $A(\omega)$ gives the probability of finding the system at the frequency ω in the interval $d\omega$. Then the normalized probability $P(E)$ of finding the system at the energy E, in the interval dE is

$$P(E) = \frac{\Gamma}{\hbar^2} |A(\omega)|^2 = \frac{1}{\pi} \frac{\Gamma/2}{(E_0-E)^2+\Gamma^2/4} \tag{5.47c}$$

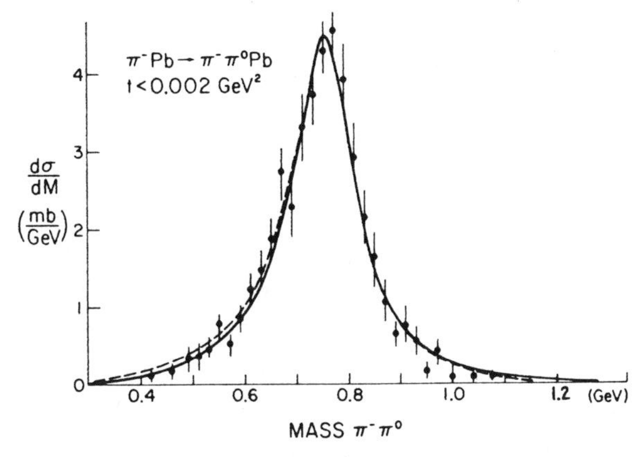

FIGURE 5.13. Distribution of the $\pi^-\pi^0$ effective mass indicates the existence of a short-lived particle, the ρ-meson. Note that the ρ-state has a finite width. [From D. Berg *et al.*, *Phys. Rev. Lett.* **44,** 706 (1980)].

Equation (5.47c) describes the shape of the energy distribution of a decaying state and is known as a Lorentzian curve. It peaks at $E=E_0$, the central value, and its full width at half maximum is $\Delta E=\Gamma$. Such a curve is shown in Fig. 5.13 for the ρ-meson, an elementary particle that decays rapidly into two π-mesons. The figure [from D. Berg *et al.*, *Phys. Rev. Lett.* **44,** 706 (1980)] shows the distribution in the energy (effective mass) of the π^- and π^0 mesons. The distribution is centered at $E_0=750$ MeV, which specifies the mass of the ρ-meson and has a full width $\Gamma \sim 150$ MeV. This indicates that the ρ-meson decays rapidly, its mean lifetime being only $\tau=\hbar/\Gamma \sim 0.5\times 10^{-23}$ sec. Similar distributions are obtained for any decaying state.

Problems

PROBLEM 1

A hydrogen atom in its ground state is subject to a time-dependent electric field of the form

$$\mathscr{E} = 0 \qquad t < 0$$
$$\mathscr{E} = \mathscr{E}_0 e^{-t/\tau}, \qquad t > 0$$

which gives rise to a perturbation Hamiltonian

$$\hat{H}' = (e\mathscr{E}) r \cos\theta$$

(*a*) Using *perturbation theory* find the probability that as $t \to \infty$ the atom is in the $n = 2$, $l = 1$, $m = 0$ state. The wavefunctions are given by Eqs. (8.32).

(*b*) Express your result in terms of the total energy of the perturbing field

$$W = \pi a^2 c \int_0^\infty |\hat{H}'|^2 \, dt$$

and discuss the relation between the time duration τ of the electric field pulse and the transition probability for fixed total power. Can you interpret this result in physical terms?

PROBLEM 2

A one-dimensional oscillator in its ground state is subjected to a perturbation of the form

$$H'(t) = C\hat{p}e^{-\alpha|t|} \cos\omega t, \qquad C, \alpha, \omega \text{ are constants}$$

Here $\hat{p}$ is the momentum operator $\hat{p} = -i(m\omega_c \hbar/2)(\hat{a} - \hat{a}^\dagger)$.

What is the probability that at $t \to \infty$ the oscillator will be found in its first excited state in first-order perturbation theory? Discuss the result as a function of α, ω, and ω_c (the oscillator frequency). The wavefunctions for the ground state and the first excited state are given by Eqs. (9.27).

PROBLEM 3

A particle is in the ground state of the one-dimensional potential

$$V(x)=\begin{cases}\infty, & x<-a\\ 0, & -a\leqslant x\leqslant a\\ \infty, & x>a\end{cases}$$

A time-dependent perturbation acts on it and has the form

$$\hat{H}'=T_0V_0\sin\frac{\pi x}{a}\,\delta(t),\qquad T_0,\ V_0\text{ constants}$$

What is the probability that the system is in the first excited state afterward? The wavefunctions for the ground state and the first excited state can be obtained from Eqs. (1.78).

PROBLEM 4

The energy of the nth stationary state of the hydrogen atom is

$$E_n=-\frac{m_e^2(e^2/4\pi\varepsilon_0)^2}{2\hbar^2}\frac{1}{n^2}$$

Show that in the limit of large n the frequency of the radiation emitted in a transition from the $(n+1)$th state to the nth state $\omega=(E_{n+1}-E_n)/\hbar$ equals the angular frequency of rotation of an electron of energy E_n moving classically in a circular orbit around the proton.

Note: This is an example of the correspondence principle.

PROBLEM 5

From the data represented by Fig. 5.11(c) and given that the total sweep width was of order $\Delta B = \pm 2.5$ G

(*a*) Estimate the width of the absorption line in gauss, and from this give the fractional inhomogeneity of the field over the area of the sample.

(*b*) From the decay of the wiggles estimate the transverse relaxation time T_2.

Chapter 6

THE COORDINATE REPRESENTATION

We have learned how to describe a quantum-mechanical system by a state in Hilbert space and have introduced the notion of operators that can act on these states. In particular, we have studied the time evolution of states and have found that the eigenstates of the energy operator are stationary. Even though we discussed several physical examples we still have not developed a systematic method for finding the Hamiltonian operator and its eigenstates for any given quantum-mechanical system. In fact, this is a very difficult problem, and there is no unique prescription for obtaining a solution.

A very useful method for finding the stationary states of a quantum-mechanical system is to use the coordinate representation, i.e., the representation in which the position eigenstates are the basis states. As discussed in Section 3.4, in this representation the amplitude for finding a system described by the state $|\psi\rangle$ at the position $\mathbf{x}$ is a continuous function of the coordinates known as the wavefunction $\psi(\mathbf{x}, t) = \langle \mathbf{x} \mid \psi \rangle$. The usefulness of the coordinate representation is related to the fact that classical systems are most frequently described by functions of the coordinates, for instance the trajectory of a particle moving freely or under the influence of forces. Thus, a fairly close analogy can be established between the classical and quantum descriptions of a physical system. Often the quantum-mechanical Hamiltonian operator can be obtained from the classical Hamiltonian by a well-defined transcription of classical variables into operators. In the coordinate representation the equation of motion, introduced in Eq. (4.12), becomes the celebrated Schrödinger equation.† This is a differential equation that the wavefunction $\psi(\mathbf{x}, t)$ must obey, and it can be thought of as an

† By extension, analogy, or habit the equation of motion in its general form of Eq. (4.12) is also often referred to as the Schrödinger equation.

eigenvalue problem. The wavefunctions for the stationary states are the solutions of this equation satisfying suitable physical boundary conditions. The Schrödinger equation has been and remains a very powerful tool for studying quantum systems. We will use it extensively in our study of physical applications. However, when the system possesses degrees of freedom that have no classical analogue, it can no longer be completely described in the coordinate representation, and we have to take recourse to more general methods.

As indicated by the title, this chapter is devoted to the coordinate representation of states and operators. We begin by showing that observables that can be simultaneously determined correspond to operators that commute. Most operators in quantum mechanics, however, are noncommuting. We then introduce the basic commutator between canonical operators $\hat{x}$ and $\hat{p}_x$, and from it deduce the coordinate representation for the momentum operator. This allows us to discuss the uncertainty relations for pairs of canonically conjugate observables. In Section 6.3 we give the coordinate representation of operators. These are either multiplicative operators such as the position operator, or they take the form of differential operators as in the case of the momentum. Next we discuss the time rate of change of the expectation values of the quantum-mechanical operators (Ehrenfest theorem) and emphasize again the correspondence principle. We then apply this theorem to several simple examples to bring out its essential character.

Next we introduce the Schrödinger equation both in its time-dependent and time-independent forms. We point out how under appropriate conditions this simple equation leads to quantized values for the energy of a bound system. We conclude the chapter with applications to two physical systems that can be adequately described by the one-dimensional Schrödinger equation. The first application is the motion of a particle in a periodic potential. This leads to band structure. In other words, there are distinct regions (bands) of energy that the particle is allowed to occupy; this has observable effects in the motion of electrons in crystalline materials. As a second application we treat the α-decay of heavy nuclei where the α-particle penetrates through a potential barrier. Classically this would have been impossible, but in quantum mechanics there is a small but finite probability for this *tunneling* to occur. Application of the Schrödinger equation to more conventional problems such as the hydrogen atom are taken up in Chapter 8 where we begin a systematic study of systems that form bound states.

6.1. COMPATIBLE OBSERVABLES

Physical observables are represented by hermitian operators, and the result of any measurement yields one of the eigenvalues of the corresponding operator. Operators can be represented by square matrices acting on column or row vectors. Furthermore, we know that in a representation where the basis states are the eigenstates of the operator $\hat{A}$, the operator $\hat{A}$ is diagonal.

In particular, the eigenstates of the energy operator $\hat{H}$ are stationary states, hence in order to describe a quantum system it is advantageous to choose the energy eigenstates as the basis states. These can be labeled by an index $1, 2, 3, \ldots, n$, corresponding to different energy values $E_1, E_2, E_3, \ldots, E_n$. However, it is possible that corresponding to a particular energy E_n there exist several states. For example, in the absence of a magnetic field the two orientations of the proton spin have the same energy but correspond to two different states. In such cases we need a second index (or more) to distinguish between the degenerate stationary states

$$|\psi_{n\alpha}\rangle$$

The index n refers to a particular eigenvalue of the energy operator. The supplemental index α must therefore correspond to some other observable of the system (for example, to the projection of the spin) and must refer to the eigenvalues of some hermitian operator other than the energy. The fact that we can label a state of the system by the eigenvalues of two (or more) different operators implies that we can measure both (or all such) observables simultaneously. Observables that can be measured simultaneously are called *compatible*.

Let $|\psi_{n\nu}\rangle$ be the simultaneous eigenstate of the operators $\hat{A}$ and $\hat{B}$ with eigenvalues a_n and b_ν. Thus

$$\begin{aligned} \hat{A}\,|\psi_{n\nu}\rangle &= a_n\,|\psi_{n\nu}\rangle \\ \hat{B}\,|\psi_{n\nu}\rangle &= b_\nu\,|\psi_{n\nu}\rangle \end{aligned} \tag{6.1a}$$

for all values of the indices n and ν. Clearly, the observables corresponding to the operators $\hat{A}$ and $\hat{B}$ are compatible. From Eq. (6.1a) we note that

$$\hat{A}\hat{B}\,|\psi_{n\nu}\rangle = \hat{A}(\hat{B}\,|\psi_{n\nu}\rangle) = \hat{A}(b_\nu\,|\psi_{n\nu}\rangle) = b_\nu\hat{A}\,|\psi_{n\nu}\rangle = a_n b_\nu\,|\psi_{n\nu}\rangle$$
$$\hat{B}\hat{A}\,|\psi_{n\nu}\rangle = \hat{B}(\hat{A}\,|\psi_{n\nu}\rangle) = \hat{B}(a_n\,|\psi_{n\nu}\rangle) = a_n\hat{B}\,|\psi_{n\nu}\rangle = a_n b_\nu\,|\psi_{n\nu}\rangle \tag{6.1b}$$

It follows that

$$(\hat{A}\hat{B} - \hat{B}\hat{A})\,|\psi_{n\nu}\rangle = 0 \tag{6.1c}$$

and since this relation holds for all values of n and ν we conclude that

$$\hat{A}\hat{B} - \hat{B}\hat{A} = [\hat{A}, \hat{B}] = 0 \tag{6.2}$$

In other words, when two operators commute the corresponding observables are compatible. If they do not commute a simultaneous eigenstate is not impossible, but is highly accidental. However, when they commute there are enough simultaneous eigenstates to form a complete set. Therefore, the basic problem in quantum mechanics lies in both establishing the eigenvalues of all compatible observables of a system and in finding the corresponding eigenstates.

Equation (6.2) has a direct interpretation when we represent operators by matrices. In general, two matrices do not commute. However, two diagonal matrices always commute. That is, if

$$\hat{M} = \begin{pmatrix} m_1 & & 0 \\ & m_2 & \\ 0 & & \ddots \end{pmatrix} \quad \text{and} \quad \hat{N} = \begin{pmatrix} n_1 & & 0 \\ & n_2 & \\ 0 & & \ddots \end{pmatrix}$$

Then

$$\hat{M}\hat{N} = \begin{pmatrix} m_1 n_1 & & 0 \\ & m_2 n_2 & \\ 0 & & \ddots \end{pmatrix} = \hat{N}\hat{M} \tag{6.3}$$

Therefore, if two operators commute there exists a representation in which the matrices for both the operators are simultaneously diagonal.

As an example, consider the Pauli matrices which can be used to represent the operator that measures the projection of the electron's spin along a particular axis. We cannot specify simultaneously the projection of the spin along, say, the X- and Y-axes since

$$\hat{\sigma}_x\hat{\sigma}_y - \hat{\sigma}_y\hat{\sigma}_x = [\hat{\sigma}_x, \hat{\sigma}_y] \neq 0 \tag{6.4}$$

On the other hand, we can specify the energy of the electron and the projection of its spin, say along the X-axis, because in the absence of a magnetic field and if the electron is at rest, then

$$\hat{H} = m_e c^2 \mathbb{1}$$

and therefore

$$[\hat{H}, \hat{\sigma}_x] = 0 \tag{6.5a}$$

If a magnetic field directed along the X-axis is present, the energy operator becomes

$$\hat{H}' = m_e c^2 \mathbb{1} - \mu B \hat{\sigma}_x \tag{6.5b}$$

However, note that

$$[\hat{H}', \hat{\sigma}_x] = 0 \tag{6.5c}$$

This shows that even in this case the energy as well as the projection of the spin along the X-axis can be simultaneously measured.

From the example above we note that when no magnetic field is present, specifying the energy of the system, does not completely determine the state of the electron because the two spin-projection states are degenerate. We must use the spin-projection eigenvalue to write

$$|\psi_{E=m_e c^2, \sigma_x=1}\rangle \quad \text{and} \quad |\psi_{E=m_e c^2, \sigma_x=-1}\rangle \tag{6.6a}$$

in order to distinguish the states. On the other hand, the magnetic field lifts the degeneracy, and we can specify the states of the system simply by giving their energy

$$|\psi_{E=m_e c^2-\mu B}\rangle \quad \text{and} \quad |\psi_{E=m_e c^2+\mu B}\rangle \tag{6.6b}$$

However, it is much more convenient and natural to retain the notation of Eq. (6.6a), where we label states by the eigenvalues of all the compatible observables of the system.

Finally, we note that compatibility of two or more observables is a property of the corresponding operators. It depends on the system under consideration only insofar as a particular operator, say the energy matrix, is different for different systems.

6.2. QUANTUM CONDITIONS AND THE UNCERTAINTY RELATION

We have argued—and this is supported by experiment—that it is not possible to determine simultaneously all the physical observables of a microscopic system. For example, we cannot determine simultaneously and with unlimited precision the position and the momentum of an electron. This implies that x and p_x are not compatible observables and, therefore, the operators $\hat{x}$ and $\hat{p}_x$, which represent the measurement of the x-component of position and the x-component of momentum, do not commute. That is,

$$[\hat{x}, \hat{p}_x] \neq 0 \tag{6.7a}$$

Furthermore, if the uncertainty in the measurement of x and p_x is given by the Heisenberg relation

$$\Delta x \, \Delta p_x \geqslant \frac{\hbar}{2}$$

then the commutator in Eq. (6.7a) must have the form

$$[\hat{x}, \hat{p}_x] = i\hbar \tag{6.7b}$$

This postulate determines the structure of quantum mechanics and is referred to as the *quantum condition.* Any physical theory that involves commutation relations such as Eq. (6.7b) is called a *quantized theory.* We note here that the factor i in Eq. (6.7b) is a result of the fact that both $\hat{x}$ and $\hat{p}_x$ are hermitian operators so that

$$[\hat{x}, \hat{p}_x]^\dagger = (\hat{x}\hat{p}_x - \hat{p}_x\hat{x})^\dagger = \hat{p}_x\hat{x} - \hat{x}\hat{p}_x = -[\hat{x}, \hat{p}_x] \tag{6.7c}$$

Thus, the rhs of Eq. (6.7b) must also change sign under hermitian conjugation. Planck's constant (divided by 2π), $\hbar$, measures the noncompatibility of x and p_x. In classical physics where the magnitude of the quantities involved is much larger, the observables x and p_x become compatible as a consequence of the smallness of $\hbar$.

The quantum condition of Eq. (6.7b) can be justified by analogy to classical mechanics. Here we will only sketch the argument. A complete derivation is given in Appendix 5. A fundamental relation of classical mechanics is the Poisson bracket: for any two functions

$A(x_i, p_i)$ and $B(x_i, p_i)$ of the canonical variables, this is defined as

$$\{A, B\} = \sum_i \left(\frac{\partial A}{\partial x_i}\frac{\partial B}{\partial p_i} - \frac{\partial B}{\partial x_i}\frac{\partial A}{\partial p_i}\right) \tag{6.8a}$$

If we now think of the functions A and B as corresponding to physical observables, then in quantum mechanics they will be represented by operators and we can attempt to express the Poisson bracket in terms of these operators. To do this we must preserve the order in which the operators act, and as shown in Appendix 5 this leads to a proportionality relation

$$[\hat{A}, \hat{B}] = (\hat{A}\hat{B} - \hat{B}\hat{A}) \propto \{\hat{A}, \hat{B}\}_Q \tag{6.8b}$$

where $\{\hat{A}, \hat{B}\}_Q$ is the quantum equivalent of the Poisson bracket of Eq. (6.8a). The proportionality constant in Eq. (6.8b) is set equal to $i\hbar$ and we assume that $\{\hat{A}, \hat{B}\}_Q$ has the same value as the corresponding classical Poisson bracket. Classically

$$\{x_r, x_s\} = 0 \qquad \{p_r, p_s\} = 0 \qquad \{x_r, p_s\} = \delta_{rs} \tag{6.9a}$$

as can be easily shown by a direct evaluation of the rhs of Eq. (6.8a). From this it follows that the quantum-mechanical operators will satisfy

$$[\hat{x}_r, \hat{x}_s] = 0 \qquad [\hat{p}_r, \hat{p}_s] = 0 \qquad [\hat{x}_r, \hat{p}_s] = i\hbar\, \delta_{rs} \tag{6.9b}$$

Once these fundamental quantum conditions are obtained, it is straightforward to calculate the commutator of two operators that depend on $\hat{x}$ and $\hat{p}_x$. Equations (6.8) and (6.9b) give a prescription for obtaining the quantum conditions in the case of operators that have a classical analogue. However, note that in quantum mechanics there exist systems with operators that do not have any classical analogue and we have to use other arguments to obtain their commutation relations. Note also that Eqs. (6.8) and (6.9b) show that classical mechanics may be thought of as a limiting case of quantum mechanics when $\hbar \to 0$.

The Uncertainty Relation

We have seen that if two operators commute, then the corresponding observables can be simultaneously measured. The uncertainty in measurement arises only when two operators do not commute. Let us consider a general commutator between two hermitian operators $\hat{A}$ and $\hat{B}$ such that

$$[\hat{A}, \hat{B}] = i\hat{C} \tag{6.10}$$

where $\hat{C}$ is another hermitian operator. The average value (the expectation value) of the operators $\hat{A}$ and $\hat{B}$ in a particular quantum state $|\psi\rangle$ will be denoted by $\langle\hat{A}\rangle$ and $\langle\hat{B}\rangle$. By analogy with statistical arguments one can also define the variance, or mean square deviation, as

$$\begin{aligned}(\Delta A)^2 &= \langle\hat{A}^2\rangle - \langle\hat{A}\rangle^2 = \langle(\hat{A} - \langle\hat{A}\rangle)^2\rangle \\ (\Delta B)^2 &= \langle\hat{B}^2\rangle - \langle\hat{B}\rangle^2 = \langle(\hat{B} - \langle\hat{B}\rangle)^2\rangle\end{aligned} \tag{6.11}$$

Furthermore, let us define the shifted operators

$$\begin{aligned}\tilde{A} &= \hat{A} - \langle\hat{A}\rangle \\ \tilde{B} &= \hat{B} - \langle\hat{B}\rangle\end{aligned}$$

which have zero average value in the state $|\psi\rangle$. Since $\langle\hat{A}\rangle$ and $\langle\hat{B}\rangle$ are numbers, their commutator is zero and hence Eq. (6.10) retains the same value for the shifted operators

$$[\tilde{A}, \tilde{B}] = i\hat{C} \tag{6.12a}$$

Furthermore, it is clear from Eq. (6.11) that $(\Delta A)^2 = \langle\tilde{A}^2\rangle$, $(\Delta B)^2 = \langle\tilde{B}^2\rangle$. Thus the product becomes

$$(\Delta A)^2(\Delta B)^2 = \langle\tilde{A}^2\rangle\langle\tilde{B}^2\rangle \tag{6.12b}$$

At this point we make use of a mathematical theorem, the Schwartz inequality, which states that for any two complex functions

$$\int |f|^2\,dx \int |g|^2\,dx \geqslant \left|\int f^* g\,dx\right|^2$$

when we apply this inequality to the expectation value of the operators $\tilde{A}^2$ and $\tilde{B}^2$ we obtain

$$\langle\tilde{A}^2\rangle\langle\tilde{B}^2\rangle \geqslant |\langle\tilde{A}\tilde{B}\rangle|^2 \tag{6.12c}$$

We can use Eq. (6.12b) in the lhs of the above result and expand the rhs, so that

$$\begin{aligned}(\Delta A)^2(\Delta B)^2 \geqslant |\langle\tilde{A}\tilde{B}\rangle|^2 &= \left|\left\langle\frac{\tilde{A}\tilde{B} + \tilde{B}\tilde{A}}{2} + \frac{\tilde{A}\tilde{B} - \tilde{B}\tilde{A}}{2}\right\rangle\right|^2 \\ &= \left|\left\langle\frac{\tilde{A}\tilde{B} + \tilde{B}\tilde{A}}{2} + i\frac{\hat{C}}{2}\right\rangle\right|^2\end{aligned} \tag{6.12d}$$

where we use the commutator of Eq. (6.12a). The first term on the rhs is a hermitian operator, and hence its expectation value is real; thus the square of the expectation value is positive definite. Therefore, we can write

$$(\Delta A)^2(\Delta B)^2 \geqslant \tfrac{1}{4}|\langle \tilde{A}\tilde{B}+\tilde{B}\tilde{A}\rangle|^2+\tfrac{1}{4}|\langle \hat{C}\rangle|^2 \geqslant \tfrac{1}{4}|\langle \hat{C}\rangle|^2$$

or

$$(\Delta A)(\Delta B) \geqslant \tfrac{1}{2}|\langle \hat{C}\rangle| \tag{6.13a}$$

We immediately see that if $\hat{A}=\hat{x}$ and $\hat{B}=\hat{p}_x$, so that $\hat{C}=\hbar\mathbb{1}$, then

$$\Delta x\, \Delta p_x \geqslant \frac{\hbar}{2} \tag{6.13b}$$

which is nothing other than Heisenberg's uncertainty relation. Thus, the quantum condition of Eq. (6.7b) implies directly the uncertainty relation of Eq. (6.13b). Simple examples of uncertainty relations were already given in Sections 1.4 and 1.5.

6.3. COORDINATE REPRESENTATION OF OPERATORS

Just as a quantum-mechanical state can be expanded in any complete set of basis states, quantum-mechanical operators can be similarly written in any representation. In particular, in the coordinate representation an arbitrary state $|\psi\rangle$ can be expanded as in Eq. (3.26a)

$$|\psi\rangle = \int dx\, |x\rangle\langle x \mid \psi\rangle = \int dx\, \psi(x, t)\, |x\rangle \tag{6.14}$$

The expansion coefficients $\psi(x, t)$ of a quantum-mechanical state are the probability amplitudes for finding the system at the position x and are functions of the coordinates and of time. $\psi(x, t)$ is the *wavefunction* corresponding to the state $|\psi\rangle$. An operator $\hat{A}$ can also be expanded in terms of a complete set of coordinate basis states. Here the expansion coefficients are nothing other than the coordinate representation of the operator $\hat{A}$. Since these expansion coefficients have to act on the wavefunction, which is a function of the coordinates and time, the operators in the coordinate representation are differential operators that act on coordinates and time. Because classical mechanics is formulated using functions of the coordinates, the closest analogy is found in the coordinate representation of quantum mechanics.

The exact form of an operator in the coordinate representation can be obtained from the quantum conditions. For example, in the coordinate representation the position operator $\hat{x}$ is simply represented by x with the meaning that all functions to the right of x are multiplied by x.

$$\hat{x} \to x \quad \text{and} \quad (\hat{x})^\dagger \to x \tag{6.15}$$

where $\hat{x}$ is a hermitian operator.

The momentum operator $\hat{p}_x$ must be chosen so that the quantum condition Eq. (6.7b) is satisfied. This determines $\hat{p}_x$ to be

$$\hat{p}_x \to -i\hbar \frac{\partial}{\partial x} \tag{6.16a}$$

where the partial derivative as usual acts on functions to the right of it. We can then explicitly evaluate the commutator in this representation

$$\begin{aligned}[\hat{x}, \hat{p}_x] &= \left[x, -i\hbar \frac{\partial}{\partial x}\right] \\ &= -i\hbar\left(x\frac{\partial}{\partial x} - \frac{\partial}{\partial x}x\right)\end{aligned} \tag{6.16b}$$

Equation (6.16b) represents an operator, and to evaluate it we let it act on an arbitrary function of the coordinates $f(x)$. That is,

$$\begin{aligned}[\hat{x}, \hat{p}_x]f(x) &= -i\hbar\left\{x\frac{\partial}{\partial x}f(x) - \frac{\partial}{\partial x}xf(x)\right\} \\ &= i\hbar f(x)\end{aligned} \tag{6.16c}$$

In other words, the effect of the commutator acting on a function $f(x)$ is to multiply it by a constant $i\hbar$. Thus, we conclude that†

$$[\hat{x}, \hat{p}_x] = i\hbar \tag{6.16d}$$

† The quantum condition does not uniquely fix the form of the momentum operator. However, the remaining arbitrariness can be absorbed into the unobservable phase of the basis states. We will choose the phase so that Eq. (6.16a) is satisfied.

The operator $\hat{p}_x$ is hermitian. This condition is satisfied by the form of Eq. (6.16a)

$$\begin{aligned} \hat{p}_x &= -i\hbar \frac{\partial}{\partial x} \\ (\hat{p}_x)^\dagger &= i\hbar \frac{\overleftarrow{\partial}}{\partial x} \end{aligned} \tag{6.17a}$$

where now the derivative acts to the *left*. We then consider the expression

$$\begin{aligned} f(x)[\hat{x}, (\hat{p}_x)^\dagger] &= f(x)\left[x, i\hbar \frac{\overleftarrow{\partial}}{\partial x}\right] \\ &= i\hbar\left\{(xf(x))\frac{\overleftarrow{\partial}}{\partial x} - f(x)\frac{\overleftarrow{\partial}}{\partial x}x\right\} \\ &= i\hbar\left\{\frac{\partial}{\partial x}(xf(x)) - \frac{\partial f}{\partial x}x\right\} \\ &= i\hbar f(x) \end{aligned} \tag{6.17b}$$

and thus we conclude that

$$[\hat{x}, (\hat{p}_x)^\dagger] = i\hbar = [\hat{x}, \hat{p}_x] \tag{6.18a}$$

Clearly

$$(\hat{p}_x)^\dagger = \hat{p}_x \tag{6.18b}$$

which shows that the momentum operator is self-adjoint and thus hermitian.

In three dimensions the momentum operator is given by

$$\hat{\mathbf{p}} = -i\hbar\boldsymbol{\nabla} \tag{6.19a}$$

a correspondence valid in any coordinate system. Similarly, the square of the momentum operator is given by

$$\hat{p}^2 = (\hat{\mathbf{p}})^2 = -\hbar^2\nabla^2 \tag{6.19b}$$

where ∇^2 is the familiar Laplace operator.

We can now construct the energy operator for a particle of mass m moving in a time-independent potential with potential energy given by $U(\mathbf{r}) = U(x, y, z)$. We know that the energy operator will correspond to the classical Hamiltonian as long as we do not need to take into account the internal degrees of freedom. For such a system the classical Hamiltonian is given by

$$H = \frac{\mathbf{p}^2}{2m} + U(x, y, z) \tag{6.20a}$$

Thus, in the coordinate representation, the quantum-mechanical energy operator is given by

$$\hat{H} = \frac{\hat{p}^2}{2m} + \hat{U} \rightarrow -\frac{\hbar^2}{2m}\nabla^2 + U(x, y, z) \tag{6.20b}$$

This is the energy operator that appears in Schrödinger equation.

We summarize our results for quantum-mechanical operators in the coordinate representation in Table 6.1. Note that we can obtain the coordinate representation of quantum-mechanical operators for all observables that are functions of the canonical coordinates and momenta by repeated application of the operators defined in the table. However, care is needed to define the product of two or more operators since the order of noncommuting operators is important in quantum mechanics.

Table 6.1. Coordinate Representation of Some Quantum-Mechanical Operators[a]

Observable	Operator	Coordinate representation
E	$\hat{H}$	$-(\hbar^2/2m)\nabla^2 + U(x, y, z)$
x	$\hat{x}$	x
y	$\hat{y}$	y
z	$\hat{z}$	z
p_x	$\hat{p}_x$	$-i\hbar(\partial/\partial x)$
p_y	$\hat{p}_y$	$-i\hbar(\partial/\partial y)$
p_z	$\hat{p}_z$	$-i\hbar(\partial/\partial z)$

[a] The energy operator refers to a particle of mass m in a region of time-independent potential with potential energy $U(x, y, z)$.

We will now construct the coordinate representation for the angular momentum operators from the basic operators we have already studied. Classically the angular momentum of a particle is given by

$$\mathbf{L} = \mathbf{r} \times \mathbf{p} \tag{6.21a}$$

and its projections along the three axes are

$$
\begin{aligned}
L_x &= yp_z - zp_y \\
L_y &= zp_x - xp_z \\
L_z &= xp_y - yp_x \\
L^2 &= \mathbf{L}^2 = L_x^2 + L_y^2 + L_z^2
\end{aligned}
\tag{6.21b}
$$

The quantum-mechanical operators, therefore, are given by

$$
\begin{aligned}
\hat{L}_x &\to -i\hbar\left(y\frac{\partial}{\partial z} - z\frac{\partial}{\partial y}\right) \\
\hat{L}_y &\to -i\hbar\left(z\frac{\partial}{\partial x} - x\frac{\partial}{\partial z}\right) \\
\hat{L}_z &\to -i\hbar\left(x\frac{\partial}{\partial y} - y\frac{\partial}{\partial x}\right)
\end{aligned}
\tag{6.21c}
$$

and

$$\hat{L}^2 = (\hat{L}_x)^2 + (\hat{L}_y)^2 + (\hat{L}_z)^2$$

We note here that construction of these operators involves no ambiguity since each component consists of products of commuting operators. Now, from the representation of Eq. (6.21*c*) we can evaluate the commutators between various components of the angular momentum operators. We simply give the results here

$$
\begin{aligned}
[\hat{L}_x, \hat{L}_y] &= i\hbar\hat{L}_z \\
[\hat{L}_y, \hat{L}_z] &= i\hbar\hat{L}_x \\
[\hat{L}_z, \hat{L}_x] &= i\hbar\hat{L}_y
\end{aligned}
\tag{6.22}
$$

and

$$[\hat{L}^2, \hat{L}_x] = [\hat{L}^2, \hat{L}_y] = [\hat{L}^2, \hat{L}_z] = 0$$

We will discuss these relations and their consequences in great detail in Chapter 7. But we note here that no two components of the angular momentum can be simultaneously determined since they do not commute. Furthermore, Eq. (6.22) represents relations between quantum-mechanical operators and is valid in any and all representations even though we have labeled the operators by x, y, and z.

6.4. TIME DEPENDENCE OF EXPECTATION VALUES: EHRENFEST'S THEOREM

We now examine the time variation of the expectation value of an operator $\hat{A}$ in a quantum-mechanical state $|\psi\rangle$. By definition, the expectation value is given by

$$\langle \hat{A} \rangle = \langle \psi | \hat{A} | \psi \rangle \tag{6.23a}$$

Therefore, the time variation is given by

$$\frac{d}{dt}\langle \hat{A} \rangle = \left(\frac{\partial}{\partial t}\langle \psi|\right)(\hat{A}\,|\psi\rangle) + \left\langle \psi \left| \frac{\partial \hat{A}}{\partial t} \right| \psi \right\rangle + (\langle \psi|\,\hat{A})\frac{\partial}{\partial t}|\psi\rangle \tag{6.23b}$$

If we use the equation of motion as given by Eq. (4.12)

$$i\hbar\frac{\partial}{\partial t}|\psi\rangle = \hat{H}\,|\psi\rangle$$

$$-i\hbar\frac{\partial}{\partial t}\langle \psi| = \langle \psi|\,\hat{H}$$

then Eq. (6.23b) becomes

$$\begin{aligned}
\frac{d}{dt}\langle \hat{A} \rangle &= -\frac{1}{i\hbar}\langle \psi|\,\hat{H}\hat{A}\,|\psi\rangle + \left\langle \psi \left| \frac{\partial \hat{A}}{\partial t} \right| \psi \right\rangle + \frac{1}{i\hbar}\langle \psi|\,\hat{A}\hat{H}\,|\psi\rangle \\
&= \frac{1}{i\hbar}\langle \psi|\,\hat{A}\hat{H} - \hat{H}\hat{A}\,|\psi\rangle + \left\langle \psi \left| \frac{\partial \hat{A}}{\partial t} \right| \psi \right\rangle \\
&= \frac{1}{i\hbar}\langle [\hat{A}, \hat{H}] \rangle + \left\langle \frac{\partial \hat{A}}{\partial t} \right\rangle
\end{aligned} \tag{6.24a}$$

Classically the time dependence of an observable A is given by

$$\frac{dA}{dt} = \{A, H\} + \frac{\partial A}{\partial t} \tag{6.24b}$$

where $\{\}$ is the classical Poisson bracket. If we now recall Eq. (6.8b), which states that the commutator of two operators is $i\hbar$ times the quantum Poisson bracket it is clear that Eq. (6.24a) is a statement of the correspondence principle. That is, the expectation values of operators in quantum mechanics obey laws similar to classical physics. This relation is known as *Ehrenfest's theorem* and has many useful applications.

It is important to remark at this point that the time dependence of states and operators in quantum mechanics can be chosen according to our convenience. The description we have followed so far is one where the operators are fixed in time and the states carry all the time dependence. This is known as the *Schrödinger picture.* In this case physical observables—that is, the expectation values of operators—acquire a time dependence because of the evolution of the state vectors.

On the other hand, we know that the time dependence of a state is given by

$$|\psi(t)\rangle = \hat{U}(t)\,|\psi(0)\rangle \tag{6.25}$$

as was shown in Chapter 4 [see, for instance, Eq. (4.1)]. Here $\hat{U}$ is a unitary operator. Thus, one can make a unitary transformation through the operator $\hat{U}^{-1}(t)$ on the quantum states in the Schrödinger picture such that the new quantum states become fixed in time. In this process, however, the operators pick up all the time dependence since the expectation value of operators are observables and are independent of the description one uses. This description of quantum mechanics is called the *Heisenberg picture.* It follows from Eq. (6.25) that since the states do not depend on time, one can write Eq. (6.24a) in this case strictly as an operator equation

$$\frac{d\hat{A}}{dt} = \frac{1}{i\hbar}[\hat{A}, \hat{H}] + \frac{\partial \hat{A}}{\partial t} \tag{6.26}$$

Although Eq. (6.26) is closer in spirit to the classical result of Eq. (6.24b), we will continue to use the Schrödinger picture.

Examples

As an application of Ehrenfest's theorem let us consider a particle moving in one dimension in a region where the position-dependent potential energy is $U(x)$. Thus

$$\hat{H} = \frac{\hat{p}_x^2}{2m} + \hat{U}(x) \tag{6.27a}$$

Noting that the operator $\hat{x}$ does not depend on time explicitly, the time rate of change of $\langle\hat{x}\rangle$ is given by

$$\frac{d}{dt}\langle\hat{x}\rangle=\frac{1}{i\hbar}\langle[\hat{x},\hat{H}]\rangle \tag{6.27b}$$

The commutator is easily evaluated to be

$$[\hat{x},\hat{H}]=\left[\hat{x},\frac{\hat{p}_x^2}{2m}+\hat{U}(x)\right]$$

$$=\left[\hat{x},\frac{\hat{p}_x^2}{2m}\right]=i\hbar\frac{\hat{p}_x}{m} \tag{6.28a}$$

Thus

$$\frac{d}{dt}\langle\hat{x}\rangle=\frac{1}{i\hbar}\left\langle i\hbar\frac{\hat{p}_x}{m}\right\rangle$$

$$=\frac{\langle\hat{p}_x\rangle}{m} \tag{6.28b}$$

a result we obtained by explicit calculation in the case of a wave packet [Eq. (1.73)]. Similarly, the time rate of change of $\langle\hat{p}_x\rangle$ is given by

$$\frac{d}{dt}\langle\hat{p}_x\rangle=\frac{1}{i\hbar}\langle[\hat{p}_x,\hat{H}]\rangle \tag{6.29a}$$

It is easy to evaluate the commutator in the coordinate representation

$$[\hat{p}_x,\hat{H}]=\left[-i\hbar\frac{\partial}{\partial x},-\frac{\hbar^2}{2m}\frac{\partial^2}{\partial x^2}+U(x)\right]$$

$$=-i\hbar\frac{\partial U}{\partial x} \tag{6.29b}$$

Substituting Eq. (6.29a) we obtain

$$\frac{d}{dt}\langle\hat{p}_x\rangle=\frac{1}{i\hbar}(-i\hbar)\left\langle\frac{\partial\hat{U}}{\partial x}\right\rangle$$

$$=\left\langle-\frac{\partial\hat{U}}{\partial x}\right\rangle \tag{6.29c}$$

The rhs of Eq. (6.29c) can be thought of as the classical force and we again note the classical correspondence.

It is interesting to point out here that if an operator does not have any explicit time dependence, then its expectation in an energy eigenstate is time independent since the quantum state in this case is stationary. That is, if $\partial\hat{A}/\partial t=0$, then for an energy eigenstate

$$\frac{d}{dt}\langle\hat{A}\rangle=\frac{1}{i\hbar}\langle[\hat{A},\hat{H}]\rangle=0 \tag{6.30}$$

Let us use this result in the case of the one-dimensional motion where the potential energy depends on position according to a simple power law. That is,

$$\hat{H}=\frac{\hat{p}_x^2}{2m}+\alpha\hat{x}^n \tag{6.31a}$$

Here α is a constant and n is an integer. Furthermore, let us evaluate the time rate of change of $\langle\hat{x}\hat{p}_x\rangle$ in an energy eigenstate. Clearly in this case

$$\frac{d}{dt}\langle\hat{x}\hat{p}_x\rangle=\frac{1}{i\hbar}\langle[\hat{x}\hat{p}_x,\hat{H}]\rangle=0 \tag{6.31b}$$

On the other hand, if we calculate the commutator in Eq. (6.31b)

$$\begin{aligned}
[\hat{x}\hat{p}_x,\hat{H}]&=\left[\hat{x}\hat{p}_x,\frac{\hat{p}_x^2}{2m}+\alpha\hat{x}^n\right]\\
&=\hat{x}\left[\hat{p}_x,\frac{\hat{p}_x^2}{2m}+\alpha\hat{x}^n\right]+\left[\hat{x},\frac{\hat{p}_x^2}{2m}+\alpha\hat{x}^n\right]\hat{p}_x\\
&=\hat{x}[\hat{p}_x,\alpha\hat{x}^n]+\left[\hat{x},\frac{\hat{p}_x^2}{2m}\right]\hat{p}_x\\
&=-i\hbar n\alpha\hat{x}^n+i\hbar\frac{\hat{p}_x^2}{m}\\
&=i\hbar(-n\hat{U}+2\hat{T})
\end{aligned} \tag{6.32a}$$

Here $\hat{U}$ is the potential energy operator and $\hat{T}$ is the kinetic energy operator. Substituting this last result into Eq. (6.31*b*) we have

$$i\hbar\langle -n\hat{U}+2\hat{T}\rangle=0$$

or

$$\langle \hat{T}\rangle=\frac{n}{2}\langle \hat{U}\rangle \tag{6.32b}$$

This is the quantum-mechanical analog of the virial theorem.

As a final example of the Ehrenfest theorem let us consider an electron at rest in a region of space where a constant magnetic field exists along the Z-axis. The Hamiltonian is

$$\hat{H}=m_e c^2 \mathbb{1}-\mu B\hat{\sigma}_z \tag{6.33a}$$

We obtain the time rate of change of the expectation value of $\hat{\sigma}_x$ through

$$\begin{aligned}\frac{d}{dt}\langle\hat{\sigma}_x\rangle &= \frac{1}{i\hbar}\langle[\hat{\sigma}_x,\hat{H}]\rangle \\ &= \frac{1}{i\hbar}\langle[\hat{\sigma}_x, mc^2\mathbb{1}-\mu B\hat{\sigma}_z]\rangle \\ &= \frac{1}{i\hbar}\langle[\hat{\sigma}_x, -\mu B\hat{\sigma}_z]\rangle \\ &= \frac{2i\mu B}{i\hbar}\langle\hat{\sigma}_y\rangle = 2\omega\langle\hat{\sigma}_y\rangle, \qquad \omega=\frac{\mu B}{\hbar}\end{aligned} \tag{6.33b}$$

If at $t=0$ the system is in an eigenstate of $\hat{\sigma}_z$, say in the $|+z\rangle$ state, then

$$|\psi(t)\rangle = e^{-im_e c^2 t/\hbar}\begin{pmatrix} e^{+i\omega t} \\ 0\end{pmatrix} \tag{6.34a}$$

Clearly then

$$\frac{d}{dt}\langle\hat{\sigma}_x\rangle = 2\omega\langle\hat{\sigma}_y\rangle=(2\omega)\begin{pmatrix} e^{-i\omega t} & 0\end{pmatrix}\begin{pmatrix} 0 & -i \\ i & 0\end{pmatrix}\begin{pmatrix} e^{i\omega t} \\ 0\end{pmatrix}=0 \tag{6.34b}$$

This result is in agreement with what we already know, since in this case

$$[\hat{\sigma}_z, \hat{H}] = 0$$

An eigenstate of $\hat{\sigma}_z$ is also an eigenstate of the Hamiltonian. As we have noted before, expectation values in a stationary state are independent of time.

Let us now consider the case where the system, initially, is in an eigenstate of $\hat{\sigma}_x$, namely, in the $|+x\rangle$ state. Then

$$|\psi(t)\rangle = \frac{e^{-im_e c^2 t/\hbar}}{2^{1/2}} \begin{pmatrix} e^{i\omega t} \\ e^{-i\omega t} \end{pmatrix} \tag{6.35a}$$

Consequently, in this case

$$\begin{aligned} \frac{d}{dt}\langle\hat{\sigma}_x\rangle &= 2\omega\langle\hat{\sigma}_y\rangle = (2\omega)\tfrac{1}{2}\,(e^{-i\omega t}, e^{i\omega t}) \begin{pmatrix} 0 & -i \\ i & 0 \end{pmatrix} \begin{pmatrix} e^{i\omega t} \\ e^{-i\omega t} \end{pmatrix} \\ &= (2\omega)\tfrac{1}{2}(-ie^{-2i\omega t} + ie^{2i\omega t}) \\ &= -2\omega \sin 2\omega t \end{aligned} \tag{6.35b}$$

To show that this result agrees with the classical analogy we first evaluate $\langle\hat{\sigma}_x\rangle$. Just as in Eq. (6.35*b*) we find

$$\langle\sigma_x\rangle = \cos 2\omega t \tag{6.35c}$$

which is consistent with our expectations. If the system at $t=0$ was in the $|+x\rangle$ state, then the probability amplitude that at time t the system would be in the $|+x\rangle$ state is given by $\langle\psi(t)\,|\,+x\rangle$

$$\tfrac{1}{2}(e^{-i\omega t}, e^{i\omega t}) \begin{pmatrix} 1 \\ 1 \end{pmatrix} = \tfrac{1}{2}(e^{-i\omega t} + e^{i\omega t}) = \cos \omega t \tag{6.36a}$$

Thus, the probability that the spin would be along $+X$-axis is

$$P(+x) = \cos^2 \omega t$$

and (6.36*b*)

$$P(-x) = 1 - P(+x) = 1 - \cos^2 \omega t = \sin^2 \omega t$$

The expectation value of $\hat{\sigma}_x$ is the weighted average for the possible eigenvalues of the $\hat{\sigma}_x$ operator. That is,

$$\langle\hat{\sigma}_x\rangle = \frac{(+1)P(+x) + (-1)P(-x)}{P(+x) + P(-x)} = \cos^2 \omega t - \sin^2 \omega t = \cos 2\omega t$$

which agrees with the direct calculation given in Eq. (6.35c). Given this result we then differentiate it with respect to time

$$\frac{d}{dt}\langle\hat{\sigma}_x\rangle = \frac{d}{dt}(\cos 2\omega t) = -2\omega \sin \omega t$$

which is in agreement with the conclusion obtained from the Ehrenfest theorem, as given in Eq. (6.35b).

6.5. THE SCHRÖDINGER EQUATION

The quantum-mechanical equation of motion given in Chapter 4 [Eq. (4.12)] is

$$i\hbar\frac{\partial}{\partial t}|\psi\rangle = \hat{H}|\psi\rangle \tag{6.37a}$$

If we take the inner product with the bra $\langle\mathbf{x}|$, then the equation becomes

$$\begin{aligned} i\hbar\frac{\partial}{\partial t}\langle\mathbf{x}\mid\psi\rangle &= \langle\mathbf{x}|\hat{H}|\psi\rangle \\ &= \int d^3x'\langle\mathbf{x}|\hat{H}|\mathbf{x}'\rangle\langle\mathbf{x}'\mid\psi\rangle \end{aligned} \tag{6.37b}$$

We have already stated in Section 6.3 that the matrix element $\langle\mathbf{x}|\hat{H}|\mathbf{x}'\rangle$ is the coordinate representation of the Hamiltonian operator. That is,

$$\langle\mathbf{x}|\hat{H}|\mathbf{x}'\rangle = \delta^3(\mathbf{x}-\mathbf{x}')\left[-\frac{\hbar^2}{2m}\nabla^2 + U(\mathbf{x})\right] \tag{6.38a}$$

where $U(\mathbf{x})$ is the potential energy of a particle of mass m. While we cannot prove Eq. (6.38a) from first principles, we will adopt it as the correct expression for the matrix element of the Hamiltonian operator. Introducing it into Eq. (6.37b) we then obtain

$$\begin{aligned} i\hbar\frac{\partial}{\partial t}\langle\mathbf{x}\mid\psi\rangle &= \int d^3x'\delta^3(\mathbf{x}-\mathbf{x}')\tilde{H}\langle\mathbf{x}'\mid\psi\rangle \\ &= \tilde{H}\langle\mathbf{x}\mid\psi\rangle \end{aligned} \tag{6.38b}$$

where

$$\tilde{H} = -\frac{\hbar^2}{2m}\nabla^2 + U(\mathbf{x})$$

Note that $\langle \mathbf{x} \mid \psi \rangle = \psi(\mathbf{x}, t)$ is nothing other than the probability amplitude for finding the particle at $\mathbf{x}$ at time t. Thus, Eq. (6.38b) actually represents an equation for the time evolution of the wavefunction

$$\boxed{i\hbar \frac{\partial \psi}{\partial t}(\mathbf{x}, t) = \left[-\frac{\hbar^2}{2m}\nabla^2 + U(\mathbf{x})\right]\psi(\mathbf{x}, t)} \tag{6.39}$$

This is the celebrated *Schrödinger equation* which when subjected to appropriate boundary conditions allows us to determine the wavefunction of a quantum system.

We are, of course, interested in the stationary states of the system. These are eigenstates of the Hamiltonian. Hence

$$\hat{H} \mid \psi_n \rangle = E_n \mid \psi_n \rangle \tag{6.40a}$$

where E_n is the energy of the nth level. Correspondingly,

$$\langle \mathbf{x} \mid \hat{H} \mid \psi_n \rangle = E_n \langle \mathbf{x} \mid \psi_n \rangle = E_n \psi_n(\mathbf{x}, t) \tag{6.41b}$$

If we substitute this result into Eq. (6.37b)

$$i\hbar \frac{\partial}{\partial t}\langle \mathbf{x} \mid \psi_n \rangle = \langle \mathbf{x} \mid \hat{H} \mid \psi_n \rangle$$

we obtain

$$i\hbar \frac{\partial}{\partial t}\psi_n(\mathbf{x}, t) = E_n \psi_n(\mathbf{x}, t) \tag{6.42}$$

Equation (6.42) is a differential equation of first-order in time, and hence the solution is easily obtained to be

$$\psi_n(\mathbf{x}, t) = \phi_n(\mathbf{x}) e^{-(i/\hbar)E_n t} \tag{6.43a}$$

The above expression gives the wavefunction for a stationary state and $\phi_n(\mathbf{x})$ is the *space part* of the wavefunction. Note that $\phi_n(\mathbf{x})$ depends only on the spatial coordinates. The time dependence of the wavefunction for a stationary state of energy E_n is given by

$$e^{-(i/\hbar)E_n t}$$

which is in agreement with our observation in Eq. (3.12).

If we substitute the form of the wavefunction given in Eq. (6.43a) into Eq. (6.38b) we obtain the equation determining the space part $\phi_n(\mathbf{x})$ of the wavefunction

$$\tilde{H}\psi_n(\mathbf{x}, t) = i\hbar \frac{\partial \psi_n}{\partial t}(\mathbf{x}, t)$$

or

$$\boxed{\left[-\frac{\hbar^2}{2m}\nabla^2 + U(\mathbf{x})\right]\phi_n(\mathbf{x}) = E_n\phi_n(\mathbf{x})} \tag{6.43b}$$

This equation is known as the time-independent Schrödinger equation, and it is an eigenvalue equation since the differential operator $\tilde{H}$ acting on $\phi_n(\mathbf{x})$ simply multiplies it by a constant. To determine $\phi_n(\mathbf{x})$ completely in addition to knowing the form of $U(\mathbf{x})$, we must impose appropriate boundary conditions. For problems where the particle is sufficiently localized, the natural boundary condition is

$$\phi_n(\mathbf{x}) \xrightarrow[|\mathbf{x}|\to\infty]{} 0 \tag{6.44}$$

With such a boundary condition, Eq. (6.43b) admits solutions $\phi_n(\mathbf{x})$ with only discrete eigenvalues E_n.

Furthermore, we note that since $\psi_n(\mathbf{x}, t)$ represents the probability amplitude for finding the particle at $\mathbf{x}$ at time t, the probability of finding the particle in a spatial volume d^3x at $\mathbf{x}$ is given by

$$\begin{aligned} P_n(\mathbf{x}, t)\, d^3x &= |\psi_n(\mathbf{x}, t)|^2\, d^3x \\ &= |\phi_n(\mathbf{x})|^2\, d^3x = P_n(\mathbf{x})\, d^3x \end{aligned} \tag{6.45a}$$

That is, the probability of finding the particle at $\mathbf{x}$ in the volume d^3x is independent of time if the particle is in a stationary state. Furthermore, since the total probability must integrate to unity

$$\begin{aligned} \int P_n(\mathbf{x}, t)\, d^3x &= \int P_n(\mathbf{x})\, d^3x \\ &= \int |\phi_n(\mathbf{x})|^2\, d^3x = 1 \end{aligned} \tag{6.45b}$$

This is the usual normalization condition.

The Schrödinger equation provides us with a powerful mathematical tool for finding the stationary states of quantum systems. We will apply it extensively to several problems in the following sections and chapters. It should be clear, however, that since Eqs. (6.39) and (6.43*b*) are expressed in the coordinate representation they can describe only those physical observables of the system that have an analogue in classical physics.

Energy Quantization

One of the most characteristic features of quantum mechanics is that the energy of certain physical systems can take only discrete values; we say that the energy is quantized. For instance, we found this to be true for a particle confined in a square well potential as discussed in Section 1.8. This is a general property of the *stationary* solutions of the Schrödinger equation [Eq. (6.39)] when supplemented by the boundary condition introduced by Eq. (6.44), namely, when the particle is localized in space. The stationary solutions obey the time-independent Schrödinger equation [Eq. (6.43*b*)], and it is interesting that such a continuous differential equation can lead to discrete solutions. In many respects this is analogous to the discrete solutions for the vibrations of a string or of an air column in classical physics.

For simplicity, we will consider motion only in one dimension and, therefore, seek the solutions of the equation

$$\left[-\frac{\hbar^2}{2m}\frac{d^2}{dx^2}+U(x)\right]\phi_n(x)=E_n\phi_n(x) \tag{6.46}$$

where the time-independent potential $U(x)$ is attractive and is of the form shown in Fig. 6.1. The total energy E_n of the particle in the nth stationary state is indicated by the dashed line. We now examine the *topology* of Eq. (6.46) and note three regions of space: I$(x<x_1)$, II$(x_1<x<x_2)$, and III$(x>x_2)$. In each of these regions our equation has the following form

$$\text{I.} \qquad x<x_1 \qquad \frac{d^2\phi}{dx^2}=\frac{2m}{\hbar^2}[U(x)-E]\phi(x)\Rightarrow(+)\phi(x)$$

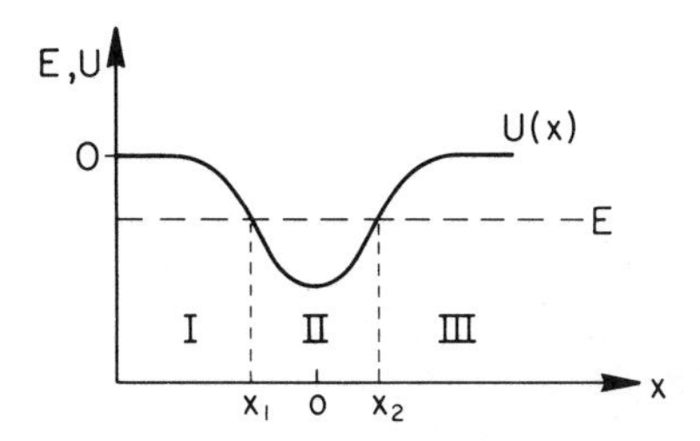

FIGURE 6.1. An attractive potential plotted as a function of x. For a range of values of the total energy E_n we recognize three distinct regions of space: (I) where $E_n < U$; (II) where $E_n > U$; and (III) where again $E_n < U$.

$$\text{II.} \quad x_1 < x < x_2 \qquad \frac{d^2\phi}{dx^2} = \frac{2m}{\hbar^2}[U(x) - E]\phi(x) \Rightarrow (-)\phi(x) \tag{6.47a}$$

$$\text{III.} \quad x_2 < x \qquad \frac{d^2\phi}{dx^2} = \frac{2m}{\hbar^2}[U(x) - E]\phi(x) \Rightarrow (+)\phi(x)$$

On the rhs of Eqs. (6.47a) the symbols (+) and (−) indicate that the second derivative $d^2\phi/dx^2$ [that is, the curvature of the function $\phi(x)$] must have the same or opposite sign as $\phi(x)$.

When $d^2\phi/dx^2 = (+)\phi(x)$ the solutions $\phi(x)$ are concave away from the X-axis: they have the same behavior as the exponentials $e^{\alpha x}$ or $e^{-\alpha x}$, where α is a real number (or function depending on x). When $d^2\phi/dx^2 = (-)\phi(x)$ the solutions are concave towards the axis: they have the same behavior as pieces of $\sin \beta x$ or $\cos \beta x$, where β is a real number (or function depending on x). If the particle is *bound* in the potential $U(x)$ the function $\phi(x)$ must vanish as $x \to \pm\infty$, as stated in Eq. (6.44), and therefore the solution must have the asymptotic form

$$\begin{aligned} \phi(x) &\to e^{+\alpha x} \qquad \text{for } x < x_1 \text{ as } x \to -\infty \\ \phi(x) &\to e^{-\alpha x} \qquad \text{for } x > x_2 \text{ as } x \to +\infty \end{aligned} \tag{6.47b}$$

In Figs. 6.2(a),(b) we have sketched the possible forms that $\phi(x)$ can take in these two regions. The heavy curves represent the only allowed forms that describe a bound state. In region II($x_1 < x < x_2$) the function $\phi(x)$ is concave towards the axis and can be oscillatory.

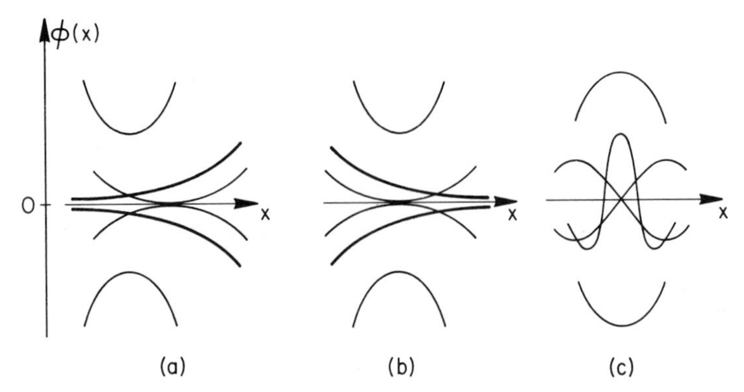

FIGURE 6.2. The permissible curvatures that the function $\phi(x)$ can take for different ranges of values of x. (a) corresponds to region I, (b) to region III, and (c) to region II. These regions are defined in the caption for Fig. 6.1.

Some possible forms are shown in Fig. 6.2(c). All of them are admissible solutions.†

We can try to obtain the complete solution of Eq. (6.46) by combining the separate solutions in each of the three regions. However, $\phi(x)$ and $d\phi/dx$ must be continuous‡ at x_1 and x_2. Suppose we pick a particular energy E_a as shown in Fig. 6.3(a). In region I, $\phi(x)$ will have the form of one of the heavy curves of Fig. 6.2(a) but is completely determined by our choice of E_a. When we continue $\phi(x)$ into region II we must use a particular piece from Fig. 6.2(c), and this is again completely determined by our choice of E_a. Finally, the continuation of $\phi(x)$ into region III is also completely determined and, in general, will *not* be of the form of the heavy curves of Fig. 6.2(b). This is shown in Fig. 6.3(a) and does not satisfy the boundary conditions. Next, we pick a different energy E_b as shown in Fig. 6.3(b). We determine $\phi(x)$ through the same procedure and $\phi(x)$ may again diverge in region III (i.e., for $x > x_2$) as shown in Fig. 6.3(b). It is only for a *particular* value of E, say E_n, that we can obtain a solution $\phi_n(x)$ that obeys Eq. (6.46) and has a form satisfying the boundary conditions. Such a solution is shown in Fig. 6.4(a).

† In drawing Fig. 6.2(c) we assumed that the potential was symmetric about $x = 0$.
‡ This is a property of the differential equation for a nonsingular $U(x)$.

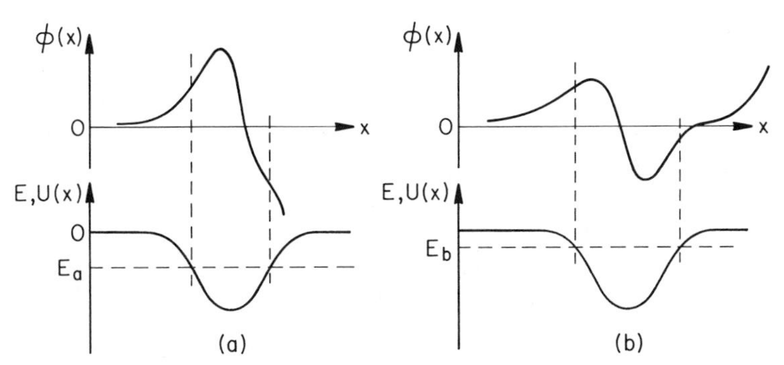

FIGURE 6.3. (a) The function $\phi(x)$ for some arbitrary total energy E_a will not satisfy the boundary condition. (b) Similarly, for some other value E_b, the boundary conditions are not satisfied. Note that the potential is also shown.

The particular value of the total energy E_n that leads to an admissible solution is a very special one because any small change from E_n will destroy the delicate balance that led to the perfect match of the three parts of the solution. Of course, for any particular potential there may be more than one (or no) values of E that give satisfactory solutions. For instance, a different possibility is shown in Fig. 6.4(b). However, the admissible energies E_n form a *discrete* spectrum. When the total energy E is larger than the potential $U(x)$ at large distances, the particle is not bound in the potential and all energies are allowed. This is the case when one considers the scattering of free particles from a potential.

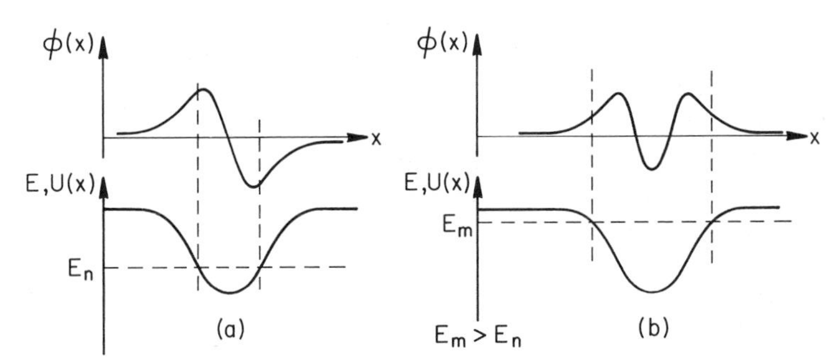

FIGURE 6.4. Only specific values of the total energy result in functions $\phi(x)$ that satisfy the boundary conditions. (a) The eigenvalue is E_n. (b) The eigenvalue $E_m > E_n$. Note that the function $\phi(x)$ in this case has more nodes than in the previous case.

We thus see that the stationary states of bound systems have a discrete spectrum of energies. This feature remains true when we solve the time-independent Schrödinger equation in three-dimensional space and will be demonstrated explicitly as we discuss specific applications.

6.6. THE PERIODIC POTENTIAL

As an interesting application of the Schrödinger equation we will examine the stationary states for a particle moving in a periodic potential. This situation arises in crystalline materials where the electrons are subject to the Coulomb potential of the nuclei or ions that are located at fixed sites. For simplicity we will consider only a one-dimensional lattice, but the method that is used is readily extended to three dimensions.

There are various forms of potentials that one could consider; for instance, square-well (as in Fig. 6.5), delta-function, Coulombic, etc. However, the general features of the solution are independent of the details of the potential. One finds that for the stationary states the energy of the electron can lie only in certain bands. It is this band structure that gives rise to the interesting electric properties of materials such as insulators and semiconductors.

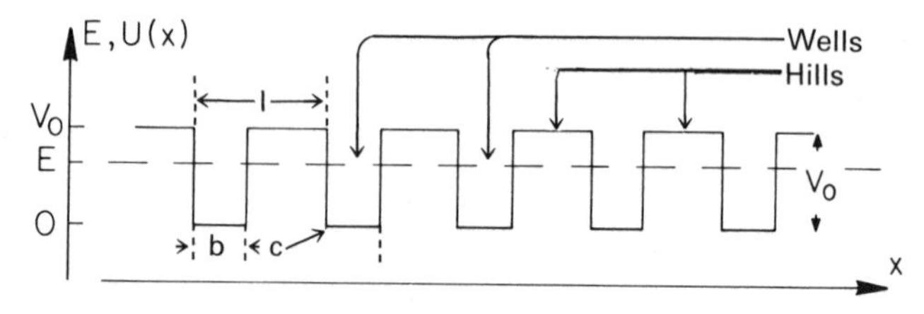

FIGURE 6.5. A periodic one-dimensional lattice of square-potential wells. The energy E of the stationary state is shown by the dashed line.

The space part of the electron's wavefunction will obey the time-independent Schrödinger equation [Eq. (6.43*b*)] in one dimension

$$\left[-\frac{\hbar^2}{2m_e}\frac{d^2}{dx^2}+U(x)\right]\psi_E(x)=E\psi_E(x) \tag{6.48}$$

where we use the subscript E to indicate the stationary state of this particular energy. The potential energy $U(x)$ is assumed periodic with a lattice spacing l, namely,

$$U(x+l)=U(x) \tag{6.49}$$

Furthermore, since we are interested in bound-state solutions, the total energy E must be less than the potential energy in certain regions of space.

Let us introduce a *translation* operation T_l such that when it acts on a function of the coordinates, it shifts the argument of the function from x to $x+l$. In other words,

$$T_l f(x)=f(x+l) \tag{6.50a}$$

It is clear that T_l acting on the Hamiltonian of Eq. (6.48) leaves it invariant. Thus, if we introduce a quantum-mechanical operator $\hat{T}_l$ to represent the translation operation of Eq. (6.50a), this operator will commute with the Hamiltonian, that is

$$[\hat{H}, \hat{T}_l]=0 \tag{6.50b}$$

We conclude from Eq. (6.2) that the two operators $\hat{H}$ and $\hat{T}_l$ are compatible, that is, the energy eigenstates must simultaneously be eigenstates of the translation operator $\hat{T}_l$.

The energy eigenstates are given by $|\psi_E\rangle$ with the wavefunction $\psi_E(x)$, and in view of our above conclusion it must hold that

$$\hat{T}_l\ |\psi_E(x)\rangle=|\psi_E(x+l)\rangle=\lambda\ |\psi_E(x)\rangle \tag{6.51a}$$

Because of the symmetry in the problem, the probability of finding the electron at x must be the same as it would at a translated point. Thus

$$|\psi_E(x)|^2=|\psi_E(x+l)|^2=|\hat{T}_l\psi_E(x)|^2 \tag{6.51b}$$

which determines that

$$\hat{T}_l\psi_E(x)=\lambda\psi_E(x) \qquad \text{where } \lambda=e^{i\delta} \tag{6.51c}$$

with δ real. Furthermore, δ must reflect the length l by which we translate the coordinates. A form of $\psi_{\mathrm{E}}(x)$ that satisfies Eqs. (6.51a)–(6.51c) is

$$\psi_{\mathrm{E}}(x)=e^{ikx}\phi_{\mathrm{E},k}(x) \tag{6.52a}$$

with

$$\phi_{\mathrm{E},k}(x+l)=\phi_{\mathrm{E},k}(x) \tag{6.52b}$$

Wavefunctions of the form of Eq. (6.52) are called Bloch functions. So far, the wave vector k is not restricted in any way, but is related to the eigenvalue λ since

$$\begin{aligned}\hat{T}_l\psi_{\mathrm{E}}(x)&=e^{ik(x+l)}\phi_{\mathrm{E},k}(x+l)\\&=e^{ikl}e^{ikx}\phi_{\mathrm{E},k}(x)=e^{ikl}\psi_{\mathrm{E}}(x)\end{aligned} \tag{6.52c}$$

We thus have shifted the problem from finding the solutions of Eq. (6.48) to one of finding the *periodic* solutions $\phi_{\mathrm{E},k}(x)$ defined by Eqs. (6.52a) and (6.52b). Introducing Eq. (6.52a) into Eq. (6.48) we obtain the differential equation obeyed by $\phi_{\mathrm{E},k}(x)$

$$\frac{d^2\phi}{dx^2}+2ik\frac{d\phi}{dx}+\frac{2m_e}{\hbar^2}\left[E-U(x)-\frac{\hbar^2k^2}{2m_e}\right]\phi(x)=0 \tag{6.53}$$

This appears to be a more complicated equation than Eq. (6.48), but in practice the periodicity of $\phi(x)=\phi(x+l)$ greatly simplifies the solution. We show this by a specific example.

We consider a square-well potential lattice as shown in Fig. 6.5. The depth of the well is V_0 and the width is b. The lattice spacing is $l=b+c$. The energy E of the stationary state is indicated by the dashed line. We introduce the notation

$$\alpha=\left(\frac{2m_e}{\hbar^2}E\right)^{1/2}\qquad\beta=\left[\frac{2m_e}{\hbar^2}(V_0-E)\right]^{1/2}$$

so that Eq. (6.53) takes the form

$$\frac{d^2\phi}{dx^2}+2ik\frac{d\phi}{dx}+(\alpha^2-k^2)\phi=0\qquad\text{in a well} \tag{6.54a}$$

$$\frac{d^2\phi}{dx^2}+2ik\frac{d\phi}{dx}-(\beta^2+k^2)\phi=0\qquad\text{in a hill} \tag{6.54b}$$

with solutions

$$\phi(x)=\begin{cases} A_n e^{i(\alpha-k)x}+B_n e^{-i(\alpha+k)x} & (6.55a) \\ C_n e^{(\beta-ik)x}+D_n e^{-(\beta+ik)x} & (6.55b) \end{cases}$$

Equation (6.55a), which is purely oscillatory, corresponds to the nth well, whereas Eq. (6.55b) is the solution in the nth hill. The $(4\times N)$ constants A_n, B_n, C_n, and D_n must be adjusted to satisfy the periodicity condition and the continuity of $\phi(x)$ and $d\phi/dx$ at the boundaries between wells and hills.

The problem is thus reduced to solving four simultaneous homogeneous equations for four constants A, B, C, and D. To obtain a consistent solution the determinant of the coefficients must vanish. This imposes an algebraic relation between the coefficients and thus relates k and E in terms of the constants V_0, b, and l. Furthermore, the algebraic condition cannot be satisfied unless E is restricted to a range of values as shown in Fig. 6.6. The discrete energy levels for a square well of the same depth and width (as the periodic wells) is also shown. We see that the energy bands correspond to a "smearing out" of the discrete energy levels. This is due to the fact that an electron is not localized in a particular well, but can move around the lattice by tunneling through the potential hills.

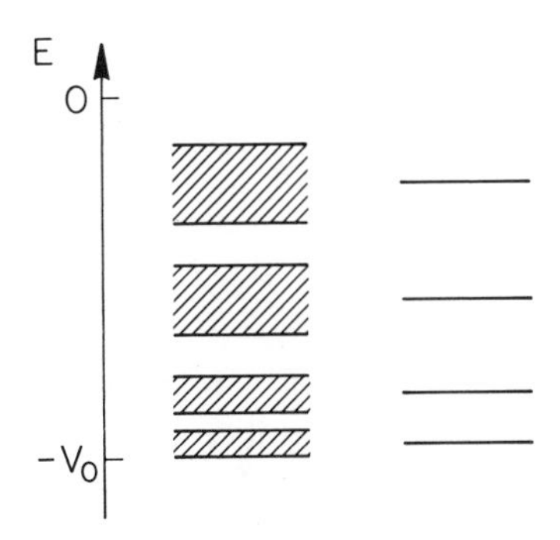

FIGURE 6.6. The bands of allowed energies for motion in a periodic lattice. To the right are shown the energy levels in a well of the same depth and width as those that form the lattice.

6.7. BARRIER PENETRATION

In this example we use the Schrödinger equation to examine the scattering of a free particle from a repulsive potential of finite width. Again we restrict ourselves to one-dimensional motion where the only possibility is that the incident particle will be either transmitted or reflected by the potential barrier, as shown in Fig. 6.7(a). We have indicated the *flux* of incident, transmitted and reflected particles by use of the symbols j_I, j_T, and j_R.

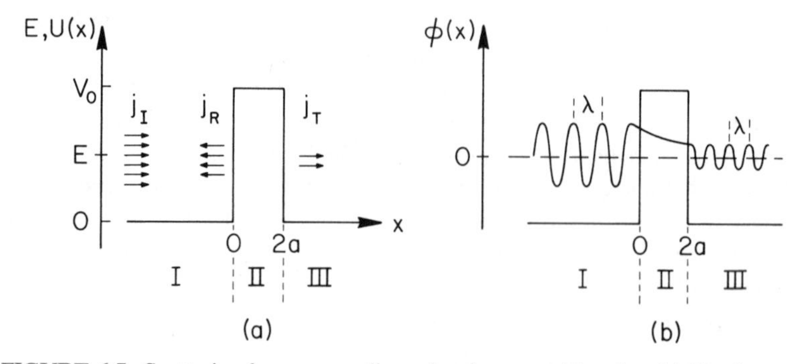

FIGURE 6.7. Scattering from a one-dimensional potential barrier. (a) The form of the potential and a schematic representation of incident, reflected, and transmitted flux. The energy E of the incident particles is indicated. (b) The function $\phi(x)$ in the three regions of space.

Flux is defined as the number of particles crossing a unit area per unit time. If the density of particles is ρ and their velocity $\mathbf{v}$, then the flux through a surface normal to $\mathbf{v}$ is

$$\text{Flux} = \rho\,|\mathbf{v}| \tag{6.56a}$$

Even though the concept of flux is related to an ensemble of particles we can define a probability current, i.e., the flux, for a single particle described by a wavefunction $\psi(\mathbf{x}, t)$ through

$$\mathbf{j}(\mathbf{x}, t) = -\frac{i\hbar}{2m}\,[\psi^*\nabla\psi - (\nabla\psi^*)\psi] \tag{6.56b}$$

Let us consider a plane wave of momentum k moving in one

dimension in the x-direction. The wavefunction is (see Section 1.6)

$$\psi(x, t) = \frac{1}{(2L)^{1/2}} e^{-i(\omega t - kx)}$$

and introducing this result in Eq. (6.56b)

$$j_x = \frac{1}{2L}\frac{\hbar k}{m} = \rho v \tag{6.56c}$$

where the density for one particle on the line $-L<x<L$ is $\rho = \frac{1}{2}L$ and $\hbar k = p = mv$.

We let the height of the barrier be V_0 and its width $2a$. The energy of the incident particle is $E < V_0$, and we seek the stationary states. For one-dimensional motion the time-independent Schrödinger equation in the three regions shown in the figure is given by

$$-\frac{\hbar^2}{2m}\frac{d^2\phi}{dx^2} = E\phi \qquad \text{(I, III)} \tag{6.57a}$$

$$-\frac{\hbar^2}{2m}\frac{d^2\phi}{dx^2} + V_0\phi = E\phi \qquad \text{(II)} \tag{6.57b}$$

Using the notation introduced in the previous section

$$\alpha = \left(\frac{2m}{\hbar^2}E\right)^{1/2} \quad \text{and} \quad \beta = \left[\frac{2m}{\hbar^2}(V_0 - E)\right]^{1/2}$$

the solutions to Eqs. (6.57) are quite generally

$$\phi_{\mathrm{I}}(x) = Ae^{i\alpha x} + Be^{-i\alpha x} \tag{6.58a}$$

$$\phi_{\mathrm{II}}(x) = Ce^{\beta x} + De^{-\beta x} \tag{6.58b}$$

$$\phi_{\mathrm{III}}(x) = Fe^{i\alpha x} + Ge^{-i\alpha x} \tag{6.58c}$$

We must now determine the six boundary coefficients A–G. First we must set $G = 0$ since it represents a wave moving toward $-x$, and no such wave can be present in region III. This leaves us with five coefficients and four constraining equations which are obtained by demanding that $\phi(x)$ and $d\phi/dx$ be continuous at $x = 0$ and $x = 2a$, the two boundaries of the barrier. These four constraints suffice to determine the four constants B, C, D, and F, in terms of the constant A. Of course, A is expressed in terms of the incident flux

$j_I=|A|^2\hbar k/m$. The graphical representation of the three functions ϕ_I, ϕ_{II}, and ϕ_{III}, joined smoothly at $x=0$ and $x=2a$, is shown in Fig. 6.7(b).

We define the transmission coefficient T (reflection coefficient R) as the ratio of transmitted (reflected) flux to the incident flux.

$$T=\frac{j_T}{j_I}=\frac{|F|^2}{|A|^2} \qquad R=\frac{j_R}{j_I}=\frac{|B|^2}{|A|^2} \tag{6.59a}$$

Performing the algebra (see Problem 6.15) we obtain

$$\begin{aligned} T&=\frac{4\alpha^2\beta^2}{4\alpha^2\beta^2+(\alpha^2+\beta^2)^2\sinh^2(2\beta a)} \\ &=\frac{1}{1+[V_0^2/4E(V_0-E)]\sinh^2(2\beta a)} \end{aligned} \tag{6.59b}$$

$$R=\frac{(\alpha^2+\beta^2)^2\sinh^2(2\beta a)}{4\alpha^2\beta^2+(\alpha^2+\beta^2)^2\sinh^2(2\beta a)}=1-T$$

Note that $R+T=1$, or in terms of flux

$$j_I=j_R+j_T \tag{6.60}$$

which is a statement of particle, or probability, conservation.

That particles can penetrate through a potential barrier is observed in the α-decay of heavy nuclei. For instance, the nucleus of thorium-232 decays by α-emission to radium-228. The energy of the α-particle is 4.05 MeV, and the observed lifetime is $\tau=1.39\times10^{10}$ years. The radius of the thorium nucleus can be taken as

$$R\simeq A^{1/3}\times1.5\text{ F}\simeq(232)^{1/3}\times1.5\text{ F}\simeq9.6\text{ F}$$

When the α-particle ($Z=2$, $A=4$) is inside the nucleus it feels an attractive potential, whereas outside the nucleus it is repelled by the Coulomb force. The potential energy as a function of radial distance is obtained using

$$U=\frac{(eZ)_{\text{radium}}(eZ)_\alpha}{(4\pi\varepsilon_0)r}=(88\times2)\frac{e^2}{(4\pi\varepsilon_0)\hbar c}\frac{\hbar c}{r}=\frac{88\times2}{137}\frac{197}{r}\quad\text{MeV}$$

(r in Fermi)

and has the form shown in Fig. 6.8. The energy of the α-particle is well below the peak of the potential barrier and classically it could never penetrate through it. However, quantum-mechanically we must use the transmission coefficient of Eq. (6.59*b*) to evaluate the probability that the α-particle will penetrate through the barrier. In an exact calculation we would include the fact that the potential extends into three dimensions and that the barrier is not square. We can approximate the real case by working in one dimension and replacing the barrier by a rectangular one of width $2a = 33$ F and height $V_0 = 14$ MeV, as shown by the dotted curve† in Fig. 6.8.

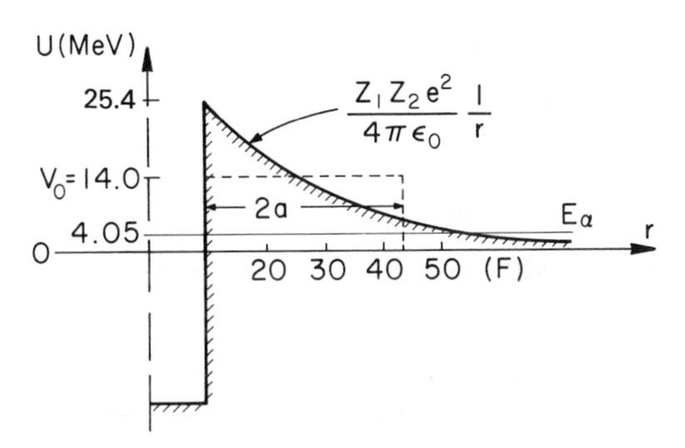

FIGURE 6.8. The Coulomb barrier in heavy nuclei. α-particles of energy E_α can tunnel through the barrier which we approximate by a square potential barrier of width $2a$ and height V_0 (dashed line).

We then obtain

$$\beta = \frac{1}{\hbar}[2m_\alpha(V_0 - E)]^{1/2} = \frac{1}{\hbar c}[2m_\alpha c^2(V_0 - E)]^{1/2}$$
$$= 1.39 \quad (\mathrm{F}^{-1}) \qquad (6.61a)$$

and $2\beta a = 45.8$, which is much larger than 1. Thus, in Eq. (6.59*b*) we can approximate $\sinh^2(2\beta a)$ as

$$\sinh^2(2\beta a) \simeq (\tfrac{1}{2}e^{2\beta a})^2 = \tfrac{1}{4}e^{4\beta a} \qquad (6.61b)$$

and therefore

$$T = \left[1 + \frac{V_0^2}{4E(V_0 - E)} \frac{e^{4\beta a}}{4}\right]^{-1} \simeq 3.29e^{-4\beta a} \qquad (6.61c)$$

† This choice corresponds to a barrier of the same area as the Coulomb potential and yields the correct result.

Finally, using $2\beta a = 45.8$ we find

$$T = 5.75 \times 10^{-40} \tag{6.62a}$$

Every time the α-particle reaches the barrier it has the very small probability of 5.75×10^{-40} to penetrate through it. However, the α-particles in the nucleus impinge on the barrier quite often. We can estimate the frequency ν with which these collisions occur by a qualitative argument

$$\nu = \frac{1}{\Delta t} \qquad \Delta t = \frac{\Delta x}{v} = \frac{\Delta x}{(2E/m)^{1/2}}$$

Using $\Delta x \simeq 10$ F and $E \simeq 4$ MeV we find

$$\nu = \frac{c}{\Delta x}\left(\frac{2E}{mc^2}\right)^{1/2} = 1.4 \times 10^{21} \text{ Hz} \tag{6.62b}$$

Therefore, the probability of α-emission per second is

$$P(\alpha_{\text{emission}}) = \nu T = 8.05 \times 10^{-19} \text{ sec}^{-1}$$

and since the lifetime is the *inverse* of the decay probability

$$\tau \simeq 1.2 \times 10^{18} \text{ sec} \simeq 3.9 \times 10^{10} \text{ years}$$

The numerical value we obtained cannot be very precise because of the simplifying assumptions we introduced. However, the approximate form of Eq. (6.61*c*), which indicates that the decay probability depends exponentially on the energy of the α-particle

$$P(\alpha_{\text{emission}}) \propto e^{-(4a/\hbar)[2m(V_0-E)]^{1/2}}$$

remains valid even in an exact calculation. Therefore, small changes in the energy of the α-particle cause very large variations in the lifetime. The lifetime for α-emission varies from 10^{-6} sec to 10^{18} sec as the energy of the emitted α-particles changes only from 9 to 5 MeV. The calculation of α-particle decay was one of the early successes of quantum mechanics.

In conclusion, we point out that the boundary conditions for this problem were quite different from those used in the discussion of the periodic potential and from the conditions applicable to bound states. As a result, even though in both cases we used the time-independent Schrödinger equation, the solutions had a very different character.

6.8. SUMMARY

We have devoted this chapter to the study of various concepts, developed earlier, in the particular representation in which coordinate states are the basis states. For many problems this is not only convenient it is also the closest analogy to the classical description of physical systems. We showed that only when two operators commute can the corresponding observables be compatible. In most cases, however, the operators do not commute. We gave then the basic commutator of the position and momentum operators and from this derived quantitatively the uncertainty relation.

The Schrödinger equation was introduced in the coordinate representation, and it was shown how the boundary conditions lead to a discrete spectrum for the energy values. The time development of the expectation values was also discussed. Finally, the formalism developed in this chapter was applied to simple physical systems that could be reasonably treated as one-dimensional, such as the band structure of a crystal and the α-decay of a heavy nucleus.

Problems

PROBLEM 1

Calculate the following Poisson brackets

$$\{j_x, j_y\}, \quad \{j_y, j_z\}, \quad \text{and} \quad \{j_z, j_x\}$$

where

$$j_x = yp_z - zp_y$$
$$j_y = zp_x - xp_z$$
$$j_z = xp_y - yp_x$$

PROBLEM 2

If a is a complex dynamical variable (complex function of q, p), a^* is its complex conjugate, and if the Poisson bracket

$$\{a, a^*\} = i$$

calculate $\{a, aa^*\}$, $\{a^*, aa^*\}$, $\{a, a^*a\}$, $\{a^*, a^*a\}$.

PROBLEM 3

Justify why the Schrödinger wavefunction $u(x)$ and its derivative $du(x)/dx$ have to be continuous at the boundary in the case of the finite square-well potential, even though the potential has a discontinuity.

PROBLEM 4

A particle moving in one dimension has a ground-state wavefunction (not normalized) given by

$$\psi_0(x) = e^{-\alpha^4 x^4/4}, \qquad (\alpha \text{ is a real constant})$$

belonging to the eigenvalue

$$E_0 = \frac{\hbar^2 \alpha^2}{m}$$

Determine the potential $U(x)$ in which the particle moves. (Do not try to normalize the wavefunction.)

PROBLEM 5

Consider a harmonic oscillator in one dimension, namely, a particle of mass m moving in a region where the potential energy is

$$U = \tfrac{1}{2}kx^2$$

Use the uncertainty principle in its *minimal* form

$$\Delta p_x \, \Delta x = \frac{\hbar}{2}$$

to find:

(*a*) The energy of the *lowest* state of the oscillator.
(*b*) Express your results in terms of the natural frequency of the s.h.o. $\omega = (k/m)^{1/2}$ and interpret the result obtained in part (*a*).

PROBLEM 6

Calculate the average kinetic and potential energy of a particle bound in a delta function potential in one dimension

$$U(x) = -\gamma\delta(x), \qquad \gamma > 0$$

(You may calculate $\langle T\rangle$ from $\langle E\rangle$ and $\langle V\rangle$.)

PROBLEM 7

Calculate $\langle T\rangle$ for Problem 6 above directly as

$$\int dx\, \psi^*(x)(-\hbar^2/2m\, d^2/dx^2)\psi(x)$$

and compare with the previous result.

PROBLEM 8

Show that in one dimension bound states cannot be degenerate.

PROBLEM 9

For a particle moving in an infinite square well of width $2a$, its wavefunction at time $t = 0$ is

$$\psi_{t=0}(x) = \frac{1}{2^{1/2}}[u_0(x) + u_1(x)]$$

where $u_0(x)$ and $u_1(x)$ are the normalized ground-state and first-excited-state wavefunctions, respectively. What is its average kinetic energy and potential energy as a function of time? Calculate Δx in this state.

PROBLEM 10

A particle moves in a potential in one dimension of the form [see the figure]

$$U(x)=\begin{cases}\infty, & x^2>a^2\\ \gamma\delta(x), & x^2<a^2, \qquad \gamma>0\end{cases}$$

For sufficiently large γ calculate the time required for the particle to tunnel from the ground state of the well extending from $x=-a$ to $x=0$ to the ground state of the well extending from $x=0$ to $x=a$.

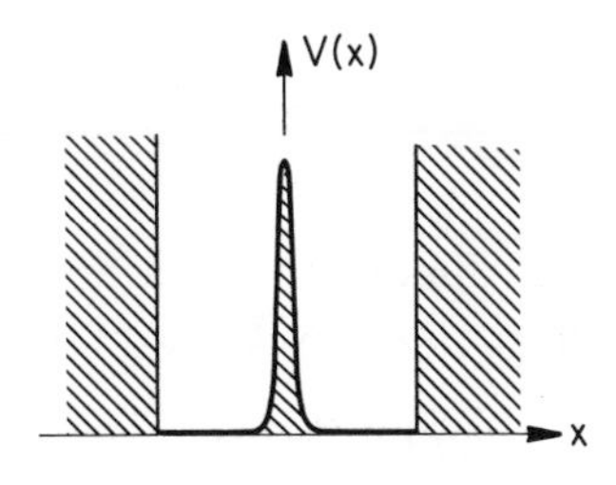

Sketch of the potential.

PROBLEM 11

The first excited state of a particle moving in a potential in one dimension is given by a wavefunction

$$\psi_1(x)=(\sinh \alpha x)\psi_0(x)$$

and belongs to the eigenvalue E_1. Here $\psi_0(x)$ is the ground-state wavefunction belonging to the eigenvalue E_0. What is the potential $U(x)$?. (α is a constant.)

PROBLEM 12

A particle moving in a potential $U(x)$ in one dimension has the wavefunction at $t=0$ given by

$$\psi(x, 0) = \left(\frac{\alpha^2}{\pi}\right)^{1/4} e^{-(\alpha^2/2)(x-a)^2 - ikx}$$

where α, a, and k are real constants.

(*a*) What are the expectation values of the position and momentum in this state at $t=0$?
(*b*) If the particle moves in a potential $V(x) = mgx$, what is the average value of its energy at $t=0$?
(*c*) What is its average energy at an arbitrary time t?

PROBLEM 13

A particle moving in one dimension has a first-excited-state eigenfunction associated with the energy E_1 given by

$$\psi_1 = x\psi_0(x)$$

where $\psi_0(x)$ is the ground-state eigenfunction associated with the energy eigenvalue E_0. Given that the potential vanishes at $x = 0$

(*a*) determine the ratio E_1/E_0.
(*b*) What is the potential $U(x)$ in which the particle moves?

PROBLEM 14

A beam of electrons of kinetic energy $E = 5\,\text{eV}$ is incident on a square-potential barrier of height 25 V and width $a = 0.52$ Å. What fraction of the beam is transmitted?

PROBLEM 15

Obtain the transmission coefficient for a particle of mass m and kinetic energy E (moving in one dimension) incident on a square-potential barrier of width $2a$ and height U_0, where $E < U_0$. That is, derive Eq. (6.59b) of the text.

PROBLEM 16

The wavefunction of a *nonrelativistic* particle moving in one dimension with well-defined momentum is (as usual)

$$\psi(x, t) = \frac{1}{(2\pi)^{1/2}} e^{-i(\omega t - kx)}$$

Observe now the same particle from a frame of reference moving with velocity $v_0 \ll c$ with respect to the original frame. The two frames are connected by the Galilean transformation $x' = x - v_0 t$. In the moving frame the wavefunction can be written as

$$\psi'(x', t) = \frac{1}{(2\pi)^{1/2}} e^{-i(\omega' t - k' x')}$$

Express $\psi'(x', t) = \psi(x, t) f(\hbar, m, v_0, x, x')$. In other words, find how the wavefunction is transformed when viewed in a moving frame. (The transformation coefficient should not contain explicitly the time t.)

Note: You can check your result by finding ψ' when $v_0 = 2v$ (where v is the velocity of the particle).

Chapter 7

SYMMETRIES AND CONSTANTS OF THE MOTION

The question of symmetry has intrigued man from the earliest times. On the one hand, symmetry is inherently manifest in every facet of nature from crystalline rocks to plant life and living organisms, even if it is not always exact. On the other hand, man has exploited and introduced symmetry in many of his activities: it is apparent in art, in music, in buildings and machines—to mention but a few examples. One might expect that as we examine simpler structures the symmetry properties would be more prevalent, and indeed this is the case. At the other end of the scale the evolution of complex systems with large numbers of constitutents, while still obeying the basic principles of symmetry, is primarily governed by statistical laws. The use of symmetries in quantum mechanics is more pronounced than in classical physics where most simple systems can be solved exactly. In quantum mechanics systems are described by state vectors belonging to a Hilbert space. In that space the symmetry operations are represented by operators with well-defined properties, and it can be easily shown that for every symmetry operation there exists a corresponding physical observable that is a *constant of the motion*. Therefore, identifying the symmetries of a system provides helpful information about the state vectors and hence about the system. In the past few decades experimental and theoretical advances have led to the identification of many (unsuspected) new symmetries of nature. But we have also learned under what conditions some symmetries, which were believed to be universally true, cease to be exact—we say then that they are "broken."

The organization of this chapter is as follows: We begin with a brief review of the concept of compatible observables and constants of the motion, which we introduced previously. We then introduce the concept of symmetry, the idea of invariance under specific transformations, and show how this leads to conservation laws.

Next, we explicitly construct unitary symmetry operators that act on quantum states. The symmetry operators and their properties are completely determined by the generators of the symmetry transformation, and we discuss this.

Having established the general properties of symmetries, we address the extremely important case of the symmetry under rotations in three-dimensional space. The generators of rotations are the angular momentum operators, and we study their properties and their eigenvalues, which are conserved. Almost all real systems in three-dimensional space possess spherical symmetry, and thus their eigenstates can be labeled by the eigenvalues of angular momentum. Section 7.4 is devoted to a study of different representations of the angular momentum operators and their eigenstates, including the coordinate representation. The composition of angular momenta is explained in the final section in terms of a simple example.

Because of the importance of angular momentum in all applications of quantum mechanics we have included two appendices that cover more difficult material. In Appendix 6 we discuss the Clebsch–Gordan coefficients that arise in the general case of the composition of arbitrary angular momenta. In Appendix 7 we show how one can evaluate or relate the matrix elements of vector operators between eigenstates of angular momentum. These results are a direct consequence of the underlying symmetry under rotations. However, even in the appendices the presentation is kept at a practical level rather than seeking mathematical rigor. One should also keep in mind that these techniques are not restricted to angular momentum but are generally applicable to the investigation of any symmetry.

7.1. COMPATIBLE OBSERVABLES AND CONSTANTS OF THE MOTION: A REVIEW

In Section 6.1 we defined two observables to be compatible if there existed a complete set of simultaneous eigenstates of the two operators representing the observables. In that case, it is possible to know with precision the eigenvalues of both operators. We also showed that two operators, $\hat{P}$ and $\hat{Q}$, represent compatible observables if, and only if, they commute, because, if the state $|\psi\rangle$ is a simultaneous eigenstate of both operators with eigenvalues p_n and q_ν, then

$$\hat{P}\,|\psi\rangle = p_n\,|\psi\rangle \qquad \hat{Q}\,|\psi\rangle = q_\nu\,|\psi\rangle \tag{7.1a}$$

and

$$\begin{aligned} \hat{Q}\hat{P}\,|\psi\rangle &= p_n\hat{Q}\,|\psi\rangle = p_n q_\nu\,|\psi\rangle \\ \hat{P}\hat{Q}\,|\psi\rangle &= q_\nu\hat{P}\,|\psi\rangle = q_\nu p_n\,|\psi\rangle \end{aligned} \tag{7.1b}$$

If this is true for a complete set of states, it implies that

$$\hat{P}\hat{Q} - \hat{Q}\hat{P} = [\hat{P}, \hat{Q}] = 0 \tag{7.1c}$$

Another important result concerns the time dependence of the expectation value of an operator. If the operator $\hat{Q}$ does not depend explicitly on time, the time derivative of the expectation value in any state $|\psi\rangle$ is given by

$$\frac{d}{dt}\langle\psi|\,\hat{Q}\,|\psi\rangle = \frac{1}{i\hbar}\langle\psi|\,\hat{Q}\hat{H} - \hat{H}\hat{Q}\,|\psi\rangle \tag{7.2a}$$

where $\hat{H}$ is the Hamiltonian, or the energy operator, of the system. Clearly, if $\hat{Q}$ commutes with $\hat{H}$, then the expectation value of $\hat{Q}$ in any arbitrary state of the system does not change with time. The operator $\hat{Q}$ in this case represents an observable that is a constant of the motion. Thus, we see that

$$\text{if } [\hat{Q}, \hat{H}] = 0, \text{ then } \hat{Q} \text{ is a constant of the motion} \tag{7.2b}$$

Comparing Eqs. (7.1c) and (7.2b) it is evident that constants of the motion are compatible with the energy operator. Thus, the eigenstates of the energy operator, i.e., the stationary states, carry, in addition to their energy eigenvalue, labels that are the eigenvalues of the constants of the motion. This is of particular importance when states corresponding to different eigenvalues of a constant of the motion are degenerate in energy. As an example, consider a proton at rest and the two states of its spin projection onto the Z-axis. In the absence of an external field the two states are degenerate and stationary. We can distinguish them by giving not only the energy eigenvalue, which in this case is just the rest mass, but also the spin projection $+1/2$ or $-1/2$ of the state. For this to be possible the spin operator $\hat{\sigma}_z$ must commute with the Hamiltonian. For a proton at rest this is true in the absence of a magnetic field. It remains true in the presence of an external field if the field is along the Z-axis.

In order to establish the complete spectrum of stationary states of a system we must find *all* the constants of the motion and label the states by the eigenvalues of these operators. Frequently, the eigenvalues of the constants of the motion are referred to as "good quantum numbers." This is analogous to classical physics where a knowledge of the constants of the motion greatly facilitates the solution of the problem.

As an example, we know from classical mechanics that when a system has spherical symmetry the angular momentum is a constant of the motion. This is also true in quantum mechanics and, therefore, the stationary states must be labeled by the values of the energy E_n *and* by the values of the angular momentum l

$$|\psi_{n,l}\rangle \tag{7.3a}$$

The projection of the angular momentum along an arbitrary axis is also a constant of the motion, and this is specified by the index m. Thus, the complete description of a stationary state of a spherically symmetric system involves the three indices n, l, and m, which specify the corresponding eigenvalues. In this case we represent a state by

$$|\psi_{n,l,m}\rangle \tag{7.3b}$$

If, in addition, the system has internal degrees of freedom, such as spin, further indices are required to label all the possible stationary states, for example, as $|\psi_{n,l,m,m_s}\rangle$. The basic problem of finding the complete set of stationary states has thus been shifted to finding the constants of the motion of the system and their eigenstates.

As a further illustration, let us consider a particle moving in a position dependent potential $U(\mathbf{x})$. In that case, as we know from Eq. (6.7b), the position coordinate and thus also the potential does not commute with the momentum operator, and therefore

$$[\hat{H}, \hat{\mathbf{p}}] \neq 0 \tag{7.3c}$$

The momentum is not a constant of the motion and cannot be used to label stationary states. On the other hand, in the absence of the potential, that is, for a free particle, the momentum operator commutes with $\hat{H}$, and therefore the familiar plane-wave states of well-defined momentum are stationary.

Finally, we recall that if two observables are compatible and the operators representing them are expressed by matrices, then there exists a representation where both operators are simultaneously diagonal. This follows from Eqs. (7.1a)–(7.1c), but is also self-evident since two diagonal matrices always commute.

7.2. SYMMETRY AND CONSERVATION LAWS IN QUANTUM MECHANICS

When we look at a geometric drawing or at an object we can discern certain *symmetries*. For instance, if we rotate a glass of water by any angle we see the same image. If we turn the drawing of a square by 90°, 180°, or 270° we perceive the same figure. Operations that leave the appearance of the object unchanged are symmetry operations for that particular object. Different objects have different symmetries. We extend this notion to a *physical system* by defining as symmetries those operations that leave the equations of motion and hence the time evolution of the system unchanged. For instance, if a system is subject *only to internal* forces, then translating it in space does not affect its time evolution, and, therefore, the operation of translation is a symmetry of this system. Similarly, if a system is subject to central forces, then a rotation of the system about the source of the force is a symmetry operation.

Symmetries are not restricted to geometric operations only, but are rather of more general nature. For instance, the exchange of identical particles discussed in Section 3.6 is clearly a symmetry operation since it leaves the physical system and its evolution unchanged. Similarly the internal degrees of freedom of elementary particles exhibit a high degree of symmetry even though they cannot be interpreted by classical geometric concepts.

So far we have considered symmetries as operations acting on the physical system. That is, the physical system undergoes the change. This is known as the *active* point of view. We can also describe a symmetry by considering an equivalent transformation of the coordinates in the inverse sense. This *passive* approach is often more convenient for calculational purposes and is widely used. We illustrate this by an example. Consider an electron moving in a Coulomb potential

$$U(x, y, z) = \frac{e^2}{4\pi\varepsilon_0 r} = \frac{e^2}{4\pi\varepsilon_0 (x^2 + y^2 + z^2)^{1/2}} \tag{7.4a}$$

If we invert the position† of the electron with respect to the origin, the electron still sees the same potential: inversion is a symmetry operation. We can view this operation also in the inverse sense, that is, we can perform a transformation that inverts the coordinates. This is known as the *parity* transformation, which we designate by P. Under such a transformation

$$x \to -x \qquad y \to -y \qquad z \to -z \tag{7.4b}$$

which leaves the Coulomb potential of Eq. (7.4*a*) unchanged or *invariant*. This conclusion is equivalent to our previous observation that the active inversion of the electron's position did not alter the potential seen by the electron.

Once the effect of a symmetry transformation on a set of variables is specified we can express the result of the symmetry operation on a function of these variables. For instance, consider the parity operation P. Its effect on an arbitrary function of the coordinates is

$$Pf(x, y, z) = f(-x, -y, -z) \tag{7.5a}$$

If the function f expressed in the transformed coordinates $f(-x, -y, -z)$, is equal to the original function, i.e., if

$$f(-x, -y, -z) = f(x, y, z) \tag{7.5b}$$

then $f(x, y, z)$ is invariant under the symmetry operation P. We can also say that the function $f(x, y, z)$ is symmetric under parity. This is the case for the Coulomb potential.

Symmetry operations are expressed in quantum mechanics as operators acting on the states of the system in Hilbert space.

† Of course, the electron is not localized at any position in space; only the probability of finding it at that position is known.

Consider a symmetry operation T and a system described by the state $|\psi\rangle$, and let the effect of T on this state be such so as to change the system to the state $|\psi'\rangle$. We then define the operator $\hat{T}$ corresponding to this particular symmetry operation through

$$\hat{T}\,|\psi\rangle=|\psi'\rangle$$

as valid for all states $|\psi\rangle$ of the system. The operation T will be a symmetry for this particular system if it leaves its time evolution unchanged, that is, if $|\psi\rangle$ and $|\psi'\rangle$ evolve in time in a similar manner.

The evolution of a quantum state is given by the time development operator [Eqs. (4.1) and (4.13b)]

$$|\psi(t_2)\rangle=\hat{U}(t_2-t_1)\,|\psi(t_1)\rangle \tag{7.6a}$$

and

$$|\psi'(t_2)\rangle=\hat{U}(t_2-t_1)\,|\psi'(t_1)\rangle \tag{7.6b}$$

where $|\psi'\rangle$ is the state resulting from the symmetry operation

$$|\psi'(t_1)\rangle=\hat{T}\,|\psi(t_1)\rangle \tag{7.7a}$$

If the two states are to evolve identically it must also hold that

$$|\psi'(t_2)\rangle=\hat{T}\,|\psi(t_2)\rangle \tag{7.7b}$$

Using Eq. (7.6a) to express $|\psi(t_2)\rangle$ and introducing the expression for $|\psi'(t_1)\rangle$ from Eq. (7.7a) into Eq. (7.6b) we obtain the identity

$$|\psi'(t_2)\rangle=\hat{T}\hat{U}(t_2-t_1)\,|\psi(t_1)\rangle=\hat{U}(t_2-t_1)\hat{T}\,|\psi(t_1)\rangle$$

Since the state $|\psi(t_1)\rangle$ is arbitrary, our result implies that

$$\hat{T}\hat{U}(t_2-t_1)=\hat{U}(t_2-t_1)\hat{T} \tag{7.8}$$

provided that $\hat{T}$ is a symmetry operation for the system.

As the time interval (t_2-t_1) becomes infinitesimal, the time development operator can be expressed by its differential form [Eq. (4.9b)]

$$\hat{U}(\Delta t)=\mathbb{1}-\frac{i}{\hbar}\hat{H}\,\Delta t, \qquad \Delta t\to 0$$

where $\hat{H}$ is the Hamiltonian for the system. Consequently, Eq. (7.8) implies that

$$\hat{T}\hat{U}=\hat{T}-\frac{i}{\hbar}\Delta t\hat{T}\hat{H}=\hat{U}\hat{T}=\hat{T}-\frac{i}{\hbar}\Delta t\hat{H}\hat{T}$$

or that $\hat{T}$ must commute with the Hamiltonian of the system.

$$\hat{T}\hat{H} - \hat{H}\hat{T} = 0$$

Thus, if

$$\boxed{[\hat{T}, \hat{H}] = 0} \tag{7.9}$$

then the transformation induced by the operator $\hat{T}$ is a symmetry operation of the system with Hamiltonian $\hat{H}$. Equation (7.9) defines symmetry transformations for quantum systems and is one of the most profound and beautiful relations of quantum mechanics.

Invariance of the Hamiltonian

An immediate physical interpretation of Eq. (7.9) is obtained by examining the effect of the transformation T on the Hamiltonian. Assuming that the inverse transformation exists and is represented by the operator $\hat{T}^{-1}$, then Eq. (7.9) is equivalent to

$$\hat{H}' = \hat{T}\hat{H}\hat{T}^{-1} = \hat{H} \tag{7.10}$$

In other words, the transformed Hamiltonian operator $\hat{H}'$ is equal to its value before the transformation was applied: If $\hat{T}$ represents a symmetry operation of the system, it must leave the Hamiltonian *invariant.*

We can illustrate this point by the example of the electron moving in the Coulomb potential of Eq. (7.4*a*). In the coordinate representation the Hamiltonian can be written [see Eq. (6.20*b*)] as

$$\hat{H} = -\frac{\hbar^2}{2m}\left(\frac{\partial^2}{\partial x^2} + \frac{\partial^2}{\partial y^2} + \frac{\partial^2}{\partial z^2}\right) - \frac{e^2/4\pi\varepsilon_0}{(x^2 + y^2 + z^2)^{1/2}} \tag{7.11a}$$

Inverting the position of the electron with respect to the origin is equivalent to the inversion of the coordinates as indicated by the transformation given in Eq. (7.4*b*). We represent this transformation by the operator $\hat{P}$, and it is clear that it leaves the Hamiltonian of Eq. (7.11*a*) invariant. We write

$$\hat{H}' = \hat{P}\hat{H}\hat{P}^{-1} = \hat{H} \tag{7.11b}$$

and conclude that the parity transformation is a symmetry for an electron moving in a Coulomb potential. Therefore, the energy eigenstates must also be eigenstates of the parity operator; they must be states of definite parity. The invariance of the Hamiltonian provides us with a reliable and easy method for finding the symmetries of quantum systems.

Conservation Laws

Comparison of Eq. (7.9), which defines the symmetry transformations of the system, with Eq. (7.2*b*), which defines the constants of the motion, shows that they are identical statements. Thus, for every symmetry of the system there must exist a constant of the motion. The existence of constants of the motion is often expressed as a *conservation law*. When we say that momentum is conserved in all interactions we imply that the total momentum of the system is a constant of the motion. The connection between *conserved quantities* and corresponding symmetry operations is a fundamental concept of both quantum and classical mechanics.†.

In the study of natural phenomena certain conservation laws appear to have universal validity. They therefore acquire extreme importance and must reflect fundamental symmetries of nature. In Table 7.1 we list some of the most important symmetries together with the corresponding conserved observables.

Table 7.1. Some of the Most Striking Symmetries of Nature

Symmetry transformation	Conservation law
Permutation of identical particles	Bose–Einstein or Fermi–Dirac statistics
Translation in space	Momentum
Rotations in space	Angular momentum
Translation in time	Energy
Reflection about a plane or point in space	Parity
Lorentz transformation	Energy-momentum relations
Gauge transformations of electromagnetic potentials	Electric charge

† This is known as Noether's theorem, which is valid for continuous symmetries.

All of the above symmetries, with the exception of the parity transformation, are presently believed to be exact symmetries of nature and are observed in all physical phenomena.

Example

We can illustrate some of these ideas by using the example of the ammonia molecule discussed in Sections 4.4 and 4.5. For this simple two-state system it is convenient to use a matrix representation for the Hamiltonian. Thus, the operation representing the symmetry transformation should be expressed by a 2×2 matrix, whereas the eigenstates will be represented by two-component column vectors. The physical configuration for the states $|1\rangle$ and $|2\rangle$ is shown in Fig. 7.1. If we perform a reflection about the plane of the three hydrogens, state $|1\rangle$ becomes state $|2\rangle$, and vice versa. We designate the *reflection* operator† by $\hat{P}_R$ so that

$$\begin{aligned} \hat{P}_R\,|1\rangle &= |2\rangle \\ \hat{P}_R\,|2\rangle &= |1\rangle \end{aligned} \tag{7.12a}$$

We represent the states $|1\rangle$ and $|2\rangle$ by the column vectors

$$|1\rangle = \begin{pmatrix} 1 \\ 0 \end{pmatrix} \qquad |2\rangle = \begin{pmatrix} 0 \\ 1 \end{pmatrix} \tag{7.12b}$$

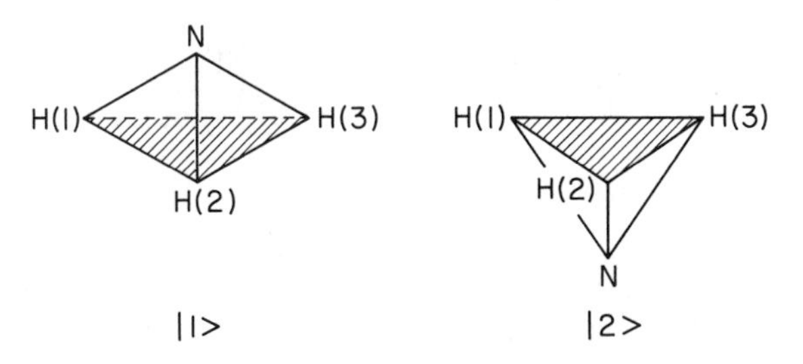

FIGURE 7.1. The two states of the ammonia molecule.

† This is not the same as the parity operator introduced in Eqs. (7.4*b*) and (7.11*b*), which represents a complete inversion about the origin.

in which case the operator $\hat{P}_R$ has the matrix representation

$$\hat{P}_R = \begin{pmatrix} 0 & 1 \\ 1 & 0 \end{pmatrix} \tag{7.12c}$$

which follows from the defining equation [Eq. (7.12*a*)].

In the $|1\rangle, |2\rangle$ representation the energy matrix is given by Eq. (4.49*a*)

$$\hat{H} = \begin{pmatrix} E_0 & -A \\ -A & E_0 \end{pmatrix} \tag{7.13a}$$

We can check whether $\hat{P}_R$ is a symmetry operation for the ammonia molecule by forming the products

$$\begin{aligned} \hat{P}_R\hat{H} &= \begin{pmatrix} 0 & 1 \\ 1 & 0 \end{pmatrix}\begin{pmatrix} E_0 & -A \\ -A & E_0 \end{pmatrix} = \begin{pmatrix} -A & E_0 \\ E_0 & -A \end{pmatrix} \\ \hat{H}\hat{P}_R &= \begin{pmatrix} E_0 & -A \\ -A & E_0 \end{pmatrix}\begin{pmatrix} 0 & 1 \\ 1 & 0 \end{pmatrix} = \begin{pmatrix} -A & E_0 \\ E_0 & -A \end{pmatrix} \end{aligned} \tag{7.13b}$$

This shows that

$$\hat{P}_R\hat{H} = \hat{H}\hat{P}_R \tag{7.13c}$$

Therefore, $\hat{P}_R$ commutes with the Hamiltonian, showing that reflection about the plane of the hydrogens is a symmetry of the ammonia molecule.

Since $\hat{P}_R$ commutes with the Hamiltonian its eigenstates must be simultaneous eigenstates of the energy operator. The stationary states of the ammonia molecule, states $|\mathrm{I}\rangle$ and $|\mathrm{II}\rangle$, expressed in the $|1\rangle, |2\rangle$ representation, are given by Eqs. (4.53*b*)

$$\begin{aligned} |\mathrm{I}\rangle &= \frac{1}{2^{1/2}}[|1\rangle + |2\rangle] = \frac{1}{2^{1/2}}\begin{pmatrix} 1 \\ 1 \end{pmatrix} \\ |\mathrm{II}\rangle &= \frac{1}{2^{1/2}}[|1\rangle - |2\rangle] = \frac{1}{2^{1/2}}\begin{pmatrix} 1 \\ -1 \end{pmatrix} \end{aligned} \tag{7.14a}$$

If we operate on these states with $\hat{P}_R$ using the matrix representation of Eq. (7.12*c*) we obtain

$$\begin{aligned} \hat{P}_R\,|\mathrm{I}\rangle &= \begin{pmatrix} 0 & 1 \\ 1 & 0 \end{pmatrix}\frac{1}{2^{1/2}}\begin{pmatrix} 1 \\ 1 \end{pmatrix} = \frac{1}{2^{1/2}}\begin{pmatrix} 1 \\ 1 \end{pmatrix} = |\mathrm{I}\rangle \\ \hat{P}_R\,|\mathrm{II}\rangle &= \begin{pmatrix} 0 & 1 \\ 1 & 0 \end{pmatrix}\frac{1}{2^{1/2}}\begin{pmatrix} 1 \\ -1 \end{pmatrix} = \frac{1}{2^{1/2}}\begin{pmatrix} -1 \\ 1 \end{pmatrix} = -|\mathrm{II}\rangle \end{aligned} \tag{7.14b}$$

which shows that the stationary states are eigenstates of $\hat{P}_R$; they have eigenvalues $+1$ and -1, respectively.

The reflection operation has the special property that when applied twice on a state it must reproduce the original state† from which it follows that

$$(\hat{P}_R)^2 = \hat{P}_R\hat{P}_R = \mathbb{1} \tag{7.15a}$$

Therefore, if a state $|\psi\rangle$ is an eigenstate of $\hat{P}_R$ with the eigenvalue α, then

$$\hat{P}_R\,|\psi\rangle = \alpha\,|\psi\rangle \quad \text{and} \quad \hat{P}_R\hat{P}_R\,|\psi\rangle = \hat{P}_R(\alpha\,|\psi\rangle) = \alpha(\hat{P}_R\,|\psi\rangle) = \alpha^2\,|\psi\rangle$$

But we also know that

$$\hat{P}_R\hat{P}_R\,|\psi\rangle = |\psi\rangle$$

Therefore, we conclude that

$$\alpha^2 = 1 \quad \text{or} \quad \alpha = \pm 1 \tag{7.15b}$$

which indicates that the only possible eigenvalues of the reflection symmetry operator $\hat{P}_R$ are $+1$ and -1. This fact is borne out by Eqs. (7.14*b*).

Since $\hat{P}_R$ commutes with $\hat{H}$ as shown in Eq. (7.13*c*), there must exist a representation in which both $\hat{H}$ and $\hat{P}_R$ are simultaneously diagonal. Of course, $\hat{H}$ is diagonal in the representation where the basis states are the stationary states $|\text{I}\rangle$ and $|\text{II}\rangle$. This representation is obtained from the $|1\rangle, |2\rangle$ representation by the unitary transformation S, which we have used before. It is given by Eq. (4.55*b*)

$$S = \frac{1}{\sqrt{2}} \begin{array}{c|cc} & |1\rangle & |2\rangle \\ \hline |I\rangle & 1 & 1 \\ |II\rangle & 1 & -1 \end{array} \quad \text{and} \quad S^{-1} = \frac{1}{\sqrt{2}} \begin{array}{c|cc} & |1\rangle & |2\rangle \\ \hline |I\rangle & 1 & 1 \\ |II\rangle & 1 & -1 \end{array} \tag{7.16}$$

Under this transformation the Hamiltonian of Eq. (7.13*a*) becomes diagonal and has the form given by Eq. (4.55*a*)

$$\hat{H}' = S\hat{H}S^{-1} = \begin{pmatrix} E_0 - A & 0 \\ 0 & E_0 + A \end{pmatrix} \tag{7.17a}$$

† The same is true for the symmetry under the exchange of two identical particles discussed in Section 3.6.

The operator $\hat{P}_R$ transforms according to the same prescription

$$\hat{P}'_R = S\hat{P}_R S^{-1} = \frac{1}{2^{1/2}}\begin{pmatrix} 1 & 1 \\ 1 & -1 \end{pmatrix}\begin{pmatrix} 0 & 1 \\ 1 & 0 \end{pmatrix}\frac{1}{2^{1/2}}\begin{pmatrix} 1 & 1 \\ 1 & -1 \end{pmatrix} = \begin{pmatrix} 1 & 0 \\ 0 & -1 \end{pmatrix} \tag{7.17b}$$

As expected, in the representation of the stationary states $|I\rangle$ and $|II\rangle$ of the system, the symmetry operator $\hat{P}'_R$ is indeed diagonal and has the eigenvalues $+1$ and -1 as we have already found in Eqs. (7.14*b*).

These manipulations illustrate the power of transformation theory. For instance, it would be difficult to establish directly the form of $\hat{P}'_R$ in the $|I\rangle, |II\rangle$ representation. On the other hand, it was straightforward to establish $\hat{P}_R$ in the $|1\rangle, |2\rangle$ representation from the physical structure of the ammonia molecule. As a further exercise the reader should verify that in the presence of an electric field, the reflection operation $\hat{P}_R$ *ceases* to be a symmetry of the system. In that case the energy matrix is given in the $|1\rangle, |2\rangle$ representation by Eq. (4.58*a*)

$$\hat{H}_1 = \begin{vmatrix} E_0 + \mu\mathscr{E} & -A \\ -A & E_0 - \mu\mathscr{E} \end{vmatrix} \tag{7.18a}$$

Using the representation given by Eq. (7.12*c*) for $\hat{P}_R$ one finds immediately that

$$\hat{P}_R\hat{H}_1 - \hat{H}_1\hat{P}_R \neq 0 \tag{7.18b}$$

Such a situation arises frequently in nature. An isolated physical system may have a high degree of symmetry; however, when it interacts with another system, or if a perturbation is applied, the symmetry or part of it may no longer be valid.

7.3. SYMMETRY TRANSFORMATIONS AND THEIR GENERATORS

We have seen that quantum systems admit certain symmetry operations that do not alter their evolution. The results of performing such a symmetry operation on the system are equivalent to those obtained by an appropriate transformation of the coordinates. Such transformations can be represented by a corresponding operator acting on the Hilbert space. As an example we considered the operation of reflection with respect to a plane and applied it to a simple two-state system. In that case the relevant operator $\hat{P}_R$ could be obtained from direct physical arguments. We would now like to develop a systematic approach for constructing the operators for different symmetries. Furthermore, we should be able to give their explicit form in certain specific representations.

Our first observation is that since a symmetry transformation leaves the evolution of the system unchanged, it must not affect the normalization of a state. Therefore, symmetry transformations must be *unitary*. Designating a symmetry transformation by the operator $\hat{T}$, it must hold that

$$\hat{T}^\dagger \hat{T} = \hat{T}\hat{T}^\dagger = \mathbb{1} \tag{7.19}$$

Next we show that the eigenvalues of a unitary matrix have modulus 1. Note that any matrix can be written in the form

$$\hat{T} = \frac{\hat{T}+\hat{T}^\dagger}{2} + i\frac{\hat{T}-\hat{T}^\dagger}{2i} \tag{7.20a}$$

where the matrices $(\hat{T}+\hat{T}^\dagger)/2$ and $(\hat{T}-\hat{T}^\dagger)/2i$ are evidently hermitian. Furthermore, they commute since $\hat{T}$ commutes with $\hat{T}^\dagger$. Thus, they can have simultaneous eigenfunctions. If we choose these eigenfunctions as basis states, the matrices $(\hat{T}+\hat{T}^\dagger)/2$ and $(\hat{T}-\hat{T}^\dagger)/2i$ are simultaneously diagonal and so is the matrix $\hat{T}$. We designate the eigenvalues of $\hat{T}$ by α, the eigenvalues of $\hat{T}^\dagger$ are then given by α^*. In view of Eq. (7.19) the eigenvalues $\hat{T}\hat{T}^\dagger$ are 1. Thus

$$\alpha_s \alpha_s^* = 1 \quad \text{or} \quad \alpha_s = e^{i\delta_s} \tag{7.20b}$$

with δ_s real.

If $\hat{T}$ is a symmetry of a particular quantum system it must commute with the Hamiltonian. Therefore, the stationary states are also eigenstates of $\hat{T}$, and $\hat{T}$ is diagonal in the representation where the stationary states are chosen as the basis states. It follows that

$$\left.\begin{array}{c}\hat{T}\,|\psi_s\rangle = \alpha_s\,|\psi_s\rangle = e^{i\delta_s}\,|\psi_s\rangle \\ \text{where } |\psi_s\rangle \text{ is a stationary state and} \\ \hat{T} \text{ is a symmetry of the system}\end{array}\right\} \tag{7.20c}$$

The second observation is that repeated application of a transformation must be equivalent to a single transformation from the initial configuration to the final one. For instance, if $\hat{T}_x(a)$ represents a translation of the system along the X-axis by a distance a, then

$$\hat{T}_x(b)\hat{T}_x(a) = \hat{T}_x(a+b) \tag{7.21}$$

Unitary operators that satisfy a relation such as that given by Eq. (7.21) are of the general form

$$\hat{U}_Q(\lambda) = e^{i\hat{Q}\lambda} \tag{7.22}$$

where $\hat{Q}$ is a *hermitian* operator; that is, $\hat{Q} = \hat{Q}^\dagger$ and λ is a *real* parameter. By the very construction of Eq. (7.22) we have assumed that the transformation $\hat{U}_Q(\lambda)$ will be unitary since

$$\hat{U}_Q(\lambda)\hat{U}_Q^\dagger(\lambda) = e^{i\hat{Q}\lambda}e^{-i\hat{Q}\lambda} = \mathbb{1} = \hat{U}_Q^\dagger(\lambda)\hat{U}_Q(\lambda)$$

The hermitian operator $\hat{Q}$ in this case is called the *generator* of the symmetry transformation.

In what follows we construct the unitary transformations for translations and rotations in space. We show that the generators of these transformations are the momentum and angular momentum operators, respectively. It is now clear that if $\hat{U}_Q(\lambda)$ is a symmetry transformation of the system, that is, if

$$[\hat{U}_Q, \hat{H}] = 0 \tag{7.23a}$$

then the generator $\hat{Q}$ must commute with the Hamiltonian

$$[\hat{Q}, \hat{H}] = 0 \tag{7.23b}$$

This follows because Eq. (7.22) can be expressed as an expansion in powers of $\hat{Q}$

$$\hat{U}_Q(\lambda) = \mathbb{1} + i\lambda\hat{Q} + \frac{(i\lambda)^2}{2!}(\hat{Q})^2 + \cdots \tag{7.23}$$

Thus, Eq. (7.23*a*) can be satisfied if, and only if Eq. (7.23*b*) is valid. We therefore reach the important conclusion that *for any symmetry* *of a quantum system the physical observable represented by the generator of the symmetry is a conserved quantity.* It is this result that provides the connection between symmetries and conservation laws, examples of which were given in Table 7.1.

Translations

We consider a specific symmetry transformation, the translation of the coordinates. Let $f(x, y, z)$ be an arbitrary function of the coordinates. If we translate the system described by this function along the X-axis by a distance ε, the function $f(x, y, z)$ will change to $f(x+\varepsilon, y, z)$. We can Taylor-expand $f(x+\varepsilon, y, z)$ about the point (x, y, z)

$$f(x+\varepsilon, y, z) = f(x, y, z) + \varepsilon \left.\frac{\partial f}{\partial x}\right|_{x,y,z} + \cdots$$

If $\varepsilon \to 0$, only the term linear in ε is important and

$$f(x+\varepsilon, y, z) = [\hat{T}_x(\varepsilon)] f(x, y, z) = \left(1 + \varepsilon \frac{\partial}{\partial x}\right) f(x, y, z) \tag{7.24a}$$

Thus, for $\varepsilon \to 0$ the translation operator $\hat{T}_x(\varepsilon)$ is given by

$$\hat{T}_x(\varepsilon) = 1 + \varepsilon \frac{\partial}{\partial x}, \qquad \varepsilon \to 0 \tag{7.24b}$$

In the coordinate representation $\partial/\partial x = (i/\hbar)\hat{p}_x$, where $\hat{p}_x$ is the momentum operator. Thus, we can write the operator for infinitesimal translations in the form

$$\hat{T}_x(\varepsilon) = 1 + \frac{i}{\hbar} \hat{p}_x \varepsilon, \qquad \varepsilon \to 0 \tag{7.24c}$$

A finite translation by the distance a along the X-axis can be generated by the repeated application of infinitesimal translations

$$\hat{T}_x(a) = [\hat{T}_x(\varepsilon)]^n = \left[1 + \frac{i}{\hbar} \hat{p}_x \varepsilon\right]^n = \left[1 + \frac{i\hat{p}_x a}{\hbar n}\right]^n = e^{(i/\hbar)\hat{p}_x a} \tag{7.25a}$$

where $a = n\varepsilon$ with $n \to \infty$ and $\varepsilon \to 0$. That the above expression does indeed induce a finite translation of the coordinates can be verified by a direct expansion of the exponential

$$\hat{T}_x(a) = 1 + a\frac{i}{\hbar}\hat{p}_x + \frac{a^2}{2!}\left(\frac{i}{\hbar}\hat{p}_x\right)^2 + \frac{a^3}{3!}\left(\frac{i}{\hbar}\hat{p}_x\right)^3 + \cdots \quad (7.25b)$$

or

$$\hat{T}_x(a) = 1 + a\frac{\partial}{\partial x} + \frac{a^2}{2!}\frac{\partial^2}{\partial x^2} + \frac{a^3}{3!}\frac{\partial^3}{\partial x^3} + \cdots \quad (7.25c)$$

Clearly, when $\hat{T}_x(a)$ as given by Eq. (7.25*c*) acts on a function of the coordinates $f(x, y, z)$, it will yield the same function translated to the point $(x+a)$, as follows from the Taylor theorem. Thus

$$[\hat{T}_x(a)]f(x, y, z) = f[(x+a), y, z] \quad (7.25d)$$

The above result can be generalized to a translation along an arbitrary direction **a** where

$$\mathbf{a} = a_x\mathbf{u}_x + a_y\mathbf{u}_y + a_z\mathbf{u}_z$$

In this case the desired translation can be achieved by a successive application of translations along each of the three axes, where we recall that the momentum operators along different directions commute. Thus

$$\hat{T}(\mathbf{a}) = \hat{T}_x(a_x)\hat{T}_y(a_y)\hat{T}_z(a_z) = e^{(i/\hbar)\hat{p}_x a_x} e^{(i/\hbar)\hat{p}_y a_y} e^{(i/\hbar)\hat{p}_z a_z}$$

or

$$\boxed{\hat{\mathbf{T}}(\mathbf{a}) = e^{(i/\hbar)\hat{\mathbf{p}}\cdot\mathbf{a}}} \quad (7.26)$$

Even though we used the coordinate representation to relate $\partial/\partial x$, $\partial/\partial y$, and $\partial/\partial z$, to the momentum operators $\hat{p}_x$, $\hat{p}_y$, and $\hat{p}_z$, Eq. (7.26) is a *relation between operators* and, therefore, remains valid in any representation.

As an example consider a free particle of mass m. In the coordinate representation the Hamiltonian is given by

$$\hat{H} = \frac{\hat{\mathbf{p}}^2}{2m} = -\frac{\hbar^2}{2m}\nabla^2 \quad (7.27a)$$

$\hat{H}$ commutes with $\hat{T}(\mathbf{a})$ because the operator $\hat{\mathbf{p}}$ commutes with $\hat{p}^2$. Thus, translation of the coordinates is a symmetry operation for the free particle, and consequently the free-particle stationary states must be eigenstates of $\hat{T}(\mathbf{a})$. They must also be eigenstates of $\hat{\mathbf{p}}$, which is the generator of $\hat{T}(\mathbf{a})$. We can check this in the coordinate representation where the stationary state wavefunctions for the free particle are given by plane waves

$$\psi_{\mathbf{p}}(\mathbf{x}, t) = \frac{1}{V^{1/2}} e^{-(i/\hbar)(Et-\mathbf{p}\cdot\mathbf{x})} \tag{7.27b}$$

Operating on this wavefunction with $\hat{T}(\mathbf{a})$ as given by Eq. (7.26) we obtain†

$$\hat{T}(\mathbf{a})\psi_{\mathbf{p}}(\mathbf{x}, t) = (e^{(i/\hbar)\mathbf{p}\cdot\mathbf{a}})\psi_{\mathbf{p}}(\mathbf{x}, t) \tag{7.27c}$$

In the above expression the vector $\mathbf{p}$ appearing in the exponent is no longer an operator, but is the eigenvalue of the momentum corresponding to the plane-wave wavefunction $\psi_{\mathbf{p}}(\mathbf{x}, t)$. We have confirmed that the free-particle wavefunctions are eigenstates of $\hat{T}(\mathbf{a})$ with the eigenvalue

$$e^{(i/\hbar)\mathbf{p}\cdot\mathbf{a}}$$

This eigenvalue has modulus 1 in accordance with our general conclusion about the eigenvalues of unitary matrices [Eq. (7.20*b*)].

Rotations

As the next symmetry transformation, we consider rotations about a fixed axis. Let $f(x, y, z)$ be a function of the coordinates and let us perform an infinitesimal rotation by the angle ω around the Z-axis on the system described by the function f. Then

$$\hat{R}_z(\omega)f(x, y, z) = f(x-\omega y, y+\omega x, z) \tag{7.28a}$$

We Taylor-expand the function on the rhs of the above equation, and since $\omega \to 0$ we keep only the terms linear in ω

$$f(x-\omega y, y+\omega x, z) = f(x, y, z) - \omega y\frac{\partial f}{\partial x} + \omega x\frac{\partial f}{\partial y} + \text{terms in } \omega^2 \tag{7.28b}$$

† Think in terms of the expansion indicated by Eq. (7.25*c*).

or

$$\hat{R}_z(\omega)f(x, y, z) = \left[1 + \omega\left(x\frac{\partial}{\partial y} - y\frac{\partial}{\partial x}\right)\right]f(x, y, z), \qquad \omega \to 0 \tag{7.28c}$$

We recognize that the differential operator

$$\left(x\frac{\partial}{\partial y} - y\frac{\partial}{\partial x}\right) = \frac{i}{\hbar}\hat{L}_z$$

is the coordinate representation of the operator for the projection of the angular momentum onto the Z-axis as established by Eqs. (6.21c). Therefore, we write

$$\hat{R}_z(\omega)f(x, y, z) = \left(1 + \frac{i}{\hbar}\omega\hat{L}_Z\right)f(x, y, z), \qquad \omega \to 0$$

or

$$\hat{R}_z(\omega) = 1 + \frac{i}{\hbar}\omega\hat{L}_Z, \qquad \omega \to 0 \tag{7.28d}$$

A finite rotation by an angle α about the Z-axis is obtained by repeated application of the operator for infinitesimal rotations leading to

$$\boxed{\hat{R}_z(\alpha) = e^{(i/\hbar)\hat{L}_z\alpha}} \tag{7.29a}$$

As before, this result is an operator relation and, therefore, *independent* of any particular representation. By the same procedure we find the symmetry transformation for rotations about the X- and Y-axes as

$$\hat{R}_x(\beta) = e^{(i/\hbar)\hat{L}_x\beta} \tag{7.29b}$$

$$\hat{R}_y(\gamma) = e^{(i/\hbar)\hat{L}_y\gamma} \tag{7.29c}$$

We must now exercise some care because the generators of rotations $\hat{L}_x$, $\hat{L}_y$, and $\hat{L}_z$, do not commute with one another and, therefore, neither will the rotation matrices $\hat{R}_x$, $\hat{R}_y$, or $\hat{R}_z$. Thus, the order in which successive rotations about different axes are performed is very important.

We see that the angular momentum operators are the generators of rotations. If the system is spherically symmetric, then rotations about any axis leave the system invariant, and hence the three operators

$$\hat{L}_x, \hat{L}_y, \text{ and } \hat{L}_z \tag{7.30a}$$

commute with the Hamiltonian. In addition, the operator

$$\hat{L}^2 = \hat{L}_x^2 + \hat{L}_y^2 + \hat{L}_z^2 \tag{7.30b}$$

commutes with the Hamiltonian as well as with each of the operators $\hat{L}_x$, $\hat{L}_y$, and $\hat{L}_z$. Therefore, the magnitude of the total angular momentum and its projection along the three axes are constants of the motion for spherically symmetric systems. However, as we have seen before, the three operators $\hat{L}_x$, $\hat{L}_y$, and $\hat{L}_z$, do not commute among themselves, and, hence, only the magnitude of the angular momentum and its projection along any one axis can be measured simultaneously. That is, only the three operators

$$\hat{H}, \hat{L}^2 \qquad (\text{and one of } \hat{L}_x, \hat{L}_y, \hat{L}_z)$$

can be made simultaneously diagonal. It is customary, but of course not necessary, to choose a representation in which $\hat{L}_z$ is diagonal. Then the stationary states of a spherically symmetric system are labeled by the eigenvalues of $\hat{H}$, $\hat{L}^2$, and $\hat{L}_z$. An electron moving in the Coulomb potential of Eq. (7.4a) is an example of a system with spherical symmetry, since the Hamiltonian [see Eq. (7.11a)] remains invariant under rotations.

In the coordinate representation the angular momentum operator $\hat{L}_z$ can be expressed as a differential operator

$$\hat{L}_z = -i\hbar\left(x\frac{\partial}{\partial y} - y\frac{\partial}{\partial x}\right) = -i\hbar\frac{\partial}{\partial \phi} \tag{7.31a}$$

where ϕ is the angle of rotation about the Z-axis, the azimuth angle. The amplitudes

$$\psi_m(\phi) = \frac{1}{(2\pi)^{1/2}} e^{im\phi}, \qquad m \text{ real} \tag{7.31b}$$

are clearly eigenfunctions of $\hat{L}_z$ with eigenvalue $\hbar m$. Therefore, they also will be eigenfunctions of the rotation operator about the Z-axis, as given by Eq. (7.29a). Consider a finite rotation by the angle α about the Z-axis

$$\hat{R}_z(\alpha)\psi_m(\phi) = e^{(i/\hbar)\hat{L}_z\alpha}\,\psi_m(\phi)$$

$$= e^{\alpha(\partial/\partial\phi)}\frac{1}{(2\pi)^{1/2}}\,e^{im\phi} = e^{i\alpha m}\psi_m(\phi) \qquad (7.32a)$$

We see that $\psi_m(\phi)$ is a representation of the eigenfunctions of $\hat{R}_z(\alpha)$ with eigenvalue $e^{i\alpha m}$, as expected for unitary operators. This result is completely analogous to that obtained in Eq. (7.27c) for translations.

On physical grounds we expect that a rotation by an angle $\alpha = 2\pi$ will return the system to its original configuration. Then, from Eq. (7.32a) we have

$$e^{i\alpha m}\big|_{\alpha=2\pi} = e^{i2m\pi} = 1 \qquad (7.32b)$$

which implies that

$$m = 0, \pm 1, \pm 2, \ldots \qquad (7.32c)$$

This result establishes the possible eigenvalues of the projection of the *orbital* angular momentum operator onto the Z-axis.† However, if we consider angular momentum in general, including spin, the eigenvalues of $\hat{L}_z$ are $\hbar m$ with

$$m = 0, \pm\tfrac{1}{2}, \pm 1, \pm\tfrac{3}{2}, \pm 2, \ldots \qquad (7.33)$$

We will show in the next section how this conclusion follows from the general properties of the angular momentum operator.

7.4. ANGULAR MOMENTUM OPERATORS AND THEIR EIGENVALUES

The angular momentum operators were defined in the coordinate representation by Eqs. (6.21c). This definition was obtained by

† As already pointed out the amplitude for a quantum state need not be invariant under transformations that correspond classically to the identity. This is because the amplitude itself is not an observable. Thus, the assumption of Eq. (7.32c) is too restrictive unless we are dealing with an observable that has an exact classical analogue.

analogy with the classical expression

$$\mathbf{L} = \mathbf{r} \times \mathbf{p} \tag{7.34a}$$

and we introduced the three operators $\hat{L}_x$, $\hat{L}_y$, and $\hat{L}_z$. Furthermore, the sum of the squares of these operators defines the total angular momentum squared operator as given by Eq. (7.30*b*). The most crucial step is to recognize that the orbital angular momentum operators obey the commutation relations given by Eqs. (6.22)

$$\begin{aligned} [\hat{L}_x, \hat{L}_y] &= i\hbar\hat{L}_z \\ [\hat{L}_y, \hat{L}_z] &= i\hbar\hat{L}_x \\ [\hat{L}_z, \hat{L}_x] &= i\hbar\hat{L}_y \end{aligned} \tag{7.34b}$$

These relations were established by direct calculation in the coordinate representation. However, since they are operator relations they must be generally valid independent of any particular representation. We will, therefore, use Eqs. (7.34*b*) as the *defining relations* for a triplet of operators

$$\hat{J}_x, \hat{J}_y, \hat{J}_z \tag{7.35a}$$

which we call the *angular momentum operators*. We use the symbol $\hat{J}$ to indicate that we are *not* restricted to orbital angular momentum, which is derived from Eq. (7.34*a*). Furthermore, we write the commutation relations in compact form†

$$\boxed{[\hat{J}_i, \hat{J}_j] = i\hbar\hat{J}_k, \qquad i, j, k \text{ cyclic}} \tag{7.35b}$$

where i, j, and k stand for x, y, and z. We then introduce the operator $\hat{J}^2$ through

$$\hat{J}^2 = \hat{J}_x^2 + \hat{J}_y^2 + \hat{J}_z^2 \tag{7.36a}$$

† The triplet of angular momentum operators can be expressed more compactly by introducing the equivalent of vector notation

$$\hat{\mathbf{J}} = \hat{J}_x\mathbf{u}_x + \hat{J}_y\mathbf{u}_y + \hat{J}_z\mathbf{u}_z$$

in which case the commutation relations of Eq. (7.35*b*) can be written as the single equation

$$\hat{\mathbf{J}} \times \hat{\mathbf{J}} = i\hbar\hat{\mathbf{J}}$$

which commutes with the operators $\hat{J}_i$ so that

$$[\hat{J}^2, \hat{J}_i] = 0 \tag{7.36b}$$

as can be easily proved by using Eq. (7.35*b*).

We are now in a position to obtain the eigenvalues of the angular momentum operators. We will do this without recourse to any one particular representation, using only the defining commutation relations of Eq. (7.35*b*). This is possible because the operators in question are generators of a symmetry transformation, and symmetry transformations form a group. In our case, Eq. (7.35*b*) is identical to the algebra of the continuous $SU(2)$ group. The derivation is rather mathematical, but the results are of such wide applicability in quantum mechanics that one should become completely familiar with them.

We consider a representation in which $\hat{J}^2$ and $\hat{J}_z$ are diagonal. Such a representation exists in view of Eq. (7.36*b*). Next we introduce the two nonhermitian operators

$$\begin{aligned} \hat{J}_+ &= \hat{J}_x + i\hat{J}_y \\ \hat{J}_- &= \hat{J}_x - i\hat{J}_y \end{aligned} \tag{7.37a}$$

so that

$$\hat{J}_+^\dagger = \hat{J}_- \quad \text{and} \quad \hat{J}_-^\dagger = \hat{J}_+ \tag{7.37b}$$

This follows from the fact that $\hat{J}_x, \hat{J}_y, \hat{J}_z$ are hermitian. It can be easily shown that the operators $\hat{J}_+$ and $\hat{J}_-$ obey the commutation relations

$$[\hat{J}_z, \hat{J}_+] = \hbar\hat{J}_+ \qquad [\hat{J}_z, \hat{J}_-] = -\hbar\hat{J}_- \tag{7.38a}$$

and

$$[\hat{J}_+, \hat{J}_-] = 2\hbar\hat{J}_z \tag{7.38b}$$

In particular, they obey the algebra

$$\begin{aligned} \hat{J}_+\hat{J}_- &= \hat{J}_x^2 + i(\hat{J}_y\hat{J}_x - \hat{J}_x\hat{J}_y) + \hat{J}_y^2 = \hat{J}^2 - \hat{J}_z^2 + \hbar\hat{J}_z \\ \hat{J}_-\hat{J}_+ &= \hat{J}_x^2 + i(\hat{J}_x\hat{J}_y - \hat{J}_y\hat{J}_x) + \hat{J}_y^2 = \hat{J}^2 - \hat{J}_z^2 - \hbar\hat{J}_z \end{aligned} \tag{7.39}$$

The operators $\hat{J}_+$ and $\hat{J}_-$ are *raising* and *lowering* operators, respectively, analogous to the $\hat{\sigma}_+$ and $\hat{\sigma}_-$ matrices defined by Eqs. (4.74*a*). To prove this we consider a state $|\psi_{\beta,\mu}\rangle$, which is a simultaneous eigenstate of $\hat{J}^2$ and $\hat{J}_z$ with the eigenvalues $\hbar^2\beta$ and $\hbar\mu$, respectively. Namely,

$$\hat{J}^2 \,|\psi_{\beta,\mu}\rangle = \hbar^2\beta \,|\psi_{\beta,\mu}\rangle \tag{7.40a}$$

$$\hat{J}_z \,|\psi_{\beta,\mu}\rangle = \hbar\mu \,|\psi_{\beta,\mu}\rangle \tag{7.40b}$$

From Eq. (7.38*a*) it follows that

$$\hat{J}_z\hat{J}_+ - \hat{J}_+\hat{J}_z = \hbar\hat{J}_+ \quad \text{or} \quad \hat{J}_z\hat{J}_+ = \hat{J}_+(\hat{J}_z + \hbar)$$

and therefore

$$\hat{J}_z\hat{J}_+ \,|\psi_{\beta,\mu}\rangle = \hat{J}_+(\hat{J}_z + \hbar)\,|\psi_{\beta,\mu}\rangle$$

If we now use Eq. (7.40*b*) to express $\hat{J}_z \,|\psi_{\beta,\mu}\rangle$ we obtain

$$\boxed{\hat{J}_z(\hat{J}_+ \,|\psi_{\beta,\mu}\rangle) = \hbar(\mu + 1)(\hat{J}_+ \,|\psi_{\beta,\mu}\rangle)} \tag{7.41}$$

This is the first important result. It shows that the state $(\hat{J}_+ \,|\psi_{\beta,\mu}\rangle)$ is an eigenstate of $\hat{J}_z$ with the eigenvalue $\hbar(\mu + 1)$. Therefore, $\hat{J}_+$ operating on an eigenstate of $\hat{J}^2$ and $\hat{J}_z$, $|\psi_{\beta,\mu}\rangle$ produces the eigenstate $|\psi_{\beta,\mu+1}\rangle$ with the eigenvalue of $\hat{J}_z$ equal to $\hbar(\mu + 1)$. Correspondingly, one can show that the operator $\hat{J}_-$ operating on $|\psi_{\beta,\mu}\rangle$ produces the eigenstate $|\psi_{\beta,\mu-1}\rangle$ with the eigenvalue of $\hat{J}_z$ equal to $\hbar(\mu - 1)$.

We express this result by writing

$$\hat{J}_+ \,|\psi_{\beta,\mu}\rangle = c_{\beta,\mu} \,|\psi_{\beta,\mu+1}\rangle \tag{7.42a}$$

$$\hat{J}_- \,|\psi_{\beta,\mu}\rangle = d_{\beta,\mu} \,|\psi_{\beta,\mu-1}\rangle \tag{7.42b}$$

where the coefficients $c_{\beta,\mu}$ and $d_{\beta,\mu}$ are constants to be determined.

If $\hat{J}_+$ and $\hat{J}_-$ are raising and lowering operators, then $\hat{J}_+\hat{J}_-$ or $\hat{J}_-\hat{J}_+$ are diagonal operators as is also evident from Eqs. (7.39). Let us now calculate the expectation value of $\hat{J}_-\hat{J}_+$

$$\langle\psi_{\beta,\mu}\,|\hat{J}_-\hat{J}_+\,|\,\psi_{\beta,\mu}\rangle = |c_{\beta,\mu}|^2\,\langle\psi_{\beta,\mu+1}\,|\,\psi_{\beta,\mu+1}\rangle$$

or

$$\langle\psi_{\beta,\mu}|\,\hat{J}^2 - \hat{J}_z^2 - \hbar\hat{J}_z\,|\psi_{\beta,\mu}\rangle = |c_{\beta,\mu}|^2$$

or (7.43*a*)

$$(\hbar^2\beta - \hbar^2\mu^2 - \hbar^2\mu)\,\langle\psi_{\beta,\mu}|\,\psi_{\beta,\mu}\rangle = |c_{\beta,\mu}|^2$$

or

$$|c_{\beta,\mu}|^2 = \hbar^2[\beta - \mu(\mu + 1)]$$

where we have assumed that the states are properly normalized. We can choose the constants to be real so that

$$c_{\beta,\mu} = c^*_{\beta,\mu} = \hbar[\beta - \mu(\mu+1)]^{1/2} \tag{7.43b}$$

Similarly, the expectation value of $\hat{J}_+\hat{J}_-$ gives

$$\langle\psi_{\beta,\mu}|\hat{J}_+\hat{J}_-|\psi_{\beta,\mu}\rangle = |d_{\beta,\mu}|^2\langle\psi_{\beta,\mu-1}|\psi_{\beta,\mu-1}\rangle$$

or

$$\langle\psi_{\beta,\mu}|\hat{J}^2 - \hat{J}_z^2 + \hbar\hat{J}_z|\psi_{\beta,\mu}\rangle = |d_{\beta,\mu}|^2 \tag{7.44a}$$

or

$$|d_{\beta,\mu}|^2 = \hbar^2[\beta - \mu(\mu-1)]$$

And if we choose the constants to be real, then

$$d_{\beta,\mu} = d^*_{\beta,\mu} = \hbar[\beta - \mu(\mu-1)]^{1/2} \tag{7.44b}$$

Note that since β corresponds to the eigenvalue of a positive operator $\hat{J}^2$, $\beta \geqslant 0$. Furthermore, note from Eqs. (7.43*a*) and (7.44*a*) that both

$$|c_{\beta,\mu}|^2 \quad \text{and} \quad |d_{\beta,\mu}|^2 \geqslant 0 \tag{7.45}$$

This implies that although the eigenvalue μ of a state can be raised and lowered, it cannot be done so indefinitely. In fact, from Eqs. (7.43*a*), (7.44*a*), and (7.45), we see that there must exist a state with a maximum μ value μ_{max} such that

$$\beta - \mu_{max}(\mu_{max}+1) = 0 \tag{7.46a}$$

Similarly, there must exist a state with a minimum eigenvalue μ_{min} such that

$$\beta - \mu_{min}(\mu_{min}-1) = 0 \tag{7.46b}$$

It is quite simple to see that the above relations must be true, for if such states did not exist, we could apply the raising or the lowering operator to raise or lower the value of μ so that the positivity condition of Eq. (7.45) would be violated.

Equations (7.46*a*) and (7.46*b*) have two solutions. One which gives

$$\mu_{min} = \mu_{max} + 1 \tag{7.47}$$

has to be rejected since it says $\mu_{min} > \mu_{max}$. The second solution has the form

$$\mu_{max} = -\mu_{min} = j$$

and

$$\beta = j(j+1) \qquad \text{with } j \geqslant 0 \tag{7.48}$$

The final step is to show that j cannot take arbitrary values. To this effect we know from Eq. (7.42*a*) that we can obtain the state $|\psi_{\beta,\mu_{max}}\rangle$ from the state $|\psi_{\beta,\mu_{min}}\rangle$ by repeated application of the $\hat{J}_+$ operator (or vice versa by repeated application of $\hat{J}_-$). Every time $\hat{J}_+$ operates it increases the eigenvalue μ by one unit: therefore, the *difference* between μ_{max} and μ_{min} must be a positive *integer* or zero

$$\mu_{max} - \mu_{min} = j - (-j) = 2j = \text{positive integer or zero}$$

Thus j can be a positive half-integer, or integer, or zero! We summarize these important results and display them below:

$$\boxed{\begin{array}{l} \text{The eigenvalues of } \hat{J}^2 \text{ are} \\ \qquad \hbar^2 j(j+1) \qquad \text{with } j = 0, \tfrac{1}{2}, 1, \tfrac{3}{2}, 2, \ldots \\ \text{For given } j \text{ the eigenvalues of } \hat{J}_z \text{ are} \\ \qquad \hbar m \qquad \text{with } m = -j, -(j-1), \ldots, (j-1), j \end{array}} \tag{7.49}$$

The coefficients $c_{\beta,\mu}$ and $d_{\beta,\mu}$ are obtained from Eqs. (7.46)

$$c_{j,m} = \hbar[j(j+1) - m(m+1)]^{1/2} = \hbar[(j-m)(j+m+1)]^{1/2}$$
$$d_{j,m} = \hbar[j(j+1) - m(m-1)]^{1/2} = \hbar[(j+m)(j-m+1)]^{1/2}$$

leading to the important relations

$$\begin{aligned} \hat{J}_+ |\psi_{j,m}\rangle &= \hbar[(j-m)(j+m+1)]^{1/2} |\psi_{j,m+1}\rangle \\ \hat{J}_- |\psi_{j,m}\rangle &= \hbar[(j+m)(j-m+1)]^{1/2} |\psi_{j,m-1}\rangle \end{aligned} \tag{7.50}$$

We have found that the eigenstates of the angular momentum operators are characterized by two eigenvalues, that of $\hat{J}^2$ and $\hat{J}_z$. For any given value of the integer j there exists in the Hilbert space a subspace with $(2j+1)$ eigenstates that correspond to different eigenvalues of $\hat{J}_z$: these are labeled by the index m. It is therefore reasonable to use the analogy where the eigenvalue of $\hat{J}^2$, $\hbar^2 j(j+1)$, is a measure of the magnitude of the angular momentum of the state. Then the eigenvalues of $\hat{J}_z$, $\hbar m$, are interpreted as the projections of that angular momentum onto the Z-axis. Consequently, the projections of the angular momentum of a system, or, say, of its magnetic moment, on any one axis are quantized. The total number of such states depends on the value of j, that is, on the total angular momentum (or spin) of the system. We have already discussed the experimental observation of these effects on several occasions. For instance, for electrons or protons, $j = \frac{1}{2}$, and thus there are only two projections onto any axis. They are labeled by $m = \pm\frac{1}{2}$.

7.5. REPRESENTATIONS OF THE ANGULAR MOMENTUM OPERATORS AND THEIR EIGENSTATES

The results that we obtained in the previous section and the ones given in Eqs. (7.49) and (7.50) are completely general and independent of any particular representation. We will now discuss two particular representations for the angular momentum operators in order to facilitate their use in specific problems.

Matrix Representation

Let us choose a representation in which $\hat{J}^2$ and $\hat{J}_z$ are diagonal matrices. Then the eigenstates are labeled by the corresponding eigenvalue indices j and m, and we indicate them by the bra $|j, m\rangle$. If we "ignore" $\hat{J}_z$, the matrix for $\hat{J}^2$ would be of the form

$$\frac{1}{\hbar^2}\hat{J}^2 = \begin{array}{c|ccccc} j = & |0\rangle & |1/2\rangle & |1\rangle & |3/2\rangle & \cdots \\ \hline |0\rangle & 0 & 0 & 0 & 0 & \\ |1/2\rangle & 0 & 3/4 & 0 & 0 & \\ |1\rangle & 0 & 0 & 2 & 0 & \\ |3/2\rangle & 0 & 0 & 0 & 15/4 & \\ \vdots & & & & & \ddots \end{array}$$

This matrix is infinite since j can be as large as we wish, as long as it remains a half-integer or an integer. When we take $\hat{J}_z$ into account we note that to each value of j correspond $(2j+1)$ states labeled by their m-values. Thus, the complete form of the hermitian $\hat{J}^2$ matrix is

$$\frac{1}{\hbar^2}\hat{J}^2 \quad \begin{array}{r|c:cc:ccc:l}
 & |0,0\rangle & |1/2,+1/2\rangle & |1/2,-1/2\rangle & |1,+1\rangle & |1,0\rangle & |1,-1\rangle & \cdots \\
\hline
|0,0\rangle & 0 & \multicolumn{2}{c:}{0} & \multicolumn{3}{c:}{0} & \\
\hdashline
|1/2,+1/2\rangle & & 3/4 & 0 & & & & \\
 & 0 & & & & 0 & & 0 \\
|1/2,-1/2\rangle & & 0 & 3/4 & & & & \\
\hdashline
|1,+1\rangle & & & & 2 & 0 & 0 & \\
|1,0\rangle & 0 & \multicolumn{2}{c:}{0} & 0 & 2 & 0 & 0 \\
|1,-1\rangle & & & & 0 & 0 & 2 & \\
\hdashline
\vdots & 0 & \multicolumn{2}{c:}{0} & \multicolumn{3}{c:}{0} & \text{etc.}
\end{array} \tag{7.51a}$$

The dashed lines are used to separate the *degenerate* parts of the $\hat{J}^2$ matrix, that is, the states that correspond to the same value of j.

In this representation the hermitian $\hat{J}_z$ matrix is also diagonal, and in view of Eqs. (7.49) it has the form

$$\frac{1}{\hbar}\hat{J}_z = \begin{array}{r|c:cc:ccc:l}
 & |0,0\rangle & |1/2,+1/2\rangle & |1/2,-1/2\rangle & |1,+1\rangle & |1,0\rangle & |1,-1\rangle & \cdots \\
\hline
|0,0\rangle & 0 & \multicolumn{2}{c:}{0} & \multicolumn{3}{c:}{0} & \\
\hdashline
|1/2,+1/2\rangle & & +1/2 & 0 & & & & \\
 & 0 & & & & 0 & & \\
|1/2,-1/2\rangle & & 0 & -1/2 & & & & \\
\hdashline
|1,+1\rangle & & & & +1 & 0 & 0 & \\
|1,0\rangle & 0 & \multicolumn{2}{c:}{0} & 0 & 0 & 0 & \\
|1,-1\rangle & & & & 0 & 0 & -1 & \\
\hdashline
\vdots & & & & & & & \text{etc.}
\end{array} \tag{7.51b}$$

The eigenstates are represented by infinite column vectors with a 1 at the appropriate entry and zeros for all other entries. For instance

$$|0,0\rangle = \begin{pmatrix} 1 \\ 0 \\ 0 \\ 0 \\ 0 \\ 0 \\ \vdots \end{pmatrix} \qquad |1/2,-1/2\rangle \begin{pmatrix} 0 \\ 0 \\ 1 \\ 0 \\ 0 \\ 0 \\ \vdots \end{pmatrix} \qquad |1,0\rangle = \begin{pmatrix} 0 \\ 0 \\ 0 \\ 0 \\ 1 \\ 0 \\ \vdots \end{pmatrix} \qquad \text{etc.} \tag{7.51c}$$

To construct the matrices for $\hat{J}_x$ and $\hat{J}_y$, we make use of the definition of Eq. (7.37*a*)

$$\hat{J}_x = \frac{1}{2}(\hat{J}_+ + \hat{J}_-) \qquad \hat{J}_y = \frac{1}{2i}(\hat{J}_+ - \hat{J}_-) \tag{7.52a}$$

Since $\hat{J}_+$ and $\hat{J}_-$ do not change the value of j, the matrix elements of these operators cannot connect states with different j value. In other words, the matrices $\hat{J}_+$ and $\hat{J}_-$ are *block-diagonal* between the dashed lines of the matrices shown in Eqs. (7.51*a*) and (7.51*b*). It is therefore customary to write out the submatrices for any given value of j separately. The matrix elements are obtained directly from Eqs. (7.50)

$$\begin{aligned} \langle j, m+1|\, J_+ \,|j, m\rangle &= \hbar[(j-m)(j+m+1)]^{1/2} \\ \langle j, m-1|\, J_- \,|j, m\rangle &= \hbar[(j+m)(j-m+1)]^{1/2} \end{aligned} \tag{7.52b}$$

and all other matrix elements are zero because of the orthogonality of the eigenstates.

Using the above results we find for the case $j = 1/2$

$\dfrac{\hat{J}_+}{\hbar} =$

	$\lvert 1/2, +1/2\rangle$	$\lvert 1/2, -1/2\rangle$
$\lvert 1/2, +1/2\rangle$	0	1
$\lvert 1/2, -1/2\rangle$	0	0

$\dfrac{\hat{J}_-}{\hbar} =$

	$\lvert 1/2, +1/2\rangle$	$\lvert 1/2, -1/2\rangle$
$\lvert 1/2, +1/2\rangle$	0	0
$\lvert 1/2, -1/2\rangle$	1	0

From these non-hermitian matrices we obtain $\hat{J}_x$ and $\hat{J}_y$ using Eqs. (7.52*a*). Thus, the $\hat{J}_x$, $\hat{J}_y$, $\hat{J}_z$, and $\hat{J}^2$ submatrices for the case $j = 1/2$ have the following forms

$$\hat{J}_x = \frac{\hbar}{2}\begin{pmatrix} 0 & 1 \\ 1 & 0 \end{pmatrix} \qquad \hat{J}_y = \frac{\hbar}{2}\begin{pmatrix} 0 & -i \\ i & 0 \end{pmatrix}$$

$$\hat{J}_z = \frac{\hbar}{2}\begin{pmatrix} 1 & 0 \\ 0 & -1 \end{pmatrix} \qquad \hat{J}^2 = \tfrac{3}{4}\hbar^2\begin{pmatrix} 1 & 0 \\ 0 & 1 \end{pmatrix} \tag{7.53}$$

Note that the Pauli matrices of Eqs. (4.68*c*) differ from the angular momentum submatrices for the case $j = 1/2$ only by a numerical constant, it holds that†

$$\hat{\mathbf{J}}(j = 1/2) = \frac{\hbar}{2}\hat{\boldsymbol{\sigma}}$$

By the same procedure we find the submatrices for $j = 1$. They are

$$\hat{J}_x = \frac{\hbar}{\sqrt{2}}\begin{pmatrix} 0 & 1 & 0 \\ 1 & 0 & 1 \\ 0 & 1 & 0 \end{pmatrix} \qquad \hat{J}_y = \frac{\hbar}{\sqrt{2}}\begin{pmatrix} 0 & -i & 0 \\ i & 0 & -i \\ 0 & i & 0 \end{pmatrix}$$

$$\hat{J}_z = \hbar\begin{pmatrix} 1 & 0 & 0 \\ 0 & 0 & 0 \\ 0 & 0 & -1 \end{pmatrix} \qquad \hat{J}^2 = 2\hbar^2\begin{pmatrix} 1 & 0 & 0 \\ 0 & 1 & 0 \\ 0 & 0 & 1 \end{pmatrix} \tag{7.54}$$

This process can be repeated for any arbitrary value of j, yielding submatrices of dimensionality $(2j+1)$. Some remarks are appropriate at this point. First, we note that the dimensionality of the matrix representation depends on the value of j, namely, it is $(2j+1)$. Secondly, the representations that we have given in Eqs. (7.53) and (7.54) are not unique since one can perform a similarity transformation to a new set of basis states. However, they are the only ones where $\hat{J}_z$ and $\hat{J}^2$ are simultaneously diagonal. One should keep in mind that the matrix elements of Eqs. (7.49) and (7.50) define the

† Here we used the shorthand notation given in the footnote to p. 281.

properties of the operators, the matrices being only a convenient tabulation of these matrix elements.

Coordinate Representation

We now turn our attention to the coordinate representation where the amplitudes will be given by functions of the coordinates. The angular momentum operators are then represented by differential operators acting on the coordinates. Clearly, we can only talk here about the orbital angular momentum since spin refers to an internal degree of freedom not connected with the coordinates. We have made use of this representation in Chapter 6 [see Eqs. (6.21c)]. In the coordinate representation the basis states are the continuous set of eigenstates of the position operator $\hat{\mathbf{x}}$. Since the angular momentum operators generate rotations, they will not change the length of the vector $\mathbf{x}$ but only its direction. It therefore suffices to consider only the basis states that describe all points on a unit sphere centered at the origin of the coordinates. For this purpose, we label the basis states by the two angles θ and ϕ, i.e., $|\theta, \phi\rangle$. These basis states form a continuous orthonormal set, and therefore a particular eigenstate $|l, m\rangle$ of the angular momentum operators can be described by the amplitude

$$\langle \theta, \phi \mid l, m\rangle \tag{7.55}$$

This represents the wavefunction for the eigenstate $|l, m\rangle$ in exact analogy to Eqs. (3.25) and (3.26). Note that we are denoting the eigenvalues of orbital angular momentum by l. The absolute square of the amplitude of Eq. (7.55) gives the probability of finding a particle, which is in the state $|l, m\rangle$ at the position (θ, ϕ) on the unit sphere.

The amplitude $\langle \theta, \phi \mid l, m\rangle = \psi_{l,m}(\theta, \phi)$ is a continuous function of (θ, ϕ), and the operators $\hat{L}_x$, $\hat{L}_y$, $\hat{L}_z$, and $\hat{L}^2$ will be expressed as differential operators acting on this function. We already know the form of these operators in the coordinate representation, as given by Eqs. (6.21c). In terms of the (θ, ϕ) coordinates, the orbital angular momentum operators take the form (see Appendix 8)

$$\hat{L}_z = -i\hbar \frac{\partial}{\partial \phi} \tag{7.56a}$$

$$\hat{L}_\pm = \hat{L}_x \pm i\hat{L}_y = \pm \hbar e^{\pm i\phi}\left[\frac{\partial}{\partial \theta} \pm i \cot \theta \frac{\partial}{\partial \phi}\right] \tag{7.56b}$$

$$\hat{L}^2 = -\hbar^2\left[\frac{1}{\sin \theta}\frac{\partial}{\partial \theta}\left(\sin \theta \frac{\partial}{\partial \theta}\right) + \frac{1}{\sin^2 \theta}\frac{\partial^2}{\partial \phi^2}\right] \tag{7.56c}$$

We now seek functions of (θ, ϕ) that are simultaneous eigenfunctions of $\hat{L}_z$ and $\hat{L}^2$ subject to the boundary condition $\psi(\theta, \phi + 2\pi) = \psi_{l,m}(\theta, \phi)$. Note from Eq. (7.56$a$) that the ϕ dependence of the eigenfunction of $\hat{L}_z$ would be of the form

$$\psi_{l,m}(\theta, \phi) \propto e^{im\phi}$$

so that

$$\hat{L}_z \psi_{l,m}(\theta, \phi) = \hbar m \psi_{l,m}(\theta, \phi)$$

Furthermore, the boundary condition

$$\psi_{l,m}(\theta, \phi + 2\pi) = \psi_{l,m}(\theta, \phi)$$

imposes the condition that the m-quantum numbers must be integers. Correspondingly, the eigenvalues of $\hat{L}^2, \hbar^2 l(l+1)$ [see Eqs. (7.49)] can only involve integer l values. The complete eigenfunctions of orbital angular momentum are known as the *spherical harmonics* and are designated by

$$Y_{l,m}(\theta, \phi)$$

where

$$l = 0, 1, 2, \ldots$$

and

$$m = -l, -l+1, \ldots, 0, \ldots, (l-1), l$$

The $Y_{l,m}$'s are expressed in terms of the associated Legendre polynomials as

$$Y_{l,m}(\theta, \phi) = \varepsilon\left[\frac{2l+1}{4\pi}\frac{(l-m)!}{(l+m)!}\right]^{1/2} P_l^m(\cos \theta) e^{im\phi}$$

where $\varepsilon = (-1)^m$ if $m > 0$ and $\varepsilon = 1$ otherwise [see also Appendix 9].

The first few normalized spherical harmonics are

$$
\begin{aligned}
Y_{00} &= \frac{1}{(4\pi)^{1/2}} \\
Y_{1,\pm 1} &= \mp\left(\frac{3}{8\pi}\right)^{1/2} \sin\theta\, e^{\pm i\phi} \\
Y_{1,0} &= \left(\frac{3}{4\pi}\right)^{1/2} \cos\theta \\
Y_{2,\pm 2} &= \frac{1}{4}\left(\frac{15}{2\pi}\right)^{1/2} \sin^2\theta\, e^{\pm 2i\phi} \\
Y_{2,\pm 1} &= \mp\left(\frac{15}{8\pi}\right)^{1/2} \sin\theta\cos\theta\, e^{\pm i\phi} \\
Y_{2,0} &= \left(\frac{5}{4\pi}\right)^{1/2} \left(\tfrac{3}{2}\cos^2\theta - \tfrac{1}{2}\right)
\end{aligned}
\tag{7.57}
$$

The normalization is such that

$$\int Y^*_{l',m'} Y_{l,m}\, d(\cos\theta)\, d\phi = \delta_{ll'}\delta_{mm'}$$

where the integration is over the surface of the unit sphere [$\cos\theta(-1 \to +1)$ and $\phi(0 \to 2\pi)$]. By direct application of Eqs. (7.56a)–(7.56c) on the above spherical harmonics one finds that they obey the fundamental properties of Eqs. (7.49) and (7.50). For arbitrary l and m the proof follows from the mathematical properties of the Legendre equation. [The Legendre equation is equivalent to the form of Eq. (7.56c).]

The spherical harmonics are also eigenstates of the parity operator (inversion of the coordinates: $\theta \to \pi - \theta$, $\phi \to \phi + \pi$) with eigenvalue $(-1)^l$

$$Y_{l,m}(\pi - \theta, \phi + \pi) = (-1)^l Y_{l,m}(\theta, \phi) \tag{7.58a}$$

and it can be shown that

$$Y_{l,-m} = (-1)^l Y^*_{l,m} \tag{7.58b}$$

Note that the $Y_{l,m}$'s can be expressed in terms of the cartesian coordinates using the substitutions

$$\cos\theta = \frac{z}{r} \qquad \sin\theta\cos\phi = \frac{x}{r} \qquad \sin\theta\sin\phi = \frac{y}{r} \tag{7.59a}$$

As an example we give the useful expressions

$$Y_{1,0} = \left(\frac{3}{4\pi}\right)^{1/2}\frac{z}{r} \qquad Y_{1,1} = -\left(\frac{3}{8\pi}\right)^{1/2}\frac{x+iy}{r}$$

$$Y_{1,-1} = \left(\frac{3}{8\pi}\right)^{1/2}\frac{x-iy}{r} \tag{7.59b}$$

7.6. COMPOSITION OF ANGULAR MOMENTA

We are familiar with the addition of two or more vectors to a resultant vector. Similarly, in classical mechanics, when a spinning top is subject to an external torque, we vectorially add the two angular momentum vectors to obtain the direction along which the axis of spin will be pointing. An analogous situation exists in quantum mechanics. For instance, a hydrogen atom consists of an electron and a proton bound to each another. Both particles can be in a given eigenstate of angular momentum, but we can also think of the two particles as forming a single system that as a whole is in a particular angular momentum eigenstate.

The problem of composition of angular momenta can then be stated as follows: given two components of a system, each in an eigenstate $|j, m\rangle$ of angular momentum, what are the possible eigenstates $|J, M\rangle$ of the combined system? The answer follows from a direct application of the general properties of the angular momentum operators [Eqs. (7.35*b*), (7.49), and (7.50)]. If we designate by $\hat{\mathbf{J}}^{(1)}$ and $\hat{\mathbf{J}}^{(2)}$ the angular momentum operators acting on each of the two components, we define the *total* angular momentum operator $\hat{\mathbf{J}}$ for the system as†

$$\hat{\mathbf{J}} = \hat{\mathbf{J}}^{(1)} + \hat{\mathbf{J}}^{(2)} \tag{7.60a}$$

† In this section we freely use the notation $\hat{\mathbf{J}}$ for the triplet of operators $\hat{J}_x$, $\hat{J}_y$, and $\hat{J}_z$, as discussed in the footnote to p. 281. We do so because we want to take advantage of the compactness of notation.

The operator $\hat{\mathbf{J}}$ has the commutation relations of angular momentum. By using Eq. (7.35b) for $\hat{\mathbf{J}}^{(1)}$ and $\hat{\mathbf{J}}^{(2)}$ and the fact that $\hat{\mathbf{J}}^{(1)}$ commutes with $\hat{\mathbf{J}}^{(2)}$ it can be shown that

$$\hat{\mathbf{J}}\times\hat{\mathbf{J}}=i\hbar\hat{\mathbf{J}} \tag{7.60b}$$

Furthermore $\hat{\mathbf{J}}^{(1)}$ acts only on the state $|j_1m_1\rangle$ of the first component and leaves the state $|j_2m_2\rangle$ of the second component unchanged. The operator $\hat{\mathbf{J}}^{(2)}$ acts on $|j_2m_2\rangle$ but leaves $|j_1m_1\rangle$ unchanged. From these definitions one can obtain the eigenstates $|J, M\rangle$ as indicated in more detail in Appendix 6. Here, instead, we will discuss a simple example where two constituents with spin 1/2 are combined.

Consider the hydrogen atom in its ground state. It consists of an electron and a proton which both are spin-1/2 particles. In the ground state there is no orbital angular momentum. The spin projection of the proton onto the Z-axis can take only the two values $m_p=\pm 1/2$ and similarly for the electron $m_e=\pm 1/2$. Thus, the atom can be found in one of the following four states

$$|m_p=+1/2\rangle\,|m_e=+1/2\rangle \tag{7.61a}$$

$$|m_p=+1/2\rangle\,|m_e=-1/2\rangle \tag{7.61b}$$

$$|m_p=-1/2\rangle\,|m_e=+1/2\rangle \tag{7.61c}$$

$$|m_p=-1/2\rangle\,|m_e=-1/2\rangle \tag{7.61d}$$

We state without proof† that the projection of the total angular momentum (that is, the eigenvalues of $\hat{J}_z$) for these four states correspond to the M-values

$$M=+1,\quad M=0,\quad M=0,\quad \text{and} \quad M=-1$$

The question is now whether the four *product* states of Eqs. (7.61a)–(7.61d) are also eigenstates of $(\hat{\mathbf{J}})^2$, and if so what the corresponding eigenvalues are. The states given by Eqs. (7.61a) and (7.61d) must belong to $J\geqslant 1$ since according to Eq. (7.49) $|M|\leqslant J$. In fact, they belong to $J=1$. The states given by Eqs. (7.61b) and (7.61c) with $M=0$ are not eigenstates of $(\hat{\mathbf{J}})^2$, but their linear combinations are. The four product states are then combined into

† To show this, simply act with $\hat{J}_z=\hat{J}_z^{(1)}+\hat{J}_z^{(2)}$ on the states given by Eqs. (7.61), keeping in mind how $\hat{J}_z^{(1)}$ and $\hat{J}_z^{(2)}$ act.

eigenvalues of $(\hat{\mathbf{J}})^2$ and $\hat{J}_z$ as follows

$$|J=1; M=+1\rangle = |m_p=+1/2; m_e=+1/2\rangle$$

$$|J=1; M=0\rangle = \frac{1}{\sqrt{2}}\{|m_p=+1/2; m_e=-1/2\rangle + |m_p=-1/2; m_e=+1/2\rangle\} \quad (7.62a)$$

$$|J=1; M=-1\rangle = |m_p=-1/2; m_e=-1/2\rangle$$

and

$$|J=0; M=0\rangle = \frac{1}{\sqrt{2}}\{|m_p=+1/2; m_e=-1/2\rangle - |m_p=-1/2; m_e=+1/2\rangle\} \quad (7.62b)$$

While we do not give the proof here, the reader can use Eq. (A6.4) of Appendix 6 to show that the states of Eq. (7.62*a*) are eigenstates of $(\hat{\mathbf{J}})^2$ with eigenvalue $\hbar^2 J(J+1)$, where $J=1$, whereas for the state of Eq. (7.62*b*) $J=0$.

We have found that two angular momenta with $j_1=1/2$ and $j_2=1/2$ combine into total angular momentum with either $J=0$ or $J=1$. We can represent this result by the sketches of Fig. 7.2 where j_1, j_2, and J are shown symbolically as collinear vectors. We refer to the configuration in (a) of the figure as the *stretched* case and to the one in (b) as the *jackknife*. This picture can be extended to values of j_1, j_2 different from 1/2. The maximum value of J is obtained always in the stretched case and equals $J_{\max}=j_1+j_2$. The minimum value of J corresponds to the jackknife and equals $J_{\min}=|j_1-j_2|$. The total angular momentum can now take all the values

$$J = J_{\min}, (J_{\min}+1), (J_{\min}+2), \ldots, (J_{\max}-1), J_{\max} \quad (7.63a)$$

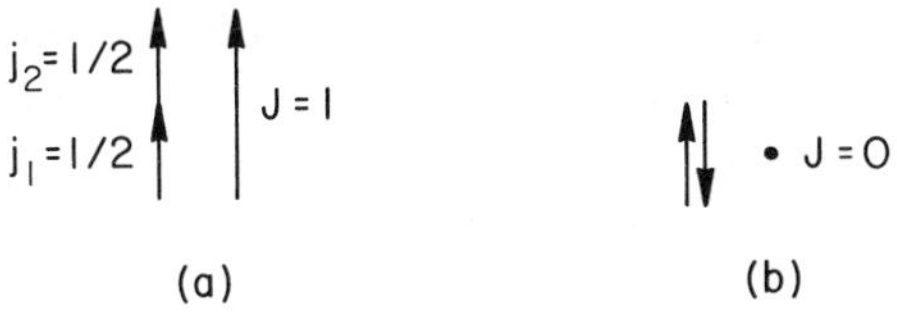

FIGURE 7.2. Vector model representation of the combination of two angular momenta with $j_1=1/2$, $j_2=1/2$. (a) The $J=1$ case (stretched). (b) The $J=0$ case (jackknife).

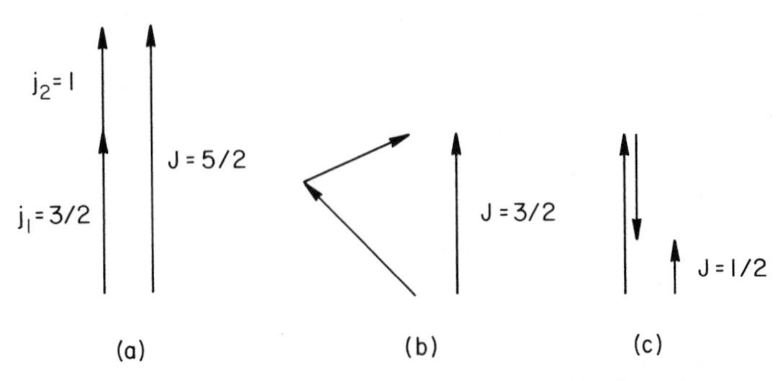

FIGURE 7.3. Vector model representation of the combination of two angular momenta with $j_1 = 3/2$, $j_2 = 1$. (a) The $J = 5/2$ case (stretched). (b) The $J = 3/2$ case. (c) The $J = 1/2$ case (jackknife).

This is shown in Fig. 7.3 for $j_1 = 3/2$ and $j_2 = 1$. Another important conclusion we draw from Eqs. (7.62) is that for all states

$$M = m_1 + m_2 \tag{7.63b}$$

This result generalizes to arbitrary values of j_1, j_2 and is always valid when we compose two angular momenta. Furthermore, three or more angular momenta can be added by combining two at a time by the methods of this section.

Example

Here we show that the number of states for the total angular momentum equals the number of product states formed from $j_1 = 3/2$ and $j_2 = 1$. We have already seen this result for the case $j_1 = 1/2$ and $j_2 = 1/2$ where there are four possible states as given by Eqs. (7.61) and (7.62). For $j_1 = 3/2$ there are $2j_1 + 1 = 4$ possible states, hereas for $j_2 = 1$ there are $2j_2 + 1 = 3$ possible states. The total number of product states is therefore $3 \times 4 = 12$. From Fig. 7.3 the combined angular momentum can take the values $J = 5/2$, $3/2$, and $1/2$ and the corresponding number of states is $2J + 1$ for each case

$$(5+1) + (3+1) + (1+1) = 12$$

The result is generally valid, as can be easily shown by summing the series for arbitrary values of j_1 and j_2.

As an application of angular momentum composition we will consider an electron moving in a fixed potential† and assume that it is in a state with orbital angular momentum characterized by $l = 1$. In addition, the electron has spin $s = 1/2$. We can compose (add) these two angular momenta, that is, the orbital angular momentum and the spin. We then see that the electron can be in a state where its total angular momentum is given by

$$J = 1/2 \quad \text{or} \quad J = 3/2$$

These two states differ in energy by a small amount and give rise to a close doublet in atomic spectra. This is particularly pronounced in sodium.

The assignments of the total angular momentum can be verified by applying a magnetic field to the atom. In the presence of the field, states with different angular momentum projections onto the field direction acquire different energy. There are four such states for the $J = 3/2$ level and two states for $J = 1/2$; this is sketched in Fig. 7.4 and is fully confirmed by experiment.

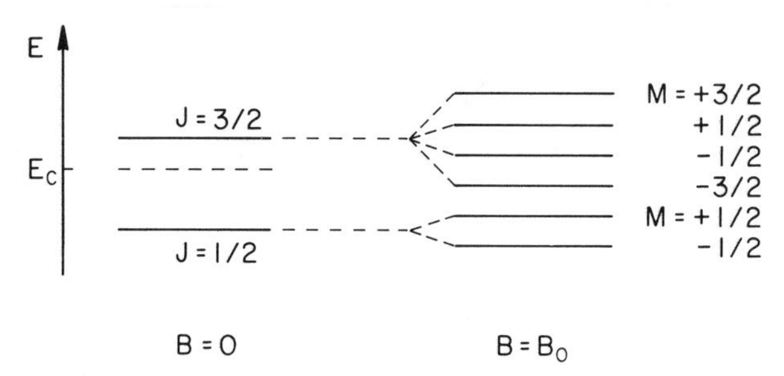

FIGURE 7.4. Splitting of angular momentum multiplets when an atom is placed in a magnetic field. The energy of states with $J = 1/2$ and $J = 3/2$ is shown in the absence and in the presence of a magnetic field.

† This is the case for the excited state of the hydrogen or sodium atom if we ignore the interaction of the electron with the magnetic moment of the nucleus, i.e., if we ignore the nuclear spin.

Problems

PROBLEM 1

Calculate the expectation value of $(d/dt)\langle\hat{x}\rangle$ for a relativistic particle moving freely in the x-direction. The Hamiltonian for this particle is $\hat{H}=(c^2\hat{p}_x^2+m_0^2c^4)^{1/2}$, with m_0 the rest mass of the particle. The wavefunction for this particle is

$$\psi(x,t)=\frac{1}{V^{1/2}}e^{-i/\hbar(Et-p_x x)}$$

where $E=(c^2p_x^2+m_0^2c^4)^{1/2}$. However, it is best to work with the Hamiltonian directly. You will need to make an expansion because of the $(\;)^{1/2}$, and then "undo" the expansion. The desired result is

$$\frac{d}{dt}\langle x\rangle=c^2\left\langle\frac{p_x}{E}\right\rangle$$

(*a*) How does this compare with the classical result for a slow particle?
(*b*) For a relativistic particle?

PROBLEM 2

The rotation matrices can be obtained from the general expression

$$\hat{R}(\mathbf{n},\theta)=e^{i/\hbar(\hat{\mathbf{J}}\cdot\mathbf{n})\theta}$$

where $\mathbf{n}$ is a unit vector along the axis of rotation, θ is the finite rotation angle, and $\hat{\mathbf{J}}$ is the angular momentum operator.

(*a*) Find the rotation matrices for spin 1/2 for a rotation about the Z-axis and for a rotation about the Y-axis. For spin-1/2 particles the angular momentum operator $\hat{\mathbf{J}}$ has components

$$\hat{J}_x = \frac{\hbar}{2}\hat{\sigma}_x \qquad \hat{J}_y = \frac{\hbar}{2}\hat{\sigma}_y \qquad \hat{J}_z = \frac{\hbar}{2}\hat{\sigma}_z$$

along the three axes.

(*b*) Find the rotation matrix about the X-axis.

PROBLEM 3

Repeat the calculations of Problem 2 but for spin 1.

PROBLEM 4

Clasically the quadrupole moment tensor is given by

$$Q_{ij} = \frac{1}{e}\int (3x_i x_j - \delta_{ij} r^2)\rho(r)\, d^3r, \qquad i, j, k = x, y, z$$

where $\rho(r)$ is the charge density of the system, such that $\int \rho(r)\, d^3r = e$.
Quantum-mechanically

$$\hat{Q}_{ij} = \frac{1}{e}\int r^2[\tfrac{3}{2}(\hat{J}_i\hat{J}_j + \hat{J}_j\hat{J}_i) - \delta_{ij}\hat{J}^2]\rho(r)\, d^3\mathbf{r}$$

The *quadrupole moment* Q_0 for a stationary state $|n, j\rangle$ is defined as the expectation value of $\hat{Q}_{zz}$ evaluated in the state $m = j$.

(*a*) Evaluate

$$Q_0 = \langle n, j, m = j|\, \hat{Q}_{zz}\, |n, j, m = j\rangle$$

in terms of j and $\overline{r^2} = \langle n, j|\, r^2\, |n, j\rangle$.

(*b*) Can the proton ($j = 1/2$) have a quadrupole moment? the deuteron ($j = 1$)?

(*c*) Evaluate $\langle n, j, m|\, \hat{Q}_{xy}\, |n, j, m\rangle$.

(*d*) The quantum-mechanical expression for the *electric* dipole moment is

$$p_0 = \langle n, j, m = j | \frac{r}{e} \hat{J}_z | n, j, m = j \rangle$$

Can an eigenstate of a Hamiltonian with a central potential have an *electric* dipole moment?

PROBLEM 5

Use the raising and lowering operators for eigenstates of angular momentum [Eqs. (7.50)] to *derive* the Clebsch–Gordan matrix for the case $j_2 = 1/2$, $j_1 =$ arbitrary. [See Appendix 6]

PROBLEM 6

(*a*) Show that for angular momentum states with $j = 1$ the operator

$$\hat{E}_1 = a\hat{J}_+^2 \hat{J}_-^2$$

is (for a suitable choice of a) a projection operator onto the state $|j = 1, m = 1\rangle$. Find the expression for a.

(*b*) Construct the projection operators onto the other two states $|j = 1, m = 0\rangle$ and $|j = 1, m = -1\rangle$.

(*c*) Generalize your results to construct the operator that projects an arbitrary state of angular momentum j onto the state $|j, m\rangle$.

Note: The definition of a projection operator $\hat{P}_\alpha$ which projects onto the state $|\alpha\rangle$ (or projects out of the state $|\phi\rangle$ its content $|\alpha\rangle$) is

$$\hat{P}_\alpha |\phi\rangle = \langle \alpha | \phi \rangle |\alpha\rangle$$

Explain the physical interpretation of this definition.

PROBLEM 7

A particle with angular momentum eigenvalue $j=5/2$ is described within a six-dimensional space. Choose a basis that diagonalizes $\hat{J}_z$ and $\hat{J}^2$.

(*a*) Show that in this basis, the operator $(\hat{J}_x^2-\hat{J}_y^2)$ is block-diagonal when the states are properly ordered. It separates into two 3×3 matrices. Explain why this happens.
(*b*) Find the eigenvalues of $(\hat{J}_x^2-\hat{J}_y^2)$.

PROBLEM 8

Let the Hamiltonian describing only the spin of two spin-1/2 particles be

$$\hat{H}=\hat{H}_0+\hat{H}'$$

where

$$\hat{H}_0=\frac{4A}{\hbar^2}\mathbf{S}_1\cdot\mathbf{S}_2 \quad \text{and} \quad \hat{H}'=\frac{2B}{\hbar}\hat{S}_{1z}$$

where A and B are constants.

(*a*) Determine the eigenvalues and eigenstates of $\hat{H}_0$.
(*b*) Calculate the lowest-order effect of $\hat{H}'$ on the energy eigenvalues.

PROBLEM 9

Prove the Jacobi identity for Poisson brackets and show that the corresponding identity for *noncommuting* operators is

$$[\hat{A},[\hat{B},\hat{C}]]+[\hat{B},[\hat{C},\hat{A}]]+[\hat{C},[\hat{A},\hat{B}]]=0$$

PROBLEM 10

In a state denoted by $|l, m\rangle$ calculate $\langle \hat{L}_x \rangle$, $\langle \hat{L}_y \rangle$, $\langle \hat{L}_x^2 \rangle$, and $\langle \hat{L}_y^2 \rangle$.

PROBLEM 11

Show that

$$e^{-i\beta \hat{L}_y} = e^{i(\pi/2)\hat{L}_x} e^{-i\beta \hat{L}_z} e^{-i(\pi/2)L_x}$$

PROBLEM 12

Three 2×2 matrices satisfy the relations

$$[\hat{Q}_x, \hat{Q}_y] = -i\hat{Q}_z$$
$$[\hat{Q}_y, \hat{Q}_z] = -i\hat{Q}_x$$
$$[\hat{Q}_z, \hat{Q}_x] = -i\hat{Q}_y$$

Determine such a set of matrices. (Note that these are not the same as the commutation relations satisfied by Pauli matrices.)

PROBLEM 13

Two particles each with angular momentum $j_1 = j_2 = 1$ form a state with total angular momentum $J = 2$ and the z-component of total angular momentum $M = 1$. Construct such a state from linear combinations of states $|j_1, m_1; j_2, m_2\rangle$.

PROBLEM 14

Let $\hat{\mathbf{A}}$ be a vector operator, namely, a triplet of operators $\{\hat{A}_i, \hat{A}_j, \hat{A}_k\}$ such that

$$[\hat{J}_i, \hat{A}_j] = i\varepsilon_{ijk}\hat{A}_k, \qquad \hbar \equiv 1$$

You are *given* the relation (see Problem 16 below)

$$[\hat{J}^2, [\hat{J}^2, \hat{\mathbf{A}}]] = 2(\hat{J}^2\hat{\mathbf{A}} + \hat{\mathbf{A}}\hat{J}^2) - 4\hat{\mathbf{J}}(\hat{\mathbf{J}} \cdot \hat{\mathbf{A}})$$

Expand the commutators on the lhs and then form the matrix elements of *both* sides between eigenstates of angular momentum $\langle j, m|$ and $|j, m'\rangle$ corresponding to the *same* eigenvalue j. Show that

$$j(j+1)\,\langle j, m|\,\hat{\mathbf{A}}\,|j, m'\rangle = \langle j, m|\,\hat{\mathbf{J}}\,|j, m'\rangle\langle j, m'|\,\hat{\mathbf{J}} \cdot \hat{\mathbf{A}}\,|j, m'\rangle$$

Note: Recall that $(\hat{\mathbf{A}} \cdot \hat{\mathbf{J}})$ is a scalar operator and thus has only diagonal elements in the j, m representation.

PROBLEM 15

Consider a system consisting of an electron and a proton. The basis states can be labeled as $|p+, e+\rangle$, $|p+, e-\rangle$, $|p-, e+\rangle$, and $|p-, e-\rangle$. The operators

$$\hat{\boldsymbol{\sigma}}_e \quad \text{and} \quad \hat{\boldsymbol{\sigma}}_p$$

operate on the electron and proton spin, respectively. Define a *spin exchange* operator $\hat{P}_{\text{spin exch}}$ that has the property of interchanging the spin projections of the electron and the proton

$$\hat{P}_{\text{spin exch}}\,|p+, e-\rangle = |p-, e+\rangle, \qquad \text{etc.}$$

Prove the following identity obeyed by this operator

$$\hat{\boldsymbol{\sigma}}_e \cdot \hat{\boldsymbol{\sigma}}_p = 2\hat{P}_{\text{spin exch}} - 1$$

PROBLEM 16

Let $\hat{\mathbf{A}}$ be a vector operator that satisfies the commutation relation

$$[\hat{J}_i, \hat{A}_j] = i\hat{A}_k \varepsilon_{ijk}, \qquad i, j, k \text{ cyclic}$$

where $\hat{\mathbf{J}}$ is the angular momentum operator.

Prove the relations

$$[\hat{J}^2, \hat{\mathbf{A}}] = i[(\hat{\mathbf{A}} \times \hat{\mathbf{J}}) - (\hat{\mathbf{J}} \times \hat{\mathbf{A}})]$$

and

$$[\hat{J}^2, [\hat{J}^2, \hat{\mathbf{A}}]] = 2(\hat{J}^2\hat{\mathbf{A}} + \hat{\mathbf{A}}\hat{J}^2) - 4\hat{\mathbf{J}}(\hat{\mathbf{J}} \cdot \hat{\mathbf{A}})$$

Chapter 8

BOUND STATES: PART I

When the forces between two or more particles (objects) are attractive, it is possible for them to become bound to each other and form a single system. For example, classically we know of the planetary system. Microscopically we know that an atom consists of a number of electrons bound to a nucleus. Nuclei are bound systems of neutrons and protons; two or more atoms bind into molecules, and so on. In these microscopic systems, however, the momentum multiplied by the distance (size) is of the order of $\hbar$, Planck's constant. Therefore, the uncertainty relation is nonnegligible in these cases, and hence such systems must be analyzed using the rules of quantum mechanics. In fact, quantum mechanics arose out of attempts to correctly describe such bound systems.

Bulk matter, such as a stone, a crystal, or a liquid drop, can also be considered as a bound system. However, such systems consist of a very large number of particles and will be discussed later. In this and the chapter following, we will consider only simple bound systems which consist of two particles and, by analogy, similar systems with a small number of constituents. The nature of a bound system depends on the force between the particles (in quantum mechanics we prefer to talk about the corresponding potential). Exact analytic solutions can be obtained only for a few special forms of the potential. In this chapter we will discuss the Coulomb potential $-\alpha/r$. The harmonic oscillator potential $\frac{1}{2}kr^2$ will be discussed in Chapter 9. Both these potentials allow for exact solutions. The behavior of many other systems can be understood by analogy with these potentials, and in these cases detailed solutions must be obtained by approximate and/or numerical methods.

To describe a bound system in quantum mechanics we look for its stationary states. A bound state is a localized state and hence must satisfy the boundary condition that the wavefunction vanishes as the distance from the origin tends to infinity. As we have seen before, this leads to a discrete spectrum of energy eigenvalues. Besides the energy, we are also interested in the expectation values (average values) of operators corresponding to physical observables. We would also like to know the matrix elements of the operators that cause transitions between different stationary states. All this is easily calculated in the coordinate representation since, in this case, the wavefunction contains all the information about the system. Thus, the problem is reduced to finding the solutions of the time-independent Schrödinger equation for a given potential. Note that both the Coulomb potential as well as the harmonic oscillator potential have spherical symmetry. That is, rotations leave the Hamiltonian invariant, and consequently the stationary states are simultaneously eigenstates of the angular momentum operators.

We begin by introducing the relevant kinematics for a two-particle system and show that for translation-invariant potentials, the two-body motion effectively reduces to a one-particle equation. For spherically symmetric potentials, the angular part of the Schrödinger equation separates conveniently, and we discuss the form of the radial equation in general. We solve the radial equation for a Coulomb potential, and then the solution is applied to the case of the hydrogen atom. We discuss the consequence of the electron also possessing a spin angular momentum. The fine structure and the hyperfine structure in the energy levels of the hydrogen atom are also explained. We introduce the variational method as an approximate method for evaluating bounds on the ground-state energy of a system. This is then applied to the case of the linear potential, and the bound state of a quark–antiquark system (quarkonium) is discussed. A brief presentation of the Laguerre polynomials and their properties can be found in Appendix 10.

8.1. SEPARATION OF THE CENTER-OF-MASS MOTION

Let us consider the motion of two particles, 1 and 2, which are subject to a force that depends on their positions $\mathbf{r}_1$ and $\mathbf{r}_2$, as shown in Fig. 8.1(a). The potential is $U(\mathbf{r}_1, \mathbf{r}_2)$ and, therefore, the Hamiltonian in the coordinate representation is given by

$$\hat{H} = \frac{\hat{p}_1^2}{2m_1} + \frac{\hat{p}_2^2}{2m_2} + \hat{U}$$

$$= -\frac{\hbar^2}{2m_1}\nabla_1^2 - \frac{\hbar^2}{2m_2}\nabla_2^2 + U(\mathbf{r}_1, \mathbf{r}_2) \tag{8.1}$$

Let the potential depend only on the separation between the two particles. That is,

$$U(\mathbf{r}_1, \mathbf{r}_2) = U(|\mathbf{r}_1 - \mathbf{r}_2|) \tag{8.2}$$

We call such a potential *central.* It remains invariant under the translation and rotation of the coordinate system.

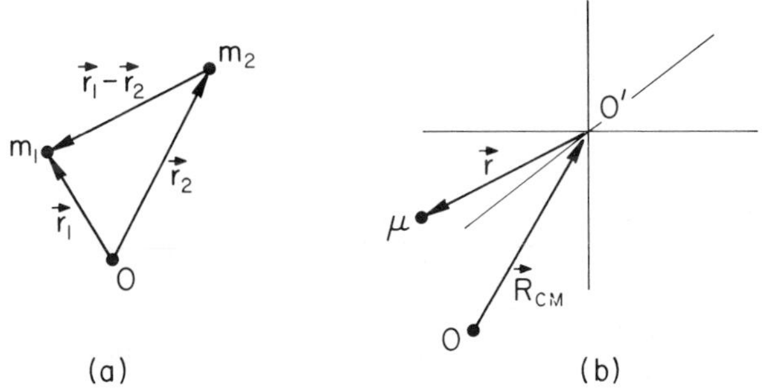

FIGURE 8.1. (a) Coordinates of two particles with respect to a fixed origin O. (b) The center-of-mass coordinate system with origin at O'.

We now introduce the relative and center-of-mass (cm) coordinates as shown in Fig. 8.1(b).

$$\mathbf{r} = \mathbf{r}_1 - \mathbf{r}_2$$

$$\mathbf{R}_{CM} = \frac{m_1\mathbf{r}_1 + m_2\mathbf{r}_2}{m_1 + m_2} \tag{8.3a}$$

The expressions for the relative momentum and the center-of-mass momentum operators are given as

$$\hat{\mathbf{p}} = \frac{m_2\hat{\mathbf{p}}_1 - m_1\hat{\mathbf{p}}_2}{m_1 + m_2}$$

$$\hat{\mathbf{P}}_{CM} = \hat{\mathbf{p}}_1 + \hat{\mathbf{p}}_2 \tag{8.3b}$$

The expression for these operators in the coordinate representation are

$$\hat{\mathbf{p}} = -i\hbar\,\boldsymbol{\nabla} = -i\hbar\left(\mathbf{n}_x\frac{\partial}{\partial x}+\mathbf{n}_y\frac{\partial}{\partial y}+\mathbf{n}_z\frac{\partial}{\partial z}\right)$$

$$\hat{\mathbf{P}}_{CM} = -i\hbar\,\boldsymbol{\nabla}_{CM} = -i\hbar\left(\mathbf{n}_X\frac{\partial}{\partial X_{CM}}+\mathbf{n}_Y\frac{\partial}{\partial Y_{CM}}+\mathbf{n}_Z\frac{\partial}{\partial Z_{CM}}\right) \tag{8.3c}$$

where $\mathbf{n}_x(\mathbf{n}_X)$, etc., are the unit vectors to which the relative, (CM), motion are referred to. Remembering that

$$\frac{\partial}{\partial x}=\frac{\partial x_1}{\partial x}\frac{\partial}{\partial x_1}+\frac{\partial x_2}{\partial x}\frac{\partial}{\partial x_2}, \qquad \text{etc.}$$

we find that

$$\frac{1}{m_1^2}\nabla_1^2+\frac{1}{m_2^2}\nabla_2^2=\frac{1}{m_1+m_2}\nabla_{CM}^2+\frac{1}{\mu}\nabla^2 \tag{8.4}$$

where

$$\mu=\frac{m_1m_2}{m_1+m_2}$$

is known as the reduced mass of the system. The Hamiltonian in the new coordinates is given by

$$\hat{H}=-\frac{\hbar^2}{2(m_1+m_2)}\nabla_{CM}^2-\frac{\hbar^2}{2\mu}\nabla^2+U(r) \tag{8.5}$$

Looking at Eq. (8.5) we note that the center-of-mass variables do not couple to the relative coordinate variables. Hence, the wave function for the system, that is, the solution of the Schrödinger equation, must be separable in the two variables $\mathbf{r}$ and $\mathbf{R}_{CM}$. Furthermore, the Hamiltonian does not depend on $\mathbf{R}_{CM}$. Consequently, the center-of-mass momentum must remain constant. Thus, we can write the wavefunction for the system as a superposition of plane-wave states in the center-of-mass coordinates

$$\left[-\frac{\hbar^2}{2(m_1+m_2)}\nabla_{CM}^2-\frac{\hbar^2}{2\mu}\nabla^2+U(r)\right]u(\mathbf{r},\mathbf{R}_{CM})=Eu(\mathbf{r},\mathbf{R}_{CM}) \tag{8.6}$$

And we can write

$$u(\mathbf{r}, \mathbf{R}_{CM}) = \sum_k \phi_k(\mathbf{r}) e^{i\mathbf{k}\cdot\mathbf{R}_{CM}} \tag{8.7}$$

so that Eq. (8.6) reduces to

$$\left[-\frac{\hbar^2}{2\mu}\nabla^2 + U(\mathbf{r})\right]\phi_k(\mathbf{r}) = \left[E - \frac{(\hbar\mathbf{k})^2}{2(m_1+m_2)}\right]\phi_k(\mathbf{r}) \tag{8.8}$$

We note here that $\hbar\mathbf{k}$ is the momentum associated with the center-of-mass motion, and since this motion is free the energy associated with the center-of-mass motion is

$$E_{CM} = \frac{(\hbar\mathbf{k})^2}{2(m_1+m_2)} \tag{8.9}$$

Therefore, the energy associated with the relative motion of the two particles is

$$E_{\text{rel}} = E - E_{CM} = E - \frac{(\hbar\mathbf{k})^2}{2(m_1+m_2)} \tag{8.10}$$

Clearly we can rewrite Eq. (8.8) as

$$\left[-\frac{\hbar^2}{2\mu}\nabla^2 + U(r)\right]\phi_k(\mathbf{r}) = E_{\text{rel}}\,\phi_k(\mathbf{r}) \tag{8.11}$$

This is, of course, the Schrödinger equation for the motion of a single particle, and this decomposition has been possible only because the potential for the two-particle system is central. Furthermore, since the motion of the center-of-mass is free, we only have to study the equation for the relative motion of the two particles. Therefore, the problem of studying a two-particle system interacting through a central potential reduces effectively to a single-particle problem with a mass μ given by the reduced mass of the system.

8.2. SPHERICALLY SYMMETRIC POTENTIALS

For a central potential we are no longer concerned with the center-of-mass motion as explained in Section 8.1. Therefore, from now on, in studying Eq. (8.11) for the relative motion, we drop all subscripts. Both the Coulomb potential and the harmonic oscillator potential are spherically symmetric. That means the Hamiltonian in this case is invariant under rotations. Hence, the eigenstates of the Hamiltonian are simultaneous eigenstates of the angular momentum operators $\hat{L}^2$ and $\hat{L}_z$ (see Section 7.4 and the following). In this case it is advantageous to work in spherical coordinates since the wavefunction separates in the radial and angular variables. In this section we show how the separation is achieved.

First, note that the momentum operator in spherical coordinates is given by

$$\hat{\mathbf{p}} = -i\hbar\,\boldsymbol{\nabla} = -i\hbar\left(\mathbf{n}_r \frac{\partial}{\partial r} + \mathbf{n}_\theta \frac{1}{r}\frac{\partial}{\partial \theta} + \mathbf{n}_\phi \frac{1}{r\sin\theta}\frac{\partial}{\partial \phi}\right) \tag{8.12a}$$

Furthermore, realizing that the Laplacian in spherical coordinates has the form [see Appendix 8]

$$\nabla^2 = \left[\frac{1}{r^2}\frac{\partial}{\partial r}\left(r^2\frac{\partial}{\partial r}\right) + \frac{1}{r^2\sin\theta}\frac{\partial}{\partial\theta}\left(\sin\theta\frac{\partial}{\partial\theta}\right) + \frac{1}{r^2\sin^2\theta}\frac{\partial^2}{\partial\phi^2}\right]$$

we have

$$\begin{aligned}
\hat{\mathbf{p}}^2 &= -\hbar^2\nabla^2 \\
&= -\hbar^2\left[\frac{1}{r^2}\frac{\partial}{\partial r}\left(r^2\frac{\partial}{\partial r}\right) + \frac{1}{r^2\sin\theta}\frac{\partial}{\partial\theta}\left(\sin\theta\frac{\partial}{\partial\theta}\right) + \frac{1}{r^2\sin^2\theta}\frac{\partial^2}{\partial\phi^2}\right] \\
&= -\hbar^2\left[\frac{1}{r^2}\frac{\partial}{\partial r}\left(r^2\frac{\partial}{\partial r}\right) - \frac{\hat{L}^2}{\hbar^2 r^2}\right]
\end{aligned} \tag{8.12b}$$

This follows from the fact that the last two terms in the parentheses correspond to $-\hat{L}^2/\hbar^2 r^2$ as given in Eq. (7.56c). (For an alternate way of deriving this relation see Problem 8.7.)

We see that, for a spherically symmetric potential, the Hamiltonian operator takes the form

$$\begin{aligned}
\hat{H} &= \frac{\hat{\mathbf{p}}^2}{2\mu} + \hat{U} \\
&= -\frac{\hbar^2}{2\mu}\left[\frac{1}{r^2}\frac{\partial}{\partial r}\left(r^2\frac{\partial}{\partial r}\right) - \frac{\hat{L}^2}{\hbar^2 r^2}\right] + U(r)
\end{aligned} \tag{8.13a}$$

Therefore, the time-independent Schrödinger equation is

$$\left\{-\frac{\hbar^2}{2\mu}\left[\frac{1}{r^2}\frac{\partial}{\partial r}\left(r^2\frac{\partial}{\partial r}\right)-\frac{\hat{L}^2}{\hbar^2 r^2}\right]+U(r)\right\}\psi_n(r,\theta,\phi)=E_n\psi_n(r,\theta,\phi)$$

or

$$\left\{\frac{1}{r^2}\frac{\partial}{\partial r}\left(r^2\frac{\partial}{\partial r}\right)-\frac{\hat{L}^2}{\hbar^2 r^2}+\frac{2\mu}{\hbar^2}[E_n-U(r)]\right\}\psi_n(r,\theta,\phi)=0 \quad (8.13b)$$

Since the wave function $\psi_n(r, \theta, \phi)$ has to be an eigenstate of the angular momentum operators, we can write

$$\psi_{nlm}(r,\theta,\phi)=R_{nl}(r)Y_{lm}(\theta,\phi) \quad (8.14a)$$

where $Y_{lm}(\theta, \phi)$ are the coordinate representation of the eigenstates of $\hat{L}^2$ and $\hat{L}_z$ [see Eqs. (7.57)] so that

$$\hat{L}^2 Y_{lm}(\theta,\phi)=\hbar^2 l(l+1)Y_{lm}(\theta,\phi) \quad (8.14b)$$

Thus, Eq. (8.13b) becomes

$$\left\{\frac{1}{r^2}\frac{d}{dr}\left(r^2\frac{d}{dr}\right)+\frac{2\mu}{\hbar^2}\left[E_{nl}-U(r)-\frac{\hbar^2 l(l+1)}{2\mu r^2}\right]\right\}R_{nl}=0 \quad (8.15)$$

Note here that we have dropped the $Y_{lm}(\theta, \phi)$ in writing Eq. (8.15) since the operator in the parentheses does not depend on the angular coordinates (θ, ϕ).

The reader will note here that the *separation* procedure is similar to that used for the center-of-mass motion of Eqs. (8.7) and (8.8). In fact, whenever an operator (in this case the Hamiltonian) contains additive terms that act on different subspaces, the eigenfunction can be taken as a product of the eigenfunctions of the different subspaces.

The operator in Eq. (8.15) depends only on the radial coordinate r, and hence Eq. (8.15) is known as the radial equation for a general spherically symmetric potential. We see that a system with spherical symmetry must have wavefunctions that are labeled by the eigenvalues l and m of the angular momentum operators. In addition, they depend on a quantum number n which emerges from the solution of the radial equation. Furthermore, the radial solutions $R_{nl}(r)$ must be labeled by n and l since l appears explicitly in Eq. (8.15) in the term $\hbar^2 l(l+1)/2\mu r^2$.

Let us now study the radial equation, Eq. (8.15), in some detail. Since the spherical harmonics $Y_{lm}(\theta, \phi)$ are properly normalized, the normalization of the total wavefunction implies that

$$\int |\psi_{nlm}(r, \theta, \phi)|^2 \, d^3r = 1$$

$$\int |R_{nl}(r)|^2 \, |Y_{lm}(\theta, \phi)|^2 \, r^2 \, dr \, d\Omega = 1 \tag{8.16}$$

$$\int_0^\infty |R_{nl}(r)|^2 \, r^2 \, dr = 1$$

Since we are seeking a bound-state solution the system must be localized and the probability of finding its two constituents at infinite relative separation must vanish. Thus the radial solution $R_{nl}(r)$ must obey the boundary condition

$$rR_{nl}(r) \to 0 \quad \text{as} \quad r \to \infty \tag{8.17a}$$

This is also necessary in order to satisfy Eq. (8.16). Less obvious is the requirement for $r \to 0$

$$rR_{nl}(r) \to 0 \quad \text{as} \quad r \to 0 \tag{8.17b}$$

This is necessary because otherwise the integrated probability over a sphere of vanishingly small radius would remain finite. We have assumed here that $rU(r)$ is nonsingular. We can then redefine the radial solution as

$$R_{nl}(r) = \frac{u_{nl}(r)}{r} \tag{8.18a}$$

in which case the boundary conditions of Eqs. (8.17*a*) and (8.17*b*) become

$$u_{nl}(r) \xrightarrow[r \to \infty]{} 0$$

$$u_{nl}(r) \xrightarrow[r \to 0]{} 0 \tag{8.18b}$$

In terms of the new functions u_{nl} Eq. (8.15) simplifies and is given by

$$\frac{d^2 u_{nl}}{dr^2} + \frac{2\mu}{\hbar^2}\left[E_{nl} - U(r) - \frac{\hbar^2 l(l+1)}{2\mu r^2}\right] u_{nl} = 0 \tag{8.19}$$

Note that this is exactly the same as the one-dimensional Schrödinger equation except for the boundary condition at the origin and the additional repulsive potential $\hbar^2 l(l+1)/2\mu r^2$. This potential term is known as the *centrifugal barrier* and its origin can be understood physically as follows.

Consider a classical particle moving in a circular orbit of radius r as shown in Fig. 8.2(a). The centrifugal force is directed radially and is of magnitude $F = mv^2/r = p^2/mr = L^2/mr^3$, where $L = rp$ is the angular momentum in the circular orbit. Since $F = -\partial U/\partial r$ the corresponding potential is $U = L^2/2mr^2$ and is repulsive, as shown in Fig. 8.2(b). But this is exactly the term in Eq. (8.19) when the eigenvalue of the operator $\hat{L}^2$ is introduced in place of the classical value of L^2.

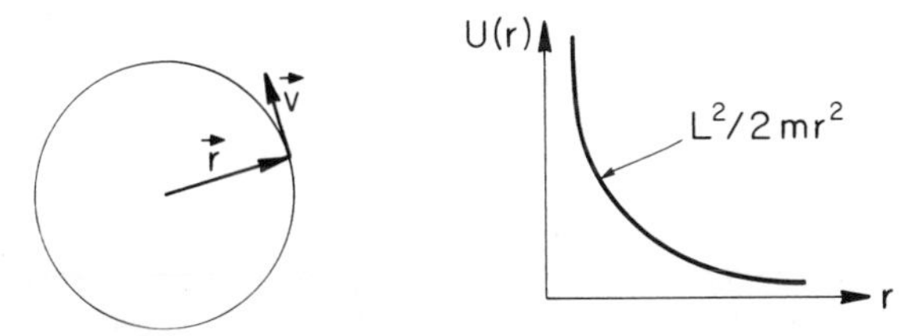

FIGURE 8.2. A particle moving in a circular orbit with angular momentum L is subject to a repulsive potential of magnitude $L^2/2mr^2$.

The nature of the solutions of Eq. (8.19), of course, depends on the value of the total energy. Let U_∞ stand for the asymptotic value of the potential. Then, if $E_{nl} < U_\infty$, the solutions represent bound states, and the spectrum of eigenvalues is discrete. If, on the other hand, $E > U_\infty$ the solutions represent free particles with a continuous spectrum of energy eigenvalues scattering from the potential. For bound states in the case of the Coulomb potential the general behavior of $u_{nl}(r)$ is shown in Fig. 8.3. For values of r such that $E_{nl} < [U(r) + \hbar^2 l(l+1)/2\mu r^2]$ the function $u_{nl}(r)$ will be concave away from the r-axis. When $E_{nl} > [U(r) + \hbar^2 l(l+1)/2\mu r^2]$, on the other hand, $u_{nl}(r)$ will be oscillatory and may contain nodes (i.e., cross the r-axis one or more times). In Fig. 8.3 we have chosen the Coulomb potential as an example. Part (a) of the figure refers to the case where $l = 0$ and hence where the centrifugal barrier is absent; part (b) refers to the case where $l \neq 0$ and shows the effect of the

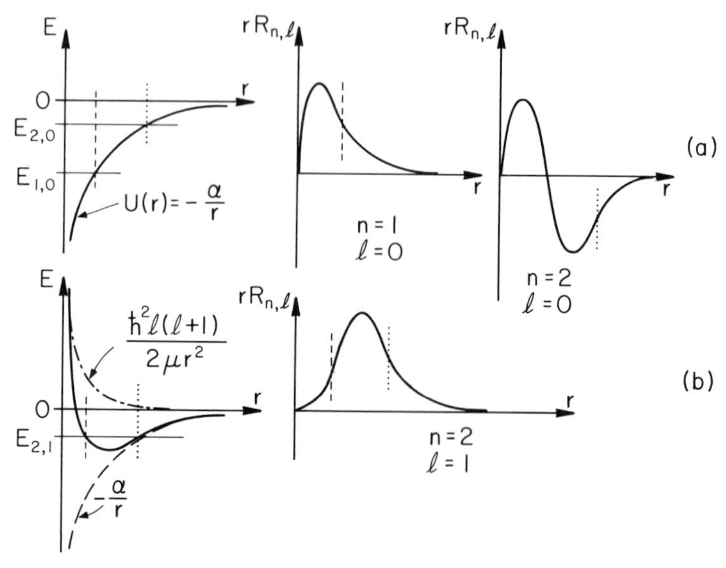

FIGURE 8.3. The effect of the angular momentum barrier. (a) The effective potential for states with $l = 0$; the lowest and next to lowest energy states and their corresponding radial functions are shown. (b) The effective potential for a state with $l \neq 0$; the lowest energy state and its radial function is also shown.

centrifugal barrier. The resulting effective potential looks like a well, and the centrifugal term pushes the particles further away from the origin.

The energy spectrum consists of discrete states of progressively higher energy which become spaced closer to one another as $E \to 0$. For long-range potentials (such as the Coulomb potential) the number of discrete states is infinite. For short-range potentials (such as the square well) the number of bound states is finite and may even be zero.

8.3. SOLUTION OF THE RADIAL EQUATION FOR THE COULOMB POTENTIAL

We consider now the hydrogen atom, which consists of a proton ($m_p = 938\ \mathrm{MeV}/c^2$) and an electron ($m_e = 0.511\ \mathrm{MeV}/c^2$). This is a two-particle system and the force between the two particles is given by the attractive Coulomb potential

$$U(r) = -\frac{e^2}{4\pi\varepsilon_0 r} \tag{8.20a}$$

The potential is central and thus, is rotationally invariant. Hence, all the formalism we have developed so far can be directly applied. First, note that the reduced mass for this system is given by

$$\mu = \frac{m_e m_p}{m_e + m_p} \simeq m_e \tag{8.20b}$$

The radial equation for this system [Eq. (8.19)] can be written as

$$\left\{\frac{d^2}{dr^2} + \frac{2m_e}{\hbar^2}\left[E_{nl} + \frac{e^2}{4\pi\varepsilon_0}\frac{1}{r} - \frac{\hbar^2 l(l+1)}{2m_e r^2}\right]\right\} u_{nl}(r) = 0 \tag{8.20c}$$

To simplify the notation we will rescale distances and energies in terms of units that are natural to the problem at hand so that all variables become dimensionless. For example, realizing that the Bohr radius is given by

$$a_0 = 4\pi\varepsilon_0 \frac{\hbar^2}{m_e e^2} = 0.5291771(8) \times 10^{-8}\ \text{cm} \tag{8.21a}$$

and that the Rydberg is defined to be†

$$\text{Ry} = \frac{m_e e^4}{2\hbar^2}\frac{1}{(4\pi\varepsilon_0)^2} = 13.60583(15)\ \text{eV} \tag{8.21b}$$

we introduce the dimensionless variables

$$\begin{aligned} \rho &= \frac{r}{a_0} \\ \varepsilon_{nl} &= \frac{E_{nl}}{\text{Ry}} \end{aligned} \tag{8.21c}$$

It is clear that in terms of these variables the radial equation [Eq.

† The accepted symbol for the Rydberg is R; we use Ry to avoid any possibility of confusion with radial distances or functions.

(8.20*c*)] becomes

$$\left\{\frac{d^2}{d\rho^2}+\left[\varepsilon_{nl}+\frac{2}{\rho}-\frac{l(l+1)}{\rho^2}\right]\right\}u_{nl}(\rho)=0 \tag{8.22}$$

We are looking for the bound-state solutions of this equation which implies that

$$\varepsilon_{nl}=-\lambda^2<0, \qquad \lambda \text{ real} \tag{8.23}$$

In the limit of $\rho\to\infty$, the radial equation takes the form†

$$\left(\frac{d^2}{d\rho^2}+\varepsilon_{nl}\right)u_{nl}(\rho)=\left(\frac{d^2}{d\rho^2}-\lambda^2\right)u_{nl}(\rho)=0$$

The solutions have the form

$$u_{nl}(\rho)\xrightarrow[\rho\to\infty]{} e^{\pm\lambda\rho}$$

The boundary condition at $r\to\infty$ allows only the negative exponential

$$u_{nl}(\rho)\xrightarrow[\rho\to\infty]{} e^{-\lambda\rho} \tag{8.24a}$$

Similarly, when $\rho\to 0$ the radial equation becomes

$$\left(\frac{d^2}{d\rho^2}-\frac{l(l+1)}{\rho^2}\right)u_{nl}(\rho)=0$$

Thus, the solutions have the form

$$u_{nl}(\rho)\xrightarrow[\rho\to 0]{} \rho^{l+1} \quad \text{or} \quad \rho^{-l}$$

However, the boundary condition that the function $u_{nl}(\rho)$ must vanish at the origin further restricts its behavior at $r\to 0$, to

$$u_{nl}(\rho)\xrightarrow[\rho\to 0]{}\rho^{l+1} \tag{8.24b}$$

† The solution of the simpler case of Eq. (8.22) when $l=0$ is easily obtained by the same method as outlined here for the general case. We leave this as an exercise for the reader.

Thus, the general solution of Eq. (8.22) can be chosen as

$$u_{nl}(\rho)=\rho^{l+1}e^{-\lambda\rho}\sum_{s=0}^{\infty} a_s\rho^s \tag{8.24c}$$

If we substitute Eq. (8.24*c*) into Eq. (8.22), it takes the form

$$\sum_{s=0}^{\infty}(s+l+1)(s+l)a_s\rho^{s+l-1}-2\lambda\sum_{s=0}^{\infty}(s+l+1)a_s\rho^{s+l}$$
$$+2\sum_{s=0}^{\infty}a_s\rho^{s+l}-l(l+1)\sum_{s=0}^{\infty}a_s\rho^{s+l-1}=0$$

or

$$\sum_{s=0}^{\infty}s(s+2l+1)a_s\rho^{s+l-1}+2\sum_{s=0}^{\infty}[1-\lambda(s+l+1)]a_s\rho^{s+l}=0$$

or

$$\sum_{s=0}^{\infty}\{(s+1)(s+2l+2)a_{s+1}+2[1-\lambda(s+l+1)]a_s\}\rho^{s+l}=0 \tag{8.25}$$

For this expression to be true the coefficient of each term must vanish, which leads to the recursion relation

$$(s+1)(s+2l+2)a_{s+1}+2[1-\lambda(s+l+1)]a_s=0$$

or

$$\frac{a_{s+1}}{a_s}=\frac{2[\lambda(s+l+1)-1]}{(s+1)(s+2l+2)} \tag{8.26}$$

It is obvious that unless the power series in Eq. (8.24*c*) terminates, as we go to higher-order terms for a fixed value of l, this ratio behaves like

$$\frac{a_{s+1}}{a_s}\xrightarrow[s\to\text{large}]{}\frac{2\lambda}{s+1}$$

But this is the behavior of an exponential series, and unless it terminates the power series in Eq. (8.24*c*) has the asymptotic form

$$\sum_{s=0}^{\infty}a_s\rho^s\xrightarrow[\rho\to\infty]{}e^{2\lambda\rho}$$

If this were true, the radial solution

$$u_{nl}(\rho)=\rho^{l+1}e^{-\lambda\rho}\sum_{s=0}^{\infty}a_s\rho^s \xrightarrow[\rho\to\infty]{} e^{\lambda\rho}$$

and hence the boundary condition Eq. (8.24a), would be violated. Thus, the series has to terminate, and this is possible only if for some value of $s=s_{\max}$ we have

$$\lambda(s_{\max}+l+1)-1=0$$

or

$$\lambda=\frac{1}{s_{\max}+l+1}$$

Since both $s_{\max}$ and l take only integer values we can rewrite this as

$$\lambda=\frac{1}{n}$$

or

$$\varepsilon_{nl}=-\lambda^2=-\frac{1}{n^2}, \qquad n=1, 2, 3, \ldots \tag{8.27a}$$

Furthermore, we note that since

$$l, s_{\max}\geqslant 0$$

and

$$n=s_{\max}+l+1$$

it follows that

$$l=n-s_{\max}-1\leqslant n-1 \tag{8.27b}$$

To get an idea of the quantum numbers involved, we note that for

$$\begin{aligned}
n=1:&\quad l=0,\ s_{\max}=0\\
n=2:&\quad l=0,\ s_{\max}=1\\
&\quad l=1,\ s_{\max}=0\\
n=3:&\quad l=0,\ s_{\max}=2\\
&\quad l=1,\ s_{\max}=1\\
&\quad l=2,\ s_{\max}=0
\end{aligned} \tag{8.27c}$$

and so on.

If we assume that $a_0=a_{s=0}=1$, then the higher coefficients in the power series in Eq. (8.24c) can be obtained through the recursion relation given by Eq. (8.26). Therefore, the radial wavefunctions can simply be constructed from the definitions of Eqs. (8.24c) and

(8.27a). Thus, for example,

$$\begin{aligned} u_{1,0}(\rho) &= \rho e^{-\rho} \\ u_{2,0}(\rho) &= \rho(1-\tfrac{1}{2}\rho)e^{-\rho/2} \\ u_{2,1}(\rho) &= \rho^2 e^{-\rho/2} \\ u_{3,0}(\rho) &= \rho(1-\tfrac{2}{3}\rho+\tfrac{2}{27}\rho^2)e^{-\rho/3} \end{aligned} \tag{8.28}$$

and so on.

Note that these solutions are correct only up to a normalization constant. They satisfy the boundary conditions both at spatial infinity as well as at the origin. The number of nodes in the radial function increases as we go to higher and higher values of n, the number being given by $n-l-1$. Since $l_{\max}=n-1$, it is clear that the solution with the maximum value of the angular momentum for a given n is nodeless.

An interesting feature of the solutions is that the energy of the stationary states depends only on the index n and not on l. That is, for $n>1$ we have states with different values of the angular momentum corresponding to the same n value, and hence these states are degenerate in energy. This is peculiar since rotational invariance only leads to a degeneracy in the m-quantum numbers, i.e., in the projections of the angular momentum along a particular axis. The degeneracy in the l-quantum number is a special feature of the hydrogen atom and is known as an *accidental degeneracy*.†

8.4. FURTHER DISCUSSION OF THE SOLUTIONS OF THE HYDROGEN ATOM

From Eq. (8.27) we know that the energy levels of the hydrogen atom are given by

$$E_{nl} = -\frac{\text{Ry}}{n^2} = -\frac{m_e}{2\hbar^2}\left(\frac{e^2}{4\pi\varepsilon_0}\right)^2\frac{1}{n^2} \tag{8.29a}$$

† In the hydrogen atom, because of the special form of the Hamiltonian, the system has a much larger rotational symmetry than we expect. Mathematically one says that the symmetry group associated with the hydrogen atom is O(4) rather than the usual O(3). The degeneracy in the l-quantum number is a consequence of this larger symmetry.

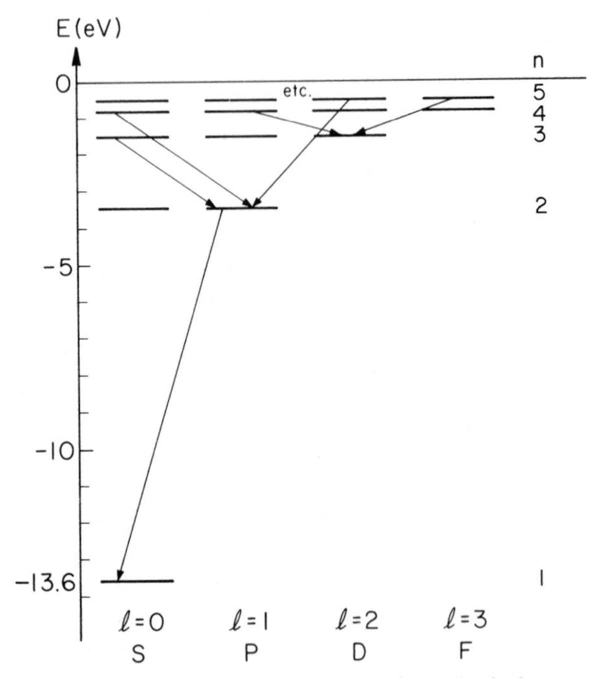

FIGURE 8.4. The spectrum of the stationary states of the hydrogen atom in the idealized case where spin is ignored. Some electric dipole transitions ($\Delta l = \pm 1$) are shown.

The energy spectrum is shown in Fig. 8.4. We note that states with different l-values but the same value of the principal quantum number n are degenerate. The result of Eq. (8.29a) is in numerical agreement with the Bohr theory† and is confirmed by experiment.

The spectral lines emitted by the hydrogen atom arise from transitions between stationary states for which $\Delta l = \pm 1$[see the discussion in Appendix 7, and in particular Eq. (A7.9)]. The frequency of a line emitted in the transition from a state with the principal quantum number n_2 to one with quantum number n_1 is

† In the Bohr theory the $n = 1$ state is assigned an angular momentum $\hbar$, whereas in reality it is a state with $l = 0$.

given by

$$\nu = \frac{E_{n_2} - E_{n_1}}{h} = \frac{\text{Ry}}{h}\left(\frac{1}{n_1^2} - \frac{1}{n_2^2}\right) \tag{8.29b}$$

Transitions in the case $n_1 = 1$ give rise to the Lyman series, whereas when $n_1 = 2$ the emitted lines are said to form the Balmer series, and so on.

The lines of the hydrogen atom have been measured to good accuracy, and using Eq. (8.29*b*) one obtains the value for the Rydberg (Ry) experimentally as†

$$\frac{1}{hc}\text{Ry}_{\text{exp}} = 109,677.58 \text{ cm}^{-1} \tag{8.30a}$$

whereas the theoretical value obtained from Eq. (8.21*b*) is

$$\frac{1}{hc}\text{Ry}_{\text{th}} = \frac{1}{hc}\frac{m_e}{2\hbar^2}\left(\frac{e^2}{4\pi\varepsilon_0}\right)^2 = 109,727.31 \text{ cm}^{-1} \tag{8.30b}$$

The difference between these two values is of the order of 0.05%, which is precisely the error between the mass of the electron and the reduced mass for the hydrogen atom, i.e.,

$$\mu = \frac{m_e m_p}{m_e + m_p} \simeq m_e\left(1 - \frac{m_e}{m_p}\right) = m_e\left(1 - \frac{0.511}{938}\right) \tag{8.30c}$$

Note that the lowest energy of the hydrogen atom is $-$Ry. Consequently, a Rydberg of energy is necessary to free the electron from the atom. This is known as the binding energy of hydrogen and has a value of 13.6 eV.

In spherical coordinates the wavefunctions for the stationary states are given by the product of the radial wavefunction $R_{nl}(r)$ and the spherical harmonics $Y_{lm}(\theta, \phi)$, which are eigenfunctions of the

---

† The Rydberg is expressed here in *wavenumbers*, $\bar{\nu}$, a unit commonly used in optical spectroscopy. The wavenumber is the inverse of the wavelength of the radiation of a given energy. Thus

$$\nu = \frac{E}{h} \qquad \lambda = \frac{c}{\nu} = \frac{hc}{E} \qquad \bar{\nu} = \frac{1}{\lambda} = \frac{E}{hc}$$

The value of the Rydberg in eV was given in Eq. (8.21*b*).

angular momentum operator. The completely normalized eigenfunctions of the hydrogen atom can then be expressed as

$$\psi_{nlm}(r,\theta,\phi)=-\varepsilon\left(\frac{2r}{na_0}\right)^l\left\{\left(\frac{2}{na_0}\right)^3\frac{(n-l-1)!}{2n[(n+l)!]^3}\right\}^{1/3}$$

$$\times e^{-r/na_0}L_{n+l}^{2l+1}\left(\frac{2r}{na_0}\right)\left\{\frac{2l+1}{4\pi}\frac{(l-|m|)!}{(l+|m|)!}\right\}^{1/2}P_l^m(\cos\theta)e^{im\phi} \quad (8.31)$$

Here $L_{n+l}^{2l+1}(2r/na_0)$ are the associated Laguerre polynomials (discussed in Appendix 10), $P_l^m(\cos\theta)$ are the associated Legendre polynomials, and ε is a signature factor which is $+1$ for $m\leqslant 0$ and is equal to $(-1)^m$ for $m>0$. We list the first few wavefunctions below.

$$\psi_{n=1,l=0,m=0}(r,\theta,\phi)=\frac{1}{(\pi a_0^3)^{1/2}}e^{-r/a_0}$$

$$\psi_{n=2,l=0,m=0}(r,\theta,\phi)=\frac{1}{4(2\pi a_0^3)^{1/2}}\left(2-\frac{r}{a_0}\right)e^{-r/2a_0} \quad (8.32)$$

$$\psi_{n=2,l=1\begin{Bmatrix}m=+1\\m=0\\m=-1\end{Bmatrix}}(r,\theta,\phi)=\frac{1}{8(\pi a_0^3)^{1/2}}\frac{r}{a_0}e^{-r/2a_0}\begin{Bmatrix}-\sin\theta\, e^{i\phi}\\ \sqrt{2}\cos\theta\\ \sin\theta\, e^{-i\phi}\end{Bmatrix}$$

For historical reasons it is common practice to use the letters *S*, *P*, *D*, *F*, . . . to designate states with $l=0,1,2,3,\ldots$

$l=0$ *S* (sharp)
$l=1$ *P* (principal)
$l=2$ *D* (diffuse)
$l=3$ *F* (fine)

This nomenclature is related to the appearance of the series of spectral lines ending at the corresponding state.

The reader should check that the expressions in Eqs. (8.32) follow from the general definition of Eq. (8.31) and agree with the form obtained earlier in Eq. (8.28). Note also that the wavefunction is finite (different from zero) at the origin *only* for $l=0$ states.

The probability for finding the electron at r,θ,ϕ in the differential volume element $d^3r=r^2\,dr\,d\Omega$ is

$$|\psi_{nlm}(r,\theta,\phi)|^2 r^2\,dr\,d\Omega=(|R_{nl}(r)|^2 r^2\,dr)(|Y_{lm}(\theta,\phi)|^2\,d\Omega) \quad (8.33a)$$

The probability for finding the electron at a distance r in the interval dr is

$$\text{Prob}_{nl}(r)\,dr = |R_{nl}(r)|^2\,r^2\,dr \tag{8.33b}$$

This radial probability density is shown in Fig. 8.5 for different values of n and l. The distance r is plotted in units of r/a_0. Note the presence of $(n-l-1)$ nodes and the effect of the centrifugal barrier in pushing the mean position of the electron further away from the origin.

The probability for finding the electron at θ, ϕ in the element $d\Omega$ is

$$\begin{aligned}\text{Prob}_{lm}(\theta, \phi)\,d\Omega &= |Y_{lm}(\theta, \phi)|^2\,d\Omega \\ &= \left[\frac{2l+1}{4\pi}\frac{(l-|m|)!}{(l+|m|)!}\right]|P_l^m \cos\theta|^2\,d(\cos\theta)\,d\phi \end{aligned}\tag{8.33c}$$

We see that the angular probability density is independent of the azimuthal angle ϕ and is of increasing complexity as l increases. The functions $|P_l^m \cos\theta|^2$ are shown in Fig. 8.6 for a few values of l and m. When l is large, the electron is found for $|m| = l$ near $\theta = \pi/2$. This corresponds classically to orbits in the x-y plane, namely, orbits for which the angular momentum vector is along the Z-axis. The opposite is true for $m = 0$.

The wavefunctions for the stationary states of the hydrogen atom with different values for the quantum numbers n, l, and m are orthogonal (see Eq. (8.31)) This can be verified simply in the following way. When $n = n'$ and $l = l'$, but $m \neq m'$, the orthogonality follows from the properties of the functions $e^{im\phi}$. When $n = n'$ and $m = m'$ but $l \neq l'$, then the properties of the Legendre polynomials make them orthogonal. Finally, when $l = l'$ and $m = m'$ but $n \neq n'$ the orthogonality follows from the properties of the Laguerre polynomials.

Central potentials are invariant under the parity operation or under reflection of any coordinate. Therefore, it must be possible to label the stationary states of the hydrogen atom by the eigenvalues of the parity operator. We know that under the parity operation

$$Y_{lm}(\theta, \phi) \xrightarrow{\text{parity}} Y_{lm}(\pi-\theta, \pi+\phi) = (-1)^l Y_{lm}(\theta, \phi)$$

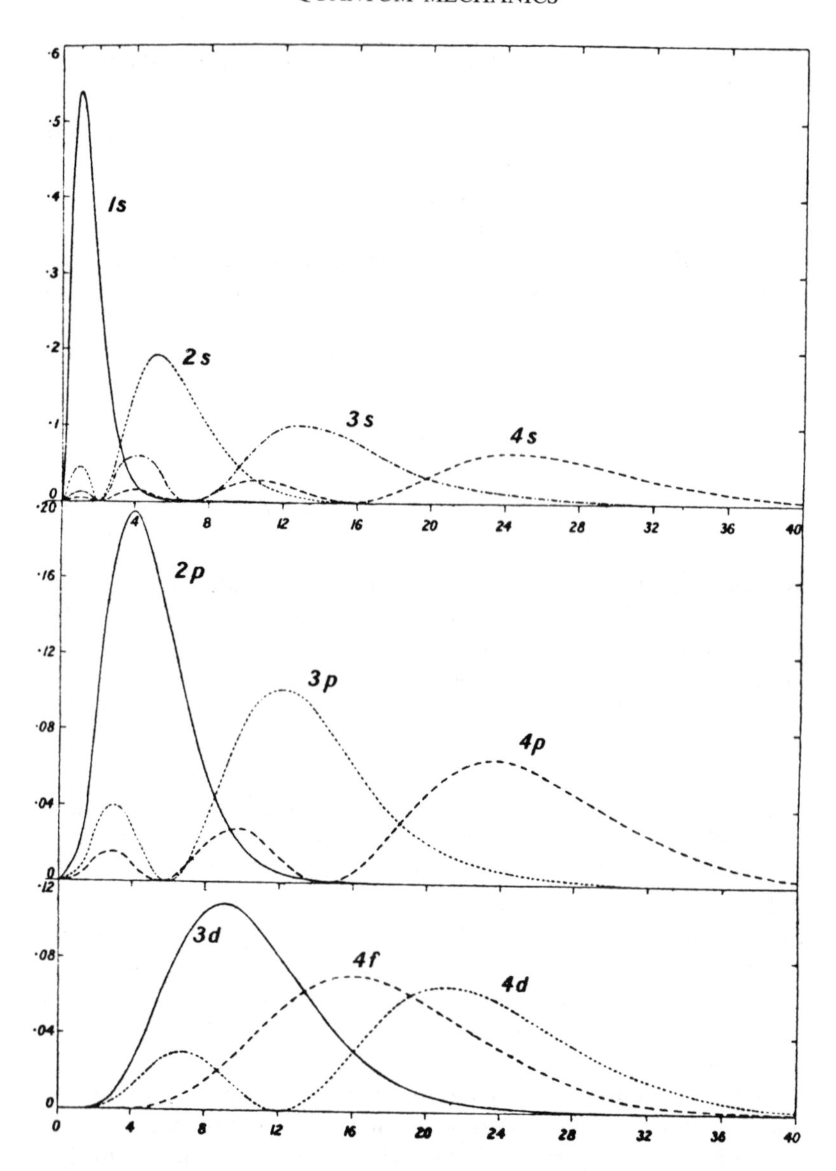

FIGURE 8.5. The radial probability distribution function $|rR_{nl}|^2$ for several values of the quantum numbers n, l. [From E. U. Condon and G. H. Shortley, *The Theory of Atomic Spectra*, Cambridge University Press, Cambridge (1953) by permission].

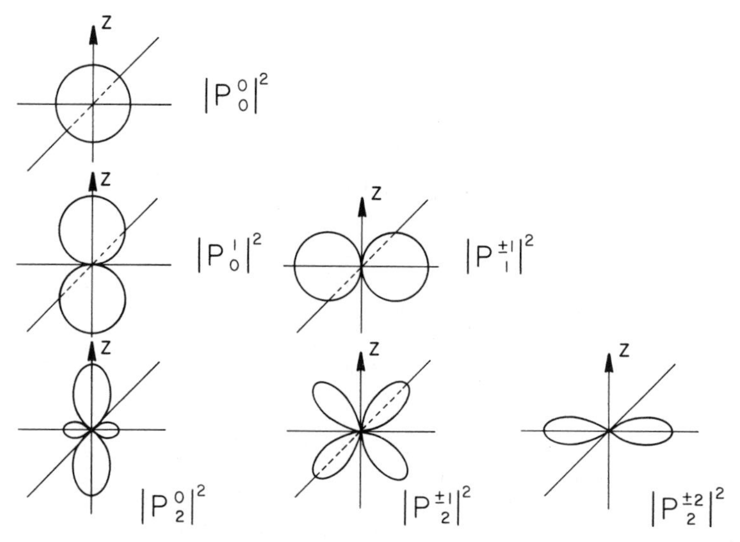

FIGURE 8.6. The probability distribution in the polar angle θ is given by $|P_l^m \cos\theta|^2$. The angle θ is measured from the Z-axis and the distributions have axial symmetry around the Z-axis.

That is, the wavefunctions of the hydrogen atom are definite parity states, the parity quantum number being completely determined by the orbital angular momentum l.

Once the wavefunctions are known, it is possible to calculate the matrix elements of operators between stationary states. Given an operator $\hat{Q}$ with a coordinate representation, the matrix element is defined to be

$$\langle n', l', m'| \hat{Q} |n, l, m\rangle = \int d^3r \psi^*_{n'l'm'}(\mathbf{r})\hat{Q}(\mathbf{r})\psi_{nlm}(\mathbf{r}) \qquad (8.34)$$

Of particular interest are the average (expectation) values of the radial position operator $\hat{r}$ and its powers.† These can be obtained in closed form. In Appendix 11 we indicate how they can be conveniently calculated. For the present we simply list some of the relations most often encountered. We follow the notation

$$\langle n, l, m| \hat{r}^s |n, l, m\rangle = \langle r^s\rangle_{n,l}$$

† In this section we use a caret to indicate the position operator $\hat{r}$. This is not necessary in the coordinate representation where $\hat{r} \to r$, and in the future we will simply write r.

and then

$$\langle \hat{r} \rangle_{n,l} = n^2 a_0 \left\{ 1 + \frac{1}{2} \left[1 - \frac{l(l+1)}{n^2} \right] \right\}$$
$$\langle \hat{r}^2 \rangle_{n,l} = n^4 a_0^2 \left\{ 1 + \frac{3}{2} \left[1 - \frac{l(l+1) - 1/3}{n^2} \right] \right\} \tag{8.35a}$$

$$\left\langle \frac{1}{\hat{r}} \right\rangle_{n,l} = \frac{1}{a_0 n^2}$$
$$\left\langle \frac{1}{\hat{r}^2} \right\rangle_{n,l} = \frac{1}{a_0^2 n^3} \frac{1}{l + 1/2} \tag{8.35b}$$

$$\left\langle \frac{1}{\hat{r}^3} \right\rangle_{n,l} = \frac{1}{a_0^3 n^3} \frac{1}{l(l+1/2)(l+1)} \tag{8.35c}$$

Equation (8.35a) are useful in determining the average distance of the electron from the center, and the reader should compare them with the density functions of Fig. 8.5. It is easy to see from Eqs. (8.35a) and (8.35b) that

$$\langle \hat{r} \rangle_{1,0} = \tfrac{3}{2} a_0$$
$$\left\langle \frac{1}{\hat{r}} \right\rangle_{1,0} = \frac{1}{a_0} \tag{8.36}$$

That is, although

$$\left\langle \frac{1}{\hat{r}} \right\rangle_{n,l} \neq \frac{1}{\langle \hat{r} \rangle_{n,l}}$$

they are of the same order of magnitude. Therefore, one can think of the Bohr radius a_0 as giving the approximate size of the hydrogen atom. Note also that Eq. (8.35c) is undefined for $l = 0$ states. This is not surprising since for S-wave states the wavefunction is finite at the origin and hence

$$\int \frac{1}{r^3} |R_{n,0}(r)|^2 \, r^2 \, dr \to \infty \quad \text{as} \quad r \to 0$$

As a further application of the above formulas let us verify the virial theorem (which we discussed in Section 6.5) in the case of the hydrogen atom. Let

$$\hat{H} = \hat{E} = \frac{\hat{p}^2}{2m_e} + \hat{U}(r) = \hat{T} + \hat{U}(r)$$

Thus

$$\hat{T} = \hat{E} - \hat{U}(r) = \hat{E} + \frac{e^2}{4\pi\varepsilon_0}\frac{1}{\hat{r}} \tag{8.37a}$$

We know that

$$\langle \hat{E} \rangle_{n,l} = -\frac{\text{Ry}}{n^2} = -\frac{e^2}{4\pi\varepsilon_0}\frac{1}{2a_0 n^2} \tag{8.37b}$$

Using the result of Eq. (8.35b) we have

$$\langle \hat{U} \rangle_{n,l} = -\frac{e^2}{4\pi\varepsilon_0}\left\langle\frac{1}{\hat{r}}\right\rangle_{n,l} = -\frac{e^2}{4\pi\varepsilon_0}\frac{1}{a_0 n^2} \tag{8.37c}$$

It is now straightforward to see that

$$\begin{aligned}\langle \hat{T} \rangle_{n,l} &= \langle \hat{E} \rangle_{n,l} - \langle \hat{U} \rangle_{n,l} \\ &= -\frac{e^2}{4\pi\varepsilon_0}\frac{1}{2a_0 n^2} + \frac{e^2}{4\pi\varepsilon_0}\frac{1}{a_0 n^2} \\ &= \frac{e^2}{4\pi\varepsilon_0}\frac{1}{2a_0 n^2} = -\tfrac{1}{2}\langle \hat{U} \rangle_{n,l}\end{aligned} \tag{8.38}$$

This is, of course, what we expected from our discussion of the virial theorem applied to the case of a Coulomb potential [see Eqs. (6.32)].

8.5. EFFECT OF SPIN

Although the results of the calculations for the hydrogen atom in the previous sections agree quite well with the experimental observations, there are minor differences. Part of this discrepancy can be recovered by using the true reduced mass of the system rather than the mass of the electron since the proton is not infinitely heavy. There are, however, still other effects that we have to consider.

First, note that we have treated the electron as a spinless particle in our entire calculation. In practice we know that the electron has spin 1/2 [the eigenvalue of $\hat{S}^2$ is $\hbar^2(1/2)(1/2+1)$] and associated with it a magnetic moment of magnitude†

$$|\boldsymbol{\mu}_e| = \frac{e\hbar}{2m_e} = \mu_B \tag{8.39a}$$

where $\mu_B = 9.274 \times 10^{-24}$ A-m^2 is the Bohr magneton (see Section 3.3). The direction of $\boldsymbol{\mu}_e$ is opposite to that of the spin because the electron has negative charge and the quantum-mechanical operator expressing the magnetic moment has the form [recall Eqs. (5.42)]

$$\hat{\boldsymbol{\mu}}_e = -\mu_B\left(\frac{2}{\hbar}\hat{\mathbf{S}}\right) = -\mu_B\hat{\boldsymbol{\sigma}} \tag{8.39b}$$

where we have used the Pauli matrices of Eq. (4.68*c*) to express the spin operator.

In the hydrogen atom, the electron is not at rest. Consequently, in its rest frame, the proton (the nucleus) is moving. Since a moving charge produces a magnetic field‡, the electron "sees" a magnetic field due to the motion of the proton, and the field is given by

$$\mathbf{B} = -\frac{1}{c^2}(\mathbf{v} \times \vec{\mathscr{E}}) = -\frac{1}{m_e c^2}(\mathbf{p} \times \mathbf{r})\frac{|\mathscr{E}|}{r} \tag{8.40a}$$

The vector relations are shown in Fig. 8.7 where **B** is directed out of

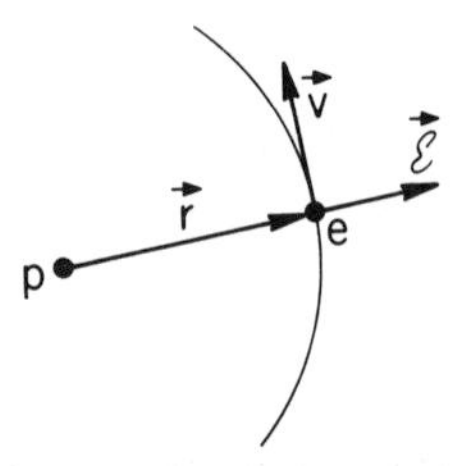

FIGURE 8.7. An electron moving with velocity **v** in the electric field $\vec{\mathscr{E}}$ of a proton is subject, in its own rest frame, to a magnetic field **B** directed out of the paper.

† The exact value is $\mu_e = 1.0015\mu_B$; see the discussion following Eq. (3.23*b*) and in Section 8.6.

‡ See J. D. Jackson, *Classical Electrodynamics*, Wiley, N.Y., 1962. First Edition p. 380. Note that we use MKS units.

the paper. Furthermore, since

$$\mathbf{L} = \mathbf{r} \times \mathbf{p}$$

and

$$|\vec{\mathscr{E}}| = \left|\frac{dV}{dr}\right|$$

we obtain the operator expression

$$\hat{\mathbf{B}} = \frac{1}{m_e c^2} \frac{1}{r} \left|\frac{dV}{dr}\right| \hat{\mathbf{L}} \tag{8.40b}$$

The magnetic moment of the electron interacts with this magnetic field, which leads to an interaction energy of the form

$$\hat{U}_{\text{int}} = -\hat{\boldsymbol{\mu}}_e \cdot \hat{\mathbf{B}} = \frac{2\mu_B}{\hbar m_e c^2} \frac{1}{r} \left|\frac{dV}{dr}\right| (\hat{\mathbf{S}} \cdot \hat{\mathbf{L}}) \tag{8.41a}$$

If we use the form of the electric potential for this case, namely, $V(r) = e/4\pi\varepsilon_0 r$ and the definition of the Bohr magneton, then the interaction energy has the form

$$\hat{U}_{\text{int}} = \frac{e^2}{4\pi\varepsilon_0} \frac{1}{m_e^2 c^2} \frac{1}{r^3} (\hat{\mathbf{L}} \cdot \hat{\mathbf{S}}) \tag{8.41b}$$

This simple-minded calculation of the interaction energy is not correct because the electron is not moving in an inertial frame. The correct expression for the interaction was calculated by Thomas† and differs from the one given in Eq. (8.41*b*) by a factor of $\frac{1}{2}$. Thus

$$\hat{U}_{\text{int}} = \frac{e^2}{4\pi\varepsilon_0} \frac{1}{2m_e^2 c^2 r^3} (\hat{\mathbf{L}} \cdot \hat{\mathbf{S}}) \tag{8.42}$$

Let us note in passing that it was through experimental observations of these effects that it was understood that the electron has spin 1/2 and magnetic moment μ_B. The doublet structure of the Zeeman effect showed that the electron had spin 1/2, and the spin precession experiments further established that the magnetic moment of the electron equalled μ_B (instead of $\frac{1}{2}\mu_B$). Theoretically, both the magnetic moment and the spin–orbit interaction follow as consequences of the relativistic Dirac equation for spin 1/2 pointlike parti-

† See J. D. Jackson, op. cit. p. 364.

cles. This additional interaction energy represents a spin–orbit interaction energy and must be added to the Hamiltonian of the hydrogen atom. Thus

$$\hat{H} = \hat{H}_0 + \hat{U} + \hat{U}_{\text{int}}$$
$$= -\frac{\hbar^2}{2m_e}\nabla^2 - \frac{e^2}{4\pi\varepsilon_0 r} + \frac{e^2}{4\pi\varepsilon_0}\frac{1}{2m_e^2c^2r^3}(\hat{\mathbf{L}}\cdot\hat{\mathbf{S}}) \qquad (8.43a)$$

To get a feeling for the order of magnitude of this additional energy, we note that $(\hat{\mathbf{L}}\cdot\hat{\mathbf{S}}) \sim \hbar^2$, and assuming that the distances involved are of the size of the Bohr radius, we have roughly

$$\hat{U}_{\text{int}} \sim \frac{e^2}{4\pi\varepsilon_0}\frac{1}{2m_e^2c^2a_0^3}\hbar^2$$
$$= \frac{1}{2}\frac{e^2}{4\pi\varepsilon_0}\frac{1}{a_0}\frac{\hbar^2}{m_e^2c^2a_0^2}$$
$$= \text{Ry}\left(\frac{e^2}{4\pi\varepsilon_0\hbar c}\right)^2 = \alpha^2\,\text{Ry} \qquad (8.43b)$$

where

$$\boxed{\alpha = \frac{e^2}{4\pi\varepsilon_0\hbar c} = \text{fine-structure const} \simeq \tfrac{1}{137}}$$

Thus we see that the spin–orbit interaction energy is about 10^4 times smaller than the energy of the stationary states, and hence we can use the time-independent perturbation theory developed in Chapter 4 to calculate its effects.

Although both $\hat{\mathbf{L}}$ and $\hat{\mathbf{S}}$ are angular momentum operators, they operate on different Hilbert spaces. $\hat{\mathbf{L}}$ operates only on the orbital angular momentum of the electron, while $\hat{\mathbf{S}}$ operates on the spin angular momentum, which truly is an intrinsic degree of freedom. Furthermore, note that neither $\hat{L}_i$ nor $\hat{S}_i$ commute with the Hamiltonian any more, and hence their eigenvalues cannot be constants of motion. We discussed in Section 7.6 how two angular momentum operators can be added to form a total angular momentum [see Eq. (7.60a) and following text, and Appendix 6]

$$\hat{\mathbf{J}} = \hat{\mathbf{L}} + \hat{\mathbf{S}} \qquad (8.44a)$$

so that

$$\hat{\mathbf{L}}\cdot\hat{\mathbf{S}} = \tfrac{1}{2}(\hat{\mathbf{J}}^2 - \hat{\mathbf{L}}^2 - \hat{\mathbf{S}}^2) \qquad (8.44b)$$

Furthermore, one can easily show that all the components $\hat{J}_i$ commute with the Hamiltonian, and hence one can choose as a basis for the unperturbed Hamiltonian the states $|l, s; j, m_j|\rangle$ such that

$$\begin{aligned}
\hat{\mathbf{S}}^2 |l, s; j, m_j\rangle &= \hbar^2 s(s+1) |l, s; j, m_j\rangle \\
\hat{\mathbf{L}}^2 |l, s; j, m_j\rangle &= \hbar^2 l(l+1) |l, s; j, m_j\rangle \\
\hat{\mathbf{J}}^2 |l, s; j, m_j\rangle &= \hbar^2 j(j+1) |l, s; j, m_j\rangle \\
\hat{J}_z |l, s; j, m_j\rangle &= \hbar m_j |l, s; j, m_j\rangle
\end{aligned} \tag{8.45a}$$

where j takes values

$$|l-s| \leqslant j \leqslant |l+s|$$

Therefore, for our problem j only takes two values

$$j = l \pm \tfrac{1}{2} \tag{8.45b}$$

In this basis, the expectation value of the operator $\hat{\mathbf{L}} \cdot \hat{\mathbf{S}}$ becomes

$$\begin{aligned}
&\langle l', s'; j', m_j| \hat{\mathbf{L}} \cdot \hat{\mathbf{S}} |l, s; j, m_j\rangle \\
&\quad = \langle l', s'; j', m'_j| \tfrac{1}{2}(\hat{\mathbf{J}}^2 - \hat{\mathbf{L}}^2 - \hat{\mathbf{S}}^2) |l, s; j, m_j\rangle \\
&\quad = \frac{\hbar^2}{2} [j(j+1) - l(l+1) - s(s+1)] \delta_{jj'}\, \delta_{m_j m'_j}\, \delta_{ll'}\, \delta_{ss'}
\end{aligned} \tag{8.46}$$

We are now ready to calculate the perturbative shift in the energy levels due to this additional interaction. Denoting the complete stationary states of the hydrogen atom by $|n, l, s; j, m_j\rangle$ we have the first-order change in energy as

$$\begin{aligned}
&\langle n, l, \tfrac{1}{2}; j, m_j| \hat{U}_{\text{int}} |n, l, \tfrac{1}{2}; j, m_j\rangle \\
&\quad = \frac{e^2}{4\pi\varepsilon_0} \frac{1}{2m_e^2c^2} \langle n, l, \tfrac{1}{2}; j, m_j| \frac{1}{r^3} (\hat{\mathbf{L}} \cdot \hat{\mathbf{S}}) |n, l, \tfrac{1}{2}; j, m_j\rangle \\
&\quad = \frac{e^2}{4\pi\varepsilon_0} \frac{1}{2m_e^2c^2} \left\langle \frac{1}{r^3} \right\rangle_{nl} \frac{\hbar^2}{2} [j(j+1) - l(l+1) - \tfrac{3}{4}] \\
&\quad = \frac{e^2\hbar^2}{4\pi\varepsilon_0 4m_e^2c^2} \left\langle \frac{1}{r^3} \right\rangle_{nl} \begin{Bmatrix} l \\ -(l+1) \end{Bmatrix} \quad \begin{matrix} \text{if } j = l + \frac{1}{2} \\ \text{if } j = l - \frac{1}{2} \end{matrix}
\end{aligned}$$

Using the value of $\langle 1/r^3 \rangle_{nl}$ from Eq. (8.35c) we have

$$\begin{aligned}
\langle \hat{U}_{\text{int}} \rangle &= \frac{e^2\hbar^2}{4\pi\varepsilon_0 4m_e^2c^2} \frac{1}{a_0^3 n^3} \\
&\quad \times \frac{1}{l(l+\frac{1}{2})(l+1)} \begin{Bmatrix} l \\ -(l+1) \end{Bmatrix} \quad \begin{matrix} \text{if } j = l + \frac{1}{2} \\ \text{if } j = l - \frac{1}{2} \end{matrix}
\end{aligned}$$

The constants in the above result can be rearranged to read

$$\frac{e^2\hbar^2}{4\pi\varepsilon_0 4m_e^2c^2}\frac{1}{a_0^3n^3}=\frac{e^2\hbar^2}{4\pi\varepsilon_0 4m_e^2c^2}\left(\frac{m_e e^2}{4\pi\varepsilon_0\hbar^2}\right)^3$$

$$=\frac{1}{2}\frac{1}{2}\frac{m_e e^4}{(4\pi\varepsilon_0)^2\hbar^2}\frac{1}{n^2}\left(\frac{e^2}{4\pi\varepsilon_0\hbar c}\right)^2\frac{1}{n}=-\tfrac{1}{2}\alpha^2E_n\frac{1}{n}$$

so that the final result is

$$\langle\hat{U}_{\text{int}}\rangle=-\tfrac{1}{2}\alpha^2E_n\frac{1}{nl(l+\frac{1}{2})(l+1)}\begin{Bmatrix} l \\ -(l+1)\end{Bmatrix}\quad\begin{matrix}\text{if } j=l+\frac{1}{2}\\ \text{if } j=l-\frac{1}{2}\end{matrix}\tag{8.47}$$

where E_n are the energy levels of hydrogen given in Eq. (8.29*a*). This gives the change in the energy levels because of the spin–orbit interaction and is called a *fine-structure splitting* since the shift is proportional to the square of the fine-structure constant.†

8.6. FINE AND HYPERFINE STRUCTURE OF ATOMIC SPECTRA

The spin–orbit interaction leads to a fine structure in the energy levels of the hydrogen atom. However, there is another effect that should also be taken into account. Namely, we have treated the electron in the hydrogen atom as a nonrelativistic particle. In reality, however, the electron is quite relativistic, as the following simple calculation shows.

Consider the electron to be in the ground state of hydrogen. From the virial theorem we know [see Eq. (8.38)]

$$\langle\hat{T}\rangle_0=-\tfrac{1}{2}\langle\hat{U}\rangle_0\tag{8.48a}$$

On the other hand, the ground-state energy is given by

$$\langle\hat{T}\rangle_0+\langle\hat{U}\rangle_0=E_0=-13.6\text{ eV}$$

† In principle $\langle 1/r^3\rangle_{n,l}$ is divergent for $l=0$ states, as we remarked in the previous section. However, Eq. (8.47) remains finite for $l=0$ states and gives the correct shift in energy even though the matrix element of $(\mathbf{L}\cdot\mathbf{S})$ is zero(!) in that case. These peculiarities are a consequence of using perturbation theory and are absent when treating the hydrogen atom by the Dirac equation.

or

$$\langle \hat{T} \rangle_0 - 2\langle \hat{T} \rangle_0 = -\langle \hat{T} \rangle_0 = -13.6 \text{ eV}$$

and assuming that $\hat{T} = \hat{p}^2/2m_e = \frac{1}{2}m_e\hat{v}^2$, we have

$$\left\langle \left(\frac{v}{c}\right)^2 \right\rangle_0 = \frac{2 \times 13.6 \text{ eV}}{m_e c^2} \simeq \frac{27.2 \text{ eV}}{0.5 \text{ MeV}} = 54.4 \times 10^{-6}$$

or

$$\left\langle \frac{v}{c} \right\rangle_0 \simeq 7 \times 10^{-3} = O(\alpha) \tag{8.48b}$$

where $\alpha \simeq 1/137$ is the fine-structure constant.

Therefore, we have to correct for the relativistic nature of the electron's motion, and we do this in the following way. The kinetic energy, relativistically defined, is

$$\begin{aligned} T = E - m_e c^2 &= (p^2c^2 + m_e^2c^4)^{1/2} - m_e c^2 \\ &= m_e c^2\left(1 + \frac{p^2}{m_e^2c^2}\right)^{1/2} - m_e c^2 \\ &= m_e c^2\left[1 + \frac{1}{2}\frac{p^2}{m_e^2c^2} - \frac{1}{8}\left(\frac{p^2}{m_e^2c^2}\right)^2 + O(p^6)\right] - m_e c^2 \\ &= \frac{p^2}{2m_e} - \frac{p^4}{8m_e^3c^2} + O(p^6) \end{aligned} \tag{8.49a}$$

Here we have neglected higher-order terms in momentum because they would lead to corrections only of higher-order in α. Under this approximation (we ignore the spin–orbit interaction for this calculation)

$$\hat{H} = \hat{T} + \hat{U} = \frac{\hat{p}^2}{2m_e} - \frac{e^2}{4\pi\varepsilon_0 r} - \frac{\hat{p}^4}{8m_e^3c^2} = \hat{H}_0 + \hat{H}'$$

where

$$\hat{H}_0 = \frac{\hat{p}^2}{2m_e} - \frac{e^2}{4\pi\varepsilon_0 r}$$

and

$$\hat{H}' = -\frac{\hat{p}^4}{8m_e^3c^2} \tag{8.49b}$$

Note that $\hat{H}'$ is rotationally invariant. Thus, it is diagonal in the $|nlm\rangle$ basis. That is,

$$\langle n'l'm'|\, \hat{H}'\, |nlm\rangle = 0 \qquad \text{unless } n = n',\, l = l',\, m = m'$$

Consequently, even though the energy levels are degenerate one can still apply nondegenerate perturbation theory since the potentially troublesome off-diagonal matrix elements vanish.

Thus, the first-order change in the energy levels is calculated from

$$E_n^{(1)} = \langle nlm|\, \hat{H}'\, |nlm\rangle$$
$$= -\frac{1}{8m_e^3c^2}\langle nlm|\, \hat{p}^4\, |nlm\rangle \tag{8.50}$$

The expectation value in Eq. (8.50) can be calculated by noting that

$$\hat{H}_0 = \frac{\hat{p}^2}{2m_e} - \frac{e^2}{4\pi\varepsilon_0 r}$$

or

$$\hat{p}^2 = 2m_e\left(\hat{H}_0 + \frac{e^2}{4\pi\varepsilon_0 r}\right)$$

or

$$\hat{p}^4 = 4m_e^2\left(\hat{H}_0 + \frac{e^2}{4\pi\varepsilon_0 r}\right)\left(\hat{H}_0 + \frac{e^2}{4\pi\varepsilon_0 r}\right) \tag{8.51a}$$

We also recognize that

$$\langle nlm|\, \hat{H}_0\, |nlm\rangle = E_n = -\frac{e^2}{4\pi\varepsilon_0}\frac{1}{2n^2a_0} = -\frac{1}{2}\frac{m_e e^4}{(4\pi\varepsilon_0)^2\hbar^2 n^2}$$
$$= -\frac{\alpha^2 m_e c^2}{2n^2} \tag{8.51b}$$

Clearly the first-order change in energy becomes

$$E_n^{(1)} = -\frac{1}{8m_e^3c^2}4m_e^2\langle nlm|\left(\hat{H}_0 + \frac{e^2}{4\pi\varepsilon_0 r}\right)\left(\hat{H}_0 + \frac{e^2}{4\pi\varepsilon_0 r}\right)|nlm\rangle$$
$$= -\frac{1}{2m_ec^2}\left[E_n^2 + 2E_n\frac{e^2}{4\pi\varepsilon_0}\left\langle\frac{1}{r}\right\rangle_{nl} + \left(\frac{e^2}{4\pi\varepsilon_0}\right)^2\left\langle\frac{1}{r^2}\right\rangle_{nl}\right]$$

Furthermore, if we use the expectation values given in Eq. (8.35*b*) we obtain

$$E_n^{(1)} = -\frac{1}{2m_ec^2}\left[E_n^2 + 2E_n\frac{e^2}{4\pi\varepsilon_0}\frac{1}{a_0n^2} + \left(\frac{e^2}{4\pi\varepsilon_0}\right)^2\frac{1}{a_0^2n^3}\frac{1}{l+\frac{1}{2}}\right]$$
$$= -\frac{E_n}{2m_ec^2}\left(E_n + 2\frac{e^2}{4\pi\varepsilon_0}\frac{1}{a_0n^2} - \frac{e^2}{4\pi\varepsilon_0}\frac{2}{a_0n}\frac{1}{l+\frac{1}{2}}\right)$$

$$= -\frac{E_n}{2m_e c^2}\left(-3E_n - \frac{2}{n}\frac{e^2}{4\pi\varepsilon_0}\frac{m_e e^2}{4\pi\varepsilon_0 \hbar^2}\frac{1}{l+\frac{1}{2}}\right)$$

$$= -\frac{E_n}{2m_e c^2}\left(\frac{3\alpha^2 m_e c^2}{2n^2} - \frac{2}{n}\alpha^2 m_e c^2 \frac{1}{l+\frac{1}{2}}\right)$$

$$= \alpha^2 E_n\left(-\frac{3}{4n^2} + \frac{1}{n}\frac{1}{l+\frac{1}{2}}\right) \tag{8.52}$$

Therefore, the total change in the energy levels, if we include relativistic corrections as well as the spin–orbit interaction, is given by the sum of Eqs. (8.47) and (8.52)

$$\Delta E_{n,l}^{\text{Tot}} = -\frac{\alpha^2 E_n}{2}\frac{1}{nl(l+\frac{1}{2})(l+1)}\begin{Bmatrix} l \\ -(l+1) \end{Bmatrix} \quad \begin{matrix} \text{if } j = l+\frac{1}{2} \\ \text{if } j = l-\frac{1}{2} \end{matrix}$$

$$+ \alpha^2 E_n\left(-\frac{3}{4n^2} + \frac{1}{n}\frac{1}{l+\frac{1}{2}}\right)$$

$$= \alpha^2 E_n\left(-\frac{3}{4n^2} + \frac{1}{n}\frac{1}{j+\frac{1}{2}}\right) \qquad \text{here } j = l \pm \tfrac{1}{2} \tag{8.53}$$

If we apply this formula to the $n = 2$ levels of hydrogen we obtain

$$\Delta E(2P_{3/2}) = \alpha^2 E_2(-\tfrac{3}{4}\tfrac{1}{4} + \tfrac{1}{2}\tfrac{1}{2}) = \frac{\alpha^2}{16}E_2$$

$$\Delta E(2P_{1/2}) = \alpha^2 E_2(-\tfrac{3}{4}\tfrac{1}{4} + \tfrac{1}{2}\,1) = \frac{5\alpha^2}{16}E_2 \tag{8.54}$$

$$\Delta E(2S_{1/2}) = \alpha^2 E_2(-\tfrac{3}{4}\tfrac{1}{4} + \tfrac{1}{2}\,1) = \frac{5\alpha^2}{16}E_2$$

The results of Eqs. (8.53) and (8.54) are intriguing in the sense that although both the spin–orbit interaction and the relativistic corrections individually remove the l-degeneracy in the hydrogen levels, when they are combined the total shift depends only on the quantum number j. That is, the levels with the same principal quantum number and the same j value are degenerate in energy even though they may possess different l-quantum numbers. This is evident in Eq. (8.54) for the two $j = 1/2$ states. In spite of our perturbative calculation, this result is generally true for the $1/r$ potential and follows directly from the Dirac equation. It shows that spin and relativistic effects are intimately connected.

In hydrogen and other atoms such as sodium which have only one electron outside a closed shell, states with $l \neq 0$ are split into doublets. As a consequence, the spectral lines that are emitted during transitions between stationary states also appear as doublets. This effect is more pronounced in atoms with large nuclear charge and is clearly evident in the yellow lines of sodium. Figure 8.8(a) shows the energy levels of sodium ($Z = 11$), and we immediately notice that states with the same principal quantum number n but different angular momentum are not degenerate anymore. This is because the *effective* potential seen by the electron outside the closed shells is not of the Coulomb type. This effect is discussed in

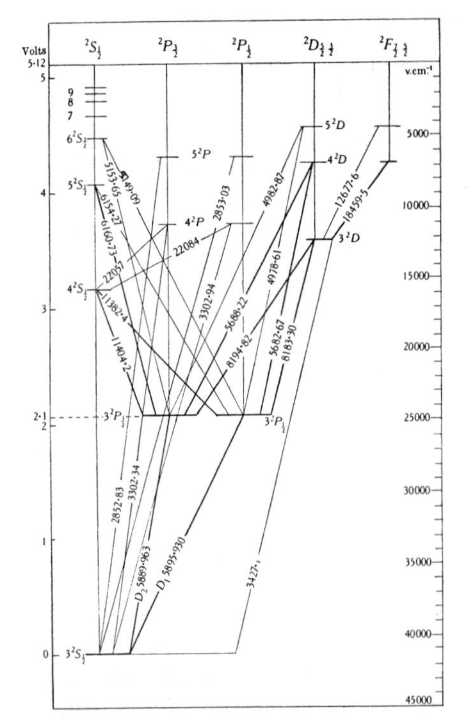

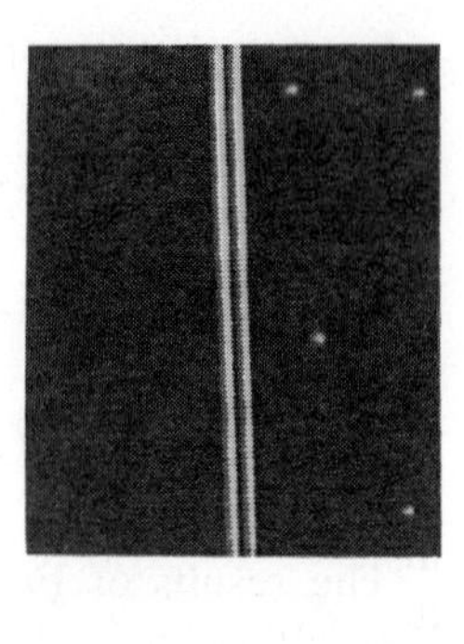

FIGURE 8.8. (a) The energy levels of sodium. The most prominent transitions are indicated. [From M. Born, *Atomic Physics*, Hafner, New York, sixth edition 1957 by permission]. (b) The sodium yellow D-lines when viewed through a simple transmission grating imaged onto a TV monitor. These lines are labeled D_1 and D_2 in the energy level diagram of part (a).

more detail in Section 10.2. The spectral lines of interest are due to the transitions

$$|n=3, l=1, j=3/2\rangle \rightarrow |n=3, l=0, j=1/2\rangle$$
$$|n=3, l=1, j=1/2\rangle \rightarrow |n=3, l=0, j=1/2\rangle$$

Note that $\Delta l = -1$, which is in agreement with the selection rule $\Delta l = \pm 1$, but $\Delta j = -1$ in the first case, whereas $\Delta j = 0$ in the second case. It is customary to designate states and transitions by the spectroscopic notation introduced in Section 8.5 with the addition of a subscript indicating the j-value. Thus, the two lines are described by the transitions

$$3P_{3/2} \rightarrow 3S_{1/2}$$
$$3P_{1/2} \rightarrow 3S_{1/2}$$

The wavelengths of the yellow lines are

$$\lambda_1 = 5889.963 \text{ Å}$$
$$\lambda_2 = 5895.930 \text{ Å}$$

so that

$$\Delta E = h\Delta\nu = 2\pi\hbar \frac{c}{\lambda}\frac{|\Delta\lambda|}{\lambda} = 2.13 \times 10^{-3} \text{ eV}$$

If we calculate the energy splitting using Eq. (8.53) it agrees with the experimental value. Figure 8.8(b) gives an enlarged view of the yellow lines obtained with a transmission grating.

When the energy levels of hydrogen are examined with even higher resolution, it was found by Lamb and Retherford in 1947 that even the $2P_{1/2}$ and $2S_{1/2}$ states of hydrogen split. This is known as the *Lamb shift* and arises from the self-interaction of the electron. Figure 8.9 shows the effect of fine structure and the Lamb shift on the $n = 2$ states of hydrogen. This effect can be calculated to a high degree of accuracy using the theory of Quantum Electrodynamics (QED). The value of the Lamb shift is

$$\Delta E = (1057.13 \pm 0.13) \text{ MHz} \quad (\sim 4.37 \times 10^{-6} \text{ eV}) \quad \text{observed}$$
$$= (1057.77 \pm 0.10) \text{ MHz} \quad \text{calculated}$$

QED also predicts the deviation of the magnetic moment from μ_B,

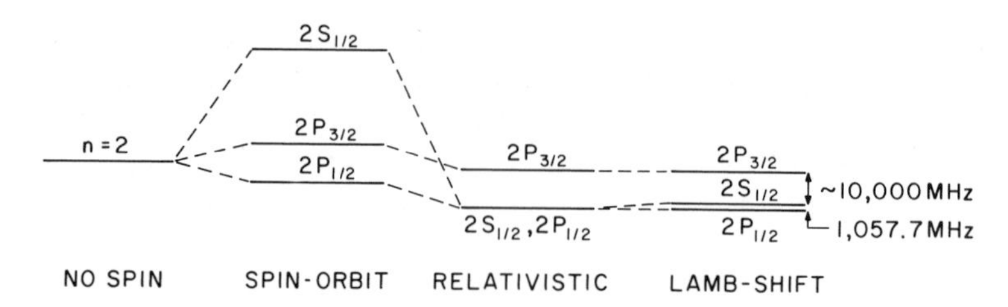

FIGURE 8.9. The fine structure of the $n = 2$ states of the hydrogen atom.

the value given by the Dirac equation. The numbers are

$$\frac{\mu_e}{\mu_B} = 1.00165922(9) \qquad \text{observed}$$

$$= 1.00165921(10) \qquad \text{calculated}$$

In our calculations so far we have treated the proton as if it were a spinless point particle. However, the proton has spin 1/2 and associated with it a magnetic moment μ_P. The magnitude of μ_P is of the order of the *nuclear magneton* (see Section 5.6)

$$\mu_P = 2.91\mu_N = 2.91\frac{e\hbar}{2m_P}$$

which is about 2000 times smaller than μ_B. Consequently, the interaction energy between the nuclear moment and the electron's orbital motion is much smaller than the fine-structure energy. It gives rise to a hyperfine structure of the stationary states. The interaction energy is proportional to $(\hat{\mathbf{I}} \cdot \hat{\mathbf{J}})$ where $\hat{\mathbf{I}}$ is the angular momentum operator for the nucleus. One proceeds as in the case of the $\hat{\mathbf{L}} \cdot \hat{\mathbf{S}}$ coupling and the resulting total angular momentum operator is designated by $\hat{\mathbf{F}} = \hat{\mathbf{I}} + \hat{\mathbf{J}}$. The hyperfine splitting of the $n = 2$ states of the hydrogen atom is of the same order of magnitude but smaller than the Lamb shift.

The fact that $\mu_P \neq \mu_N$ indicates that the proton cannot be considered as a point particle. In fact, as has been deduced from many experiments, the proton has a finite size, and this results in a deviation of the potential from the exact Coulomb form for $r \to 0$. This gives rise to a small shift (but no splitting) of the energy levels of the same order as the hyperfine structure. We speak of an *isotope shift* of the energy levels. However, for light atoms, the shifts between the spectra of different isotopes are dominated by the effective reduced masses rather than by the finite size of the nucleus [see Problem 8.5].

8.7. VARIATIONAL METHOD: APPLICATION TO THE LINEAR POTENTIAL

Sometimes the interaction potential for a system may be complicated enough so that an exact solution is not possible. The form of the potential may not also yield readily to a perturbative treatment. Furthermore, perturbation theory may not be suitable for the kinds of questions we are interested in here. As an example, let us consider the Hamiltonian with a linear potential:

$$\begin{aligned} \hat{H} &= -\frac{\hbar^2}{2\mu}\nabla^2 + k\hat{r} \\ &= \hat{H}_0 + \hat{U} \end{aligned} \tag{8.55}$$

It is clear that if the strength of the interaction k is large, then there will exist bound states of the system. This potential does not lend itself in general to an exact solution. If, on the other hand, we try to treat the potential as a perturbation on the free Hamiltonian $\hat{H}_0$, it is clear we will run into difficulty. This is because the eigenstates of the free Hamiltonian are the plane-wave states, and we have seen in Chapter 1 that for such states

$$\langle \hat{r} \rangle \to \infty$$

Furthermore, it is clear that one cannot obtain a bound-state wavefunction from a plane-wave solution through any perturbative method. In such cases where the perturbation techniques are not quite applicable (we are really looking for nonperturbative effects) one looks for alternate approximate methods. One such method is the variational method, and it is quite useful for obtaining upper bounds on the bound-state energy levels of a system. The variational method rests mainly on two theorems.

Theorem 1

The expectation value of the Hamiltonian is stationary in the neighborhood of its eigenstates.

To show this let us assume that $|\psi\rangle$ is the state in which the expectation value of the Hamiltonian is evaluated. Furthermore, let us modify the state infinitesimally to

$$|\psi\rangle \to |\psi\rangle + |\delta\psi\rangle \tag{8.56a}$$

The expectation value of the Hamiltonian is given by

$$\langle \hat{H} \rangle = \frac{\langle \psi | \hat{H} | \psi \rangle}{\langle \psi \mid \psi \rangle} \qquad \text{(assuming the states are not normalized)}$$

or

$$\langle \psi \mid \psi \rangle \langle \hat{H} \rangle = \langle \psi | \hat{H} | \psi \rangle \tag{8.56b}$$

Under the infinitesimal change of the states, therefore,

$$\langle \psi \mid \psi \rangle\, \delta\langle \hat{H} \rangle + \langle \delta\psi \mid \psi \rangle \langle \hat{H} \rangle + \langle \psi \mid \delta\psi \rangle \langle \hat{H} \rangle = \langle \delta\psi | \hat{H} | \psi \rangle + \langle \psi | \hat{H} | \delta\psi \rangle$$

or

$$\langle \psi \mid \psi \rangle\, \delta\langle \hat{H} \rangle = \langle \delta\psi | \hat{H} - \langle \hat{H} \rangle | \psi \rangle + \langle \psi | \hat{H} - \langle \hat{H} \rangle | \delta\psi \rangle \tag{8.56c}$$

If the expectation value is stationary, i.e., $\delta\langle \hat{H} \rangle = 0$, then the rhs must vanish for any arbitrary $|\delta\psi\rangle$. In particular, if we choose

$$|\delta\psi\rangle = \varepsilon\, |\psi\rangle \qquad (\varepsilon \text{ infinitesimal parameter})$$

then Eq. (8.56c) would lead to a stationary value for $\langle \hat{H} \rangle$ only if

$$(\hat{H} - \langle \hat{H} \rangle)\, |\psi\rangle = 0$$

or

$$|\psi\rangle \text{ is an eigenstate of } \hat{H}$$

This shows that the expectation value of the Hamiltonian is stationary in the neighbourhood of its eigenstates.

Theorem 2

The expectation value of the Hamiltonian in an arbitrary state is greater than or equal to the ground-state energy, i.e.,

$$\frac{\langle \psi | \hat{H} | \psi \rangle}{\langle \psi \mid \psi \rangle} \geqslant E_0$$

where E_0 is the energy of the ground state.

Although this theorem is true, in general, for simplicity of proof we assume that the Hamiltonian has a discrete spectrum such that

$$\hat{H}\,|n\rangle = E_n\,|n\rangle \tag{8.57a}$$

Hence, we can expand the state $|\psi\rangle$ as

$$|\psi\rangle = \sum_{n=0}^{\infty} C_n\,|n\rangle$$

Consequently,

$$\langle\psi\mid\psi\rangle = \sum_{n=0}^{\infty} |C_n|^2$$

and (8.57b)

$$\langle\psi|\,\hat{H}\,|\psi\rangle = \sum_{n=0}^{\infty} E_n\,|C_n|^2$$

Hence

$$\frac{\langle\psi|\,\hat{H}\,|\psi\rangle}{\langle\psi\mid\psi\rangle} = \frac{\sum_n E_n\,|C_n|^2}{\sum_n |C_n|^2} \geqslant \frac{\sum_n E_0\,|C_n|^2}{\sum_n |C_n|^2} = E_0 \tag{8.57c}$$

where the last step follows from the fact that E_0 is smaller than E_n for all values of n in the series; that is, E_0 is the lowest energy of the system.

With these two theorems in mind we can now discuss the variational method. Here we choose a wavefunction that resembles the ground state of the system as much as is possible. That is, the wavefunction must possess all the symmetries of the system and satisfy the necessary boundary conditions but should depend on a few undetermined parameters. We can then calculate the expectation value of the Hamiltonian using this trial wavefunction. The expectation value will be a function of the unknown parameters, and we can minimize the expression so as to determine the best values for these parameters. When we substitute these best values into the expression for the expectation value of the Hamiltonian we obtain an upper bound on the ground-state energy. We can try to improve our bound by adding more and more unknown parameters and using the minimization procedure. When there is difficulty in lowering the bound any further we can think of it as being very close to the true ground state.

Furthermore, we can also estimate energy eigenvalues of the excited states by this method. We simply have to choose a trial wavefunction that is orthogonal to the ground-state wavefunction. The trial wavefunction must have the right symmetry properties and must satisfy the correct boundary conditions. If we again follow the minimization procedure described above we obtain an upper bound for the first excited state. The method becomes laborious and less accurate as we go to higher states.

The variational method gives a very good estimate of the energy eigenvalues. But it does not determine the wavefunction accurately. Consequently the expectation values of observables calculated by this method, other than the energy, are not reliable.

Let us apply all these ideas to the case of the linear potential introduced in Eq. (8.55). The potential is attractive and is of infinite range (like the Coulomb potential). Therefore, the system will have bound states with energies becoming progressively more closely spaced as shown in Fig. 8.10.

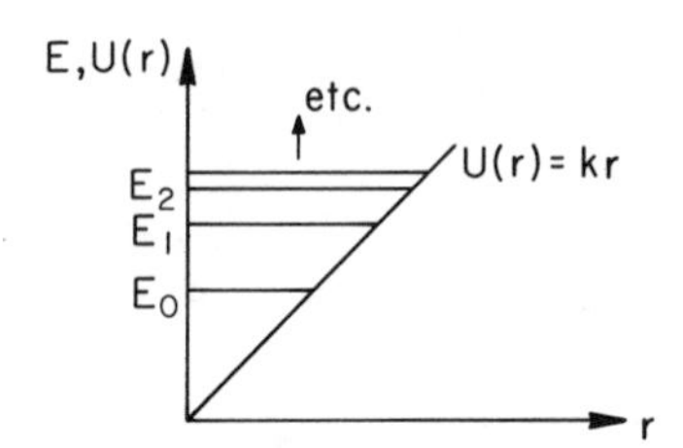

FIGURE 8.10. The linear potential $U(r) = kr$ and some of the energy levels for such a potential.

The radial equation for this system becomes [see Eq. (8.19)]

$$\left\{\frac{d^2}{dr^2} + \frac{2\mu}{\hbar^2}\left[E_{nl} - kr - \frac{\hbar^2 l(l+1)}{2\mu r^2}\right]\right\} u_{nl}(r) = 0 \qquad (8.58a)$$

This equation can be solved exactly for $l = 0$ states. We will evaluate the exact solution for this case so as to be able to compare it with the value obtained from the variational method. When $l = 0$ the equation reduces to

$$\left[\frac{d^2}{dr^2} + \frac{2\mu}{\hbar^2}(E_{nl} - kr)\right] u_{nl}(r) = 0 \qquad (8.58b)$$

The wavefunction has to satisfy the usual boundary conditions.

$$u_{nl}(r) \to 0 \quad \text{as} \quad r \to 0 \quad \text{and} \quad r \to \infty \tag{8.58c}$$

Let us now make the following redefinition:

$$x = (k\gamma)^{1/3}(r - r_0) = (k\gamma)^{1/3}(r - E/k)$$

or

$$r = (k\gamma)^{-1/3}x + E/k, \qquad \text{where } \gamma = \frac{2\mu}{\hbar^2} \tag{8.59a}$$

The range of x, therefore, becomes $-E(\gamma/k^2)^{1/3} \leqslant x \leqslant \infty$, and in terms of this variable Eq. (8.58*b*) becomes

$$\left(\frac{d^2}{dx^2} - x\right)u(x) = 0 \tag{8.59b}$$

with the boundary conditions

$$u\left[x = -E\left(\frac{\gamma}{k^2}\right)^{1/3}\right] = 0 \quad \text{and} \quad u(x \to \infty) = 0 \tag{8.59c}$$

The solutions of Eq. (8.59*b*) are known as the Airy functions and are denoted by Ai(x).† The Airy function which tends to zero as $x \to \infty$ is shown in Fig. 8.11(a). It is monotonically decaying for $x > 0$ and is oscillatory for $x < 0$. The values of the bound-state energy are obtained from the zeros of the Airy function since it must satisfy the boundary condition of Eq. (8.59*c*). Thus

$$x = -E_n\left(\frac{\gamma}{k^2}\right)^{1/3} = a_n$$

$$E_n = -\left(\frac{k^2}{\gamma}\right)^{1/3} a_n \tag{8.60a}$$

where the a_n's give the zeros of the Airy function. The numerical values of the first few zeros are

$$a_1 = -2.338 \qquad a_2 = -4.088 \qquad a_3 = -5.521 \qquad a_4 = -6.787 \tag{8.60b}$$

† See *Handbook of Mathematical Functions*, by M. Abramowitz and I. Stegun, Dover publication N.Y. 1965, p. 446, for a tabulation of the functions and of their zeros.

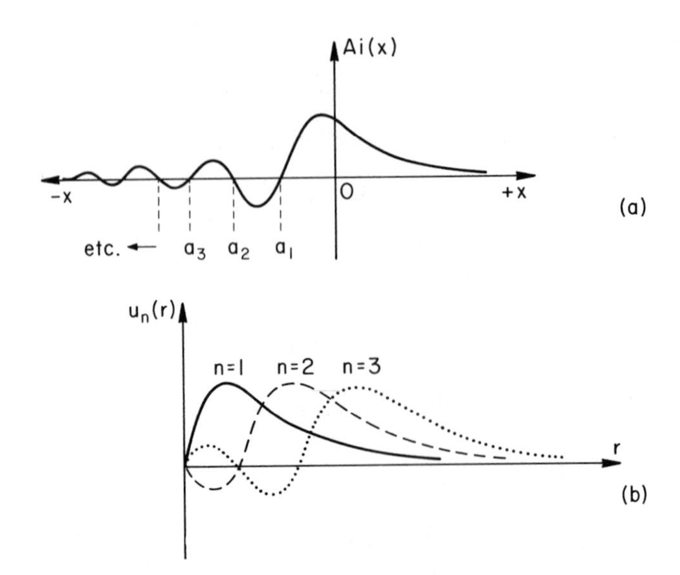

FIGURE 8.11. (a) The Airy function of the first kind with real argument. (b) The radial eigenfunctions $rR_{n,l=0}(r)$ for the linear potential.

Equation (8.60a) gives the bound-state energies for the $l=0$ states of the linear potential. The radial wavefunctions $u_{n0}(r)$ are obtained from the Airy functions subject to the transformation of its argument as indicated by Eqs. (8.59a). They are shown in Fig. 8.11(b) for the low values of n.

Next we evaluate the energy eigenvalues by the variation method. Let us first calculate the $l=0$ state. The ground-state wavefunction must be nodeless and must satisfy the boundary conditions of Eq. (8.58c). Thus, the simplest trial wavefunction has the form

$$u(r)=re^{-\lambda r}$$

or

$$R(r)=e^{-\lambda r} \qquad (\lambda \text{ is the unknown parameter}) \tag{8.61a}$$

Thus

$$\langle\psi\mid\psi\rangle=\int|\psi|^2\,r^2\,dr\,d\Omega=4\pi\int_0^\infty e^{-2\lambda r}r^2\,dr$$

$$=4\pi\frac{1}{4\lambda^3}=\frac{\pi}{\lambda^3} \tag{8.61b}$$

Furthermore,

$$\frac{2\mu}{\hbar^2}\langle\psi|\,\hat{H}\,|\psi\rangle=\frac{2\mu}{\hbar^2}\int\psi^*\hat{H}\psi r^2\,dr\,d\Omega$$

$$=\int e^{-\lambda r}\left[\left(-\frac{d^2}{dr^2}-\frac{2}{r}\frac{d}{dr}+k\gamma r\right)e^{-\lambda r}\right]r^2\,dr\,d\Omega$$

$$=4\pi\int_0^\infty(-\lambda^2r^2+2\lambda r+k\gamma r^3)e^{-2\lambda r}\,dr$$

$$=4\pi\left(-\frac{\lambda^2}{4\lambda^3}+\frac{2\lambda}{4\lambda^2}+k\gamma\frac{3}{8\lambda^4}\right)$$

$$=\frac{\pi}{\lambda^3}\left(\lambda^2+\frac{3}{2}\frac{k\gamma}{\lambda}\right) \tag{8.61c}$$

Therefore, the expression that we must minimize becomes

$$\langle\hat{H}\rangle=\frac{\langle\psi|\,\hat{H}\,|\psi\rangle}{\langle\psi\mid\psi\rangle}=\frac{\hbar^2}{2\mu}\left(\lambda^2+\frac{3}{2}\frac{k\gamma}{\lambda}\right) \tag{8.62a}$$

Now

$$\frac{d\langle\hat{H}\rangle}{d\lambda}=0=\frac{\hbar^2}{2\mu^2}\left(2\lambda-\frac{3}{2}\frac{k\gamma}{\lambda^2}\right) \tag{8.62b}$$

yields

$$\lambda_{\min}=(\tfrac{3}{4}k\gamma)^{1/3}$$

Introducing this value into Eq. (8.62a) we obtain the minimum energy

$$E_{\min}=\langle\hat{H}\rangle_{\min}=\frac{\hbar^2}{2\mu\lambda}(\lambda^3+\tfrac{3}{2}k\gamma)$$

$$=\frac{\hbar^2}{2\mu\lambda}\frac{9}{4}k\gamma=\frac{9\hbar^2}{8\mu}k\gamma(\tfrac{3}{4}k\gamma)^{-1/3}$$

$$=2.48\left(\frac{k^2}{\gamma}\right)^{1/3}=2.48\left[\frac{(\hbar k)^2}{2\mu}\right]^{1/3} \tag{8.62c}$$

Comparison of the result of the variation method with the exact ground-state energy given in Eqs. (8.60a) and (8.60b) shows that the discrepancy is only 6% and that $E_{\min}>E_0$, as it should.

The same technique can be extended to obtain the energy of the excited states as discussed earlier. For the $l=0$ states with $n>1$ we must choose a trial wavefunction with the appropriate number of nodes, whereas for the $n = 2$, $l = 1$ and $n = 3$, $l = 2$ states it suffices to choose a nodeless trial wavefunction, but including an additional factor of r^l. The results of the variational calculation are given in Table 8.1 where they are also compared with the exact results when

Table 8.1. Energy Eigenvalues for the Linear Potential in units of $[(\hbar k)^2/2\mu]^{1/3}$

State	Exact solution	Variation method
1*S*	2.338	2.48
2*P*	Cannot be obtained	3.48
2*S*	4.088	4.29
3*S*	5.521	
3*D*	Cannot be obtained	4.35
4*S*	6.787	

these can be calculated. The energy spectrum is shown graphically in Fig. 8.12.

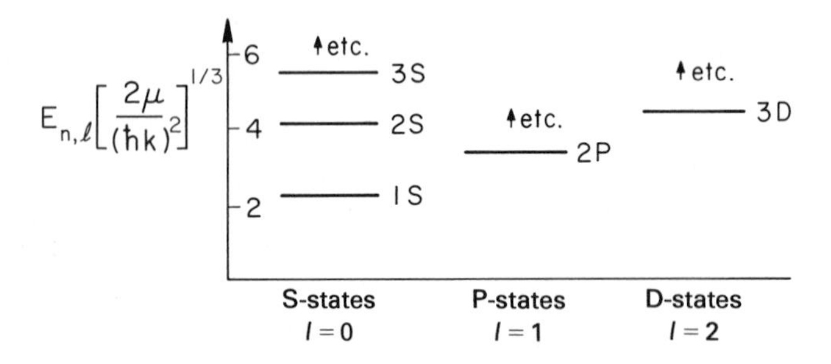

FIGURE 8.12. The energy level diagram for the low-lying levels of the linear potential.

8.8. QUARKONIUM

We have studied the linear potential because it has interesting applications to physical systems. For example, the interaction potential between the constituents of the proton can be phenomenologically represented by a linear potential, and it agrees quite well with experimental observations which we describe below.

It has only recently been established that matter consists of quarks. Quarks have a fractional electric charge of $-1/3$ or $2/3$ of the proton's charge, and the corresponding antiquarks have charges $1/3$ or $-2/3$. The quarks have spin $1/2$ and consequently the nucleons (i.e., protons and neutrons) are composites of three quarks, whereas the mesons are composites of a quark and an antiquark. No free quarks have been observed, and it is believed that free quarks cannot exist in nature. This peculiar phenomenon can be explained only if quarks inside elementary particles are bound by a force that increases as their separation increases. At large separations the potential energy becomes large enough that it exceeds the threshold for the production of a quark–antiquark pair so that a new meson can be produced without the appearance of a free quark. We say that the quarks are confined. There are many potential models of confinement and the linear potential of Eq. (8.55) is one of them.

At present five different varieties (flavors) of quarks have been observed. They are

u-quark (up) and d quark (down)

c-quark (charm) and s quark (strange)

b-quark (bottom) and ?

A sixth quark labeled t (for top) is postulated to exist as the partner of the bottom quark but has not yet been observed.† In Table 8.2 we give the charge and mass of the quarks as inferred from many experiments.

Table 8.2. Charges and Masses of Quarks

Quark	Charge	Mass (mc^2)
u	2/3	~few MeV
d	−1/3	~few MeV
s	−1/3	~0.4 GeV
c	2/3	1.15 GeV
b	−1/3	4.35 GeV

† A recent experiment (1984) has claimed preliminary evidence for the observation of the top quark with $30 < m_t < 50$ GeV. (Private communication).

We have already mentioned that a system consisting of a quark and an antiquark can be observed in nature in the form of short-lived particles. For instance, the $u\bar{d}$ system is the π^+-meson, while the $\bar{u}d$ system is the π^--meson, $u\bar{s}$ the K^+-meson, etc. In particular, systems consisting of a quark and its own antiquark have been observed such as

$$\pi^0 = \frac{1}{2^{1/2}}(u\bar{u} + d\bar{d})$$

$s\bar{s}$ is the ϕ-meson

$c\bar{c}$ are the J/Ψ-mesons

$b\bar{b}$ are the Υ-mesons

If the potential energy between the $q\bar{q}$ pair obeys Eq. (8.55), then at *small* distances the $q-\bar{q}$ force is weak and the system resembles the hydrogen atom. In systems consisting of heavy quarks such as the J/Ψ and Υ-mesons the quarks move with low velocity so that it is possible to apply the Schrödinger equation and treat relativistic effects as a small correction.

From our general arguments we would expect that such $q\bar{q}$ systems would have an infinite number of bound states. This is not true because as the potential energy increases an additional pair of quarks can be produced, and the $q\bar{q}$ system can immediately decay into two mesons, for instance,

$$c\bar{c} \to c\bar{u} + \bar{c}u \Rightarrow D^0\overline{D^0}$$

where $c\bar{u}$ is the D^0 meson and $\bar{c}u$ and $\bar{D}^0$. Thus, the observable spectrum of $q\bar{q}$ stationary states is limited to energies below the threshold for production of the corresponding mesons.

In Fig. 8.13 we show the spectrum of some of the states observed in the $c\bar{c}$ system (J/Ψ mesons) and in the $b\bar{b}$ system (Υ-mesons; pronounced upsilon). The energy indicated on the lhs scale gives the rest-mass Mc^2 of the $q\bar{q}$ system in that particular state. If the mass of the quark (and antiquark) is m_q, then

$$(Mc^2)_{n,l} = 2(m_q c^2) + E_{n,l}$$

where $E_{n,l}$ is the sum of the kinetic and potential energy of the $q\bar{q}$ system in the state $|n, l\rangle$. Assuming a linear potential the values of $E_{n,l}$ are as obtained in the previous section and summarized in Table

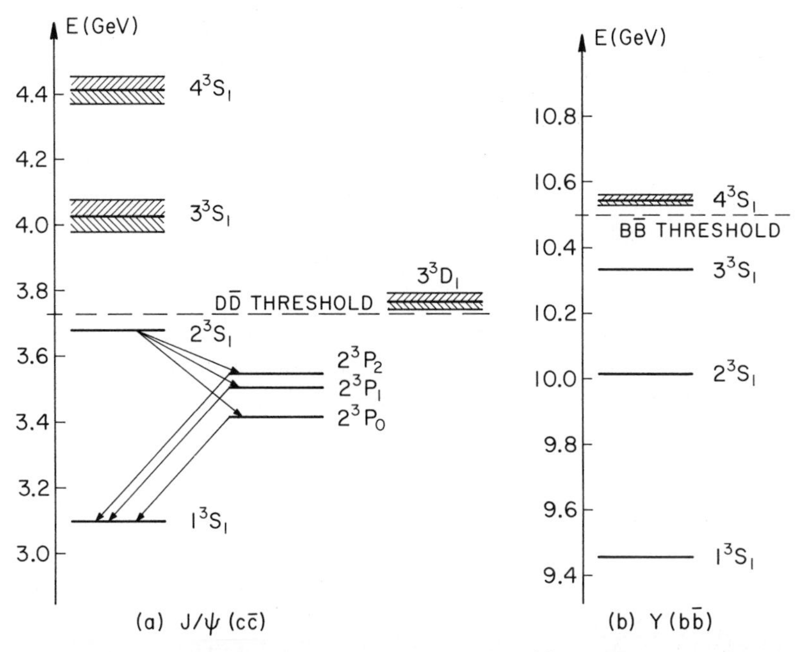

FIGURE 8.13. The observed spectrum of quarkonium (a) for the $c\bar{c}$ system known as the J/ψ mesons, and (b) for the $b\bar{b}$ mesons known as the Υ mesons. Note that the total energy of the J/ψ and Υ states is due in large part to the quark mass. Above the threshold for production of $D\bar{D}$ (or $B\bar{B}$) mesons the states acquire a measurable width. Some of the radiative transitions in the J/ψ system are shown.

8.1. The states of the $c\bar{c}$ and $b\bar{b}$ mesons are observed in high-energy collisions of elementary particles as, for instance, in πp and pp collisions, but are most easily produced in e^+e^- annihilations at high energies.

In Fig. 8.13 we have also indicated the threshold energy for the production of $D\bar{D}$ and $B\bar{B}$ mesons. Above this energy the $q\bar{q}$ states are not any more exactly stationary because they can decay into a pair of mesons. Thus, the energy of the $q\bar{q}$ states above threshold is not sharp but has considerable width as indicated. We note that not only S-states but also P and D states have been identified. We draw attention to the three closely spaced P-states indicated for the $c\bar{c}$ system. The splitting between these states is due to the spin–orbit interaction in analogy to the discussion of Section 8.5. The quarks have spin 1/2, but in contrast to the hydrogen atom q and $\bar{q}$ have the same mass, and thus magnetic moments of the same magnitude. Therefore, we must compose the two spins S_1 and S_2 into a single spin (angular momentum) vector

$$\hat{\mathbf{S}} = \hat{\mathbf{S}}_1 + \hat{\mathbf{S}}_2$$

where $\hat{\mathbf{S}}$ will have eigenvalues $s = 0$, and 1. The total angular momentum will be

$$\hat{\mathbf{J}} = \hat{\mathbf{L}} + \hat{\mathbf{S}}$$

where the eigenvalues of $\hat{\mathbf{J}}$ will be $j = l-1, l, l+1$ when $l \neq 0$, and $j = 0, 1$ when $l = 0$. To distinguish these states we add a superscript $(2S+1)$ to the spectroscopic notation; for instance,

$$\begin{aligned} 2^3P_2 &\quad \text{means} \quad n=2, l=1, s=1, j=2 \\ 1^1S_0 &\quad \text{means} \quad n=1, l=0, s=0, j=0 \\ 1^3S_1 &\quad \text{means} \quad n=1, l=0, s=1, j=1, \qquad \text{etc.} \end{aligned}$$

The arrows in Fig. 8.13 indicate transitions between the states that proceed by the emission of a photon. Note that they obey the selection rule $\Delta j = 0, \pm 1$ ($j = 0 \rightarrow j = 0$ not permitted).

We can now compare the data with our calculation. Since q and $\bar{q}$ have the same mass, the reduced mass μ of the system is given by

$$\mu = \frac{m_q m_{\bar{q}}}{m_q + m_{\bar{q}}} = \tfrac{1}{2} m_q$$

Thus, the total energy of the states is

$$(E_T)_{n,l} = (Mc^2)_{n,l} = 2(m_q c^2) + \left[\frac{(\hbar k)^2}{m_q}\right]^{1/3} a_{n,l} \qquad (8.63)$$

where the $a_{n,l}$ are given in Table 8.1. There are two unknowns in Eq. (8.63), m_q and the strength of the linear potential k. The total energy of the observed $c\bar{c}$ and $b\bar{b}$ states is given in Table 8.3. Using the $(c\bar{c})$ 1^3S_1 and 2^3S_1 states we find

$$(E_{2,0}) - (E_{1,0}) = 0.588 \text{ GeV} = \left[\frac{(\hbar k)^2}{m_c}\right]^{1/3} (4.29 - 2.48)$$

$$(E_{1,0}) = 3.097 \text{ GeV} = 2(m_c c^2) + \frac{0.588}{1.81} \times 2.48$$

Table 8.3. Masses (Energies) of the $J/\Psi(c\bar{c})$ and $\Upsilon(b\bar{b})$ states (in GeV)

State	J/Ψ	Υ	
1^3S_1	3.097	9.458	
$2^3P_{0,1,2}$	3.514		(Center of gravity)
2^3S_1	3.685	10.016	
3^3S_1	4.730	10.323	
$3^3D_{1,2,3}$	3.770		(3^3D_1)
4^3S_1	4.414	10.547	

which yield

$$m_c c^2 \simeq 1.146\ \text{GeV}; \qquad \frac{(\hbar k)^2}{m_c} = 0.0343\ \text{GeV}$$

We write

$$\frac{(\hbar k)^2}{m_c} = \frac{(\hbar c)^2}{m_c c^2} k^2 \quad \text{and therefore} \quad k^2 = m_c c^2 \frac{0.0343}{(\hbar c)^2}$$

with $m_c c^2 = 1.15$ GeV and $\hbar c = 0.197$ GeV-F

$$k = 1.006\ \text{GeV/F} \tag{8.64}$$

We have found the strength of the attractive force between quarks!!†

To get a feeling for the magnitude of k we can compare it to the force that binds the electron to the proton in the hydrogen atom. The potential energy in that case is

$$U(r) = \frac{e^2}{4\pi\varepsilon_0} \frac{1}{r} = \frac{e^2}{4\pi\varepsilon_0 \hbar c} \frac{1}{r} \frac{\hbar c}{r} = \alpha \frac{\hbar c}{r}$$

Since $r \simeq a_0 \simeq 10^{-8}$ cm $= 10^5$ F

$$U_{ep}(r \simeq 10^5\ \text{F}) = \frac{1}{137} \frac{0.2}{10^5} \simeq 10^{-8}\ \text{GeV}$$

† In published work k is usually expressed in dimensionless units using the proton mass as a scale. Thus

$$k_s = k \frac{\hbar c}{m_p c^2} \simeq 1.02 \frac{0.2}{0.938} \simeq 0.2$$

At such a distance of 10^5 F the $q\bar{q}$ force would result in a potential energy

$$U_{q\bar{q}}(r=10^5\ \text{F}) \sim 10^5\ \text{GeV}$$

As a further comparison, consider the nuclear force that binds a proton and neutron into a deuteron. The potential energy is of the order of 40–60 MeV and the separation $r \sim 1$ F. At the same separation the $q\bar{q}$ potential energy is ~20 times stronger.

We can also obtain a feeling for the size (i.e., the average separation) between the q and $\bar{q}$ quarks in these bound states. It can be easily shown that

$$\langle r\rangle_{n,l} \simeq \frac{E_{n,l}}{k}$$

Since $E_n \sim 0.32a_n$ GeV

$$\langle r_n\rangle \sim 0.32a_n\ \text{F} \qquad \text{or} \qquad \langle r\rangle_{n=1} \simeq 0.8\ \text{F}$$

As n increases the size of the system grows.

When we apply the linear potential model to the $b\bar{b}$ system we find using the energy of the $l=0$, $n=1, 2$ states that

$$m_b c^2 \simeq 4.347\ \text{GeV}$$
$$k \simeq 0.869\ \text{GeV/F}$$

We note that the force between the b and $\bar{b}$ quarks is roughly equal to that between the c and $\bar{c}$ quark. In fact, the slight decrease in the strength of the potential energy between heavier quarks is a fundamental prediction of the current theory of quark–quark interactions (this theory is known as Quantum Chromo Dynamics—QCD, for short). Furthermore the model predicts the energy difference between the $3S-1S$ and $4S-1S$ states with good accuracy, as the reader can check using the solution given in Table 8.1.

The $c\bar{c}$ and $b\bar{b}$ states exist only for a short time, the lifetime of the state being typically

$$\tau \simeq 10^{-18}\ \text{sec}$$

This is a very short time interval by macroscopic standards, but it is 10^3 times longer than the lifetime of other mesons in this mass range. From the uncertainty principle the width of a state (the uncertainty in its energy) is

$$\Gamma = \frac{\hbar}{\tau} = \frac{\hbar c}{c\tau}$$

which for $\tau = 10^{-18}$ sec yields $\Gamma \simeq 60$ keV. Therefore, $\Gamma/m_{c\bar{c}} \approx 2 \times 10^{-5}$ indicating that the relative width of these states is fairly narrow.

Below the threshold for meson production the $q\bar{q}$ system decays because the q and $\bar{q}$ annihilate one another *electromagnetically*. The probability for annihilation depends on the charge of the quarks, their mass, and the probability that they are found at the same position in space at the same time. The latter probability is proportional to the square of the wavefunction at the origin, $|\psi(r=0)|^2$. For S-states the wavefunction at the origin is finite, and therefore S-states can decay through annihilation. On the contrary, the P-states will have a very small probability for annihilation since the wavefunction at the origin is zero. This is indeed observed experimentally, the P-states making a transition to the S-states by the emission of a photon (see Fig. 8.13). These transitions are exactly analogous to atomic transitions, but instead of visible light, γ-rays of a few 100 MeV energy are emitted.

In conclusion we have seen that bound states of heavy $q\bar{q}$ pairs are very similar to the hydrogen atom. The size of these systems is 10^5 times smaller than the Bohr radius, and the energies of the states are correspondingly 10^5 times larger. For a further discussion of these systems the reader can consult an interesting article by K. Gottfried ("Comments in Nuclear and Particle Physics", Vol. **9,** 141, 1981).

8.9. SUMMARY

We have indicated the method for obtaining bound-state solutions for simple physical systems. For a central potential, we have shown that the motion of the center of mass can be separated out just as in classical mechanics. Furthermore, if the potential is rotationally invariant, the angular solutions separate and are simply the eigenstates of the angular momentum. We have solved the radial equation for the Coulomb potential in the hydrogen atom. The effect of the electron's spin and the relativistic nature of the electron's motion were treated perturbatively. This enabled us to discuss the fine structure in the energy levels of hydrogen in detail. One of the approximate methods for studying bound-state problems is the variational method. We discussed this method and applied it to the case of the linear potential. This potential is particularly interesting since it phenomenologically explains several features of quark binding in mesons and nucleons. We concluded with an application to quarkonium and compared the theoretical results with the experimental observations.

Problems

PROBLEM 1

(*a*) Plot to scale—but choose your units in a reasonable way—the probability of finding the electron at a distance r from the proton for the $1S$, $2S$, and $2P$ states of the hydrogen atom. (See Fig. 8.5).

(*b*) Plot the probability of finding the electron at an angle θ with respect to the quantization axis for the five $3D$ states of the hydrogen atom (see Fig. 8.6). Also indicate the most probable direction of the orbital angular momentum.

PROBLEM 2

Consider the ground state of the hydrogen atom. As in Problem 1 above one can plot to scale the probability of finding the electron at a distance $|\mathbf{r}|$ (within the interval $|d\mathbf{r}|$) from the proton; it is best to plot r/a_0.

For the same state plot *to scale* the probability that the electron has momentum $|\mathbf{p}|$ in the interval $|d\mathbf{p}|$; it is best to plot $p(a_0/\hbar)$.

Note: You must evaluate the wavefunction $\phi(k)$ in momentum space.

PROBLEM 3

Plot *to scale* the potential $V(r)$ for the hydrogen atom and the angular momentum barrier for $l=5$. From the plot estimate the lowest value of n, the principal quantum number for this state. You may use for instance Eq. (8.29a).

PROBLEM 4

From reference tables in the library find the frequency of atomic transitions in hydrogen and deuterium. Establish the energy of the corresponding states and show explicitly the effect of the reduced mass.

PROBLEM 5

Because of the finite size of the nucleus, the potential seen by an electron can be approximated by the form shown in the figure.

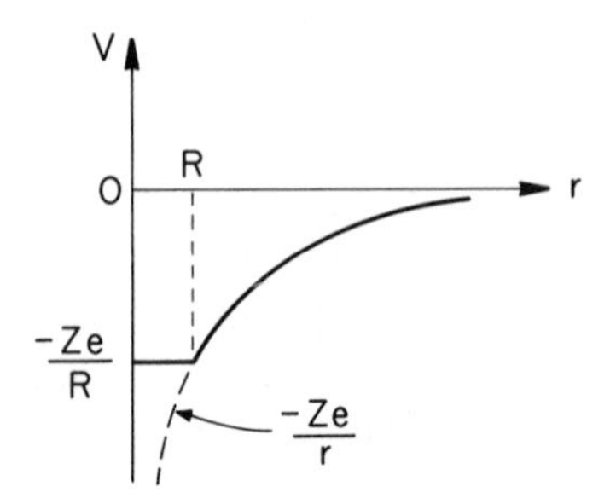

Modification of the Coulomb potential due to the finite size of the nucleus.

(a) Calculate this effect in first-order perturbation for the ground state of the hydrogen atom and the perturbation

$$\hat{H}' = \frac{e^2}{r} - \frac{e^2}{R}, \qquad r \leqslant R$$

$$= 0, \qquad r > R$$

(b) Explain the above choice of $\hat{H}'$.

(c) You can expand in powers of $R/a_0 \ll 1$, but be careful.

(d) Evaluate numerically your result for $R = 1\,\text{F}$ and $R = 100\,\text{F}$.

(*e*) Give the fractional shift of the energy of the ground state.

Note: This effect is known as the isotope shift and appears in the spectral lines of the heavy elements.

PROBLEM 6

Consider the operator

$$\hat{M}_i = \frac{1}{2\mu}\varepsilon_{ijk}(\hat{p}_j\hat{L}_k - \hat{L}_j\hat{p}_k) - \frac{e^2}{r}\hat{r}_i$$

and the Hamiltonian for the hydrogen atom

$$\hat{H} = \frac{\hat{\mathbf{p}}^2}{2\mu} - \frac{e^2}{r}$$

Show that

(*a*) $[\hat{M}_i, \hat{H}] = 0$
(*b*) $[\hat{M}_i, \hat{L}_j] = i\hbar\varepsilon_{ijk}\hat{M}_k$
(*c*) $[\hat{M}_i, \hat{M}_j] = -(2i\hbar/\mu)\varepsilon_{ijk}\hat{H}\hat{L}_k$

Here $\hat{L}_i$ are the orbital angular momentum operators.

PROBLEM 7

Prove the following operator relation

$$\hat{p}^2 = \frac{1}{r^2}[\hat{L}^2 + (\hat{\mathbf{r}}\cdot\hat{\mathbf{p}})^2 - i\hbar(\hat{\mathbf{r}}\cdot\hat{\mathbf{p}})]$$

starting from the commutation relations [Eqs. (6.9*b*)]

$$[\hat{x}, \hat{p}_x] = [\hat{y}, \hat{p}_y] = [\hat{z}, \hat{p}_z] = i\hbar$$
$$[\hat{x}, \hat{y}] = [\hat{x}, \hat{p}_y] = 0 = \cdots, \text{etc.}$$

If you have difficulty you can express $\hat{L}^2$ in cartesian coordinates.

PROBLEM 8

A particle in a spherically symmetric potential is in a state (*not* a stationary state) described by the wavefunction

$$\psi(x, y, z) = C(xy + yz + zx)e^{-\alpha r^2}$$

What is the probability that a measurement of $\hat{L}^2$ will yield zero? that it will yield $6\hbar^2$? If the value of l is found to be 2 what are the relative probabilities for $m = 2, 1, 0, -1, -2$?

Hint: Expand the wavefunction in eigenstates of $\hat{L}^2$, $\hat{L}_z$ according to Eqs. (7.59).

PROBLEM 9

The exact solution for the energy levels of the hydrogen atom as obtained from the Dirac equation is

$$E_n = \mu c^2 \left\{ 1 + \left[\frac{(e^2/4\pi\varepsilon_0\hbar c)}{n - (j+\frac{1}{2}) + \sqrt{(j+\frac{1}{2})^2 - \left(\frac{e^2}{4\pi\varepsilon_0\hbar c}\right)^2}} \right]^2 \right\} - \mu c^2$$

Expand this result through *fourth* order in $\alpha \equiv e^2/4\pi\varepsilon_0\hbar c$ and show that it agrees with the results obtained for the $n = 2$ levels as given by Eqs. (8.54) of the text.

PROBLEM 10

Consider a μ^- meson bound to a copper nucleus. Using hydrogen wavefunctions find the probability that in the ground state the μ^- meson is *inside* the nucleus. Take the radius of the copper nucleus to be $R_0 = A^{1/3} \times 1.3\text{ F} = 5.2\text{ F}$; $A = 64$, $Z = 29$; $m_\mu c^2 = 105\text{ MeV}$; $\hbar c = 200\text{ MeV-F}$, $\alpha \equiv e^2/4\pi\varepsilon_0\hbar c = 1/137$.

PROBLEM 11

(*a*) From the energy-level diagram for the sodium atom given in Fig. 8.8 calculate the effective nuclear charge Z_{eff}, such that

$$E_{n,l} = -\frac{me^4 Z_{\text{eff}}^2}{2\hbar^2 n^2}$$

(*b*) Give a numerical value for the fine-structure splitting of the yellow lines of sodium, and make an order-of-magnitude comparison of this effect with the result given by Eq. (8.47) of the text.

PROBLEM 12

Because of the interaction between the proton and electron spins the ground state of the hydrogen atom has a *hyperfine* structure. The energy matrix is of the form

$$H = \begin{array}{c|cccc} & |1\rangle & |2\rangle & |3\rangle & |4\rangle \\ \hline |1\rangle & A & 0 & 0 & 0 \\ |2\rangle & 0 & -A & 2A & 0 \\ |3\rangle & 0 & 2A & -A & 0 \\ |4\rangle & 0 & 0 & 0 & A \end{array}$$

where the basis states are

$$|1\rangle = |e+, p+\rangle$$
$$|2\rangle = |e+, p-\rangle$$
$$|3\rangle = |e-, p+\rangle$$
$$|4\rangle = |e-, p-\rangle$$

The notation $(e+)$ means electron spin along the $+Z$-axis, etc.

(*a*) Find the energy of the stationary states.

(*b*) Express the stationary states as appropriate linear combinations of the basis states $|1\rangle$, $|2\rangle$, $|3\rangle$, and $|4\rangle$.

(*c*) What would happen to the energy of the stationary states if a weak magnetic field were applied along the Z-axis? Can you give the value of the projections of the spin along the Z-axis and the value of the *total* spin of the atom in each of the stationary states?

PROBLEM 13

A hydrogen atom is placed in a uniform electric field of strength $\mathscr{E}_z$ along the z direction. Choose as a trial wavefunction

$$\psi_\alpha(x, y, z) = \left(\frac{1}{\pi a_0^3}\right)^{1/2} e^{-r/a_0}(1 + \alpha z),$$

$$[r = (x^2 + y^2 + z^2)^{1/2};\ a_0 = \text{Bohr radius}]$$

and calculate the ground-state energy using the variational method. Can you justify the choice of this wavefunction? (Neglect higher powers than $\mathscr{E}_z^2$. *Note*: This is the second-order Stark effect in hydrogen.)

PROBLEM 14

A particle of mass m moves in a potential in one dimension of the form

$$V(x) = 0, \qquad \text{for } 0 \leqslant x \leqslant a$$
$$= \infty, \qquad \text{everywhere else}$$

Inside the well it is subjected to another potential of the form $A(x - a/2)$, where A is a constant. What is the change in the ground-state energy due to this potential? (Keep terms only up to A^2. Can you guess at a suitable trial wavefunction from Problem 13?)

PROBLEM 15

Calculate the ground-state energy of the hydrogen atom by the variational method using the following wavefunction

$$\psi_\alpha(r) = \left(1 - \frac{r}{\alpha}\right), \qquad r \leqslant \alpha$$
$$= 0, \qquad r > \alpha$$

How is $\alpha_{\min}$ related to the Bohr radius?

PROBLEM 16

Calculate by the variational method the ground state energy of a particle of mass m in one dimension in the potential

$$V(x) = 0, \qquad -a \leqslant x \leqslant a$$
$$= \infty, \qquad \text{everywhere else}$$

Choose the trial wavefunction as

$$\psi_\alpha(x) = (a^2 - x^2)(a^2 - \alpha x^2), \qquad \text{for } -a \leqslant x \leqslant a$$
$$= 0, \qquad \text{everywhere else}$$

The minimization leads to more than one root. Show that one root gives very good agreement with the true eigenvalue. What can you say about the other root?

PROBLEM 17

The Hamiltonian for the one-dimensional s.h.o in units $\hbar = 1$, $\omega = 1$, and $m = 1$, is given by

$$\hat{H} = \tfrac{1}{2}\hat{p} + \tfrac{1}{2}\hat{x}^2 = -\frac{1}{2}\frac{d^2}{dx^2} + \tfrac{1}{2}x^2$$

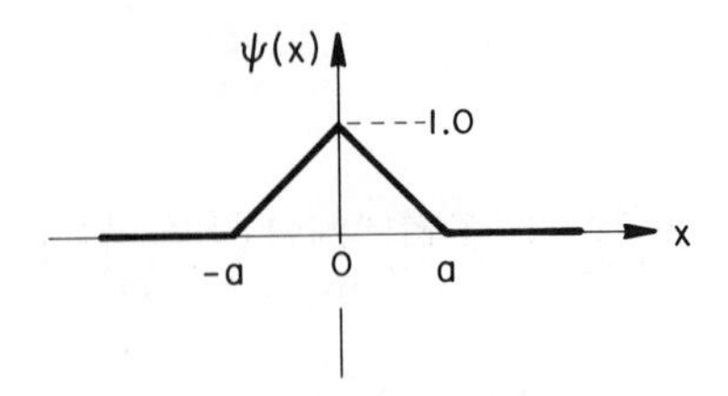

Sketch of the trial wavefunction.

Use as a trial wavefunction $u(x)$ the piecewise continuous function graphed above

$$u(x) = \begin{cases} 1 - x/a, & 0 < x < a \\ 1 + x/a, & -a < x < 0 \end{cases}$$

and apply the variational method to find the ground-state energy of the s.h.o. Compare with the exact value $E_0 = 0.500$.

Note:

$$\frac{d^2u(x)}{dx^2} = -\frac{2}{a}\delta(x) + \frac{1}{a}[\delta(x+a) + \delta(x-a)]$$

Chapter 9

BOUND STATES: PART II

In Chapter 8 we examined the properties of a system of two particles interacting through an attractive potential. We found that under certain conditions the system is bound, in which case the stationary states have a discrete energy spectrum. We paid special attention to the Coulomb potential and were able to find the wave functions for the stationary states of the hydrogen atom. The effects of spin and other corrections were calculated using perturbation theory. In this chapter we continue the study of bound states with special attention to the simple harmonic oscillator (s.h.o.) potential.

The model of the s.h.o. plays an important role in quantum mechanics, as it does in classical physics. We encounter the mass spring system early in the study of mechanics where $F = m(d^2x/dt^2) = -kx$ has the simple solution $x = A\sin(\omega t + \phi)$. Similarly, in quantum mechanics the problem of the s.h.o. has an exact solution which can be easily obtained and which clearly illustrates the methods and principles used. Apart from its pedagogical virtues, the s.h.o. potential is an excellent approximation for any system near its equilibrium position, and thus has a large range of physical applications. The structure of the Hamiltonian for the s.h.o. coincides with that of most field theories including the electromagnetic field, and hence the quantum oscillator description can be easily carried over to such cases. Historically, the one-dimensional s.h.o. was the first problem solved by Heisenberg using the matrix methods of quantum mechanics.

In our discussion of the s.h.o. we will stress operator techniques rather than the coordinate representation because of the generality of the former approach. We then proceed to a brief discussion of molecular band spectra where the vibrational motion of diatomic molecules can be correctly described as one-dimensional s.h.o. motion. The s.h.o. in three dimensions is next introduced and its usefulness in describing the nucleus in terms of a shell model is presented in the following section.

The last two sections are devoted to a description of a simple short-range potential, namely, the square well in three dimensions. First, we discuss the solutions for a well of finite depth and width, and then apply them to the problem of the deuteron. The special functions needed to describe the radial dependence of the wavefunctions can be found in Appendices 10 and 12.

9.1. THE SIMPLE HARMONIC OSCILLATOR IN ONE DIMENSION

We consider a particle of mass m moving in one dimension in a potential of the form

$$U(x)=\tfrac{1}{2}kx^2 \tag{9.1}$$

where k is a real, positive constant. Thus, the Hamiltonian operator is

$$\hat{H}=\frac{1}{2m}\hat{p}^2+\tfrac{1}{2}k\hat{x}^2 \tag{9.2}$$

The form of the potential energy is sketched in Fig. 9.1(a); x can take positive or negative values in the range $-\infty<x<+\infty$. Since $U(x)$ is positive, the eigenvalues of the Hamiltonian would be *positive semidefinite*† ($E_n \geqslant 0$). Furthermore, since asymptotically the potential energy becomes infinite, the solutions represent bound states and they have a discrete spectrum.

For large values of x the s.h.o. potential tends to infinity and thus becomes unrealistic. However, it provides a good approximation for any system of two particles that are in equilibrium at some separation x_0. At the equilibrium separation the potential energy is at a minimum, $(dU/dx)_{x=x_0}=0$, as shown in Fig. 9.1(b). If we expand $U(x)$ in a Taylor series around x_0

$$U(x)=U(x_0)+\left(\frac{dU}{dx}\right)_{x_0}(x-x_0)+\frac{1}{2}\left(\frac{d^2U}{dx}\right)_{x_0}(x-x_0)^2+\cdots \tag{9.3}$$

† We say *semidefinite* to include the possibility that $E=0$.

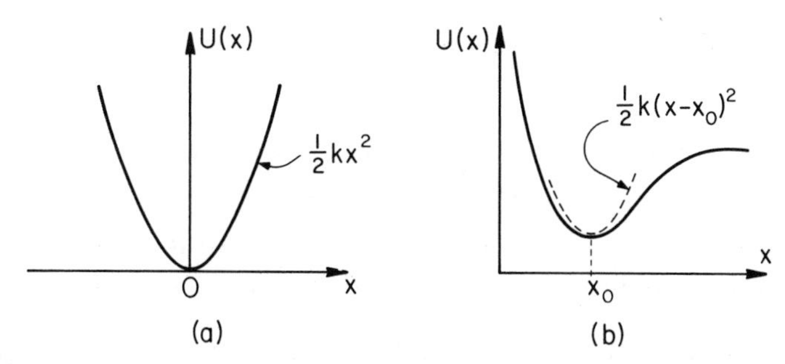

FIGURE 9.1. (a) The form of the potential $U(x)=(1/2)kx^2$ for the simple harmonic oscillator in one dimension. (b) Any potential with a minimum can be approximated in that region by a s.h.o. potential.

The first term is a constant and corresponds to an arbitrary shift of the energy scale. The second term is zero at the equilibrium point. Thus the leading term in the expansion given in Eq. (9.3) is the s.h.o. potential. Note that we can identify $(d^2U/dx^2)_{x_0}=k>0$ since the extremum is a minimum.

We now proceed to find the stationary states. As usual they will obey the eigenvalue equation

$$\hat{H}\,|u_n\rangle = E_n\,|u_n\rangle \tag{9.4a}$$

with the normalization requirement

$$\langle u_n \mid u_m\rangle = \delta_{mn} \tag{9.4b}$$

We could obtain the wavefunctions for the states $|u_n\rangle$ by solving the time-independent Schrödinger equation using the Hamiltonian of Eq. (9.2). However, it is much more instructive to use operator methods to obtain the energy eigenvalues. It is convenient to introduce dimensionless variables

$$\hat{X}=\left(\frac{m\omega}{2\hbar}\right)^{1/2}\hat{x} \qquad \hat{P}=\frac{1}{(2m\hbar\omega)^{1/2}}\,\hat{p} \tag{9.5a}$$

where

$$\omega=(k/m)^{1/2} \tag{9.5b}$$

is the natural frequency of the s.h.o. In terms of the new variables the Hamiltonian of Eq. (9.2) is expressed compactly as

$$\hat{H}=\hbar\omega(\hat{P}^2+\hat{X}^2) \tag{9.6}$$

The operators $\hat{X}$ and $\hat{P}$ are related through the commutation relation

$$[\hat{X}, \hat{P}] = \frac{1}{2\hbar}[\hat{x}, \hat{p}] = \frac{1}{2\hbar} i\hbar = \frac{i}{2} \tag{9.7}$$

which follows from the quantum condition [Eqs. (6.7b)].

Next we introduce the *non-Hermitian* operators

$$\begin{aligned} \hat{a} &= \hat{X} + i\hat{P} \\ \hat{a}^\dagger &= \hat{X} - i\hat{P} \end{aligned} \tag{9.8}$$

where $\hat{X}$ and $\hat{P}$ are hermitian. In view of Eq. (9.7) the operators $\hat{a}$ and $\hat{a}^\dagger$ obey

$$\begin{aligned} \hat{a}\hat{a}^\dagger &= (\hat{X} + i\hat{P})(\hat{X} - i\hat{P}) = \hat{X}^2 + \hat{P}^2 + \tfrac{1}{2} \\ \hat{a}^\dagger\hat{a} &= (\hat{X} - i\hat{P})(\hat{X} + i\hat{P}) = \hat{X}^2 + \hat{P}^2 - \tfrac{1}{2} \end{aligned}$$

Thus, the *algebra* of the operators $\hat{a}$, $\hat{a}^\dagger$ is defined by the commutation relation

$$\boxed{[\hat{a}, \hat{a}^\dagger] = 1} \tag{9.9}$$

and we see that the Hamiltonian for the s.h.o. can be written as

$$\hat{H} = \hbar\omega(\hat{a}^\dagger\hat{a} + \tfrac{1}{2}) \tag{9.10}$$

Equations (9.9) and (9.10) completely define the s.h.o. system.

The energy levels can be obtained by solving the operator eigenvalue equation, and to do this let us define $|u_n\rangle$ to be an eigenstate of $\hat{H}$ with the eigenvalue E_n as in Eq. (9.4a). Thus, in the notation of Eq. (9.10),

$$\hat{H}\,|u_n\rangle = E_n\,|u_n\rangle$$

or

$$(\hat{a}^\dagger\hat{a} + \tfrac{1}{2})\,|u_n\rangle = \frac{E_n}{\hbar\omega}\,|u_n\rangle = \varepsilon_n\,|u_n\rangle \tag{9.11a}$$

with $\varepsilon_n = E_n/\hbar\omega$, a dimensionless number. We operate from the left with $\hat{a}$ to find

$$(\hat{a}\hat{a}^\dagger + \tfrac{1}{2})(\hat{a}\,|u_n\rangle) = \varepsilon_n(\hat{a}\,|u_n\rangle)$$

In view of the commutator of Eq. (9.9), $\hat{a}\hat{a}^\dagger = \hat{a}^\dagger\hat{a} + 1$, we can rewrite our result as

$$(\hat{a}^\dagger\hat{a} + \tfrac{3}{2})(\hat{a}\,|u_n\rangle) = \varepsilon_n(\hat{a}\,|u_n\rangle)$$

or

$$(\hat{a}^\dagger\hat{a} + \tfrac{1}{2})(\hat{a}\,|u_n\rangle) = (\varepsilon_n - 1)(\hat{a}\,|u_n\rangle) \tag{9.11b}$$

Comparison of Eqs. (9.11a) and (9.11b) shows that

$$\hat{a}\,|u_n\rangle$$

is also an eigenstate of the Hamiltonian with the energy eigenvalue

$$E = \hbar\omega(\varepsilon_n - 1) = E_n - \hbar\omega \tag{9.12a}$$

Thus, $\hat{a}$ acting on an eigenstate $|u_n\rangle$ with energy eigenvalue E_n, produces an eigenstate with eigenvalue $E_n - \hbar\omega$. That is, it lowers the energy eigenvalue and hence is called a *lowering operator*. If we operate on Eq. (9.11a) with $\hat{a}^\dagger$ from the left we find

$$\begin{aligned}(\hat{a}^\dagger\hat{a}^\dagger\hat{a} + \tfrac{1}{2}\hat{a}^\dagger)\,|u_n\rangle &= [\hat{a}^\dagger(\hat{a}\hat{a}^\dagger - 1) + \tfrac{1}{2}\hat{a}^\dagger]\,|u_n\rangle \\ &= (\hat{a}^\dagger\hat{a} - \tfrac{1}{2})(\hat{a}^\dagger\,|u_n\rangle) = \varepsilon_n(\hat{a}^\dagger\,|u_n\rangle)\end{aligned}$$

or

$$(\hat{a}^\dagger\hat{a} + \tfrac{1}{2})(\hat{a}^\dagger\,|u_n\rangle) = (\varepsilon_n + 1)(\hat{a}^\dagger\,|u_n\rangle) \tag{9.12b}$$

Therefore, $\hat{a}^\dagger$ is a *raising operator*, i.e., acting on an eigenstate $|u_n\rangle$ with energy eigenvalue E_n, it produces an eigenstate with a higher eigenvalue $E_n + \hbar\omega$.

We have assumed that $|u_n\rangle$ is an eigenstate of the Hamiltonian corresponding to the eigenvalue E_n. If we operate on $|u_n\rangle$ repeatedly with the lowering operator $\hat{a}$, E_n will decrease and eventually become negative. This is not possible because as we have said earlier, the eigenvalues must be positive semidefinite. Thus, there must exist an eigenstate $|u_0\rangle$ corresponding to the *lowest eigenvalue* E_0 such that

$$\hat{a}\,|u_0\rangle = 0 \tag{9.13a}$$

We can find the eigenvalue E_0 in the following way. Let us operate on Eq. (9.13a) with $\hat{a}^\dagger$ to obtain

$$\hat{a}^\dagger\hat{a}\,|u_0\rangle = 0 \tag{9.13b}$$

The lowest energy eigenvalue is obtained from the equation

$$\hat{H}\,|u_0\rangle = E_0\,|u_0\rangle$$

or

$$\hbar\omega(\hat{a}^\dagger\hat{a}+\tfrac{1}{2})\,|u_0\rangle = E_0\,|u_0\rangle \tag{9.13c}$$

which gives $E_0 = \hbar\omega/2$ in view of Eq. (9.13*b*). Furthermore, by repeated application of $\hat{a}^\dagger$ on the state $|u_0\rangle$ we can construct the higher eigenstates $|u_n\rangle$

$$|u_1\rangle = \hat{a}^\dagger\,|u_0\rangle \qquad |u_n\rangle = \underbrace{\hat{a}^\dagger\hat{a}^\dagger\cdots\hat{a}^\dagger}_{n\text{ times}}\,|u_0\rangle$$

The energy eigenvalue for these states follows from Eq. (9.12*b*) and is

$$\boxed{E_n = \hbar\omega(n+\tfrac{1}{2})} \tag{9.14}$$

where n is a *positive integer or zero.*

Equation (9.14) is a fundamental result. It shows that the energy levels of the s.h.o. in one dimension are equally spaced as shown in Fig. 9.2. Furthermore, the lowest energy state, the ground state, has energy $E_0 = \frac{1}{2}\hbar\omega$. This is in contrast to the classical problem where the s.h.o. can be found at rest in the bottom of the potential well. We will return to a discussion of this point later. But Eq. (9.14) distinctively shows that the spectrum of the energy eigenvalues in the case of s.h.o. has to be strictly *positive definite.*

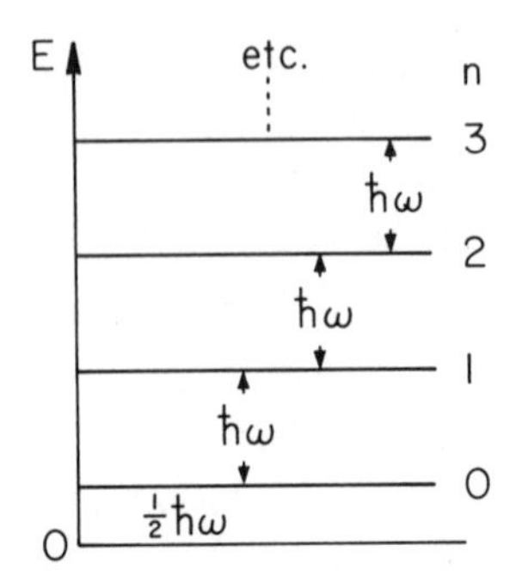

FIGURE 9.2. The spectrum of the energy levels of the one-dimensional s.h.o. Note that the levels are equally spaced.

To complete the derivation of Eq. (9.14) we must show that these indicated eigenvalues are the only ones possible. To do so let us suppose that there exists an eigenstate with eigenvalue

$$E_i = \hbar\omega(\rho_i + \tfrac{1}{2})$$

where ρ_i is positive but *not* an integer. The repeated application of the lowering operator $\hat{a}$ would then not lead to the state $|u_0\rangle$ [as defined by Eqs. (9.13) and (9.13c)] and thus we would reach negative eigenvalues. Since these are not allowed, the only possible values of ρ_i are those where ρ_i is a positive integer. It can also be shown that all the eigenstates of the one-dimensional s.h.o. are nondegenerate and orthogonal to one another.

The reader should note that the operator methods that enabled us to find the eigenvalues of the Hamiltonian of Eqs. (9.2) or (9.10) are identical to those used in Chapter 7 to find the eigenvalues of the angular momentum operators, except that the algebra obeyed by the operators was different in the two cases. We will now continue using operator techniques to find the matrix elements of the operators $\hat{a}$ and $\hat{a}^\dagger$ between the eigenstates of the Hamiltonian.

We label the eigenstates corresponding to the eigenvalue E_n by the index n [as in Eq. (9.14) n is a positive integer or zero]. The effect of the operators $\hat{a}$ and $\hat{a}^\dagger$ on these states can, then, be written as

$$\begin{aligned} \hat{a}^\dagger\,|u_n\rangle &= c_n\,|u_{n+1}\rangle \\ \hat{a}\,|u_n\rangle &= d_n\,|u_{n-1}\rangle \end{aligned} \qquad (9.15a)$$

where c_n and d_n are constants to be determined. First, we note from Eq. (9.15a) that since

$$\hat{a}^\dagger\,|u_n\rangle = c_n\,|u_{n+1}\rangle$$

hermitian conjugation gives

$$\langle u_n\,|\hat{a} = c_n^*\langle u_{n+1}|$$

and thus the norm

$$\langle u_n|\,\hat{a}\hat{a}^\dagger\,|u_n\rangle = |c_n|^2\,\langle u_{n+1}\,|\,u_{n+1}\rangle = |c_n|^2$$

Using the commutator of Eq. (9.9) we rewrite this as

$$\begin{aligned}\langle u_n|\, \hat{a}^\dagger a + 1\, |u_n\rangle &= \langle u_n|\, (\hat{a}^\dagger \hat{a} + \tfrac{1}{2}) + \tfrac{1}{2}\, |u_n\rangle \\ &= \varepsilon_n + \tfrac{1}{2} = (n + \tfrac{1}{2}) + \tfrac{1}{2} = (n+1) = |c_n|^2\end{aligned}$$

or

$$c_n = c_n^* = (n+1)^{1/2} \tag{9.15b}$$

Here we have chosen the constant to be real. Similarly, we can look at the norm

$$\begin{aligned}\langle u_n|\, \hat{a}^\dagger \hat{a}\, |u_n\rangle &= |d_n|^2 \langle u_{n-1} \mid u_{n-1}\rangle = |d_n|^2 \\ &= \langle u_n|\, [(\hat{a}^\dagger \hat{a} + \tfrac{1}{2}) - \tfrac{1}{2}]\, |u_n\rangle \\ &= \varepsilon_n - \tfrac{1}{2} = (n + \tfrac{1}{2}) - \tfrac{1}{2} = n = |d_n|^2\end{aligned}$$

or

$$d_n = d_n^* = n^{1/2} \tag{9.15c}$$

where we have again chosen the phase of the constant to be real. We can now express the effect of the operators $\hat{a}$ and $\hat{a}^\dagger$ on the eigenstates as

$$\begin{aligned}\hat{a}^\dagger\, |u_n\rangle &= (n+1)^{1/2}\, |u_{n+1}\rangle \\ \hat{a}\, |u_n\rangle &= n^{1/2}\, |u_{n-1}\rangle\end{aligned} \tag{9.16a}$$

In view of the orthogonality of the eigenstates, we can obtain the matrix elements of the raising and lowering operators for the one-dimensional s.h.o. If we denote the normalized eigenstates $|u_n\rangle$ by the simplified notation $|n\rangle$, then the matrix elements become

$$\boxed{\begin{aligned}\langle n+1|\, \hat{a}^\dagger\, |n\rangle &= (n+1)^{1/2} \\ \langle n-1|\, \hat{a}\, |n\rangle &= n^{1/2}\end{aligned}} \tag{9.16b}$$

which is in agreement with our conclusions that $\hat{a}^\dagger$ raises the system to a higher level, whereas $\hat{a}$ lowers it. Note that when $\hat{a}$ acts on the ground state, i.e., $n = 0$, the result is zero as defined by Eq. (9.13*a*).

It should be clear that although both $\hat{a}$ and $\hat{a}^\dagger$ are off-diagonal in the energy basis, the operator $\hat{a}^\dagger \hat{a}$ is diagonal, and when acting on the state $|n\rangle$ has the eigenvalue n. That is,

$$\hat{a}^\dagger \hat{a}\, |n\rangle = n\, |n\rangle \tag{9.17}$$

which follows directly from Eqs. (9.16*a*). The operator $\hat{a}^\dagger \hat{a}$ is frequently referred to as the *number operator*.† We note that when the state $|u_0\rangle$ is normalized as

$$\langle 0 \mid 0 \rangle = 1$$

the normalized eignestates $|n\rangle$ can be written in terms of $|0\rangle$ as

$$|n\rangle = \frac{1}{(n!)^{1/2}} (\hat{a}^\dagger)^n\, |0\rangle \tag{9.18}$$

9.2. REPRESENTATIONS OF THE ONE-DIMENSIONAL S.H.O.

The results of Section 9.1, in particular the eigenvalue spectrum given by Eq. (9.14) and the matrix elements given by Eqs. (9.16*b*), provide the complete solution of the problem of the s.h.o. in one dimension. They are, of course, valid in any representation. However, to appreciate the structure of the s.h.o. it is instructive to consider specific representations of the system and calculate a few of its properties, as we shall do in this section.

Let us calculate the matrix elements of the physically observable (hermitian) operators $\hat{x}$ and $\hat{p}$. In view of the defining Eqs. (9.5) and (9.8)

$$\begin{aligned} \hat{x} &= \left(\frac{2\hbar}{m\omega}\right)^{1/2} \hat{X} = \left(\frac{\hbar}{2m\omega}\right)^{1/2} (\hat{a}^\dagger + \hat{a}) \\ \hat{p} &= (2m\hbar\omega)^{1/2} \hat{P} = i\left(\frac{m\hbar\omega}{2}\right)^{1/2} (\hat{a}^\dagger - \hat{a}) \end{aligned} \tag{9.19a}$$

† This is related to the fact, mentioned in the introduction to this chapter, that the s.h.o. algebra can be used to represent the number of quanta or particles in a particular state. Thus, the state $u_n(\mathbf{k})$ represents n particles in the state of momentum $|\mathbf{k}\rangle$. In this case $\hat{a}^\dagger$ is a creation operator and $\hat{a}$ a destruction operator.

and using Eqs. (9.16b) we can write

$$\langle n+1|\,\hat{x}\,|n\rangle = \left(\frac{\hbar}{2m\omega}\right)^{1/2} \langle n+1|\,\hat{a}^\dagger\,|n\rangle = (n+1)^{1/2}\left(\frac{\hbar}{2m\omega}\right)^{1/2}$$

$$\langle n-1|\,\hat{x}\,|n\rangle = \left(\frac{\hbar}{2m\omega}\right)^{1/2} \langle n-1|\,\hat{a}\,|n\rangle = n^{1/2}\left(\frac{\hbar}{2m\omega}\right)^{1/2} \tag{9.20a}$$

All other matrix elements vanish. For the momentum operator

$$\langle n+1|\,\hat{p}\,|n\rangle = i\left(\frac{\hbar m\omega}{2}\right)^{1/2} \langle n+1|\,\hat{a}^\dagger\,|n\rangle = i(n+1)^{1/2}\left(\frac{\hbar m\omega}{2}\right)^{1/2}$$

$$\langle n-1|\,\hat{p}\,|n\rangle = -i\left(\frac{\hbar m\omega}{2}\right)^{1/2} \langle n-1|\,\hat{a}\,|n\rangle = -in^{1/2}\left(\frac{\hbar m\omega}{2}\right)^{1/2} \tag{9.20b}$$

with all other matrix elements vanishing. We see that the matrix elements of the position and momentum operators connect only states for which the index n differs by one unit, $\Delta n = \pm 1$.

The eigenstates of the s.h.o. can be represented most conveniently by a *column* vector with a "1" entry in the $(n+1)$th slot and zeros everywhere else. This is indicated below, where

$$|n\rangle = \begin{pmatrix} 0 \\ 0 \\ \vdots \\ 0 \\ 1 \\ 0 \\ \vdots \end{pmatrix} \begin{matrix} n=0 \\ n=1 \\ \\ \\ \leftarrow (n+1)\text{th slot} \\ \\ \\ \end{matrix} \qquad \langle n| = \overbrace{\underbrace{0,\ 0,\ \cdots,\ 0,\ 1,\ 0,\ \cdots}}} \tag{9.21}$$

$\uparrow$ $n=0$, $\uparrow$ $n=1$, $\uparrow$ $(n+1)$th slot

The ket $|n\rangle$ is indicated by a column vector and the bra $\langle n|$ by a *row* vector. The column and row vectors are *infinite*-dimensional but this causes no complication. Note that as given in Eq. (9.21) the states are properly normalized.

$$\langle n\,|\,n\rangle = 1$$

The operators $\hat{H}, \hat{x}$, and $\hat{p}$ will, of course, be represented by matrices. Again, these must be infinite matrices and we can easily construct them from Eqs. (9.14) and (9.20) as

$$\hat{H}=\frac{\hbar\omega}{2}\begin{bmatrix} 1 & 0 & 0 & \cdots & 0 & \cdots \\ 0 & 3 & 0 & \cdots & 0 & \cdots \\ 0 & 0 & 5 & \cdots & 0 & \cdots \\ \vdots & \vdots & \vdots & \ddots & & \\ 0 & 0 & 0 & & 2n+1 & \\ \vdots & \vdots & \vdots & & & \ddots \end{bmatrix} \tag{9.22a}$$

$$\hat{x}=\left(\frac{\hbar}{2m\omega}\right)^{1/2}\begin{bmatrix} 0 & \sqrt{1} & 0 & 0 & \cdots & 0 & \cdots \\ \sqrt{1} & 0 & \sqrt{2} & 0 & \cdots & 0 & \cdots \\ 0 & \sqrt{2} & 0 & \sqrt{3} & \cdots & 0 & \cdots \\ 0 & 0 & \sqrt{3} & 0 & \cdots & 0 & \cdots \\ \vdots & \vdots & \vdots & \vdots & & \vdots & \\ 0 & 0 & 0 & 0 & 0 & \sqrt{n} & \cdots \\ \vdots & \vdots & \vdots & \vdots & \sqrt{n} & 0 & \\ \vdots & \vdots & \vdots & \vdots & & & \end{bmatrix}$$

$$\hat{p}=\left(\frac{\hbar m\omega}{2}\right)^{1/2}\begin{bmatrix} 0 & -i\sqrt{1} & 0 & 0 & \cdots & 0 & \cdots \\ i\sqrt{1} & 0 & -i\sqrt{2} & 0 & \cdots & 0 & \cdots \\ 0 & i\sqrt{2} & 0 & -i\sqrt{3} & \cdots & 0 & \cdots \\ 0 & 0 & i\sqrt{3} & 0 & \cdots & 0 & \cdots \\ \vdots & \vdots & \vdots & \vdots & & \vdots & \\ 0 & 0 & 0 & 0 & 0 & -i\sqrt{n} & \cdots \\ \vdots & \vdots & \vdots & \vdots & i\sqrt{n} & 0 & \cdots \\ \vdots & \vdots & \vdots & \vdots & & & \end{bmatrix}$$

(9.22b)

Equations (9.21) and (9.22) constitute the *matrix representation* (also referred to as the *number representation*) for the s.h.o. They do not convey any new information beyond that of the defining equations. They should be thought of as a mnemonic notation for these equations. It is a simple exercise to construct the matrices for the non-hermitian operators $\hat{a}$ and $\hat{a}^\dagger$ and show by direct matrix multiplication that they satisfy the basic commutation relation as given in Eq. (9.9). The effect of these operators on the eigenstates of Eq. (9.21) is then obvious.

Next we want to obtain the coordinate representation of the s.h.o., namely, to find the wavefunctions for the stationary states. To do this we must express the Hamiltonian in the coordinate representation, which is possible since the s.h.o. can be described classically as well. Using the expression for $\hat{p}$ in the coordinate representation $\hat{p} = -i\hbar(d/dx)$ (see Table 6.1), the wavefunction $\phi_n(x)$ will be a solution of the time-independent Schrödinger equation [see Eq. (6.43b)] with the Hamiltonian of Eq. (9.2)

$$\left(-\frac{\hbar^2}{2m}\frac{d^2}{dx^2} + \tfrac{1}{2}kx^2\right)\phi_n(x) = E_n\phi_n(x) \qquad (9.23a)$$

Furthermore, $E_n > 0$, and $\phi_n(x)$ is subject to the boundary condition

$$\phi_n(x) \to 0 \quad \text{as} \quad x \to \pm\infty \qquad (9.23b)$$

because the system is localized. The wavefunction should be normalized according to

$$\int_{-\infty}^{+\infty} \phi_n^*(x)\phi_n(x)\,dx = 1 \qquad (9.23c)$$

There are various methods for solving the eigenvalue problem defined by Eqs. (9.23). One convenient method is to factor the differential equation [Eq. (9.23a)] in terms of the operators $\hat{a}$ and $\hat{a}^\dagger$. In the coordinate representation these operators are of the form

$$\begin{aligned}\hat{a} &= \left(\frac{m\omega}{2\hbar}\right)^{1/2} x + \left(\frac{\hbar}{2m\omega}\right)^{1/2}\frac{d}{dx} \\ \hat{a}^\dagger &= \left(\frac{m\omega}{2\hbar}\right)^{1/2} x - \left(\frac{\hbar}{2m\omega}\right)^{1/2}\frac{d}{dx}\end{aligned} \qquad (9.24)$$

We now recall the result of Eq. (9.13a) about the existence of a ground state

$$\hat{a}\,|u_0\rangle = 0$$

Thus, in the coordinate representation, this condition takes the form

$$\left(\frac{\hbar}{2m\omega}\right)^{1/2}\left(\frac{d}{dx}+\frac{m\omega}{\hbar}x\right)\phi_0(x)=0 \tag{9.25a}$$

and can be solved by an elementary integration. We have

$$\frac{d\phi_0}{dx}=-\frac{m\omega}{\hbar}x\phi_0(x)$$

or

$$\frac{d\phi_0}{\phi_0(x)}=-\frac{m\omega}{\hbar}x\,dx$$

and, therefore,

$$\phi_0(x)=A\exp\left(-\frac{m\omega}{\hbar}\frac{x^2}{2}\right) \tag{9.25b}$$

This result satisfies the boundary condition of Eq. (9.23*b*) and can be normalized to yield

$$\boxed{\phi_0(x)=\left(\frac{m\omega}{\pi\hbar}\right)^{1/4}e^{-(1/2)(m\omega/\hbar)x^2}} \tag{9.25c}$$

Note that the ground-state wavefunction is nodeless, symmetric about $x=0$, and peaks at $x=0$. In fact, it is a simple Gaussian.

The wavefunctions for $n\neq 0$ are obtained from Eq. (9.25*c*) by repeated application of the $\hat{a}^\dagger$ operator. They will remain normalized if we make use of Eq. (9.18). therefore,

$$\phi_n(x)=\frac{1}{(n!)^{1/2}}(\hat{a}^\dagger)^n\phi_0(x)$$

where the differential operator $\hat{a}^\dagger$ is as given by Eq. (9.24). When this operator acts on $\phi_0(x)$ it generates the *Hermite* polynomials $H_n(x)$, which are defined through

$$H_n(x)=(-1)^n e^{x^2}\frac{d^n}{dx^n}e^{-x^2} \tag{9.26}$$

A short review of the properties of these polynomials is given in Appendix 10. In terms of these polynomials the wavefunction for the nth state is given by

$$\phi_n(y) = \frac{1}{(2^n n!)^{1/2}} \left(\frac{\beta^2}{\pi}\right)^{1/4} H_n(y) e^{-y^2/2} \qquad (9.27a)$$

where we have used the notation

$$\beta = (m\omega/\hbar)^{1/2} \quad \text{and} \quad y = \beta x \qquad (9.27b)$$

It is shown in Appendix 10 that the functions $\phi_n(x)$ are mutually orthogonal and normalized as expected. The first few polynomials are

$$\begin{aligned} H_0(y) &= 1 \\ H_1(y) &= 2y \\ H_2(y) &= 4y^2 - 2, \qquad \text{etc.} \end{aligned} \qquad (9.27c)$$

as can be easily derived from Eq. (9.26).

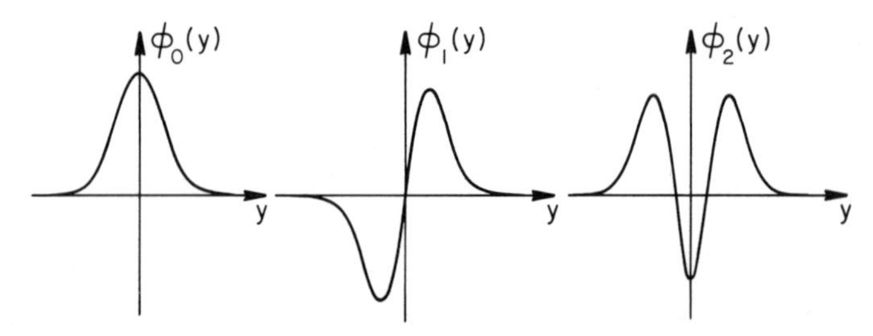

FIGURE 9.3. The wavefunction for the one-dimensional s.h.o. plotted as a function of the variable $y = (m\omega/\hbar x)^{1/2}$, is shown for $n = 0$, $n = 1$, and $n = 2$.

The wavefunctions for $n = 0$, $n = 1$, and $n = 2$ are shown in Fig. 9.3 as a function of the dimensionless parameter y. In Fig. 9.4 we show the probability distribution, i.e., $|\phi_n(y)|^2$, for $n = 10$. The dashed curve gives the probability distribution for a classical s.h.o. with the same parameters. The close proximity of the average value of the quantum-mechanical result with the classical distribution is, of course, a manifestation of the correspondence principle. We want to draw attention to the fact that the wavefunctions are also eigenstates of the parity operator, as can be clearly seen from Fig. 9.3 and as is evident from Eq. (9.27c). We note that for n even or

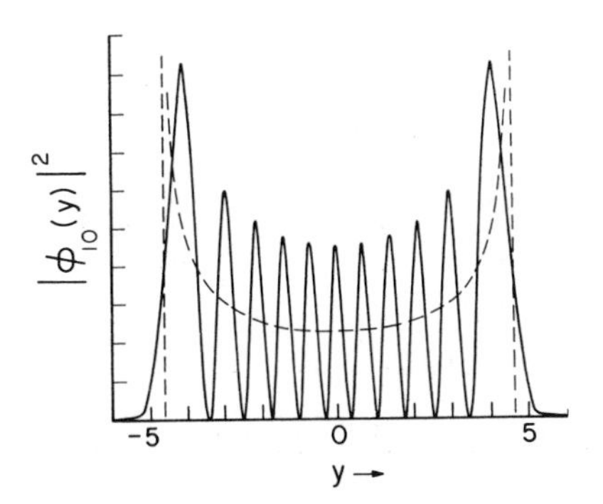

FIGURE 9.4. The probability distribution $|\phi_n(y)|^2$ for the $n=10$ state of the s.h.o. The dashed curve is the distribution for a classical oscillator of the same total energy.

zero the wavefunction is even under reflection about the origin, whereas it is odd when n is odd. This follows, in general, from the properties of the Hermite polynomials. That the wavefunctions exhibit definite parity is an immediate consequence of the invariance of the Hamiltonian of Eq. (9.2) under reflection of the coordinate. This is in agreement with the general relationship of symmetry and constants of the motion that we studied in Chapter 7.

The s.h.o. is a relatively simple system so that one can obtain explicit results in closed form for many of its properties. As an example, let us calculate the (rms) uncertainty in position $(\langle \Delta x^2 \rangle)^{1/2}$ and momentum $(\langle \Delta p^2 \rangle)^{1/2}$ for an arbitrary stationary state $|n\rangle$. These uncertainties are defined in the same way as in Eq. (6.11)

$$\langle (\Delta x)^2 \rangle = \langle \hat{x}^2 \rangle - \langle \hat{x} \rangle^2$$
$$\langle (\Delta p)^2 \rangle = \langle \hat{p}^2 \rangle - \langle \hat{p} \rangle^2$$

From Eqs. (9.20a) (9.20b) we know that

$$\langle n|\, \hat{x}\, |n\rangle = 0 = \langle n|\, \hat{p}\, |n\rangle$$

Thus

$$\begin{aligned}
\langle (\Delta x)^2 \rangle &= \langle n|\, (\hat{x})^2\, |n\rangle \\
&= \sum_m \langle n|\, \hat{x}\, |m\rangle \langle m|\, \hat{x}\, |n\rangle \\
&= \langle n|\, \hat{x}\, |n+1\rangle \langle n+1|\, \hat{x}\, |n\rangle + \langle n|\, \hat{x}\, |n-1\rangle \langle n-1|\, \hat{x}\, |n\rangle \\
&= (n+1)\frac{\hbar}{2m\omega} + n\frac{\hbar}{2m\omega} = (2n+1)\frac{\hbar}{2m\omega}
\end{aligned} \qquad (9.28a)$$

and similarly

$$\langle(\Delta p)^2\rangle = \langle n|\,(\hat{p})^2\,|n\rangle$$
$$= \sum_m \langle n|\,\hat{p}\,|m\rangle\langle m|\,\hat{p}\,|n\rangle = (2n+1)\frac{\hbar m\omega}{2} \tag{9.28b}$$

where we have used Eqs. (9.20). therefore

$$\Delta x\,\Delta p = (\langle(\Delta x)^2\rangle)^{1/2}(\langle(\Delta p)^2\rangle)^{1/2} = \hbar(n+\tfrac{1}{2}) \tag{9.28c}$$

We see that the uncertainty product grows with n and

$$\Delta x\,\Delta p \geqslant \frac{\hbar}{2}$$

which is in accordance with the general result of Eq. (6.13). Furthermore, the ground state of the s.h.o. is a state of minimal uncertainty. This is a reflection of the Gaussian form of the ground-state wavefunction (see Problem 1.13).

Let us now compare the quantum and classical solutions of the one-dimensional s.h.o. in order to establish their correspondence for large values of the quantum number n. For a classical oscillator of energy E and natural frequency ω, the solutions corresponding to the initial conditions $x(t=0)=0$ and $v(t=0)=(2E/m)^{1/2}$ are

$$x = A\sin\omega t = \left(\frac{2E}{m\omega^2}\right)^{1/2}\sin\omega t$$

$$v = A\omega\cos\omega t = \left(\frac{2E}{m}\right)^{1/2}\cos\omega t$$

If we consider a classical oscillator with energy equal to that of the quantum s.h.o. in its nth state, we must set $E=(2n+1)\hbar\omega/2$ in the above equations

$$x = (2n+1)^{1/2}\left(\frac{\hbar}{m\omega}\right)^{1/2}\sin\omega t$$
$$p = (2n+1)^{1/2}(\hbar m\omega)^{1/2}\cos\omega t \tag{9.29a}$$

Thus, the kinetic energy T and potential energy U are given by

$$\begin{aligned} T &= \frac{1}{2}\frac{p^2}{m} = \frac{2n+1}{2}\hbar\omega\cos^2\omega t \\ U &= \tfrac{1}{2}kx^2 = \frac{2n+1}{2}\hbar\omega\sin^2\omega t \end{aligned} \tag{9.29b}$$

The average values of T and U (obtained by replacing $\cos^2\omega t, \sin^2\omega t \to \frac{1}{2}$) are exactly equal to the expectation values

$$\begin{aligned} \langle n|\,\hat{T}\,|n\rangle &= \frac{1}{2m}\langle n|\,(\hat{p})^2\,|n\rangle = \frac{2n+1}{4}\hbar\omega \\ \langle n|\,\hat{U}\,|n\rangle &= \tfrac{1}{2}k\langle n|\,(\hat{x})^2\,|n\rangle = \frac{2n+1}{4}\hbar\omega \end{aligned} \tag{9.29c}$$

That exact correspondence exists for all values of n and not only when $n \to \infty$ is again a consequence of the simple structure of the s.h.o. system.

The expectation values of Eqs. (9.29c) can be calculated directly by using the matrix elements of Eqs. (9.20). They do, however, follow immediately from the virial theorem, which is given in Eq. (6.32b) as

$$\langle\hat{T}\rangle = \frac{n}{2}\langle\hat{U}\rangle$$

Since for the s.h.o. $n = 2$ [see Eqs. (6.31a) and (9.2)], it must hold $\langle\hat{T}\rangle = \langle\hat{U}\rangle = \frac{1}{2}E$.

As a final example let us obtain a description of the s.h.o. in the *momentum representation*. The basis states form a continuum $|k\rangle$ and therefore the amplitudes

$$\hat{\phi}_n(k) = \langle k \mid n\rangle \tag{9.30}$$

necessary to describe the stationary state $|n\rangle$, will be a continuous function of momentum. We are using the wave vector $k = p/\hbar$ to denote the basis states, in keeping with the notation used so far.

We can find the *momentum wavefunction* $\tilde{\phi}_n(k)$ for the nth state by performing a unitary transformation of the space wavefunction $\phi_n(x)$. From the general rule for the relation between amplitudes in different basis states

$$\tilde{\phi}_n(k) = \langle k \mid n \rangle = \int dx \langle k \mid x \rangle \langle x \mid n \rangle = \int \langle k \mid x \rangle \phi_n(x)\, dx \tag{9.31a}$$

We have expressed the transformation by an integral rather than a sum because the intermediate states $|x\rangle$ form a continuum. The transformation coefficients—in this case, a function of k and x—were given in Eq. (1.59)

$$\langle k \mid x \rangle = \frac{1}{(2\pi)^{1/2}} e^{-ikx}$$

so that

$$\tilde{\phi}_n(k) = \frac{1}{(2\pi)^{1/2}} \int e^{-ikx} \phi_n(x)\, dx \tag{9.31b}$$

As we have already noted, the wavefunctions in the two representations are related by a Fourier transformation (see also Appendix 1). In the case of the ground state, since $\phi_0(x)$ is a Gaussian, so is $\tilde{\phi}_0(k)$. One obtains

$$\tilde{\phi}_0(k) = \left(\frac{\hbar}{m\omega\pi}\right)^{1/4} e^{-(1/2)(\hbar/m\omega)k^2} \tag{9.31c}$$

The similarity in the form of the coordinate and momentum representation wavefunctions is a consequence of the symmetry of the Hamiltonian in these two variables [see Eq. (9.6)].

9.3. MOLECULAR BAND SPECTRA

A physical system that closely approximates a one-dimensional s.h.o. is that of a diatomic molecule. Diatomic molecules consist of two neutral atoms bound into a single system. We can categorize them as homonuclear, such as H_2, N_2, etc., and as heteronuclear, such as HCl, HF, etc. To get a feeling of why two neutral atoms can be attracted to one another let us consider the following crude model. At large distances the atoms maintain their identity and do not interact. The electron cloud of each atom completely screens the nuclear charge. On the other hand, at very short distances the mutual electrostatic interaction between the two nuclei tends to repel them. At some intermediate distances, we can imagine that only part of the electrons surround the respective nucleus and that there is a concentration of negative charge (electrons) between the two nuclei as shown in Fig. 9.5(a). Under these conditions the force acting on each nucleus is the sum of the electrostatic repulsion due to the other nucleus and of the electrostatic attraction due to the electron cloud. In terms of the sketch of Fig. 9.5(a) the resulting force is attractive and is given by

$$F = \frac{q^2}{r^2} + \frac{2q(-q)}{(r/2)^2} = -\frac{7q^2}{r^2}$$

In reality, the situation is more complex, nevertheless, the balance between the nuclear repulsion and the electronic interaction energy leads to a potential energy curve that has a minimum at some equilibrium distance r_e as shown in Fig. 9.5(b). At large r the potential energy varies as $1/r^6$ and gives rise to a very weak attraction. At small r the potential energy becomes very large due to the nuclear repulsion. The binding of molecules will be discussed in more detail in Section 10.4.

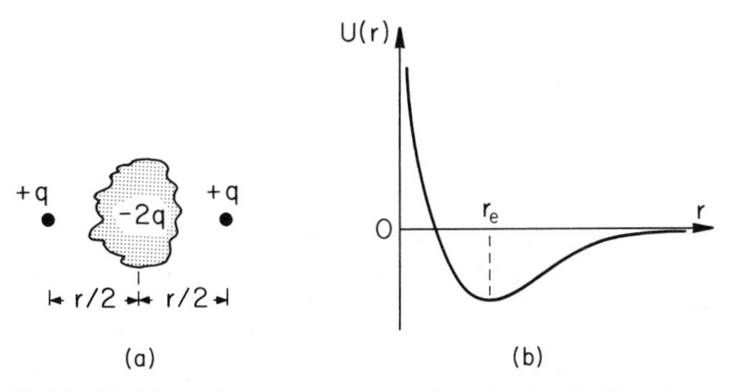

FIGURE 9.5. (a) Schematic representation of the effective binding of two positive charges due to the presence of an equal negative charge between them. (b) The effective potential binding two neutral atoms is shown as a function of the interatomic distance r. Note the existence of an equilibrium position at $r = r_e$.

Near the equilibrium separation, $r = r_e$, the potential energy curve approximates the form of the s.h.o., and we can write

$$U(r) - U(r_e) \simeq \tfrac{1}{2}k(r - r_e)^2$$

Thus, the two nuclei can be expected to vibrate along the line joining them. This motion is quantized and gives rise to discrete stationary states according to the spectrum of the one-dimensional s.h.o., and as we know, these states are equidistant, spaced by $\hbar\omega$. The molecule can undergo transitions from one to another (lower) *vibrational* state with the emission of a photon of energy $\hbar\omega$. However, the vibrational quantum number v can change only by one unit because the perturbation causing the transition is proportional to the operator $\hat{x}$ (it is due to the existence of an electric dipole moment along the line joining the nuclei†). From Eqs. (9.20) we know that the $\hat{x}$ operator has matrix elements different from zero only when $\Delta v = \pm 1$, and this gives rise to the above selection rule.

The energy of the emitted photons depends on the *spring constant* k and on μ, the reduced mass of the two nuclei. As a rough estimate we can assume that $U = \frac{1}{2}kx^2 \simeq 2$ eV—for $x \sim 2 \times 10^{-9}$ cm this implies $k = 10^{18}$ cm^{-2} eV—and if we let $\mu c^2 \simeq 10(m_p c^2) \simeq 10^{10}$ eV, then

$$\nu = \frac{\omega}{2\pi} = \frac{1}{2\pi}\left(\frac{k}{\mu}\right)^{1/2} = \frac{c}{2\pi}\left(\frac{k}{\mu c^2}\right)^{1/2}$$

$$= \frac{3 \times 10^{10}}{2\pi}\left(\frac{10^{18}}{10^{10}}\right)^{1/2} \simeq 5 \times 10^{13}\ \text{Hz}$$

or

$$\lambda = \frac{c}{\nu} \simeq 6 \times 10^{-4}\ \text{cm} = 60{,}000\ \text{Å}$$

Thus, in general, the photons emitted in transitions between vibrational states lie in the infrared.

In reality, the situation is far more complex. First of all, the deviations from the pure s.h.o. potential modify the position of the energy levels so that transitions between different v values give rise to separate lines. Second, the molecule as a whole can rotate about

† Homonuclear diatomic molecules where both atoms are in their ground state can not have an electric dipole moment because of symmetry.

an axis fixed in space. Classically, if A is the moment of inertia of the molecule about the axis of rotation, the kinetic energy is

$$T=\tfrac{1}{2}A\omega^2=\frac{1}{2}\frac{L^2}{A}$$

where $L=A\omega$ is the angular momentum, and ω is the rotational angular frequency. Since there is no potential energy associated with the rotation, the Hamiltonian operator takes the form

$$\hat{H}_{\text{rot}}=\frac{1}{2A}\hat{J}^2 \tag{9.32a}$$

where $\hat{J}$ is the angular momentum operator. Such a Hamiltonian gives rise to stationary states labeled by the eigenvalues of $\hat{J}$, and, therefore, their energy is

$$E_j=\frac{\hbar^2}{2A}j(j+1) \qquad j=0,1,2,\ldots \tag{9.32b}$$

Of course, the energy is independent of m, i.e., of the orientation of the axis of rotation. The spectrum of the rotational states is shown in Fig. 9.6(a), and transitions can occur between states such that $j'=j\pm1$ (because the perturbation is a vector operator). The frequency of the emitted photons is (for $j'=j-1$)

$$\nu=\frac{E_j-E_{j'}}{h}=\frac{\hbar}{4\pi A}[j(j+1)-(j-1)(j-1+1)]=\frac{\hbar}{2\pi A}j \tag{9.33}$$

and consists of equally spaced discrete lines as shown in Fig. 9.6(b).

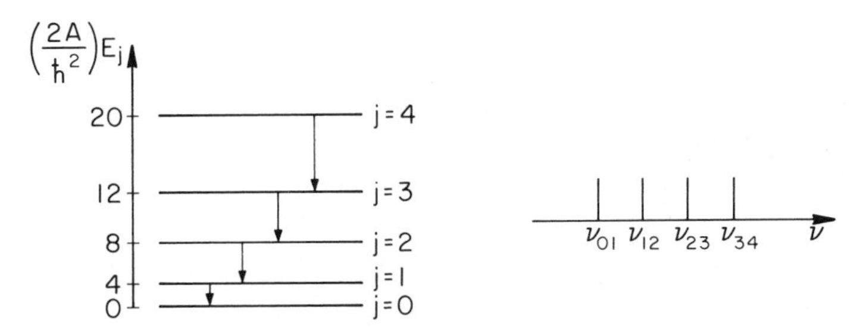

FIGURE 9.6. (a) The rotational states of a diatomic molecule. (b) The spectrum of the emitted radiation in transitions with $\Delta j=1$. This spectrum is usually in the far infrared.

We can estimate the frequency by setting

$$A = \mu r^2 \qquad \text{with } \mu c^2 \sim 10^{10}\,\text{eV},\ r \sim 10^{-8}\,\text{cm}$$

so that

$$\nu \sim c\frac{\hbar c}{2\pi A c^2} = c\frac{200\,(\text{MeV-F})}{2\pi(10^4\,\text{MeV})(10^5\,\text{F})^2} \simeq 10^{11}\,\text{Hz}$$

or

$$\lambda = \frac{c}{\nu} = 0.3\text{ cm}$$

The rotational levels have much smaller energies than the vibrational states and lie in the far infrared (millimeter microwaves). However, transitions can now occur from a rotation–vibration state $|v, j\rangle$ to a different rotation–vibration state $|v', j'\rangle$, as shown in Fig. 9.7(a). This gives rise to a *band* spectrum around the central vibrational frequency ν_{vib} which is indicaaed in (b) of the figure. We recognize two *branches* in the band. The ascending branch that corresponds to the case $j' = j - 1$ for which the photon frequencies are

$$\nu^{(+)} = \frac{\Delta E^{(+)}}{h} = \frac{1}{h}\left[\hbar\omega_v + \frac{\hbar^2}{A}j\right] \tag{9.34a}$$

and the descending branch where $j' = j + 1$ and

$$\nu^{(-)} = \frac{\Delta E^{(-)}}{h} = \frac{1}{h}\left[\hbar\omega_v - \frac{\hbar^2}{A}(j+1)\right] \tag{9.34b}$$

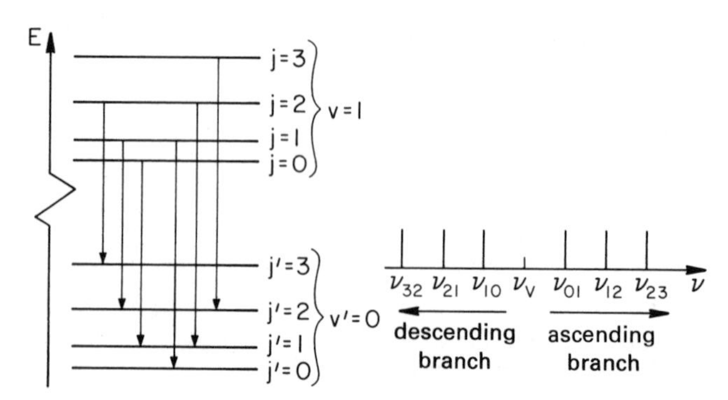

FIGURE 9.7. (a) The rotation–vibration states of a diatomic molecule. (b) The spectrum of the emitted radiation in transitions with $\Delta j = \pm 1$ and $\Delta v = 1$. This spectrum is usually in the near infrared.

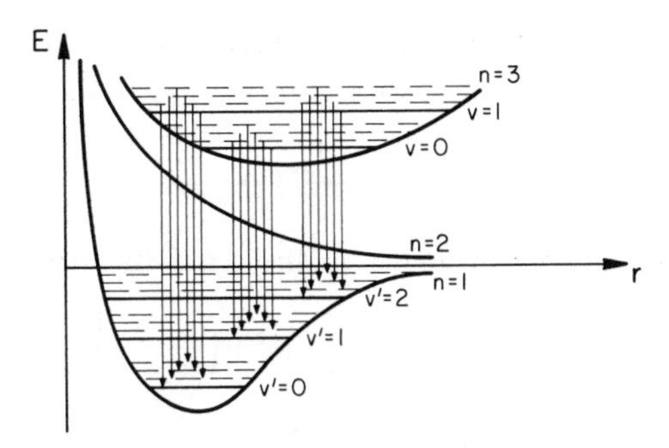

FIGURE 9.8. Schematic representation of the potential of a diatomic molecule when both atoms are in the ground state $n = 1$ and when one atom is excited. Note the transitions between rotation–vibration states belonging to different electronic excitation of the molecule. The spacing of the rotational levels is greatly exaggerated in this sketch.

Finally, it is possible that one of the atoms in the molecule becomes excited to a higher electronic state. This results in a significant change in the interaction energy, and the molecule may dissociate. If the potential energy in the excited state has a minimum, the molecule remains bound and transitions to the levels of the unexcited molecule can take place. Figure 9.8 is a semiquantitative representation of the energy levels of the hydrogen molecule showing the excited states and their rotation–vibration levels. The photons emitted in these transitions have frequencies in the visible spectrum, but instead of the single line emitted by a free atom, the whole band of the superimposed vibrational and rotational transitions manifests itself. Since the initial state lies well above the final state, transitions between vibrational levels are not restricted to $v' = v - 1$, but it holds more generally that $v' = v \pm 1$. Furthermore, the interatomic distance will be larger when the molecule is in the state where an electron is excited than when both atoms are in the ground state. Consequently, the spacing of the rotational levels is smaller in the excited state. Taking all these factors into account, one finds for the frequency of the emitted photons

$$\nu = \frac{1}{h}\left\{(E_n - E_{n'}) \pm \hbar\omega_v + \frac{\hbar^2}{2}\left[\frac{j(j+1)}{A_n} - \frac{j'(j'+1)}{A_{n'}}\right]\right\} \quad (9.35a)$$

Here A_n and $A_{n'}$ are the moments of inertia of the molecule in the initial and final state, respectively. We set $\nu_0 = [(E_n - E_{n'}) \pm \hbar\omega_v]/h$ so that

$$\nu = \nu_0 + \frac{\hbar}{2\pi}\,\nu_{j'j}$$

and we recognize three branches in the band spectrum

$$\begin{aligned}
j' &= j-1 \qquad \nu^{(+)} = \nu_0 + \frac{\hbar}{2\pi}\left[\left(\frac{1}{A_n} - \frac{1}{A_{n'}}\right)j^2 + \left(\frac{1}{A_n} + \frac{1}{A_{n'}}\right)j\right] \\
j' &= j \qquad \nu^{(0)} = \nu_0 + \frac{\hbar}{2\pi}\left[\left(\frac{1}{A_n} - \frac{1}{A_{n'}}\right)(j^2 + j)\right] \qquad (9.35b) \\
j' &= j+1 \qquad \nu^{(-)} = \nu_0 + \frac{\hbar}{2\pi}\left[\left(\frac{1}{A_n} - \frac{1}{A_{n'}}\right)(j+1)^2 - \left(\frac{1}{A_n} + \frac{1}{A_{n'}}\right)(j+1)\right]
\end{aligned}$$

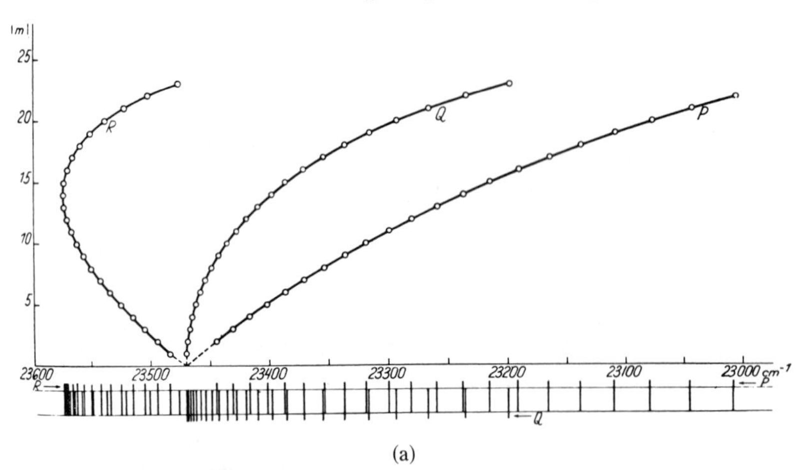

(a)

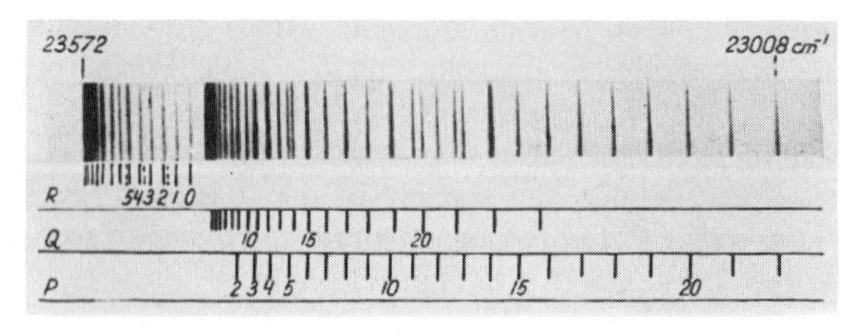

(b)

Equations (9.35b) are plotted in Fig. 9.9(a) where the abscissa gives the frequency and the ordinate gives the quantum number j. In general, the band has a distinct *head* at high (or low) frequency and becomes diffuse at the other end. A typical band spectrum is shown in Fig. 9.9(b). Rotation–vibration spectra are also clearly observed in nuclear spectra; an example is shown in Fig. 9.9(c).

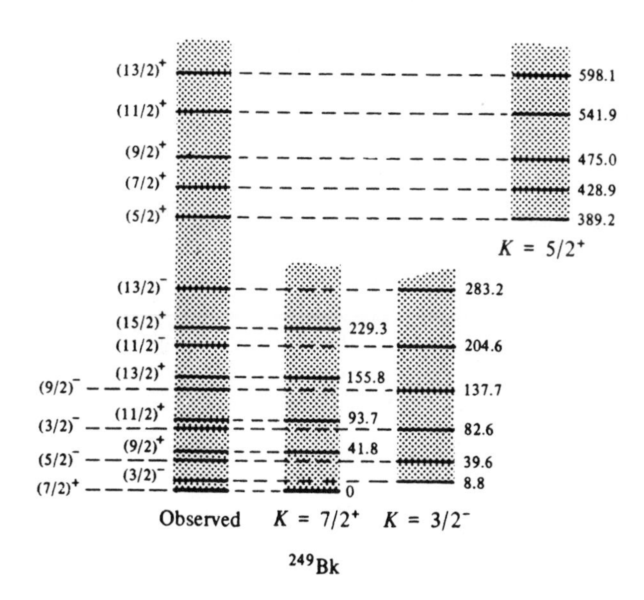

(c)

FIGURE 9.9. Band spectra of molecules. (a) The three branches, $\nu^{(+)}$, $\nu^{(0)}$, and $\nu^{(-)}$ for the *AlH* band at $\lambda = 4241$ Å. (b) The *AlH* band at $\lambda = 4241$ Å [from G. Herzberg *Molekülspektren und Molekülstruktur* Th. Steinkopff Verlag. Leipzig, 1939] (c) Example of rotational energy levels in nuclei shown here for ^{249}Bk. All levels up to 600 keV are shown on the left. They fall into three rotational bands as indicated on the right [from E. M. Henley and H. Frauenfelder "Subatomic Physics" 1974. Reprinted by permission of Prentice Hall Inc., Englewood Cliffs, NJ.]

9.4. THE S.H.O. IN THREE DIMENSIONS:

A harmonic oscillator need not be confined to one dimension but may be free to move in three dimensions. This is equivalent to the motion of a particle, in three dimensions, under the influence of a central force which is proportional to the distance $|\mathbf{r}|$ from the origin. The potential for such an *isotropic* harmonic oscillator is given by

$$U(r)=\tfrac{1}{2}kr^2=\tfrac{1}{2}k(x^2+y^2+z^2) \tag{9.36}$$

A particle moving in the above potential will remain bound and, therefore, will have a spectrum of discrete stationary states. The three-dimensional isotropic harmonic oscillator serves as one of the successful models for describing the structure of nuclei, and more generally it is useful whenever a system is subject to strong attractive short-range forces. Because of its simplicity the three-dimensional isotropic harmonic oscillator can be exactly solved in both cartesian and spherical coordinates. Thus it provides a nice example of the interconnection between two different coordinate representations of a quantum-mechanical system.

In the coordinate representation the Hamiltonian for a particle of mass m moving in the potential given by Eq. (9.36) is

$$\hat{H}(\mathbf{r})=-\frac{\hbar^2}{2m}\nabla^2+\tfrac{1}{2}(m\omega^2)r^2 \tag{9.37a}$$

where we use $m\omega^2$ to express the spring constant k. This Hamiltonian is invariant under rotations and, therefore, angular momentum is conserved. Consequently, we could label the stationary states by the eigenvalues $|l, m\rangle$ and a radial quantum number k. We will return to this representation later, but for the moment we note that the Hamiltonian is separable when expressed in cartesian coordinates

$$\hat{H}(x, y, z)=-\frac{\hbar^2}{2m}\left(\frac{d^2}{dx^2}+\frac{d^2}{dy^2}+\frac{d^2}{dz^2}\right)+\tfrac{1}{2}m\omega^2(x^2+y^2+z^2) \tag{9.37b}$$

The wavefunctions for the stationary states must obey the

Schrödinger equation

$$\hat{H}(x, y, z)\phi_n(x, y, z) = E_n\phi_n(x, y, z) \tag{9.38}$$

In view of the separable structure of $\hat{H}(x, y, z)$, the wavefunctions $\phi_n(x, y, z)$ can be written as a product of three functions, each depending on only one of the variables x, y, or z

$$\phi_n(x, y, z) = X_{n_x}(x)Y_{n_y}(y)Z_{n_z}(z) \tag{9.39a}$$

Each of the functions $X_{n_x}(x)$, $Y_{n_y}(y)$, and $Z_{n_z}(z)$ satisfies the one-dimensional s.h.o. eigenvalue problem

$$\left(-\frac{\hbar^2}{2m}\frac{d^2}{dx^2} + \tfrac{1}{2}m\omega^2x^2\right)X_{n_x}(x) = E_{n_x}X_{n_x}(x) \tag{9.39b}$$

and similarly for $Y_{n_y}(y)$, $Z_{n_z}(z)$. Therefore, the eigenvalues E_{n_x}, E_{n_y}, and E_{n_z} are given by Eq. (9.14), and the corresponding eigenfunctions are expressed in the coordinate representation by the functions $\phi_{n_x}(x)$ of Eq. (9.27*a*). We can immediately write for the solution

$$\phi_n(x, y, z) = \phi_{n_x}(x)\phi_{n_y}(y)\phi_{n_z}(z) \tag{9.39c}$$

and it follows from Eq. (9.38) that the energy eigenvalues for the three-dimensional isotropic oscillator are

$$E_n = \hbar\omega(n_x + n_y + n_z + \tfrac{3}{2}) \tag{9.40a}$$

Since n_x, n_y, and n_z take only zero or positive integer values, the quantum number

$$n = n_x + n_y + n_z \tag{9.40b}$$

also takes a value equal to zero or a positive integer.

It is clear from Eqs. (9.40*a*) and (9.40*b*) that a given energy eigenvalue $E_n = \hbar\omega(n + \frac{3}{2})$ is degenerate in general since it can be obtained for different combinations of n_x, n_y, and n_z. The ground state is nondegenerate because it corresponds to the unique assignment $n_x = n_y = n_z = 0$. However, the first excited state of the system is triply degenerate, the second excited level has six-fold degeneracy, and so on. This is indicated in Table 9.1.

Table 9.1. Classification of the Stationary States of the Three-Dimensional Isotropic Harmonic Oscillator

Quantum numbers		Number of states
n	n_x, n_y, n_z	
0	0, 0, 0	1
1	1, 0, 0; 0, 1, 0; 0, 0, 1	3
2	2, 0, 0; 0, 2, 0; 0, 0, 2 1, 1, 0; 0, 1, 1; 1, 0, 1	6
3	3, 0, 0; 0, 3, 0; 0, 0, 3 2, 1, 0; 0, 2, 1; 1, 0, 2 1, 2, 0; 0, 1, 2; 2, 0, 1 1, 1, 1	10
e.t.c		

In general, the number of degenerate states for a given value of n, is given by

$$g_n = \frac{(n+1)(n+2)}{2} \tag{9.40c}$$

as can be easily proven. The energy spectrum is shown in Fig. 9.10(a).

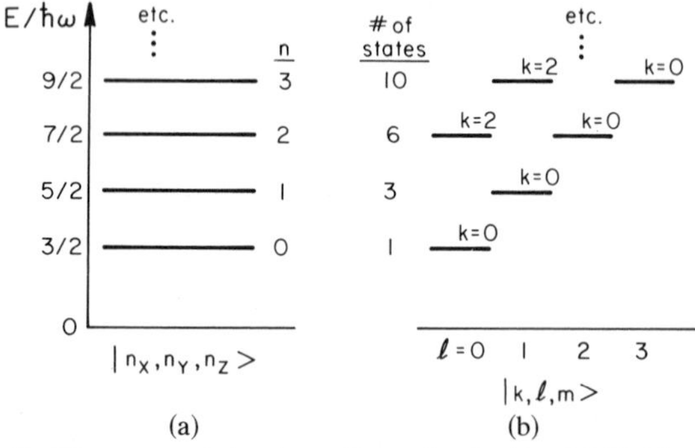

FIGURE 9.10. The states of the three-dimensional s.h.o. labeled by the $|n_x, n_y, n_z\rangle$ and $|k, l, m\rangle$ representations. Note that because of the short-range nature of the potential the spectrum does not condense as E_n increases.

We draw attention to the fact that the stationary states are also eigenstates of parity, as they must be in view of the invariance of the Hamiltonian of Eq. (9.37) under inversion of the coordinates. The eigenfunctions of the one-dimensional s.h.o. $\phi_{n_x}(x)$, etc., are eigenstates of parity with eigenvalue $(-1)^{n_x}$; therefore, the parity of

$\phi_n(x, y, z)$ is given by

$$(-1)^{n_x}(-1)^{n_y}(-1)^{n_z} = (-1)^{n_x+n_y+n_z} = (-1)^n \tag{9.41}$$

Note that we can obtain the explicit expression for $\phi_n(x, y, z)$ in terms of the Hermite polynomials using Eq. (9.27a). Here we give only the ground-state wavefunction

$$\phi_0(x, y, z) = \left(\frac{m\omega}{\pi\hbar}\right)^{3/4} e^{-(m\omega/2\hbar)r^2} \tag{9.42}$$

We will now find the stationary states in a representation where they are labeled by the angular momentum eigenvalues. Of course, these states will have the same energy eigenvalues as found above, but may be given as linear combinations of the degenerate states defined by Eq. (9.39c). We write the Hamiltonian [Eq. (9.37a)] in terms of the angular momentum operator $\hat{L}^2$ as in Eq. (8.13a)

$$\hat{H} = -\frac{\hbar^2}{2m}\left[\frac{1}{r^2}\frac{\partial}{\partial r}\left(r^2\frac{\partial}{\partial r}\right) - \frac{\hat{L}^2}{\hbar^2 r^2}\right] + \tfrac{1}{2}m\omega^2 r^2 \tag{9.43a}$$

and express the wavefunctions as a product of angular momentum eigenstates (the spherical harmonics) and a radial function labeled by the quantum numbers k and l

$$\phi_{k,l,m}(r, \theta, \phi) = R_{k,l}(r) Y_{l,m}(\theta, \phi) \tag{9.43b}$$

The solution proceeds exactly as for the hydrogen atom (see Sections 8.2 and 8.3). The radial function must obey the equation

$$\left\{\frac{1}{r^2}\frac{d}{dr}\left(r^2\frac{d}{dr}\right) + \frac{2m}{\hbar^2}\left[E_{k,l} - \tfrac{1}{2}m\omega^2 r^2 - \frac{\hbar^2 l(l+1)}{2mr^2}\right]\right\} R_{k,l}(r) = 0 \tag{9.43c}$$

the solution of which is postponed till the end of the section. One can show that the energy eigenvalues are given in this case by

$$E_{k,l} = \hbar\omega[(k+l) + \tfrac{3}{2}] \tag{9.44a}$$

where k is a positive, even integer, or zero. That is,

$$k = 0, 2, 4 \cdots \tag{9.44b}$$

To each energy $E_{k,l}$ correspond $2l+1$ states distinguished by their corresponding m value. The states in the $|k, l, m\rangle$ representation are shown in Fig. 9.10(b), and the correspondence between the two representations is evident.

Comparison of Eqs. (9.40) and (9.44) shows that the eigenvalues are the same when the quantum numbers

$$n_x + n_y + n_z = n = k + l$$

Since k is always even, we note that even values of l correspond to states with n even. According to Eq. (9.41) these states have even parity. Similarly, odd l states correspond to n odd, namely, to odd parity. Furthermore, for a given n-value, l can take values $n, n-2, n-4$, and so on. Since each l state has $(2l+1)$-fold degeneracy, the number of states found at a given value of $(k+l)$ is

$$\begin{aligned} k+l=n \quad &\text{even} \quad 2n+1+(2n+1-4)+\cdots+5+1 \\ k+l=n \quad &\text{odd} \quad 2n+1+(2n+1-4)+\cdots+7+3 \end{aligned}$$

The above arithmetic progressions can be summed directly to give the number of states

$$g_n = \tfrac{1}{2}(n+1)(n+2)$$

which is valid whether n is even or odd. As expected, the result is in agreement with Eq. (9.40*c*).

The states in the $|k, l, m\rangle$ representation are related to those in the $|n_x, n_y, n_z\rangle$ representation through a unitary transformation

$$|k, l, m\rangle = \sum_{n_x+n_y+n_z=k+l} |n_x, n_y, n_z\rangle\langle n_x, n_y, n_z\,|k, l, m\rangle \tag{9.45}$$

Of course, this transformation connects only states of the same energy. For instance, the reader can verify that for $n=1$ the following relations hold

$$\phi_{n_x=1,\, n_y=0,\, n_z=0} = \frac{1}{2^{1/2}}(\phi_{k=0,\, l=1,\, m=-1} - \phi_{k=0,\, l=1,\, m=1})$$

$$\phi_{n_x=0,\, n_y=1,\, n_z=0} = \frac{-i}{2^{1/2}}(\phi_{k=0,\, l=1,\, m=-1} + \phi_{k=0,\, l=1,\, m=1})$$

$$\phi_{n_x=0,\, n_y=0,\, n_z=1} = \phi_{k=0,\, l=1,\, m=0}$$

The degeneracy between states of different l-value but of the same value of $(k+l)$ is due to a dynamical symmetry of the isotropic harmonic oscillator potential, similar to that encountered in the case of the Coulomb potential.

The solution of the radial equation [Eq. (9.43c)] is obtained by introducing [as in Eq. (8.18a)] a modified radial function

$$u_{k,l}(r) = rR_{k,l}(r) \tag{9.46a}$$

and the notation [$\varepsilon_{k,l}$ has dimensions (length)$^{-2}$]

$$\varepsilon_{k,l} = \frac{2m}{\hbar^2} E_{k,l} \quad \text{and} \quad \beta = \left(\frac{m\omega}{\hbar}\right)^{1/2} \tag{9.46b}$$

to obtain the simpler equation

$$\left[\frac{d^2}{dr^2} + \varepsilon_{k,l} - \beta^4 r^2 - \frac{l(l+1)}{r^2}\right] u_{k,l}(r) = 0 \tag{9.46c}$$

with the boundary conditions $u_{k,l}(r) = 0$ as $r \to 0$ or $r \to \infty$. In the asymptotic limit ($r \to \infty$), the equation becomes

$$\frac{d^2 u_{kl}}{dr^2} - \beta^4 r^2 u_{kl} = 0$$

and the solution that satisfies the boundary condition is given by

$$u_{kl}(r) \xrightarrow[r \to \text{large}]{} e^{-\beta^2 r^2/2} \tag{9.47a}$$

Furthermore, near the origin, i.e., $r \to 0$, the equation becomes

$$\frac{d^2 u_{kl}}{dr^2} - \frac{l(l+1)}{r^2} u_{kl} = 0$$

The solution that vanishes at the origin has the form

$$u_{kl}(r) \xrightarrow[r \to 0]{} r^{l+1} \tag{9.47b}$$

Thus, we can write the general solution to be of the form

$$\begin{aligned} u_{kl}(r) &= e^{-\beta^2 r^2/2} r^{l+1} \sum_{s=0}^{\infty} a_s r^s \\ &= e^{-\beta^2 r^2/2} \sum_{s=0}^{\infty} a_s r^{l+s+1} \end{aligned} \tag{9.47c}$$

Substituting this into Eq. (9.46c) we have

$$\sum_{s=0}^{\infty} \{[\varepsilon_{kl} - \beta^2(2s+2l+3)]a_s r^{l+s+1} + [(l+s+1)(l+s) - l(l+1)]a_s r^{l+s-1}\} = 0$$

or,

$$\sum_{s=0}^{\infty} \{\varepsilon_{kl} - \beta^2(2s+2l+3)]a_s r^{l+s+1} + s(s+2l+1)a_s r^{l+s-1}\} = 0 \tag{9.48a}$$

Consider the lowest power of r in the series, namely, r^{l-1}. The vanishing of this term does not impose any constraint on the coefficient a_0. In fact, even if

$$a_0 \neq 0$$

the term will vanish. On the other hand, the second term in the series has the form r^l. The vanishing of this term requires that

$$2(l+1)a_1 = 0$$

which determines that $a_1 = 0$. Thus we have

$$a_0 \neq 0 \qquad a_1 = 0 \tag{9.48b}$$

Furthermore, we can write Eq. (9.48a) as

$$\sum_{s=0}^{\infty} \{[\varepsilon_{kl} - \beta^2(2s+2l+3)]a_s + (s+2)(s+2l+3)a_{s+2}\} r^{l+s+1} = 0$$

This is true only if the coefficient of each term in the series vanishes. That is,

$$[\varepsilon_{kl} - \beta^2(2s+2l+3)]a_s + (s+2)(s+2l+3)a_{s+2} = 0$$

or,

$$a_{s+2} = \frac{\beta^2(2s+2l+3) - \varepsilon_{kl}}{(s+2)(s+2l+3)} a_s \tag{9.48c}$$

Note that this is a recursion relation which connects either only the even or the odd coefficients. However, in view of Eq. (9.48b), only the even terms of the series survive. Furthermore, the series must terminate, otherwise the solution for large distances would have a behavior of the form

$$u_{kl}(r) \xrightarrow[r \to \text{large}]{} e^{\beta^2 r^2/2}$$

The series will terminate if there exists a value of s, $s_{\max}$ such that

$$\beta^2(2s_{\max} + 2l + 3) - \varepsilon_{kl} = 0$$

or,

$$\begin{aligned}\varepsilon_{kl} &= \beta^2(2s_{\max} + 2l + 3)\\ &= 2\beta^2(s_{\max} + l + \tfrac{3}{2})\end{aligned}$$

Putting in the values of ε_{kl} and β from Eq. (9.46b) we obtain

$$\begin{aligned}E_{kl} &= \hbar\omega(s_{\max} + l + \tfrac{3}{2})\\ &= \hbar\omega(k + l + \tfrac{3}{2})\end{aligned} \tag{9.49}$$

as indicated in Eq. (9.44a). Here we have identified k with $s_{\max}$, and it can only take values $k = 0, 2, 4, 6, \ldots$.

The polynomials in Eq. (9.47c) can be identified with the associated Laguerre polynomials of half-integer order. They are briefly discussed in Appendix 10. As expected, the $k = 0$ functions are nodeless, whereas for $k = 2$ there is one node, and so on (see also Fig. A10.1).

9.5. THE SHELL MODEL OF THE NUCLEUS

We do not have an exact understanding of nuclear forces and the structure of nuclei. On the other hand, a great wealth of information is available about the stationary states and the properties of nuclei. Such information enables us to construct models of the nucleus that can be used to make fairly accurate predictions. We discuss one such model in which it is assumed that the individual nucleons (neutrons or protons) move in a potential produced by all the other nucleons. The potential—due to the nuclear forces—is experimentally known to be of short-range and can be assumed, in most cases, to be central. The three-dimensional isotropic harmonic oscillator potential provides an adequate approximation to the nuclear potential and produces results in agreement with observation, so we adopt it.

We start from the well-established experimental observation that nuclei for which the number of neutrons *or* protons is

$$2, 8, 20, 28, 50, 82, 126 \qquad (9.50a)$$

show extreme stability. By this we mean that it is difficult to excite them. The sequence of numbers listed above are known, therefore, as *magic numbers*. Nuclei that are doubly magic should be exceptionally stable, as is indeed the case for

$$ {}_{2}^{2}\mathrm{He}^{4}, \quad {}_{8}^{8}\mathrm{O}^{16}, \quad {}_{20}^{20}\mathrm{Ca}^{40}, \quad {}_{20}^{28}\mathrm{Ca}^{48}, \quad {}_{82}^{126}\mathrm{Pb}^{208} $$

This would indicate that the magic numbers represent *closed shells* of nucleons so that the excitation to the next higher level requires a large amount of energy. We must remember that protons and neutrons are spin-1/2 particles. Thus, according to the Pauli exclusion principle, there can be only two particles in any one eigenstate of energy corresponding to distinct eigenvalues k, l, and m. One of the particles will be in the spin-up state $m_s = +1/2$ and the other in the spin-down state $m_s = -1/2$.

If the nuclear potential is of the form of the three-dimensional s.h.o. we would expect a closed shell whenever *all* states corresponding to a given n-value, i.e., of the same energy, are filled. According to Table 9.1 the number of states for a given n, including the two possible spin states, is [see Eq. (9.40c)]

$n=0$	$n=1$	$n=2$	$n=3$
$(l=0)$	$(l=1)$	$(l=0,2)$	$(l=1,3)$
2	6	12	20

$n=4$	$n=5$	$n=6$
$(l=0,2,4)$	$(l=1,3,5)$	$(l=0,2,4,6)$
30	42	56

and therefore the particularly stable nuclei should occur when the number of protons or neutrons is

$$2, 8, 20, 40, 70, 112, 168 \qquad (9.50\mathrm{b})$$

The first three magic numbers agree with this sequence. Beyond the third magic number the closed nuclear shells appear to correspond to

$$
\begin{aligned}
28 &= 20+8 = 20+(2\times 7/2+1)\\
50 &= 40+10 = 40+(2\times 9/2+1)\\
82 &= 70+12 = 70+(2\times 11/2+1)\\
126 &= 112+14 = 112+(2\times 13/2+1)
\end{aligned}
\tag{9.50c}
$$

nucleons.

The sequence of numbers in Eqs. (9.50*c*) is suggestive. For instance, to fill the $n=2$ shell we require $N=20$ nucleons. We now start to fill the $n=3$ shell which admits angular momentum $l=1$ and $l=3$ (P- and F-states). Since the nucleons have spin 1/2, a possible spin-orbit coupling would give rise to four values of the total angular momentum with corresponding multiplicities

$$
\begin{aligned}
&P_{1/2} \quad \text{multiplicity} \quad (2J+1) = 2\\
&P_{3/2} \qquad\qquad\qquad\qquad\quad = 4\\
&P_{5/2} \qquad\qquad\qquad\qquad\quad = 6\\
&P_{7/2} \qquad\qquad\qquad\qquad\quad = 8
\end{aligned}
$$

If the state with the highest J-value has the lowest energy it may be possible that it becomes associated with the *preceding* shell. Indeed, examination of the sequences in Eqs. (9.50*c*) shows that the shells *close* when we add to them the number of states corresponding to the highest J-value of the following shell. This is shown in Fig. 9.11, where on the lhs are indicated the (equidistant) energy levels of the three-dimensional s.h.o.; on the rhs are the levels expected from a square-well potential. In the center of the figure are given the stationary states as calculated when the spin-orbit coupling is included. This spectrum clearly reproduces the observed magic numbers and gives confidence in the shell model of the nucleus. The model was proposed independently by Mayer and by Haxel, Jensen, and Suess in 1955.

As we saw in Section 8.5 the spin–orbit coupling term arises, in principle, from the relativistic correction to the wavefunction of a particle moving in a central potential $V(r)$. According to Eq. (8.40*b*) it should be given by

$$
\frac{\hbar^2}{2m_N^2c^2}\left(\frac{1}{r}\frac{dV}{dr}\right)\hat{\mathbf{L}}\cdot\hat{\mathbf{S}}
$$

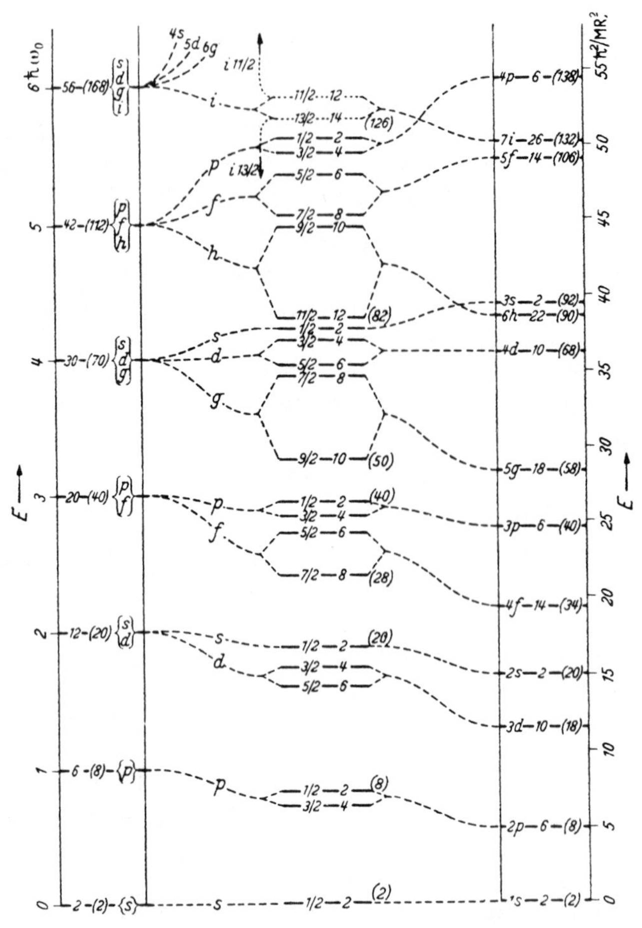

FIGURE 9.11. Energy levles of a nucleon bound in a nucleus. On the left are the oscillator levels; on the right are the square-well levels. In between are levels that are intermediate with spin–orbit coupling. The 1*s* level is taken to be the zero of energy of the system. [From Haxel, Jensen, and Suess, *Z. Phys.*, **128,** 298 (1950) by permission].

However, this term results in too small a splitting (by a factor ~30) and has the wrong sign. It requires states with larger J-values to lie at higher energies. We conclude that the spin–orbit splitting in nuclei is not due to a relativistic correction, as it is for the fine structure of atomic spectra, but is directly related to the spin-dependence of the nuclear froces. This is corroborated by scattering experiments with polarized nucleons and is in complete agreement with observation.

Nuclei consisting of closed shells must, of course, have zero total angular momentum. This is the case for ${}^{8}_{8}O^{16}$, ${}^{20}_{20}Ca^{40}$, and ${}^{126}_{82}Pb^{208}$. Nuclei with one nucleon outside a closed shell will then have the angular momentum of the *next highest* level and it is found that

$$\text{for } {}^{9}_{8}O^{17},\ I = 5/2 \qquad \text{and for } {}^{126}_{83}Bi^{209},\ I = 9/2$$

Correspondingly, for nuclei with one nucleon missing (a hole) from a closed shell, their angular momentum must be that of the level closing the shell and we find that

$$\begin{array}{lll} \text{for} & {}^{8}_{7}N^{15}, & I = 1/2 \\ \text{for} & {}^{20}_{19}K^{39}, & I = 1/2 \\ \text{for} & {}^{125}_{82}Pb^{207}, & I = 1/2 \end{array}$$

in complete agreement with the assignments indicated in Fig. 9.11.

The energy spectrum shown in Fig. 9.11 can be refined by more detailed calculations. For instance, the levels for protons differ from those of neutrons because of the electrostatic repulsion present in the former case. Furthermore, the s.h.o. wavefunctions can be used to calculate the energy of excited states and to predict the magnetic dipole and electric quadrupole moments of nuclei. However, one must remember that the shell model is most successful for nuclei having closed shells and almost closed shells. Away from closed shells the nucleons exhibit collective phenomena such as deformations and collective vibrations. These properties can be best understood in terms of different nuclear models.

9.6. THE SQUARE-WELL POTENTIAL

As we have already observed, the nuclear forces are of short range. Therefore, the simplest model for the nuclear potential is that of a square well. Such a model is more realistic than the three-dimensional s.h.o. since beyond a certain range the nucleons are free. While in reality the potential may not have sharp edges, the square-well approximation is reasonably accurate. We will also

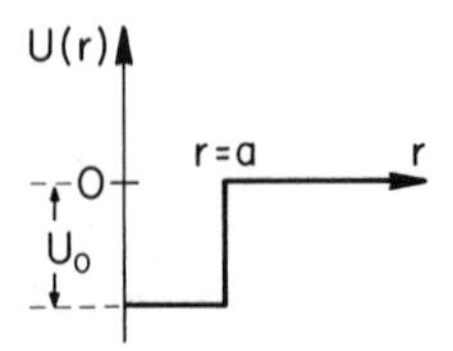

FIGURE 9.12. The square-well potential. Note that the figure shows only the radial dependence; otherwise the potential is spherically symmetric.

assume that the potential is central, i.e., it is invariant under rotations. Such a potential is shown as a function of r in Fig. 9.12. It is characterized by two numbers, its depth U_0 and its width a.

$$\begin{aligned} U(r) &= -U_0 \qquad r<a \\ &= 0 \qquad\quad\; r>a \end{aligned} \tag{9.51}$$

Note that even though we use the term "square well" the region where the potential is different from zero is bounded by a *sphere* of radius $r=a$.

Since the potential is central we immediately know that the stationary states will be eigenstates of $\hat{L}^2$ and can be labeled by $|l, m\rangle$. Because of the short-range nature of the potential we expect only a finite number of bound states. In terms of the energy scale used in Eq. (9.51), bound states will have $E<0$, so that when $r>a$ their wavefunction must be a decaying exponential. For $r<0$ the bound-state wavefunctions will be oscillatory. Only for those energies for which the logarithmic derivative of the interior ($r<a$) and exterior ($r>a$) solutions can be matched at $r=a$, will a bound state occur. This is analogous to our discussion in Section 6.5, even though in three dimensions the particle may have angular momentum about the origin. We recall that we had considered the problem of an infinitely deep potential well, but with a cubic boundary, in Section 1.8. In that case we obtained the solution in cartesian coordinates, whereas now we shall use the representation of angular momentum eigenstates.

In the coordinate representation the Hamiltonian is

$$\hat{H} = -\frac{\hbar^2}{2\mu}\left[\frac{1}{r^2}\frac{\partial}{\partial r}\left(r^2\frac{\partial}{\partial r}\right) - \frac{\hat{L}^2}{\hbar^2 r^2}\right] + U(r) \tag{9.52a}$$

with $U(r)$ given by Eq. (9.51) and μ the mass of the particle or the reduced mass as the case may be. We express the wavefunction as a product of the spherical harmonics and a radial function $R_{n,l}(r)$

$$\psi_{n,l,m}(\mathbf{r}) = R_{n,l}(r)Y_{l,m}(\theta, \phi) \tag{9.52b}$$

Thus, the radial function obeys the equation

$$\left\{\frac{\hbar^2}{2\mu}\left[\frac{1}{r^2}\frac{d}{dr}\left(r^2\frac{d}{dr}\right)-\frac{l(l+1)}{r^2}\right]+E_{n,l,m}-U(r)\right\}R_{n,l}(r)=0 \tag{9.52c}$$

and must satisfy the boundary conditions

$$[rR_{n,l}(r)] \to 0 \quad \text{as} \quad r \to 0, \infty \tag{9.52d}$$

Even though we can solve Eq. (9.52c) directly for arbitrary values of l we will first consider the special case $l = 0$. We introduce a modified radial function

$$u_{n,0}(r) = rR_{n,0}(r) \tag{9.53a}$$

and the specific form of the square-well potential $U(r)$ as given by Eq. (9.51) to obtain the two equations

$$\begin{aligned} r<a \qquad & \left[\frac{\hbar^2}{2\mu}\frac{d^2}{dr^2}+(E_n+U_0)\right]u_n(r)=0 \\ r>a \qquad & \left(\frac{\hbar^2}{2\mu}\frac{d^2}{dr^2}+E_n\right)u_n(r)=0 \end{aligned} \tag{9.53b}$$

For bound states, $E_n < 0$ and the boundary conditions are $u_n(r) \to 0$ as $r \to 0$ or $+\infty$. Thus, the solutions for $r<a$ and $r>a$ are, respectively,

$$\begin{aligned} r<a \qquad & u(r) = A\sin\alpha r \qquad & \alpha = \left[\frac{2\mu(U_0-|E|)}{\hbar^2}\right]^{1/2} \\ r>a \qquad & u(r) = Ce^{-\beta r} \qquad & \beta = \left(\frac{2\mu\,|E|}{\hbar^2}\right)^{1/2} \end{aligned} \tag{9.54a}$$

The physical solution to the problem is such that $u(r)$ and du/dr as calculated from either region have the same value at $r = a$. To match $u(r)$ the ratio of the constants must equal

$$\frac{A}{C} = \frac{e^{-\beta a}}{\sin\alpha a} \tag{9.54b}$$

Matching the derivatives du/dr at $r=a$ provides us with one more relation

$$\left[\frac{du}{dr}(r<a)\right]_{r=a} = \alpha A\cos(\alpha a) = -C\beta e^{-\beta a} = \left[\frac{du}{dr}(r>a)\right]_{r=a}$$

or

$$-\beta = \frac{A}{C}\alpha\cos(\alpha a)e^{\beta a} = \alpha\frac{\cos(\alpha a)}{\sin(\alpha a)}$$

where we have used the result of Eq. (9.54b). The relation

$$\boxed{\beta = -\alpha\cot\alpha a} \qquad (9.54c)$$

connects the constants α and β defined in Eq. (9.54a) and thus determines the possible eigenvalues $|E|$ in terms of U_0, a, and μ. Note that Eq. (9.54c) is a transcendental relation.

To solve it graphically it is convenient to introduce the dimensionless variables

$$\xi = \alpha a \qquad \eta = \beta a \qquad (9.55a)$$

so that we seek the solution of

$$\eta = -\xi\cot\xi \qquad (9.55b)$$

Furthermore, ξ and η are related through

$$\xi^2+\eta^2 = a^2(\alpha^2+\beta^2) = \frac{2\mu U_0 a^2}{\hbar^2} \qquad (9.55c)$$

Since ξ and η are positive definite, it suffices to consider their values only in the first quadrant. This is shown in Fig. 9.13 where we have plotted the function $-\xi\cot\xi$ as a function of ξ (the abscissa). For any given value of ξ, η is fixed by the condition of Eq. (9.55c). Therefore, for a specific value of U_0, η must lie on a segment of the circle of radius $(\xi^2+\eta^2)^{1/2} = \text{const}$. The intersection of this circle with the curve $-\xi\cot\xi$ gives the desired solutions. It is clear from the topology of Fig. 9.13 that when

$$0 < \frac{2\mu U_0 a^2}{\hbar^2} \leq \left(\frac{\pi}{2}\right)^2 \qquad \text{there is no solution}$$

$$\left(\frac{\pi}{2}\right)^2 < \frac{2\mu U_0 a^2}{\hbar^2} \leqslant \left(\frac{3\pi}{2}\right)^2 \qquad \text{there is one solution}$$

$$\left(\frac{3\pi}{2}\right)^2 < \frac{2\mu U_0 a^2}{\hbar^2} \leqslant \left(\frac{5\pi}{2}\right)^2 \qquad \text{there are two solutions}$$

and so on. We conclude that there will be no bound state unless

$$\left(\frac{2\mu}{\hbar^2} U_0 a^2\right)^{1/2} > \frac{\pi}{2} \quad \text{or} \quad U_0 a^2 > \frac{\hbar^2 \pi^2}{8\mu} \tag{9.56a}$$

More generally, the system will have N bound states if

$$(N - \tfrac{1}{2})\pi < \left(\frac{2\mu}{\hbar^2} U_0 a^2\right)^{1/2} \leqslant (N + \tfrac{1}{2})\pi \tag{9.56b}$$

We note that the condition for the existence of bound states depends on the product of the depth and the square of the width of the potential. This can be readily understood in terms of the uncertainty principle: when the potential is of very short range, i.e., a is small, the position uncertainty for a bound state is small. Therefore, the potential must be deep to allow for a large uncertainty in momentum. Conversely, for a long-range potential the uncertainty in the momentum of a bound state can be small, thereby allowing states of low kinetic energy to fit in a shallow well.

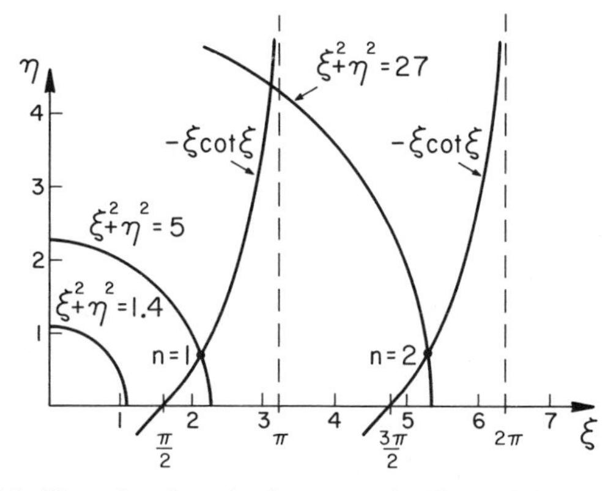

FIGURE 9.13. Plot of $-\xi \cot \xi > 0$ versus ξ. The solutions of the equation $\eta = -\xi \cot \xi$ where $\eta^2 + \xi^2 = \text{const}$ are given by the intersection of the two curves. Note that for $\eta^2 + \xi^2 = 1.4$ there is no solution. For $\eta^2 + \xi^2 = 5$ there is one solution, whereas when $\eta^2 + \xi^2 = 27$ there are two solutions.

General Case

We now consider the solution of the radial equation [Eq. (9.52c)] for the case when l can be different from zero, and for the potential of Eq. (9.51). We introduce dimensionless variables as follows. In the interior region, $r<a$, we set

$$\rho=\alpha r, \qquad \text{where} \quad \alpha=\left[\frac{2\mu(U_0-|E|)}{\hbar^2}\right]^{1/2} \tag{9.57a}$$

so that the radial equation becomes

$$\left\{\frac{d^2}{d\rho^2}+\frac{2}{\rho}\frac{d}{d\rho}+\left[1-\frac{l(l+1)}{\rho^2}\right]\right\}R(\rho)=0 \tag{9.57b}$$

with the boundary condition $\rho R(\rho)=0$ as $\rho\to 0$. Equation (9.57b) is the spherical Bessel equation and is discussed in Appendix 12. Its solutions that satisfy the boundary condition are the spherical Bessel functions $j_l(\rho)$. These are simply trigonometric functions modulated by powers of ρ; for instance, the first two are

$$\begin{aligned} j_0(\rho)&=\frac{\sin\rho}{\rho} \\ j_1(\rho)&=\frac{1}{\rho}\left(\frac{\sin\rho}{\rho}-\cos\rho\right) \end{aligned} \tag{9.57c}$$

In the exterior region $r>a$ we replace

$$\rho\to i\beta r, \qquad \text{where} \quad \beta=\left(\frac{2\mu\,|E|}{\hbar^2}\right)^{1/2} \tag{9.58a}$$

so that the radial equation reduces to the same equation as for the interior case [Eq. (9.57b)]. However, the boundary conditions are now different. It must hold that $\rho R(\rho)=0$ as $\rho\to\infty$, and there is no condition at the origin. The appropriate solutions for this case [see Appendix 12] are the spherical Hankel functions of the first kind

$$h_l^{(1)}(\rho)=j_l(\rho)+in_l(\rho) \tag{9.58b}$$

In general, these are oscillatory functions modulated by powers of ρ.

The first two are

$$
\begin{aligned}
h_0^{(1)}(\rho) &= -\frac{i}{\rho}\, e^{i\rho} \\
h_1^{(1)}(\rho) &= -\left(\frac{1}{\rho}+\frac{i}{\rho^2}\right) e^{i\rho}
\end{aligned}
\tag{9.58c}
$$

However, in view of our choice of ρ as shown by Eq. (9.58*a*), the radial functions are indeed decaying exponentials as required for a bound state

$$
\begin{aligned}
R_{n,0} &= \frac{1}{\beta r}\, e^{-\beta r} \\
R_{n,1} &= i\left[\frac{1}{\beta r}+\frac{1}{(\beta r)^2}\right] e^{-\beta r}
\end{aligned}
\tag{9.58d}
$$

Finally, the interior and exterior solutions and their derivatives must be matched at the boundary $r=a$. It suffices to match the logarithmic derivative $(1/R)(dR/dr)$. This leads to a transcendental equation as in Eq. (9.54*c*), which must be solved to obtain the eigenvalues. The algebra becomes laborious, and for example the condition for $l=1$ is

$$
\frac{\cot \xi}{\xi} - \frac{1}{\xi^2} = \frac{1}{\eta} + \frac{1}{\eta^2} \tag{9.59}
$$

Here we use the notation introduced in Eq. (9.55) and the expression given by the second of Eqs. (9.57*c*) and (9.58*d*) for the interior and exterior solutions, respectively. In the case of the square well, there is no degeneracy for states with the same radial quantum number n but different angular momentum l. This is contrary to the results found for the Coulomb potential or the three-dimensional s.h.o.

It is interesting to note that even though the form of the radial functions does not appear to depend on the quantum number n, the eigenvalues certainly do because they correspond to different solutions of the transcendental equation such as Eq. (9.54*c*) or (9.59). This, in turn, modifies the coefficients α and β that determine the scale of the radial functions, the number of nodes, etc.

Infinitely Deep Square Well

One can obtain a better feeling for the position of the energy levels of the square-well potential by considering the (unphysical) model of an infinitely deep potential well. In this case we set the zero of the energy at the bottom of the well as shown in Fig. 9.14 and, of course, the wavefunction cannot penetrate beyond $r=a$. Thus, the boundary conditions are

$$rR_{n,l}(r)=0, \qquad \text{when } r=0 \text{ and } r\geqslant a \tag{9.60a}$$

which greatly simplifies the mathematics. We know that with $\rho=\alpha r$ the solutions are of the form

$$R_{n,l}(\rho)=j_l(\rho)$$

where we can express

$$\alpha=\left(\frac{2mE_{n,l}}{\hbar^2}\right)^{1/2} \tag{9.60b}$$

where $E_{n,l}>0$ and is measured from the bottom of the well since we have set $U_0=0$.

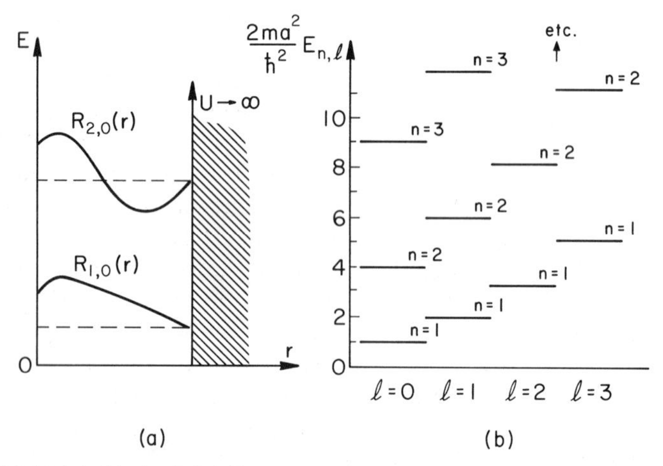

FIGURE 9.14. (a) An infinitely deep, spherically symmetric square-well potential and the radial functions for two low-lying states with $l=0$. (b) The spectrum of energy levels for the infinitely deep, spherically symmetric potential well.

To satisfy the boundary condition when $\rho = \alpha a$, it suffices that (αa) be a root of $j_l(\rho)$. We indicate these roots by $\rho_{n,l}$. Thus, the allowed values of $\alpha_{n,l}$ are given by

$$\alpha_{n,l} = \frac{\rho_{n,l}}{a} \quad \text{and} \quad E_{n,l} = \frac{\hbar^2}{2m}\left(\frac{\rho_{n,l}}{a}\right)^2 \tag{9.60c}$$

The first few roots of $j_l(\rho)$ are tabulated in Appendix 12 and it is easy to use them to construct the energy spectrum as shown in Fig. 9.14(a) (see also the rhs scale in Fig. 9.11). The reader should note the similarities of this spectrum with that obtained in Section 1.8 for an infinitely deep potential well bounded by a cubic rather than a spherical volume. In that case, of course, the system did not have rotational invariance.

9.7. THE DEUTERON

As an application of the square-well potential we will consider the deuteron. It consists of a proton (p) and a neturon (n). Since the neutron is not electrically charged the binding force cannot be electrostatic. The force between the magnetic moments of the neutron and the proton is ~ 100 times too small and the gravitational force is $\sim 10^{38}$ times too weak. Thus, we must admit the existence of a strong *nuclear force.* The nuclear force must be of short range and, therefore, we can represent it by a square-well potential. We will assume that the force is central and label the stationary states by the eigenvalues of angular momentum. Furthermore, we will assume that the ground state of the deuteron is an $l = 0$ state of orbital angular momentum. This statement is only approximately true. In reality, the ground state is a linear combination of $l = 0$ and $l = 2$ states, the amplitude for the latter being small. This conclusion is obtained directly from a study of the magnetic moment of the deuteron. Independent evidence for the noncentral nature of the nuclear forces comes from the electric quadrupole moment of the deuteron. This implies that the ground-state wavefunction deviates from spherical symmetry. Of course, the total angular momentum of the deuteron is a constant of the motion. It equals $I = 1$, and is referred to as the *spin* of the deuteron.

The binding energy of the ground state can be obtained by comparing the precise values of the deuteron, proton, and neutron masses. The deuteron and proton masses are determined from mass spectrography, whereas the neutron mass is found from the end-point of the electron energy in the β-decay of the free neutron $n \to p + e^- + \bar{\nu}_e$. One obtains

$$E_0 = [m(d) - m(p) - m(n)]c^2$$
$$= (1875.62 - 938.28 - 939.57)\ \text{MeV} = -2.23\ \text{MeV}$$

namely, the *binding* energy of the deuteron is

$$W = -E_0 = 2.33\ \text{MeV}$$

This value is confirmed by studying the threshold for the photodisintegration of the deuteron $\gamma + d \to n + p$ and by other experiments.

We can now apply the results of Section 9.6 to the deuteron. The reduced mass entering the Hamiltonian of Eq. (9.25*a*) is

$$\mu = \frac{m_p m_n}{m_p + m_n} \simeq \frac{m_N}{2}$$

where we use m_N to designate the nucleon mass. From the results of Fig. 9.13 we know that for the $l = 0$ ground state to be bound, the parameter $\xi = a\alpha$ must lie between $\pi/2$ and π. Thus, we can establish that

$$\frac{\pi^2}{4} < a^2 \frac{m_N(U_0 - W)}{\hbar^2} < \pi^2$$

or

$$\frac{1}{4}\frac{(\hbar c)^2}{m_N c^2}\pi^2 < a^2(U_0 - W) < \frac{(\hbar c)^2}{m_N c^2}\pi^2$$

with $\hbar c \simeq 200$ MeV-F, $m_N c^2 \simeq 1000$ MeV, and letting $a \simeq 2$ F we find the bound

$$25\ \text{MeV} < U_0 - W \leqslant 100\ \text{MeV}$$

namely, $U_0 \gg W$. Therefore, W is negligible as compared to U_0, which implies that $\alpha \gg \beta$ or, correspondingly, $\xi \gg \eta$ [see Eqs. (9.54*a*)]. In that case, ξ for the ground state is very close to $\pi/2$, and we can deduce that

$$a^2 U_0 \simeq \frac{(\hbar c)^2}{m_N c^2}\frac{\pi^2}{4} = 102\ \text{MeV-F}^2 \qquad (9.61a)$$

Since we used only one input, the binding energy W, it is not surprising that we cannot determine a and U_0 separately but only their product a^2U_0. The range of the nuclear force is known to be $a \simeq 1.7$ F [comparable to the Compton wavelength of the π-meson which is, $\lambda_C = \hbar/mc = (197\text{ MeV-F})/(140\text{ MeV}) = 1.41$ F] which yields

$$U_0 \simeq 35\text{ MeV}$$

The binding energy W uniquely determines the exponential falloff of the wavefunction in the exterior region since

$$\beta = \left(\frac{2\mu\,|E|}{\hbar^2}\right)^{1/2} = \left[\frac{(m_N c^2)W}{(\hbar c)^2}\right]^{1/2} \tag{9.61b}$$

This enables us to sketch the wavefunction. In the interior region the wave function is of the form $(1/r)\sin\alpha r$ where α is such that αa is just slightly larger than $\pi/2$. Thus, the radial function $u(r)$ is as shown in Fig. 9.15(a). If we decrease the range of the potential we must increase its depth so that the interior wavefunction bends faster to meet the exponential at $r = a$. This is shown in part (b) of the figure.

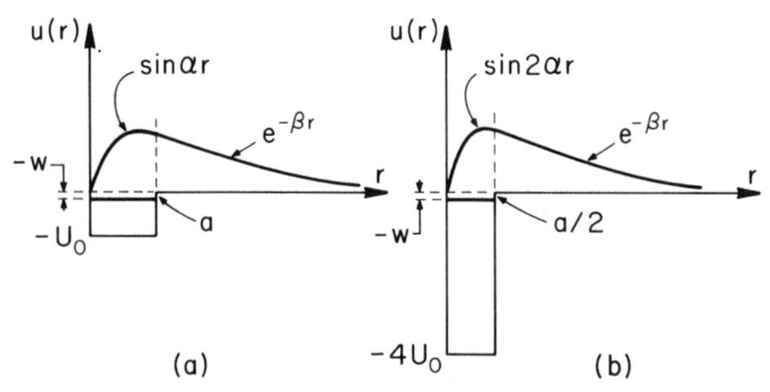

FIGURE 9.15. The radial function for a square-well potential. The indicated binding energy in this case is much smaller than the depth of the potential. This is the case for the deuteron. (b) A well of depth $-U_0$ and width a gives the same exterior wavefunction as a well of depth $-4U_0$ and width $a/2$.

The value of $\gamma = 1/\beta$ is a measure of the spatial extent of the deuteron. From Eq. (9.61b) we obtain

$$\gamma = 1/\beta = 4.31\text{ F} \tag{9.61c}$$

indicating that on average the nucleons are outside the range of the potential. It is not surprising, therefore, that the wavefunction is insensitive to the details of the potential. It can be easily checked that the deuteron has no excited bound states for $l=0$ or for higher values of l. This would necessitate a potential at least four times as deep as the one we determined. To obtain further information about the $n-p$ force and its spin dependence we must resort to scattering experiments. A brief description of such experiments is given in Chapter 11.

9.8. SUMMARY

We have examined and found solutions for several physical systems that form bound states. We tried to emphasize the great similarity between all of these systems and the common mathematical techniques used for their solution. Bound systems have, in general, a discrete spectrum of stationary states, and the number and density of these states is a reflection of the binding potential.

The systems that we considered are found in nature, and we discussed how the observational data is interpreted and related to the theoretical results. These physical systems span a large domain of energies ranging from molecules where the typical transition energies are $\sim 10^{-3}$ eV to the atomic nucleus where the energies are of the order of few MeV $\sim 10^{6}$ eV. We were able to use the coordinate representation because the dominant behavior of the systems that we analyzed depended on variables that had a classical analogue. The effects of spin were, in general, small, and therefore could be treated by perturbation theory.

The problem of the one-dimensional s.h.o. was presented and solved. It is of great theoretical importance because it serves as the model for the quantum-mechanical description of the electromagnetic field. We found that the s.h.o. system exhibited great symmetry between the position and momentum variables and that the energy spectrum was extremely simple, consisting of equidistant stationary states. Furthermore, the operator techniques that we introduced have wide application in many areas of quantum mechanics.

Problems

PROBLEM 1

Consider two harmonic oscillators with raising and lowering operators $\hat{a}^\dagger$ and $\hat{a}$ for the first, and $\hat{b}^\dagger$ and $\hat{b}$ for the second.

$$[\hat{a}, \hat{a}^\dagger] = 1 = [\hat{b}, \hat{b}^\dagger]$$
$$[\hat{a}, \hat{b}] = [\hat{a}, \hat{b}^\dagger] = [\hat{a}^\dagger, \hat{b}] = [\hat{a}^\dagger, \hat{b}^\dagger] = 0$$

From the four products $\hat{a}^\dagger\hat{b}$, $\hat{b}^\dagger\hat{a}$, $\hat{a}^\dagger\hat{a}$, and $\hat{b}^\dagger\hat{b}$ show that by linear combinations one can find operators that have the same commutation relations as the angular momentum operators. What operator plays the role of angular momentum squared?

Note: This is J. Schwinger's approach to treating angular momentum.

PROBLEM 2

Find $\langle x\rangle$, $\langle p\rangle$, $\langle x^2\rangle$, $\langle p^2\rangle$, and $\Delta x\, \Delta p$ in the state $|n\rangle$ of the s.h.o. What is the uncertainty relation for the ground state?

PROBLEM 3

If $u_m(x)$ and $u_n(x)$ are eigenfunctions for the s.h.o. in one dimension corresponding to the energy $(m+1/2)\hbar\omega$ and $(n+1/2)\hbar\omega$, respectively, use the generating function for the Hermite polynomials to calculate the integral

$$\langle p\rangle_{nm} = \int_{-\infty}^{\infty} u_n(x)\left[-i\hbar \frac{d}{dx} u_m(x)\right] dx$$

PROBLEM 4

A particle moves in three dimensions in a potential of the form

$$V(\mathbf{r}) = \begin{cases} \infty, & z^2 > a^2 \\ \frac{1}{2}m\omega^2(x^2+y^2), & \text{otherwise} \end{cases} \qquad (m, \omega \text{ constants})$$

Obtain a formula for the eigenvalues, the degeneracy of each level, and the eigenfunctions associated with them.

PROBLEM 5

For a diatomic molecule the Hamiltonian corresponding to its rotational motion can be written as

$$\hat{H} = \frac{\hbar^2}{2I}\hat{L}^2$$

where I is the moment of inertia about axes passing through the origin and perpendicular to the line joining the two atoms. (a) Find the energies of the stationary states. (b) Given that the *difference* between the frequency of *neighboring* lines in the rotational spectrum is

$$\Delta\bar{\nu} = \frac{\Delta\nu}{c} = 20.9 \text{ cm}^{-1}$$

Find the moment of inertia of the molecule and the interatomic distance if the molecule is N_2.

PROBLEM 6

Consider the $l = 2$ states of the hydrogen atom. Obtain the radial equation and make a harmonic oscillator approximation around the minimum of the equivalent potential appearing in the *radial* equation. Treat this potential as a one-dimensional s.h.o. and find the ground- and first-excited-state energy of the s.h.o. Compare with the exact solution of the Coulomb problem.

PROBLEM 7

A diatomic molecule is restricted to rotate about the Z-axis. The Hamiltonian in the coordinate representation can be taken as

$$\hat{H} = -\frac{\hbar^2}{2I}\frac{d^2}{d\phi^2}$$

where I is the moment of inertia, and ϕ is the angle around the Z-axis. The wavefunctions of the stationary states are

$$\psi_m(\phi) = \frac{1}{(2\pi)^{1/2}} e^{im\phi}$$

with $m = 0, \pm 1, \pm 2, \ldots$.

(*a*) Prove that $\psi_m(\phi)$ describe stationary states. Construct the energy matrix in a representation labeled by the stationary states.

(*b*) Assume that the rotator has a dipole moment p along the direction joining the two atoms. An electric field $\mathscr{E}$ is established along the X-axis, giving rise to a perturbation

$$H_1 = -p\mathscr{E}\cos\phi = -\tfrac{1}{2}p\mathscr{E}(e^{i\phi} + e^{-i\phi})$$

Construct the perturbation matrix.

(*c*) Use perturbation theory (through second order) to find the energy of the stationary states in the presence of the perturbation.

(*d*) Make a graph of the energy of the $m = 0, +1, -1$ states as a function of the field strength.

PROBLEM 8

Consider a particle of mass m moving in *one* dimension in the potential

$$U(x) = \lambda x^4 - kx^2$$

as shown in the figure. λ and k are positive, and $\lambda \ll (k^{3/2}m^{1/2})/4\hbar$.

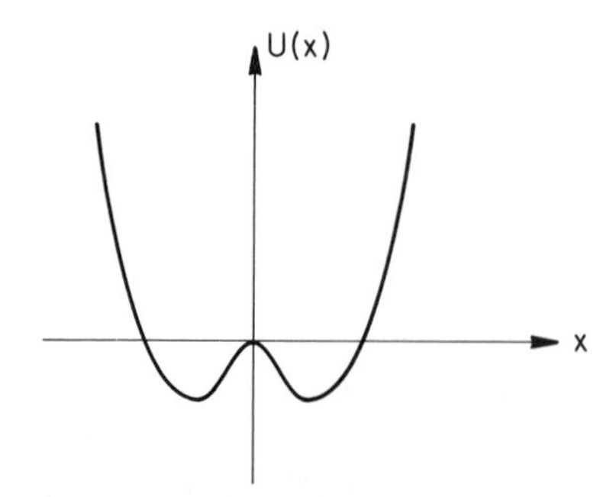

Sketch of the potential.

(*a*) Approximate the potential near the minima by a s.h.o. and find the energy of the states of lowest energy.

(*b*) Sketch the wavefunction of the state $|\psi_R\rangle$, which is defined as the state when the particle is found most of the time at $x > 0$, and of the state $|\psi_L\rangle$ when the particle is found most of the time at $x < 0$.

(*c*) Since the potential is invariant under reflection about the origin the stationary states must be eigenstates of the parity operator. Express the ground-state wavefunction in terms of $|\psi_R\rangle$ and $|\psi_L\rangle$.

PROBLEM 9

A particle moves in three dimensions in an anisotropic oscillator potential

$$V(x, y, z) = \tfrac{1}{2}m\omega^2(x^2 + 4y^2 + 9z^2)$$

(a) What is the general expression for the energy eigenvalues?
(b) What are the associated eigenfunctions (wavefunctions)? (They need not be normalized.)
(c) What are the degeneracies of the three lowest eigenvalues?

PROBLEM 10

A two-dimensional isotropic oscillator is subjected to a time-independent perturbation $\hat{H}'$ whose matrix elements vanish between two states that have the same parity in either x or y (for example, $\hat{H}' = xy$).

(a) What is the degeneracy of the unperturbed state with eigenvalue $E^{(0)} = 3\hbar\omega$?
(b) List in bra–ket notation the matrix elements of $\hat{H}'$ between eigenfunctions belonging to this value which do not vanish from symmetry.
(c) What is the first-order change in the energy level in terms of these matrix elements?

PROBLEM 11

A s.h.o. is subject to a *constant* force F so that the perturbation $\hat{H}'$ is given by

$$\hat{H}' = -Fx$$

(a) Use perturbation theory to find the energy of the nth state of the s.h.o.
(b) By a transformation of coordinates the Hamiltonian can be brought to an exactly soluble form. Show that the exact solution agrees with the perturbative result obtained in part (a) above.

PROBLEM 12

Consider a lead nucleus that captures a μ^- meson. Find the binding energy of this muonic ion using the *Coulomb wavefunctions*. What is the mean value of $\langle r \rangle$ for this state? Note that the lead nucleus can be assumed to be a uniformly, positively charged sphere of radius $r_{Pb} = A^{1/3} r_0$; $r_0 = 1.2$ F and $A = 208$.

Now consider the form of the potential *inside* the nucleus (where you will find that it is a three-dimensional oscillator potential). Find the energy of the ground state of the (lead nucleus $-\mu^-$) ion, assuming the three-dimensional s.h.o. potential. Compare this with the result from above and discuss it. (See Cohen-Tannoudji, Diu, Laloe, *Quantum Mechanics*, Wiley, N.Y. Vol. I, p. 525.)

PROBLEM 13

Compare the energy levels of an infinite spherical-potential well $U = 0, r < a$; $U = \infty, r > a$, with those for a particle confined in a cubical box of side $2L = 2a$.

Discuss equivalent levels and the observed degeneracy.

PROBLEM 14

Consider an electron bound in a three-dimensional s.h.o. potential in the $n = 1$ state. Obtain the corresponding eigenstates of the total angular momentum

$$\hat{\mathbf{J}} = \hat{\mathbf{L}} + \hat{\mathbf{S}}$$

and let the spin–orbit splitting be given by $A \ll \hbar\omega$. A uniform magnetic field B_0 is applied along the Z-axis. Find the position of the energy levels

(*a*) when B_0 is weak: $\mu B_0 \ll A$;
(*b*) when B_0 is strong: $\mu B_0 \gg A$.

PROBLEM 15

A one-dimensional square-well potential has a bound state for any positive $U_0 a^2$. The three-dimensional potential has a bound state only if $U_0 a^2 > \pi^2 \hbar^2 / 8\mu$. What is the analogous condition for a two-dimensional circularly symmetric square-well potential? What, if any, is the physical significance of these results?

PROBLEM 16

Consider the three-dimensional s.h.o. with Hamiltonian

$$\hat{H}_0 = \frac{\hat{\mathbf{p}}^2}{2\mu} + \tfrac{1}{2}\mu\omega_0^2 \mathbf{r}^2$$

Assume that the particle has a charge $e(>0)$ and is in a uniform magnetic field $\mathbf{B}$ along the Z-axis. Writing $\omega_L = -eB/2\mu$ and choosing

$$\mathbf{A}(r) = -\tfrac{1}{2}(\mathbf{r} \times \mathbf{B}) \qquad (\mathbf{A} \text{ is the vector potential})$$

one can write

$$\hat{H} = \hat{H}_0 + \hat{H}'(\omega_L)$$

where $\hat{H}'(\omega_L)$ is the sum of two terms—one which is linear in ω_L (paramagnetic), and the other which is quadratic in ω_L (diamagnetic). Show that the new stationary states of the system and their degrees of degeneracy can be determined exactly. How does the energy of the ground state vary as a function of ω_L? Is it an eigenvector of L^2? of $\hat{L}_z$? of $\hat{L}_x$?

Note: This is the Zeeman effect for the harmonic oscillator.

PROBLEM 17

By solving the transcedental equation [Eq. (9.54*c*)], find the *exact* value of α, the wave vector for the interior region of the deuteron, assuming a square-well potential and the binding energy given in Section 9.7. Plot the radial function $u(r)$ and the wavefunction $R(r)$ for the deuteron. Compare your result to the Hulthén wavefunction.

PROBLEM 18

A particle moves in one dimension in a potential $U(x)$ given by

$$U(x) = \begin{cases} \infty, & \text{for } x < 0 \\ -\gamma\delta(x-a), & \text{for } x < 0; \qquad a, \gamma > 0 \end{cases}$$

What is the minimum value of γa for which a bound state exists?

PROBLEM 19

A neutron ($mc^2 = 1$ GeV) is bound in (a) a square-well potential of width 1 F and depth $U_0 = -0.5$ GeV. Find the binding energy of the lowest state and of the first excited $l = 0$ state. (b) A s.h.o. potential that has strength $k = 0.25$ GeV/F^2. Find the energy of the ground state and of the second excited state and compare it with the result of part (a) above.

PROBLEM 20

Use perturbation theory to calculate the cross section for the photodisintegration of the deuteron

$$\gamma + d \rightarrow n + p$$

Note: This is a fairly long problem. You may wish to consult the treatment in Bethe and Morrison, *Elementary Nuclear Theory*, Wiley, New York, (1956); or a comparable text.

PROBLEM 21

The magnetic moment of a nucleus with total angular momentum **I** is defined as

$$\boldsymbol{\mu} = \frac{e\hbar}{2M_p} g\left(\frac{\mathbf{I}}{\hbar}\right)$$

where M_p is the mass of the proton and

$$\frac{e\hbar}{2M_p} = \mu_N$$

is the *nuclear magneton.* The number g is called the gyromagnetic ratio. The numerical value of the magnetic moment is its magnitude when I_z has its maximum value, $I_z = I$.

The deuteron consists of a proton and a neutron that have intrinsic (spin) magnetic moments

$$\mu_p = 2.7927\,\mu_N$$
$$\mu_n = -1.9131\mu_N$$

whereas for the deuteron one measures

$$\mu_d = 0.8574\mu_N$$

The spin of the proton and neutron is 1/2, and the deuteron has total angular momentum $I = 1$. It holds that

$$\mathbf{I} = \mathbf{L} + \mathbf{S}$$

(*a*) Find the magnetic moment of the deuteron when the neutron and proton are in the

$$^3S_1((S = 1, L = 0, I = 1) \text{ state}$$
$$^3D_1(S = 1, L = 2, I = 1) \text{ state}$$

Notes: (i) Since the neutron has no charge it cannot contribute to the magnetic moment through orbital motion. (ii) The orbital angular momentum of the proton equals one-half of the orbital angular momentum of the state. (iii) You should start with

$$\hat{\boldsymbol{\mu}}_d = \mu_n \hat{\boldsymbol{\sigma}}_n + \mu_p \hat{\boldsymbol{\sigma}}_p + \mu_N \frac{1}{\hbar} \hat{\mathbf{L}}_p$$

Does this make sense to you? Explain.

(*b*) From the experimental values find the admixture of S and D states in the deuteron.

Chapter 10

SYSTEMS WITH IDENTICAL PARTICLES

In the previous chapters we examined several physical systems which, in general, consisted of two particles and could be treated as a single particle moving in a potential. In nature, however, we encounter systems that contain several particles, and the constituents of the system interact with one another so that we cannot treat each of them independently. Furthermore, if the system contains identical particles, these are indistinguishable, and the quantum state describing the system must reflect this fact. The observable consequences of this indistinguishability are striking and have already been briefly alluded to in Section 3.6.

We begin this chapter by showing that under the exchange of any two identical particles the wavefunction must remain invariant if the particles are bosons and must change sign if the particles are fermions. This follows from both their indistinguishability and the uncertainty principle of quantum mechanics. In Section 10.2 we discuss the structure of atoms and how it depends on the antisymmetry of the many-electron wavefunction. We show that this suffices to correctly reproduce the Periodic Table.

In Section 10.3 we examine the structure of homonuclear molecules, especially as it is revealed by Raman scattering. Such molecules have a complex spectrum of energy levels and are well suited for testing the symmetry properties of the overall wavefunction of a system. Next we treat the simple example of the helium atom which is of interest because it contains only two electrons. We obtain the energy levels by a first-order perturbation calculation and, in particular, show the importance of the exchange contribution to the energy.

Section 10.5 is devoted to another simple system: the hydrogen molecule, H_2. Again, because of the exchange energy, the system is bound. In this case we must consider the mutual electrostatic interaction between all four constituents, that is, the two protons and two electrons. Reasonably good results can be obtained by simple methods such as the one first introduced by Heitler and London which we will describe. We do not perform the integrals in detail but give the final results and discuss their accuracy. We also consider the hydrogen molecule as a simple two-state system and show how this leads immediately to the general features, but not the details, of the solution.

In the final section we consider a system of identical bosons and discuss in particular an assembly of photons enclosed in a cavity. In this case the photons do not interact with one another, but they do interact with the cavity walls; thus, they can be treated as being in statistical equilibrium. We then show how the exchange symmetry for Bose particles leads directly to the Planck radiation law.

The applications that we choose to discuss in this chapter refer to important and often encountered physical systems. Furthermore, they demonstrate explicitly how the exchange symmetry must be satisfied whenever a system contains identical particles. The existence in nature of these two types of particles—bosons and fermions—is one of the most fundamental aspects of the physical world. It is, however, a purely quantum-mechanical phenomenon and has no analogue in classical physics.

10.1. INDISTINGUISHABILITY AND EXCHANGE SYMMETRY

We have already introduced the concept of exchange symmetry in Section 3.6. Here we will elaborate on these arguments from a slightly more formal point of view. Consider a system of two identical particles, 1 and 2, governed by the Hamiltonian $\hat{H}_{12}$. We assume that the system has a stationary state $|\mathrm{I}\rangle$, which can be characterized by labeling particle 1 by the index α and particle 2 by the index β. The indices α and β could, for instance, refer to the position of the particles, to their spin projection, to some other quantum number, or to a combination of these. We write

$$|\mathrm{I}\rangle = |1_\alpha 2_\beta\rangle \tag{10.1a}$$

The states of the two-particle system, e.g., the state $|\mathrm{I}\rangle$, are defined in a Hilbert space $\mathcal{H}_{1,2}$ of higher dimensionality than the Hilbert spaces $\mathcal{H}_1$ and $\mathcal{H}_2$ which span the states of the single particles 1 and 2, respectively. Mathematically we say that the space $\mathcal{H}_{1,2}$ is the *direct product* of the spaces $\mathcal{H}_1$, $\mathcal{H}_2$ and write $\mathcal{H}_{1,2} = \mathcal{H}_1 \otimes \mathcal{H}_2$.

If the particles 1 and 2 are identical, then clearly the state

$$|\mathrm{II}\rangle = |2_\alpha 1_\beta\rangle \tag{10.1b}$$

is also a stationary state of the system with the same energy as the state $|\mathrm{I}\rangle$. In fact, the state $|\mathrm{II}\rangle$ is physically *indistinguishable* from the state $|\mathrm{I}\rangle$. This is because in quantum mechanics we cannot follow the evolution of individual particles and thus cannot tell which one of two identical particles is particle 1 and which is particle 2. That the states $|\mathrm{I}\rangle$ and $|\mathrm{II}\rangle$ cannot be distinguished physically is therefore a direct and fundamental consequence of quantum mechanics. It does *not* imply that the states $|\mathrm{I}\rangle$ and $|\mathrm{II}\rangle$ are identical; in fact, they are different states of the system.

At this point it is convenient to introduce a two-particle permutation operator† $\hat{P}_{12}$, which has the property of exchanging particles 1 and 2. For instance, we immediately see that

$$\hat{P}_{12}|\mathrm{I}\rangle = |\mathrm{II}\rangle \quad \text{and} \quad \hat{P}_{12}|\mathrm{II}\rangle = |\mathrm{I}\rangle \tag{10.2a}$$

Furthermore, if we are considering indistinguishable particles, $\hat{P}_{12}$ must leave the Hamiltonian of the system invariant because otherwise one would be able to distinguish between the two identical particles. Thus, the stationary states of the system will be simultaneous eigenstates of both the Hamiltonian and $\hat{P}_{12}$. Let us look at the eigenstates of $\hat{P}_{12}$ alone. If $|\psi\rangle$ is an eigenstate of $\hat{P}_{12}$, then

$$\begin{aligned} \hat{P}_{12}|\psi\rangle &= \eta\,|\psi\rangle \\ \hat{P}_{12}\hat{P}_{12}|\psi\rangle &= \eta\hat{P}_{12}|\psi\rangle = \eta^2\,|\psi\rangle \end{aligned} \tag{10.2b}$$

On the other hand, operating with $\hat{P}_{12}$ twice is equivalent to coming back to the original state itself. Thus

$$\hat{P}_{12}\hat{P}_{12}|\psi\rangle = |\psi\rangle = \eta^2\,|\psi\rangle$$

† Not to be confused with the parity operator introduced in Chapter 8. When the possibility of confusion arises we will use $\hat{\pi}$ for the parity operator.

Hence

$$\eta^2 = 1 \quad \text{or} \quad \eta = \pm 1$$

Consequently, a physical state of two identical particles must be either symmetric or antisymmetric under their exchange. Particles that have even symmetry are called *bosons*; particles that have odd symmetry are called *fermions*.

$$\boxed{\begin{aligned} &\hat{P}_{12}|\psi\rangle = |\psi\rangle && \text{bosons} \\ &\hat{P}_{12}|\psi\rangle = -|\psi\rangle && \text{fermions} \end{aligned}} \tag{10.2c}$$

The permutation operator $\hat{P}_{12}$ is a constant of the motion since it commutes with the Hamiltonian. Thus, if a quantum system is in a symmetric (antisymmetric) state under exchange of identical particles, it will remain in a symmetric (antisymmetric) state. It is for this reason that Eqs. (10.2*c*) can be interpreted as a general classification of the two types of particles encountered in nature. If we are given an arbitrary state $|n\rangle$ of the system that does not obey the symmetry of Eqs. (10.2*c*) it cannot be used to describe the system. However, we can easily construct from the state $|n\rangle$ a suitable symmetric (antisymmetric) state in the following way

$$|12; n\rangle_s = |12; n\rangle + \hat{P}_{12}|12; n\rangle \tag{10.3a}$$

$$|12; n\rangle_a = |12; n\rangle - \hat{P}_{12}|12; n\rangle \tag{10.3b}$$

The states of Eqs. (10.3) have the appropriate symmetry and can be used to describe a system of two identical bosons (fermions).

The arguments that led to Eqs. (10.3) can be generalized to systems with an arbitrary number of identical particles. Consider first the case of three identical particles 1, 2, and 3 and a state

$$|\mathrm{I}\rangle = |1_\alpha 2_\beta 3_\gamma\rangle \tag{10.4a}$$

We can create indistinguishable states by permuting the identical particles such as

$$|\mathrm{II}\rangle = |2_\alpha 1_\beta 3_\gamma\rangle = \hat{P}_{213}|1_\alpha 2_\beta 3_\gamma\rangle \tag{10.4b}$$

$$|\mathrm{IV}\rangle = |3_\alpha 1_\beta 2_\gamma\rangle = \hat{P}_{312}|1_\alpha 2_\beta 3_\gamma\rangle \tag{10.4c}$$

In all, there are six permutation operators (including the identity $\hat{P}_{123}$) for this three-particle system. For a system of N particles there is a total of $N!$ possible permutations. It is important to recognize that any permutation is equivalent to a finite number of *transpositions*, namely, the exchange of any *two* particles. For instance the permutation of Eq. (10.4*b*) was reached from that indicated by Eq. (10.4*a*) through the single transposition of particles $1 \leftrightarrow 2$. The permutation of Eq. (10.4*c*) was reached from Eq. (10.4*a*) by the two transpositions $1 \leftrightarrow 2$ followed by $2 \leftrightarrow 3$. It could also have been reached by first transposing $1 \leftrightarrow 3$, followed by $2 \leftrightarrow 1$, etc.† Obviously the order in which the transpositions are carried out is important, but the *number* of transpositions that can be used to reach a particular permutation is always either even or odd. For instance, for three particles

$$P_{123}, P_{312}, P_{231} \quad \text{are even permutations}$$

$$P_{213}, P_{321}, P_{132} \quad \text{are odd permutations}$$

We can now define the symmetry properties under particle exchange for a system of N identical particles as follows: we say that the state is *completely* symmetric if it remains unchanged under any one of the $N!$ permutations. The state is *completely* antisymmetric if it changes sign under any one of the $N!/2$ odd permutations, but remains unchanged under the $N!/2$ even permutations. For instance, for a three-particle system using the notation of Eqs. (10.4*a*)–(10.4*c*) we can write explicitly

$$\begin{aligned} |\alpha\beta\gamma\rangle_s = &|1_\alpha 2_\beta 3_\gamma\rangle + |2_\alpha 1_\beta 3_\gamma\rangle + |3_\alpha 1_\beta 2_\gamma\rangle \\ &+ |3_\alpha 2_\beta 1_\gamma\rangle + |2_\alpha 3_\beta 1_\gamma\rangle + |1_\alpha 3_\beta 2_\gamma\rangle \end{aligned} \tag{10.5a}$$

$$\begin{aligned} |\alpha\beta\gamma\rangle_a = &|1_\alpha 2_\beta 3_\gamma\rangle - |2_\alpha 1_\beta 3_\gamma\rangle + |3_\alpha 1_\beta 2_\gamma\rangle \\ &- |3_\alpha 2_\beta 1_\gamma\rangle + |2_\alpha 3_\beta 1_\gamma\rangle - |1_\alpha 3_\beta 2_\gamma\rangle \end{aligned} \tag{10.5b}$$

for the completely symmetric and antisymmetric states of three identical particles‡. The above linear combinations are the natural extension of the two-particle states given by Eqs. (10.3*a*) and (10.3*b*).

---

† Here the action of the permutation operator is always implied to be on the ordered sequence 1, 2, 3. That is, $\hat{P}_{321}$ indicates to place the last entry into first place and the first entry into last place while leaving the middle entry unchanged.

‡ A system of particles can have partial symmetry under exchange, but this case does not appear in nature and, therefore, we will not be concerned with it further.

The indistinguishability of identical particles led us, in the case of the two-particle system, to Eqs. (10.2a)–(10.2c) according to which the state must be symmetric or antisymmetric under exchange. For the same reasons:

> The state vector describing a physical system of N identical particles must be completely symmetric or completely antisymmetric under particle exchange. If the particles are bosons the state vector must be completely symmetric, and if they are fermions it must be completely antisymmetric.

As we have pointed out in Section 3.6, particles that have zero or integer spin are bosons, whereas particles with half-integer spin are fermions. This connection, known as the *spin-statistics theorem*, has no exceptions whatsoever, but its origin is not well understood as yet. The exchange symmetry properties of the state vector are, of course, reflected in the corresponding amplitudes; this is discussed below.

Many-Particle Wavefunctions

Often it is possible to construct the many-particle state from the spectrum of single-particle states. This is true when the particles do not interact among themselves. In that case the total Hamiltonian for the system is a sum of single-particle Hamiltonians, each acting in the different Hilbert subspace of the corresponding particle

$$\hat{H}_{\mathrm{T}} = \hat{H}_1 + \hat{H}_2 + \cdots + \hat{H}_N \tag{10.6a}$$

If the particles $1, 2, \ldots, N$, are identical, then the single-particle Hamiltonians $\hat{H}_i$ must all have the same form, since $\hat{H}_{\mathrm{T}}$ must be invariant under particle exchange. We designate the stationary states of the single-particle Hamiltonians $\hat{H}_i$ by $|\alpha\rangle$, $|\beta\rangle$, $|\gamma\rangle, \ldots,$ etc. It is then clear that direct products of the form†

$$|\alpha\rangle_1 \otimes |\beta\rangle_2 \otimes |\alpha\rangle_3 \otimes \cdots \otimes |\gamma\rangle_N \tag{10.6b}$$

† Hereafter we will drop the direct product sign, the simpler expression

$$|\alpha\rangle_1 \, |\beta\rangle_2 \, |\alpha\rangle_3 \cdots |\gamma\rangle_N$$

implying a direct product.

are eigenstates of $\hat{H}_T$ with eigenvalue

$$E_T = E_\alpha + E_\beta + E_\alpha + \cdots + E_\gamma \tag{10.6c}$$

Even though the product state of Eq. (10.6*b*) is a stationary state of $\hat{H}_T$, it cannot be used to describe the physical system because it does not satisfy the symmetry conditions under particle exchange. To construct a suitable state we must symmetrize (antisymmetrize) the product state when the identical particles are bosons (fermions). This is achieved by acting on Eq. (10.6*b*) with the $N!$ permutation operators and forming the linear combination of all the resulting states. For bosons we add all the states, whereas for fermions we include a factor of $(-1)^p$, where p is the number of transpositions used to reach that particular permutation.

To better understand this procedure let us return to a system of two identical noninteracting particles which we label 1 and 2. As before, let state $|\mathrm{I}\rangle$ of the system be given by

$$|\mathrm{I}\rangle = |\alpha\rangle_1 \, |\beta\rangle_2$$

with α and β referring to different single-particle states. We can express the wavefunction for this state by using the basis states $|x_1\rangle$ and $|x_2\rangle$, which label the coordinate representation for these particles. These basis states span the single-particle Hilbert space $\mathscr{H}_1$ and $\mathscr{H}_2$ so that†

$$\begin{aligned} \psi_{\mathrm{I}}(x_1, x_2) &= \langle x_1, x_2 \,|\mathrm{I}\rangle = \langle x_1 \mid \alpha\rangle\langle x_2 \mid \beta\rangle \\ &= \psi_\alpha(x_1)\psi_\beta(x_2) \end{aligned} \tag{10.7a}$$

Of course, as written, $\psi_{\mathrm{I}}(x_1, x_2)$ does not satisfy exchange symmetry, since $\alpha \neq \beta$. To achieve this we must form a linear combination of $\psi_{\mathrm{I}}(x_1 x_2)$ and $\psi_{\mathrm{II}}(x_1, x_2)$, and obtain

$$\psi_{\alpha\beta}(x_1, x_2) = \psi_\alpha(x_1)\psi_\beta(x_2) \pm \psi_\alpha(x_2)\psi_\beta(x_1) \tag{10.7b}$$

For bosons we use the (+) sign; for fermions we use the (−) sign.

Equation (10.7*b*) is not properly normalized. If the single-particle states are properly normalized to unit probability, then as long as $\alpha \neq \beta$ it is clear that the normalization constant is $1/2^{1/2}$. However, we must exercise caution when $\alpha = \beta$. If the particles are bosons,

† This is the notation we used in Section 3.6.

Eq. (10.7*b*) reduces to

$$\psi_{\alpha\alpha}(x_1, x_2) = 2\psi_\alpha(x_1)\psi_\alpha(x_2) \quad \text{(bosons)}$$

Therefore, the normalization constant must be chosen to be 1/2. If the particles are fermions, then clearly

$$\psi_{\alpha\alpha}(x_1, x_2) = 0 \qquad \text{(fermions)}$$

That is, the wavefunction for two identical fermions in the same single-particle state vanishes identically. We recognize this as the *Pauli exclusion principle* according to which in any given system no two electrons (more generally, *identical* fermions) can be simultaneously in the same quantum state.

The symmetrization and antisymmetrization of products of single-particle states can be formally expressed for N identical particles as follows

$$|\alpha, \beta, \ldots, \mu\rangle_s = \left(\frac{N_\lambda!\,N_\nu!\cdots}{N!}\right)^{1/2} \sum_P \hat{P}(|\alpha\rangle_1|\beta\rangle_2\cdots|\mu\rangle_N) \qquad \text{(bosons)} \quad (10.8a)$$

$$|\alpha, \beta, \ldots, \mu\rangle_a = \frac{1}{(N!)^{1/2}} \sum_P (-1)^p \hat{P}(|\alpha\rangle_1|\beta\rangle_2\cdots|\mu\rangle_N) \qquad \text{(fermions)} \quad (10.8b)$$

Here $\hat{P}$ stands for the $N!$ permutation operators, and all of them must be included in the sum. The parity of the permutation is designated by p. If any of the states $|\alpha\rangle$, $|\beta\rangle$, etc., are identical, Eq. (10.8*b*) vanishes. On the other hand, in Eq. (10.8*a*) $N_\lambda, N_\nu, \ldots$, etc., represent the number of times the identical states $|\lambda\rangle, |\nu\rangle, \ldots$, etc., appear in the product. In that case permutations that lead to the same product state are to be excluded, and the normalization has been adjusted accordingly.

Whereas Eq. (10.8*b*) correctly expresses the antisymmetrization procedure, it is much more convenient to visualize it by constructing the *Slater determinant*. As an example, consider a system of three identical fermions that occupy the single-particle states α, β, and γ. The completely antisymmetric wavefunction can be obtained from products of single-particle wavefunctions by evaluating the determinant

$$\Psi_{\alpha\beta\gamma}(x_1x_2x_3)_a = \frac{1}{(3!)^{1/2}} \det \begin{vmatrix} \psi_\alpha(x_1) & \psi_\alpha(x_2) & \psi_\alpha(x_3) \\ \psi_\beta(x_1) & \psi_\beta(x_2) & \psi_\beta(x_3) \\ \psi_\gamma(x_1) & \psi_\gamma(x_2) & \psi_\gamma(x_3) \end{vmatrix} \tag{10.8c}$$

We see that the resulting wavefunction corresponds to the completely antisymmetric state described by Eq. (10.5*b*). It is also clear that if any two of the states α, β, or γ are identical, two rows of the matrix will be identical, and thus the determinant will vanish.

The use of products of single-particle wavefunctions to construct the many-particle wavefunction is strictly correct only when the particles are noninteracting. However, as long as the interaction is weak the product state provides an excellent approximation to the real many-particle state and is particularly useful as a starting point in perturbative or variational calculations.

As we have already mentioned, the effects of exchange symmetry manifest themselves only in the quantum domain, namely, when the product of momentum and distance typical of the system are of the order of $\hbar$. This can be illustrated by considering two identical noninteracting particles localized around the positions $x = a$ and $x = b$. The probability density of finding the two-particle system in this configuration is given by the square of the wavefunction

$$\begin{aligned} |\psi(x_1, x_2)|^2 &= \left| \frac{1}{2^{1/2}} [\psi(a; x_1)\psi(b; x_2) \pm \psi(a; x_2)\psi(b; x_1)] \right|^2 \\ &= \tfrac{1}{2}\{|\psi(a; x_1)|^2 \, |\psi(b; x_2)|^2 + |\psi(a; x_2)|^2 x_2)|^2 \, |(b; x_1)|^2 \\ &\quad \pm 2\,\mathrm{Re}[\psi^*(a; x_1)\psi(b; x_1)\psi^*(b; x_2)\psi(a; x_2)]\} \end{aligned}$$

Here $\psi(a; x)$ is a wavefunction describing a particle localized around $x = a$, and $\psi(b; x)$ describes a particle localized at $x = b$. If the separation between a and b is large, then the two wavefunctions have *little overlap.* As a consequence, the product $\psi^*(a; x_1)\psi(b; x_1)$ is vanishingly small for all values of x_1. Similarly, $\psi^*(b; x_2)\psi(a; x_2)$ is vanishingly small for all x_2, so that the interference term vanishes. If $(a - x_1)$ and $(b - x_2)$ are small, the term $|\psi(a; x_1)|^2 \, |\psi(b; x_2)|^2$ is significant and if $(a - x_2)$ and $(b - x_1)$ are small, the second term dominates irrespective of the separation between a and b. We see that the joint probability of finding one particle at $x = a$ and the other at $x = b$ is given by the product of the probabilities of finding particle 1 at $x = a$, particle 2 at $x = b$, *and* of finding particle 2 at $x = a$, particle 1 at $x = b$. With the inclusion of the normalization factor of $\frac{1}{2}$ this result agrees with the classical definition of the joint probability, and there is no conflict with our intuition. However, we can never know whether particle 1 is at $x = a$ and 2 at $x = b$, or vice versa.

When the separation of the two particles is in the quantum domain the interference (or exchange) term becomes important. As we will soon see it is this term that gives rise to the binding of molecules.

10.2. THE PERIODIC TABLE

It is well known that only a finite number of chemical elements are found in nature. These elements exhibit some common chemical and physical properties and can be accordingly classified in the Periodic Table. We now know that every element corresponds to a different atomic species and that atoms consist of a positively charged nucleus of charge Ze to which Z electrons are bound. Z is called the atomic number and determines the chemical properties of the atom. The atomic mass number is designated by $A = Z + N$, where N is the number of neutrons in the nucleus and Z is the number of protons in the nucleus. Thus, atoms are, in general, electrically neutral. However, we do also encounter positive (or negative) ions which are atoms with one or more electrons removed from (or added to) the neutral atom configuration.

To understand the structure of atoms we assume that all electrons move in the attractive potential of the nucleus, and we ignore their mutual electrostatic repulsion. In such a model the total Hamiltonian is a sum of single-particle Hamiltonians and the single-particle states will be described by hydrogenlike wavefunctions where the proton's charge is replaced by Ze. This is an extremely crude model and fails badly for atoms with large Z. It does, however, have a correct qualitative behavior, and we can use it to build up the atomic structure of the lighter elements.

The ground state of the atom (in this model) corresponds to the configuration where all single-particle states are in the lowest possible state. However, since electrons are fermions, the overall wavefunctions must be antisymmetric. Therefore, no two electrons can occupy the same single-particle state. We label the single-particle states by the quantum numbers of the hydrogen atom (see Sections 8.3 and 8.4)

$n = 1, 2, \ldots$	principal quantum number
$l = 0, \ldots, n-1$	orbital angular momentum
$m = -l, -(l-1), \ldots, 0, \ldots, (l-1), l$	projection of the angular momentum onto the Z-axis
$m_s = -1/2, +1/2$	projection of the spin onto the Z-axis

The ground state of hydrogen will correspond to

$$Z = 1, \quad n = 1, \quad l = 0, \quad m = 0, \quad m_s = -1/2, \quad \text{or} \quad +1/2$$

The next atom, helium, has $Z = 2$ and thus has two electrons, which will occupy the states

$$Z = 2\begin{cases} n = 1, & l = 0, \quad m = 0, \quad m_s = -1/2 \\ n = 1, & l = 0, \quad m = 0, \quad m_s = +1/2 \end{cases} \qquad \text{(helium)}$$

Next comes lithium with $Z = 3$. It is clear that the third electron cannot be accommodated in the $n = 1$ shell. Thus, the single-particle wavefunctions for lithium will be those of helium with the addition of an electron in the $n = 2$ state

$$Z = 3\begin{cases} n = 1, & l = 0, \quad m = 0, \quad m_s = -1/2 \\ n = 1, & l = 0, \quad m = 0, \quad m_s = +1/2 \\ n = 2, & l = 0, \quad m = 0, \quad m_s = -1/2, \quad \text{or} \quad +1/2 \end{cases} \qquad \text{(lithium)}$$

It is only when we get to boron ($Z = 5$) that we obtain a single-particle state with $l = 1$. The configuration is

$$Z = 5\begin{cases} n = 1, & l = 0, \quad m = 0, \quad m_s = -1/2 \\ n = 1, & l = 0, \quad m = 0, \quad m_s = +1/2 \\ n = 2, & l = 0, \quad m = 0, \quad m_s = -1/2 \\ n = 2, & l = 0, \quad m = 0, \quad m_s = +1/2 \\ n = 2, & l = 1, \quad m = -1, 0, \text{ or } +1, \\ & \qquad m_s = -1/2, \quad \text{or} \quad +1/2 \end{cases} \qquad \text{(boron)}$$

We see that as the electrons arrange themselves into the lowest energy state compatible with the exclusion principle, they progressively fill shells. For instance, the $n = 1$ shell contains two electrons. The $n = 2$ shell contains the $l = 0$ subshell with two electrons and the $l = 1$ subshell with six electrons, for a total of eight electrons. In this way we can construct the *shell structure* of atoms, and we can safely assume that whenever a shell or a subshell is closed, the atomic configuration is relatively stable. Closed shells occur for the noble gases

He, Ne, Ar, Kr, and Xe

which are indeed very stable chemically. On the other hand, if an atom has one electron outside a closed shell, the electron can be easily removed and the atoms are very active chemically. This is the case for the alkalis

Li, Na, K, Rb, and Cs

The occurrence of closed shells can also be inferred from the ionization potential of the various atoms, as shown in Fig. 10.1. The ionization potential is the energy that must be supplied to the neutral atom in order to release one electron. The location of the closed shells and subshells can be clearly seen from the graph.

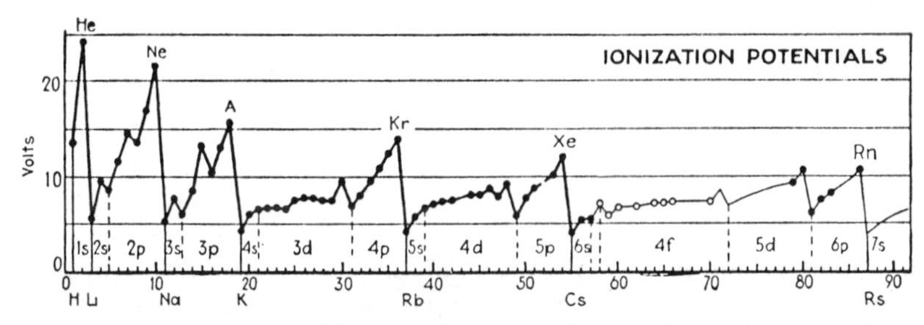

FIGURE 10.1. The potential required to ionize the neutral atom for each of the elements. Note the appearance of sharp discontinuities at closed shells. [From H. E. White, *Introduction to Atomic Spectra*, McGraw-Hill, New York (1934); by permission.]

In building up the first few elements we assumed that the $n = 2$ and $l = 0$ states are filled before the $n = 2$, $l = 1$ states. This in turn implies that the energy of the $l = 0$ state is lower than that of the

$l=1$ state, even though for hydrogenlike systems we have found that these two states are degenerate. The reason that multielectron atoms differ from hydrogen in this respect is that even in the light atoms the potential of the nucleus is modified by the presence of the other electrons. This modification can be thought of as a *screening* of the positive charge of the nucleus due to the negative charge of the electrons. Thus, the potential seen by the individual electrons is no longer of the Coulomb type, and the degeneracy between states of the same n-value but different angular momentum is lifted. As we know, for S-states the probability of finding the electron at the origin is finite, and thus the electron will, in general, be subject to a stronger potential than when it is in states with higher angular momentum. Consequently, S-states are more strongly bound and, therefore, have lower energy than the corresponding P-states, and so on. We took this into account when we listed the quantum numbers of the electrons in the ground state of lithium and boron.

At this point we need a more concise notation for describing the electronic structure of atoms. We do this in part by using the spectroscopic notation introduced in Section 8.4. For instance, for boron the electron configuration is

$$(1s)^2(2s)^2 2p \qquad \text{(boron)}$$

implying that two electrons are in the $n=1$, $l=0$ state, two more electrons in the $n=2$, $l=0$ state, and one electron in the $n=2$, $l=1$ state. We can also give the spectroscopic *term* to describe the state of the atom. This refers to the state of the electrons outside closed shells or subshells. For instance, for boron the term is

$$^{2S+1}L_J \rightarrow {}^2P_{1/2} \qquad \text{(boron)}$$

indicating that the total spin is $S=1/2$, the orbital angular momentum is $L=1$, and the total angular momentum is $J=1/2$. In Table 10.1 we give the electron configuration and spectroscopic term for the first 36 elements.

From a study of the table we see that for the first 18 elements the shells are filled exactly as predicted. For potassium, however, we would expect that after filling the configuration of argon the next electron would be in the $3d(n=3, l=2)$ state. Instead, the electron occupies the $n=4$, $l=0$ state; therefore, this state must lie lower in energy than the $n=3$, $l=2$ state. This is indeed possible because

Table 10.1. Electron Configurations of the First 36 Elements

Z	Element	W_I(eV)	Electron configuration											Spectro-scopic term
			$1s$	$2s$	$2p$	$3s$	$3p$	$3d$	$4s$	$4p$	$4d$	$4f$		
1	Hydrogen (H)	13.6	1											$^2S_{1/2}$
2	Helium (He)	24.6	2											$^1S_0$
3	Lithium (Li)	5.4	filled (2)†	1										$^2S_{1/2}$
4	Beryllium (Be)	9.3		2										$^1S_0$
5	Boron (B)	8.3		2	1									$^2P_{1/2}$
6	Carbon (C)	11.3		2	2									$^3P_0$
7	Nitrogen (N)	14.5		2	3									$^4S_{3/2}$
8	Oxygen (O)	13.6		2	4									$^3P_2$
9	Fluorine (F)	17.4		2	5									$^2P_{3/2}$
10	Neon (Ne)	21.6		2	6									$^1S_0$
11	Sodium (Na)	5.1	filled (2)	filled (8)		1								$^2S_{1/2}$
12	Magnesium (Mg)	7.6				2								$^1S_0$
13	Aluminum (Al)	6.0				2	1							$^3P_{1/2}$
14	Silicon (Si)	8.1				2	2							$^3P_0$
15	Phosphorus (P)	10.5				2	3							$^4S_{3/2}$
16	Sulfur (S)	10.4				2	4							$^3P_2$
17	Chlorine (Cl)	13.0				2	5							$^2P_{3/2}$
18	Argon (A)	15.8				2	6							$^1S_0$
19	Potassium (K)	4.3	filled (2)	filled (8)		(8)			1					$^2S_{1/2}$
20	Calcium (Ca)	6.1							2					$^1S_0$
21	Scandium (Sc)	6.5						1	2					$^2D_{3/2}$
22	Titanium (Ti)	6.8						2	2					$^3F_2$
23	Vanadium (V)	6.7						3	2					$^4F_{3/2}$
24	Chromium (Cr)	6.8						5	1					$^7S_3$
25	Manganese (Mn)	7.4						5	2					$^6S_{3/2}$
26	Iron (Fe)	7.9						6	2					$^5D_4$
27	Cobalt (Co)	7.9						7	2					$^4F_{9/2}$
28	Nickel (Ni)	7.6						8	2					$^3F_4$
29	Copper (Cu)	7.7						10	1					$^2S_{1/2}$
30	Zinc (Zn)	9.4						10	2					$^1S_0$
31	Gallium (Ga)	6.0	filled (2)	filled (8)		(18)			2	1				$^2P_{1/2}$
32	Germanium (Ge)	7.9							2	2				$^3P_0$
33	Arsenic (As)	9.8							2	3				$^4S_{3/2}$
34	Selenium (Se)	9.7							2	4				$^3P_2$
35	Bromine (Br)	11.8							2	5				$^2P_{3/2}$
36	Krypton (Kr)	14.0							2	6				$^1S_0$

† The numbers in parenthesis indicate the electrons in the corresponding closed shell.

the potential seen by the $n = 3$, $l = 2$ state is so highly screened that the state is more loosely bound than the penetrating $l = 0$ state corresponding to the next higher radial quantum number, $n = 4$. That this is indeed the case can be seen in Fig. 10.2, where the energy levels of sodium, magnesium, and aluminum are plotted. In all cases the $4s$ state lies lower than the $3d$ state.

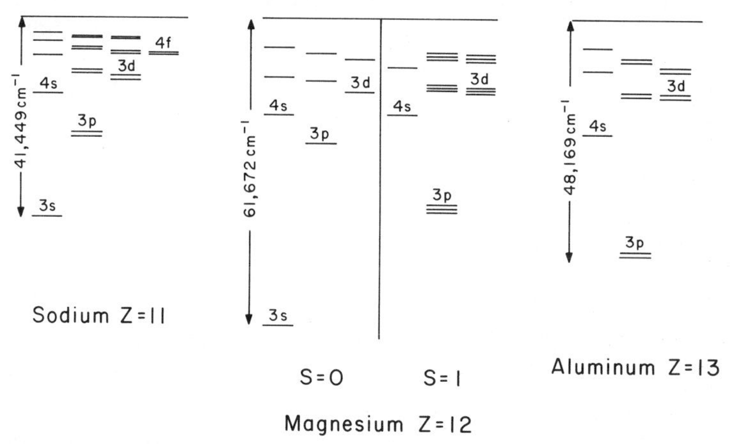

FIGURE 10.2. Atomic energy levels of sodium, magnesium, and aluminum. In all cases the $3P$ state lies lower than the $4S$ state.

From the energy level diagrams in Fig. 10.2 it is obvious that the Coulomb degeneracy is lifted even for the light atoms. The spin–orbit interaction, as for the case of hydrogen, gives rise to a fine structure. In systems with a single electron outside a closed shell, such as is the case for sodium and aluminum, the levels split into doublets, whereas in systems with two electrons outside a closed shell two types of spectra occur. When the total spin of the two electrons is composed into an $S = 0$ state, no fine structure is apparent. When they are composed into an $S = 1$ state the fine structure reveals a triplet of states. This situation will be discussed in more detail in connection with the helium atom in Section 10.4. Note that the fine structure indicated in Fig. 10.2 has been highly exaggerated in order to be made visible in the figure.

We thus see that the chemical properties of the elements follow from their atomic structure. In turn, the atomic structure observed in nature is a consequence of the antisymmetry of the atomic wavefunction under the exchange of identical particles, in this case electrons. This is a remarkable conclusion and leaves no doubt that one must use quantum mechanics in order to properly describe nature at the atomic level. Incidentally, the fact that no elements with $Z>92$ are naturally found has nothing to do with their electronic configuration. It is so only because nuclei with such large values of Z are not stable; such elements have been produced artificially.

10.3. RAMAN SPECTRA OF HOMONUCLEAR MOLECULES

In Section 9.3 we discussed the rotational and vibrational states of diatomic molecules and showed how they lead to fairly complex band spectra. We can gain further insight into molecular structure and even into properties of the nucleus by a study of the spectra of homonuclear molecules. In this case it is best to study the energy levels of the system not only from their emission spectrum but also by exciting the system with intense electromagnetic radiation and observing the shift in frequency of the scattered (really re-emitted) radiation. This process is known as the *Raman effect* and can be thought of as the inelastic scattering of photons from a quantum system that has a discrete energy spectrum.

A homonuclear diatomic molecule is a system containing two identical nuclei and several electrons. The nuclei can be either bosons or fermions depending on whether they have zero, integral spin, or half-integral spin. For instance O^{16} has spin-zero, and therefore the overall wavefunction of the O_2 molecule must be symmetric under the exchange of the two nuclei. Of course, it must be totally antisymmetric under the permutation of any electrons. On the contrary, the wavefunction for the H_2 molecule must be antisymmetric under the exchange of the two protons. These symmetry properties are clearly reflected in the spectrum of the rotational energy levels of the molecule.

The overall wavefunction can be written as a product of the wavefunctions for the different degrees of freedom

$$\psi_{\text{overall}} = \psi_{\text{electron}}\phi_{\text{vibration}}\rho_{\text{rotation}}\chi_{\text{nuclear spin}} \qquad (10.9a)$$

As an example, we consider the O_2 molecule, which contains primarily O^{16} nuclei. If $\hat{P}^N_{12}$ indicates the exchange of the two nuclei, it must hold that

$$\hat{P}^N_{12}\psi_{\text{overall}} = +\psi_{\text{overall}} \qquad (10.9b)$$

We can examine the action of $\hat{P}^N_{12}$ on each part of the wavefunction of Eq. (10.9*a*). With respect to the electron wavefunction, exchange of the nuclei corresponds to an exchange of the spatial coordinates of the electrons, but *not* of their spin wavefunction. Thus, ψ_{electron} can be either symmetric or antisymmetric under the exchange of the nuclei

$$\hat{P}^N_{12}\psi_{\text{electron}} = \pm\psi_{\text{electron}} \qquad (10.9c)$$

For the vibrational wavefunction it holds that

$$\hat{P}^N_{12}\phi_{\text{vibration}} = +\phi_{\text{vibration}}$$

since the vibrational motion depends only on the distance between the nuclei. Finally, since the O^{16} nuclei have spin-zero, $\chi_{\text{nuclear spin}}$ also is symmetric:

$$\hat{P}^N_{12}\chi_{\text{nuclear spin}} = +\chi_{\text{nuclear spin}}$$

Under these conditions Eq. (10.9*b*) can be satisfied only if the rotational wavefunction has the same symmetry as Eq. (10.9*c*)

$$\hat{P}^N_{12}\rho_{\text{rotation}} = \pm\rho_{\text{rotation}} \qquad (10.9d)$$

If we recall the model of the rigid rotator for a diatomic molecule (see Fig. 10.3) we note that permutation of the two nuclei is equivalent to an inversion of the coordinates, i.e., the parity operation $\hat{\pi}$. Thus, if

$$\rho = \rho(1_{\theta,\phi};\, 2_{\pi-\theta,\pi+\phi})$$

then

$$\hat{P}^N_{1,2}\rho = \hat{\pi}\rho = \rho(2_{\theta,\phi};\, 1_{\pi-\theta,\pi+\phi})$$

However,

$$\rho = Y_{l,m}(\theta, \phi)$$

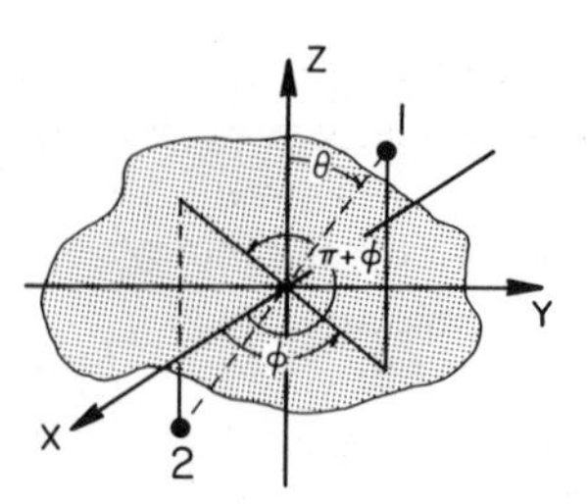

FIGURE 10.3. Coordinate system for a diatomic molecule and the effect of the parity operation.

and we know that the parity of the spherical harmonics is $(-1)^l$

$$\hat{\pi} Y_{l,m}(\theta, \phi) = (-1)^l Y_{l,m}(\theta, \phi)$$

Consequently, to satisfy Eq. (10.9*d*) the rotational levels of O^{16} must all have either positive parity, $l = 0, 2, 4, \ldots$, etc., or they must all have negative parity, $l = 1, 3, 5, \ldots$, etc., depending on the signature of Eq. (10.9*c*). For the O_2 molecule in its ground state† $\hat{P}^N_{12}\psi_{\text{electron}} = -\psi_{\text{electron}}$, and accordingly only rotational states with odd l exist.

As mentioned in Section 9.3, homonuclear molecules do not have an electric dipole moment, and therefore the rotational lines cannot be observed in the emission spectrum. However, under the influence of intense radiation an electric dipole moment can be induced and the molecule can absorb and reemit the incident radiation. This is known as the Raman effect. If the incident photon has energy $\hbar\omega$ and the scattering results in the excitation of the molecule to a level of energy higher than the initial state by ΔE_+, then the scattered photon must have energy

$$\hbar\omega' = \hbar\omega - \Delta E_+$$

If the molecule is in an excited state it is possible that in the scattering process it makes a transition to a level of energy lower than the initial state by ΔE_-; in this case the scattered photon has higher energy than the incident photon

$$\hbar\omega' = \hbar\omega + \Delta E_-$$

† The electron spin wavefunction for the O_2 molecule is symmetric so that the spatial wavefunction is antisymmetric. This is an exception to the general trend where for molecules in their ground state one finds most frequently $\hat{P}^N_{12}\psi_{\text{electron}} = +\psi_{\text{electron}}$.

To find the selection rules for Raman scattering it is important to realize that the molecule cannot resonantly absorb a photon of arbitrary energy. However, a two-step process as shown in Fig. 10.4 can take place. As long as the matrix elements between the initial state E_i and some intermediate state E_n *and* between the intermediate state E_n and the final state E_f, are different from zero, the transition $E_i \to E_f$ can take place. This is a second-order transition, and energy conservation requires that

$$\hbar\omega + E_i = \hbar\omega' + E_f$$

The transition to the intermediate state *need not conserve energy* because of the uncertainty principle. However, the closer $\hbar\omega + E_i$ is to E_n the larger the transition probability. This is because it minimizes the denominator in the expression for the transition probability as given by Eq. (5.13*c*).

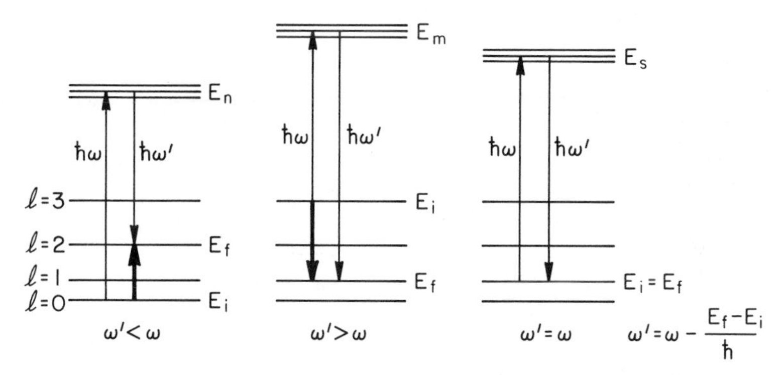

FIGURE 10.4. Some possible transitions in Raman scattering. The heavy arrows indicate the transition of the system. The light arrows indicate one of the possible transitions to an intermediate state that can connect the initial and final states.

The transitions to and from the intermediate state must have nonzero matrix elements and, therefore, the selection rule $\Delta l = \pm 1$ and parity change must be obeyed. As a consequence the selection rules for the Raman effect are

$$\Delta l = 0, \pm 2$$

When the initial and final states are members of a rotational spectrum the energy shift of the scattered photons is given by

$$\Delta E = B[l_f(l_f+1) - l_i(l_i+1)]$$

and if we set $l_f = l_i + 2$, then

$$\Delta E = B[(l_i+2)(l_i+3) - l_i(l_i+1)] = 4B(l_i+3/2)$$

As a result a band spectrum with equally spaced lines around the central frequency is observed. The spacing between lines is $4B$, where $B = \hbar/2\pi A$, with A the effective moment of inertia of the molecule about the axis of rotation [see Eqs. (9.32)].

Suppose now that all odd values of l are missing. Then the lines will be spaced by $8B$ and the difference of the first line from the center (this will be due to the transition $l_i = 0 \rightarrow l_f = 2$) will be $6B$ as shown in Fig. 10.5(a). If all the even lines are missing, then the spacing between lines is still $8B$, but the difference between the first line and the center (due to the transition $l_i = 1 \rightarrow l_f = 3$) will be $10B$ as shown in Fig. 10.5(b). Thus, one can unambiguously determine whether the molecule possesses even or odd rotational states. The Raman spectrum of the O_2 molecule is shown in Fig. 10.6(a) and its analysis indicates that only odd rotational states are present, as required if the nuclei obey Bose statistics.

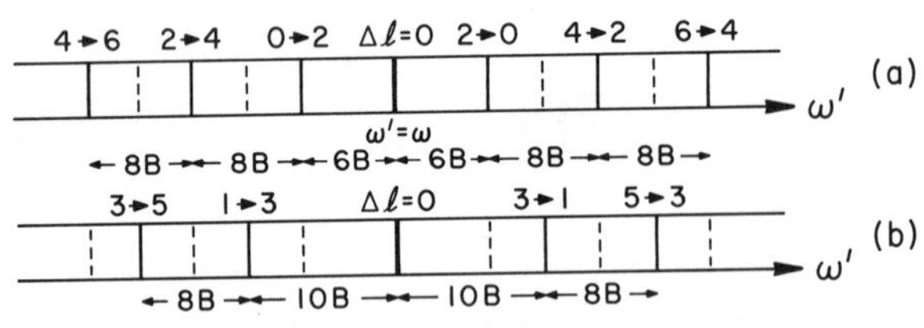

FIGURE 10.5. The Raman spectrum arising from transitions between rotational levels (a) when only even rotational levels participate; (b) when only odd rotational levels participate.

An interesting case occurs when the nuclear spin is different from zero as is the case in the N_2 molecule, where the N^{14} nucleus has spin $I = 1$. The two nuclear spins can combine into a total nuclear spin

$$T = 2, 1, 0$$

The states with $T = 2, 0$ are symmetric under the exchange of the

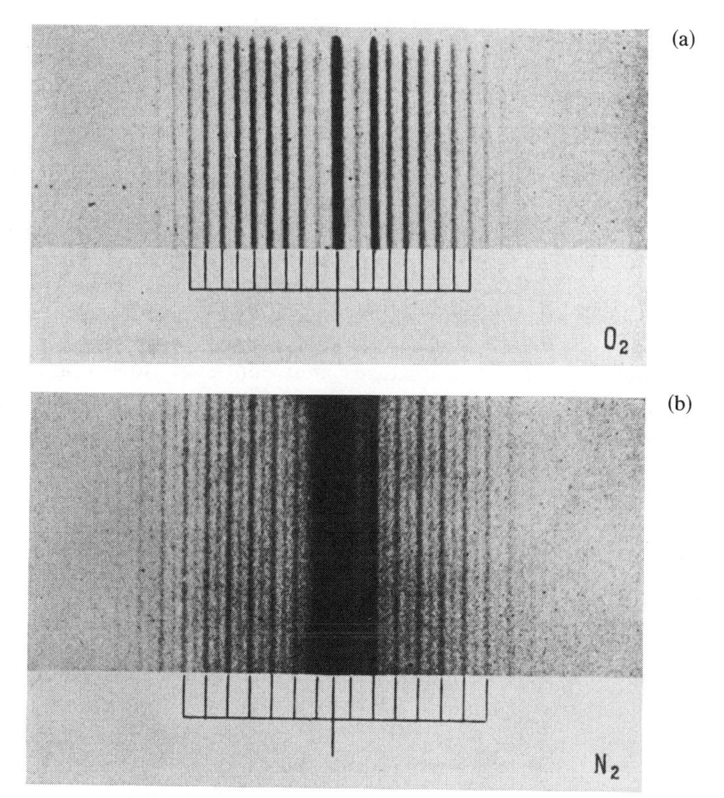

FIGURE 10.6. (a) Raman spectrum of O_2. Only odd rotational states are present. (b) Raman spectrum of N_2 excited by the 2536 Å Hg line. Note lines of alternating intensity corresponding to transitions in molecules where the total nuclear spin is even (intense lines) or odd (faint lines). [From G. Herzberg Molekülspectren und Molekülstruktur Th. Steinkopff Verlag, Leipzig, 1939].

nuclei and therefore in this case the molecule has only even rotational levels,† whereas when the nuclei are in the antisymmetric total nuclear spin state $T = 1$, only odd rotational levels appear. Therefore, we expect to see both even and odd levels. However, the *statistical weight*, that is, the number of M-sublevels, for a state of

† For the N_2 and H_2 molecules in the ground state $\hat{P}^{N}_{12}\psi_{\text{electron}} = +\psi_{\text{electron}}$, because the spin wavefunction in the ground state is antisymmetric. This is the opposite configuration to that for the O_2 molecule that we just analyzed.

angular momentum T is

$$2T+1 \qquad \text{(statistical weight of } T\text{)}$$

Thus, the nuclei can be found in any one of a total of

$$(2\times 2+1)+(2\times 0+1)=6 \qquad \text{(symmetric spin states)}$$

and

$$2\times 1+1=3 \qquad \text{(antisymmetric spin states)}$$

Therefore the Raman band spectrum of the N_2 molecule must consist of alternating lines, with an intensity ratio of $6/3=2/1$, where the lines from transitions between even levels are more intense. This is shown in Fig. 10.6(b) and such spectra can be used to determine the spin I of the nucleus since the intensity ratio is in general $(I+1)/I$.

10.4. THE HELIUM ATOM

The simplest system containing two identical fermions is the helium atom. It consists of a nucleus† with $Z=2$ and $A=4$ to which two electrons are bound. Even for such a simple system we cannot find the exact solution for the stationary states. Instead we must use approximation methods. We will choose the two-electron wavefunction as a product of single-particle wavefunctions, assuming that each electron moves independently in the central potential of the helium nucleus. We will then evaluate the mutual electrostatic repulsion of the electrons by perturbation theory. We will, however, include the effects of the electron spin from the outset, since they play a dominant role in determining the energy spectrum.

Since electrons are fermions, the overall wavefunction must be antisymmetric under the exchange of the two electrons. Furthermore, we assume that the wavefunction can be written as a product‡ of a function of the space coordinates $u(\mathbf{r}_1, \mathbf{r}_2)$ and a function

† The helium nucleus is referred to as an α-particle because it is emitted in the radioactive decay of heavy elements; see Section 6.7.

‡ This is justified because the spin–spin forces are very weak.

$\chi(s_1, s_2)$ of the spin coordinates

$$\psi = u(\mathbf{r}_1, \mathbf{r}_2)\chi(s_1, s_2) \tag{10.10}$$

The spin function $\chi(s_1, s_2)$ cannot be expressed in the coordinate representation. It can, however be represented by column vectors such as those introduced in Section 4.6.

For the spin function we will use a representation in which the operators for the *total* spin $\hat{\mathbf{S}}^2$ and $\hat{S}_z$ are diagonal. The operator $\hat{\mathbf{S}}$ is given by

$$\hat{\mathbf{S}} = \hat{\mathbf{S}}_1 + \hat{\mathbf{S}}_2$$

The combination of two spin-1/2 particles results in four eigenstates—three states corresponding to a total spin eigenvalue $S = 1$ and one state corresponding to $S = 0$. These can be expressed in terms of single-particle spin states by the methods introduced in Section 7.6. Just as in Eqs. (7.62) the four states and their eigenvalues are given by†

S	S_z	$\lvert\chi\rangle = \lvert S_1, S_2, m_{s_1}, m_{s_2}\rangle$		
1	+1	$\lvert +1/2, +1/2\rangle$	(*stretched*) triplet	$\Uparrow$
1	0	$\frac{1}{2^{1/2}}[\lvert +1/2, -1/2\rangle + \lvert -1/2, +1/2\rangle]$		
1	−1	$\lvert -1/2, -1/2\rangle$		

(10.11*a*)

0	0	$\frac{1}{2^{1/2}}[\lvert +1/2, -1/2\rangle - \lvert -1/2, +1/2\rangle]$	(*jacknife*) singlet	$\Uparrow\Downarrow$

(10.11*b*)

Note that for $S = 1$ the spin function is symmetric under $\hat{P}_{12}$ and thus can be combined only with an antisymmetric space function $u^{(a)}(\mathbf{r}_1, \mathbf{r}_2)$. For $S = 0$ the spin function is antisymmetric under $\hat{P}_{12}$ and thus can be combined only with a symmetric space function $u^{(s)}(\mathbf{r}_1, \mathbf{r}_2)$. Thus

$$\psi(1, 2) = \begin{cases} u^{(a)}(\mathbf{r}_1, \mathbf{r}_2)\chi(S = 1) \\ u^{(s)}(\mathbf{r}_1, \mathbf{r}_2)\chi(S = 0) \end{cases} \tag{10.12a}$$

† When S_1 and S_2 equal 1/2, as in the present case, we will omit these two indices when writing the total spin function.

For the space functions $u(\mathbf{r}_1, \mathbf{r}_2)$ we will use a product of single-particle wavefunctions. When symmetrized these wavefunctions are given by

$$
\begin{aligned}
u_{\alpha\beta}^{(s)}(\mathbf{r}_1, \mathbf{r}_2) &= \frac{1}{2^{1/2}}\left[u_\alpha(\mathbf{r}_1)u_\beta(\mathbf{r}_2) + u_\alpha(\mathbf{r}_2)u_\beta(\mathbf{r}_1)\right] \\
u_{\alpha\beta}^{(a)}(\mathbf{r}_1, \mathbf{r}_2) &= \frac{1}{2^{1/2}}\left[u_\alpha(\mathbf{r}_1)u_\beta(\mathbf{r}_2) - u_\alpha(\mathbf{r}_2)u_\beta(\mathbf{r}_1)\right]
\end{aligned}
\tag{10.12b}
$$

where $u_\alpha(\mathbf{r}_1)$, $u_\beta(\mathbf{r}_2)$, etc., represent hydrogenlike wavefunctions with the proton charge e replaced by Ze. The indices $\alpha, \beta, \ldots$, etc., label the eigenvalues n, l, and m of a stationary state. The energy of such a state is

$$
E_{\alpha\beta} = E_\alpha + E_\beta \tag{10.12c}
$$

and to this approximation the states $u^{(s)}(\mathbf{r}_1, \mathbf{r}_2)$ and $u^{(a)}(\mathbf{r}_1, \mathbf{r}_2)$ have the same energy.

The mutual electrostatic interaction between the two electrons is given by

$$
\hat{H}'_{12} = U_{12} = \frac{e^2}{4\pi\varepsilon_0}\frac{1}{|\mathbf{r}_1 - \mathbf{r}_2|} \tag{10.13a}
$$

and is positive since the two electrons repel each other. The definition of the coordinates $\mathbf{r}_1$ and $\mathbf{r}_2$ is shown in Fig. 10.7. The contribution of $\hat{H}'_{12}$ to the energy of the state $\psi_{\alpha\beta}$ in first-order perturbation theory is given simply by the matrix element

$$
\Delta E = \sum_{\text{spins}} \int \psi_{\alpha\beta}^*(\mathbf{r}_1, \mathbf{r}_2)\hat{H}'_{12}\psi_{\alpha\beta}(\mathbf{r}_1, \mathbf{r}_2)\, d^3r_1\, d^3r_2 \tag{10.13b}
$$

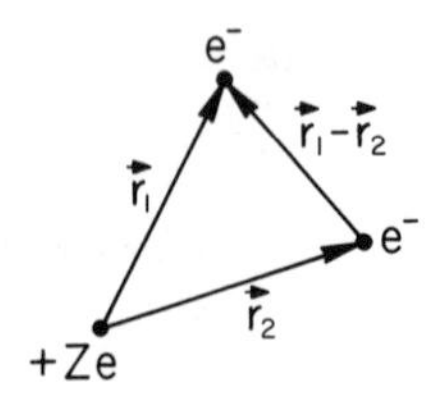

FIGURE 10.7. Coordinates used in discussing the helium atom.

Interaction Energy

Since the spin states of Eqs. (10.11) are orthonormal the summation over spins in Eq. (10.13*b*) will give

$$\langle S', S'_z \mid S, S_z \rangle = \delta_{S,S'} \delta_{S_z,S'_z}$$

and therefore the matrix element in Eq. (10.13*b*) reduces to

$$\begin{aligned} \Delta E^{(s)}_{\alpha\beta} &= \int u^{(s)*}_{\alpha\beta}(\mathbf{r}_1, \mathbf{r}_2) \hat{H}'_{12} u^{(s)}_{\alpha\beta}(\mathbf{r}_1, \mathbf{r}_2)\, d^3r_1\, d^3r_2 \\ \Delta E^{(a)}_{\alpha\beta} &= \int u^{(a)*}_{\alpha\beta}(\mathbf{r}_1, \mathbf{r}_2) \hat{H}'_{12} u^{(a)}_{\alpha\beta}(\mathbf{r}_1, \mathbf{r}_2)\, d^3r_1\, d^3r_2 \end{aligned} \qquad (10.14a)$$

Using the definition of $u^{(s)}_{\alpha,\beta}(\mathbf{r}_1, \mathbf{r}_2)$ and $u^{(a)}_{\alpha,\beta}(\mathbf{r}_1, \mathbf{r}_2)$ given by Eqs. (10.12*b*) we find

$$\begin{aligned} \Delta E_{\alpha\beta} = \frac{e^2}{4\pi\varepsilon_0} \Bigg[&\int |u_\alpha(\mathbf{r}_1)|^2 \, |u_\beta(\mathbf{r}_2)|^2 \frac{1}{|\mathbf{r}_1 - \mathbf{r}_2|}\, d^3r_1\, d^3r_2 \\ &\pm \int u^*_\alpha(\mathbf{r}_1) u^*_\beta(\mathbf{r}_2) \frac{1}{|\mathbf{r}_1 - \mathbf{r}_2|}\, u_\alpha(\mathbf{r}_2) u_\beta(\mathbf{r}_1)\, d^3r_1\, d^3r_2 \Bigg] \end{aligned} \qquad (10.14b)$$

where the (+) sign is to be used with $\Delta E^{(s)}_{\alpha,\beta}$ and the (−) sign with $\Delta E^{(a)}_{\alpha,\beta}$.

Equation (10.14*b*) is of fundamental importance. The integrals can be evaluated and are positive definite. Let us designate the values of the first integral in the rhs of Eq. (10.14*b*) by

$$J_{\alpha\beta} \qquad \text{(the } \textit{direct} \text{ integral)}$$

and the second integral by

$$K_{\alpha\beta} \qquad \text{(the } \textit{exchange} \text{ integral)} \qquad (10.15a)$$

Thus, the states with symmetric and antisymmetric space wavefunction have different energies

$$\begin{aligned} E^{(s)}_{\alpha\beta} &= E_\alpha + E_\beta + (J_{\alpha\beta} + K_{\alpha\beta}) \\ E^{(a)}_{\alpha\beta} &= E_\alpha + E_\beta + (J_{\alpha\beta} - K_{\alpha\beta}) \end{aligned} \qquad (10.15b)$$

We see that the $S = 1$ states—the spin triplets to which corresponds the antisymmetric space function—will be more tightly bound than the $S = 0$ states—the spin singlets. In other words, the relative spin orientation between the two electrons affects the energy of the system. This effect is *much* stronger than the one we obtain by calculating the mutual interaction between the magnetic moments of the two electrons, which of course would also be dependent on the orientation of the spins. Thus, the symmetrization of the wavefunction imposed in order to satisfy the exchange symmetry gives rise to an apparent spin-dependent force. Such forces, which are really *exchange forces*, appear in any system where the spins must be ordered; a striking example is ferromagnetism.

We can now construct the spectrum of the helium atom using our approximate result for ΔE. We expect to find a series of states corresponding to $S = 1$—these are referred to as *orthohelium*—and a series of states corresponding to $S = 0$—referred to as *parahelium*. These series of states are not connected by radiative transitions and thus the atom tends to find itself in one or the other configuration. This is because, to first order, the emission of electromagnetic radiation does not change the spin orientation. Thus, for electric dipole radiation the selection rule

$$\Delta S = 0$$

is strictly valid, and transitions occur only between parahelium states or only between orthohelium states.

The ground state of the system is obtained by assigning to the single-particle indices α and β their lowest value

$$\alpha = \beta \Rightarrow (n = 1, l = 0, m = 0)$$

In this case $u^{(a)}(\mathbf{r}_1, \mathbf{r}_2)$ vanishes (the exclusion principle at work) and only the parahelium state, $S = 0$, exists. Therefore†

$$u_0(\mathbf{r}_1, \mathbf{r}_2) = u_{1,0,0}(\mathbf{r}_1) u_{1,0,0}(\mathbf{r}_2) \tag{10.16a}$$

The single-particle ground-state functions $u_{1,0,0}(\mathbf{r})$ are taken to be the hydrogenlike wavefunctions given by Eq. (8.32) but with e replaced by Ze. Thus

$$u_{1,0,0}(\mathbf{r}) = \frac{1}{\pi^{1/2}} \left(\frac{Z}{a_0}\right)^{3/2} e^{-rZ/a_0} \tag{10.16b}$$

† Note that we have changed the normalization from that given in Eq. (10.12*b*) since the state is completely symmetric and refers to identical single-particle states. This was shown explicitly in Eq. (10.8*a*).

where

$$a_0 = 4\pi\varepsilon_0 \frac{\hbar^2}{m_e e^2}$$

is the Bohr radius. Therefore, Eq. (10.16a) can be written as

$$u_0(\mathbf{r}_1, \mathbf{r}_2) = \frac{1}{\pi}\left(\frac{Z}{a_0}\right)^3 e^{-r_1 Z/a_0} e^{-r_2 Z/a_0} \tag{10.16c}$$

This permits us to give an explicit expression for the interaction energy in the ground state

$$\Delta E_0^{(s)} = \frac{e^2}{4\pi\varepsilon_0} \int |u_{1,0,0}(\mathbf{r}_1)|^2 \frac{1}{|\mathbf{r}_1 - \mathbf{r}_2|} |u_{1,0,0}(\mathbf{r}_2)|^2 \, d^3r_1 \, d^3r_2 \tag{10.17a}$$

The integral can be evaluated exactly with the use of Eq. (10.16b) and yields†

$$\Delta E_0^{(s)} = \frac{5}{8} \frac{e^2}{4\pi\varepsilon_0} \frac{Z}{a_0} = \tfrac{5}{4} Z \text{ Ry} \tag{10.17b}$$

The integration is straightforward but still involves several algebraic steps. Therefore, it is presented in Appendix 13. It is, however, instructive to obtain some physical insight into the form of Eq. (10.17a). To this effect we express $\Delta E_0^{(s)}$ by a two step process

$$\Delta E_0^{(s)} = \int |u_{1,0,0}(\mathbf{r}_1)|^2 \, U(\mathbf{r}_1) \, d^3r_1 \tag{10.18a}$$

where

$$U(\mathbf{r}_1) = eV(\mathbf{r}_1) = e\int \frac{e}{4\pi\varepsilon_0} \frac{1}{|\mathbf{r}_1 - \mathbf{r}_2|} |u_{1,0,0}(\mathbf{r}_2)|^2 \, d^3r_2 \tag{10.18b}$$

Clearly $V(\mathbf{r}_1)$ defined by Eq. (10.18b) gives the electrostatic potential at the position $\mathbf{r}_1$ due to the presence of electron "2". Therefore, $\Delta E_0^{(s)}$ is just the interaction energy of electron "1" moving in the potential of electron "2". We can say that the interaction energy is calculated by smearing the electrostatic repulsion over the probability distribution of the two electrons. The evaluation of the two integrals of Eqs. (10.18) can be found in Appendix 13.

---

† As before we use Ry to designate the Rydberg; see Eqs. (8.21b).

Energy Spectrum

In our present approximation the energy of the ground state is given by

$$E_0(\text{He}) = E_{1,0,0} + E_{1,0,0} + \Delta E_0^{(s)} \tag{10.19a}$$

where $E_{1,0,0}$ is the ground-state energy of the hydrogenlike atom

$$E_{1,0,0} = -Z^2 \text{ Ry} = -4 \text{ Ry}$$

Thus

$$E_0(\text{He}) = -8 \text{ Ry} + \tfrac{5}{2} \text{ Ry} = -108.8 + 34.0 = -74.8 \text{ eV} \tag{10.19b}$$

to be compared with the experimental value of

$$E_0(\text{He}) = -78.98 \text{ eV} \tag{10.19c}$$

Note that in Eqs. (10.19) we measure the energy with reference to the energy of the doubly ionized helium ion, namely, when *both* electrons have been removed to infinity. This is shown in Fig. 10.8. In part (a) of the figure we indicate the energy levels resulting from two noninteracting electrons. In part (b) we show the modification introduced by the direct integral of the interaction energy [see Eqs. (10.15*c*)]. In part (c) we include the effect of the exchange integral, and this results in a different energy for the singlet and triplet states. Finally, in part (d) we have included the effect of the spin–orbit coupling. This has been highly magnified since it is typically of the order of 10^{-4} eV.

Returning to Eq. (10.17*b*) we note that the contribution of the mutual electrostatic interaction is positive since it corresponds to a repulsive force and amounts to a considerable correction to the ground-state energy. It is therefore not surprising that the calculation differs by ~5% from the exact value. Physically, we attribute this discrepancy to a neglect of the screening of the nuclear potential by one of the electrons. As already discussed in Section 10.2, screening of the nuclear charge is a very pronounced effect in multielectron atoms. The amount of screening present in the helium atom can be inferred from a variational calculation where one uses a hydrogenlike trial wavefunction but lets $Z = Z_{\text{eff}}$ be the variation parameter. One finds in such a case that the minimum of the

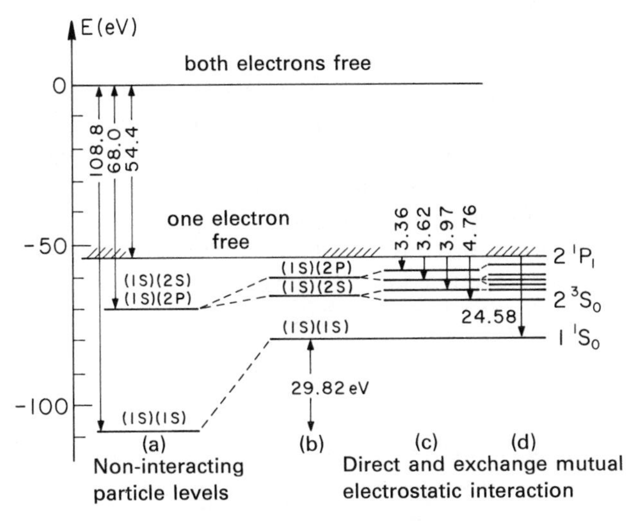

FIGURE 10.8. The low-lying energy levels of the helium atom. Here the energy scale has been chosen to be zero when both electrons are removed from the nucleus. The spin–orbit coupling gives rise to energy spacings of order 10^{-4} eV.

ground-state energy corresponds to

$$Z_{\text{eff}} = 2 - \tfrac{5}{16} = \tfrac{27}{16}$$

This is a significant screening. If one then uses the above value of Z_{eff} to evaluate the ground-state energy of the atom in the single-particle approximation the result is

$$E_0(\text{He}) = 2 \times E'_{1,0,0} = -2(Z_{\text{eff}})^2 \text{ Ry} = -77.46 \text{ eV}$$

which is in better agreement with the experimental value than obtained by perturbation theory [see Eqs. (10.19)].

The ionization potential of the helium atom is given by the energy required to remove *one* of its electrons, namely,

$$E_0(\text{He}^+) = E_0(\text{He}) + E_{\text{ionization}}$$

where $E_0(\text{He}^+)$ indicates the ground-state energy of the helium ion. The He^+ ion has the same energy levels as a hydrogen atom with $Z = 2$ or $E_0(\text{He}) = -Z^2$ Ry. Using the exact value for $E_0(\text{He})$ we find

$$E_{\text{ionization}} = -4 \text{ Ry} - E_0(\text{He}) = 24.58 \text{ eV}$$

in agreement with observation. The energy at which ionization takes place is shown in Fig. 10.8.

Excited States

A quantitative discussion of the excited states of the helium atom is beyond our scope here. We will, however, make some general remarks. First, note that the excited states in helium, and for that matter in most atoms, result from the excitation of *one* of the electrons to a higher state, since if we suppose that both electrons are excited to the next highest state, $n=2$, of their corresponding single-particle wavefunction, the energy of the state would be according to Eq. (10.19*b*)

$$E_{2,2}(\text{He}) = E_{2,0,0} + E_{2,0,0} + \Delta E_2^{(s)} = -2Z^2 \frac{\text{Ry}}{4} + \tfrac{5}{2}\,\text{Ry}$$

where we have used $\Delta E_2^{(s)} \simeq \Delta E_0^{(s)}$. The above result shows that $E_{2,2}(\text{He}) > 0$, and therefore the system, is unbound. A more precise calculation including screening effects confirms our estimate and it is found that $E_{2,2}(\text{He}) = +27.2\,\text{eV}$. While these states do not form a bound system they are nevertheless stationary states of the Hamiltonian and have *discrete* energies. They can be observed in scattering experiments.

We conclude that the spectrum of the bound states of the helium atom corresponds to single-particle excitation where one electron is in the $n=1$, $l=0$, $m=0$ state and the other electron can have $n \geqslant 1$; and $l=0, 1, \ldots, n-1$; $-l \leqslant m \leqslant l$ as usual. In general it is convenient to label the states of a multielectron system by the configuration of the corresponding single-particle states as in Table 10.1. For example, for helium we have the following states

$(1s)(1s)$	$1\,^1S_0$	ground state
$(1s)(2s)$	$2\,^1S_0$, $2\,^3S_1$	first excited state
$(1s)(2p)$	$2\,^1P_1$, $2\,^3P_{0,1,2}$	second excited state
$(1s)(3s)$	$\cdots$	
$(1s)(3p)$	$\cdots$	
$(1s)(3d)$	$3\,^1D_2$, $3\,^3D_{1,2,3}$	
etc.		

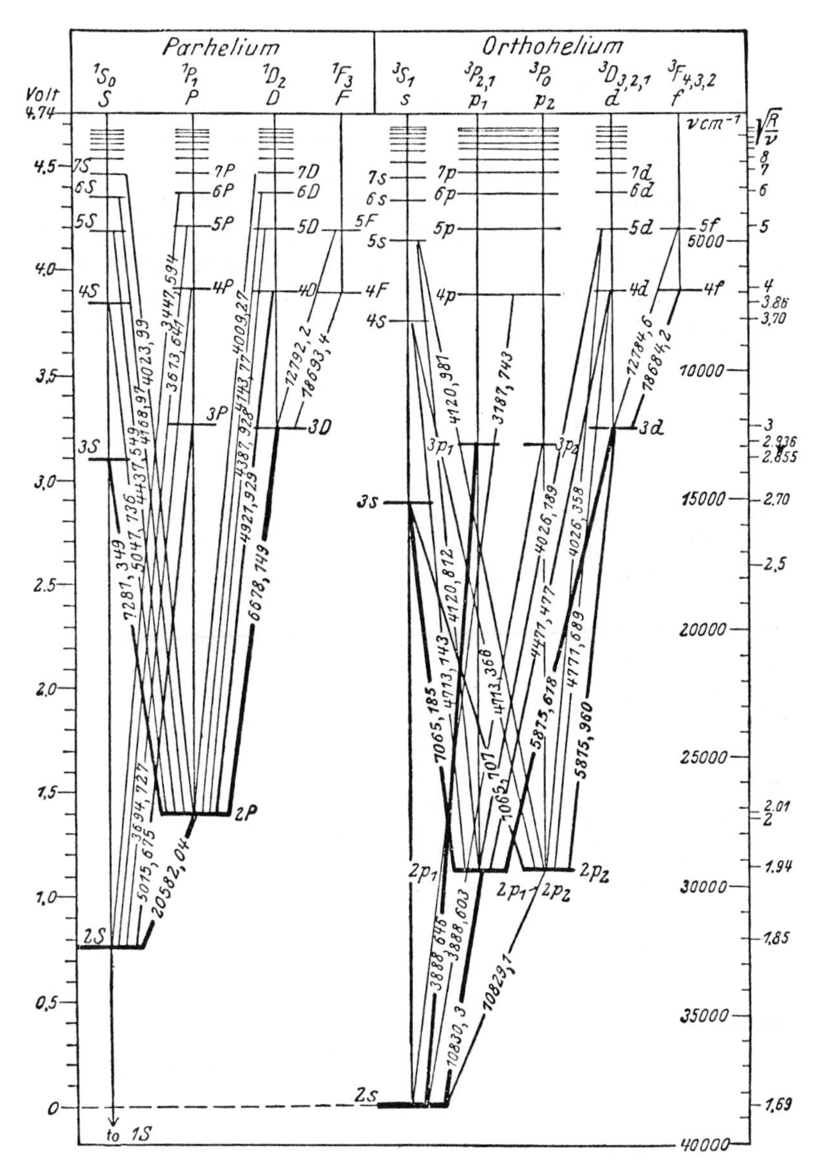

FIGURE 10.9. The energy levels of helium and the observed transitions. Note that there are no transitions between the $S=0$ and $S=1$ states. The $1S$ state of parahelium (spin singlet) lies much deeper and is indicated by the arrow. [From M. Born, *Atomic Physics*, Hafner (1957), sixth edition].

As in Table 10.1 the left-hand column gives the configuration of the individual electrons and the middle column gives the spectroscopic term. The energy level diagram of the excited states of helium is shown in Fig. 10.9 where the energy is referenced to the ionization level. Note that the ground state is too low to appear in this figure.

The parahelium and orthohelium systems are shown separately, and it is clear that they give rise to stationary states of different energy as expected from Eqs. (10.15*b*). Furthermore, the Coulomb degeneracy between states with the same n-value but different l-values has been removed because of the effects of screening. Spin–orbit coupling splits the orthohelium states for which $L \neq 0$, giving rise to fine structure. Because of the presence of two electrons, states with higher values of J lie lower, in contrast to the results for the hydrogen atom. The position of the levels is further complicated by a direct interaction between the magnetic moments of the two electrons and relativistic corrections, all of which can be calculated to substantial accuracy.† The most prominent transitions in the emission spectrum of helium are shown by the heavier lines.

10.5. THE HYDROGEN MOLECULE

As the next example of a system containing identical particles we will discuss the hydrogen molecule. This consists of two hydrogen atoms bound to each other, or more concisely of two protons and two electrons bound into a single four-particle system. The protons are much more massive than the electrons and thus move with smaller velocities. Therefore, thinking of the molecule as a system of two identical atoms is a reasonable approximation to the true physical situation and serves as a useful model. This picture is further supported by the fact that when sufficient energy is supplied to the molecule, it dissociates, most often into two neutral atoms. Since hydrogen atoms are neutral the binding mechanism cannot be due to a direct electrostatic attraction between the atoms; instead it

† See, for instance, H. Bethe and E. Salpeter, *Quantum Mechanics of One- and Two-Electron Atoms*, Academic, New York (1957).

is a consequence of the mutual electrostatic interaction of all four particles, the two electrons and the two protons. A qualitative argument along these lines was given in Section 9.3 and underscores the main effect that leads to binding. If the electrons are found in the region of space between the two protons the net force can be attractive (see Fig. 9.5). This implies that there is a finite probability for finding the two electrons in the same region of space. We say then that the electron wavefunctions overlap.

A related physical system is the hydrogen molecular ion H_2^+ which consists of two protons and a single electron, which is shared by the two protons. In this case the binding arises from the sharing of the electron. The hydrogen molecule will on occasion dissociate into H_2^+ and an electron. This indicates that at least for part of the time the hydrogen molecule is in a state where one electron is associated with both protons.

When we discussed diatomic molecules in Sections 9.3 and 10.3 we were mainly concerned with their rotational and vibrational states. These states arose from the motion of the nuclei about their equilibrium position. At present we are interested in the spectrum of the stationary states that result from the interaction of the electrons both with the nuclei and among themselves. This interaction gives rise to a binding of the molecule, and in evaluating it we treat the nuclei as being fixed in the molecule's center of mass at some equilibrium distance R one from the other. In this case the Hamiltonian in the center of mass of the molecule is relatively simple. It contains the single-particle Hamiltonian for each electron as well as additional terms due to the electrostatic interaction between the four constituents.

Of course, we cannot obtain an exact solution for the stationary states of the Hamiltonian. Instead we use approximation methods very similar to those introduced in the previous section for the solution of the problem of the helium atom. In general, we construct a many-particle wavefunction out of linear combinations of single-particle wavefunctions. The latter can be taken as hydrogenlike wavefunctions. Then the many-particle wavefunction can be used either as a zero-order approximation for perturbative calculations or as a trial wavefunction in a variational approach.

The choice of the form of the initial many-particle wavefunction is crucial and there are no exact rules for selecting one combination over another. In fact, the choice of the wavefunction depends on the type of molecule that is under consideration. For instance, for the case of the H_2 molecule we argued that it resembles a system with two identical hydrogen atoms. Thus, when the two atoms are at large separation from one another, $R \to \infty$, the proper choice of the two-electron wavefunction is

$$\psi(1, 2) = \psi_A(1)\psi_B(2) \tag{10.20a}$$

where 1 and 2 label the two electrons, and A and B label the two protons. In writing Eq. (10.20*a*) we implicitly assume that electron 1 is associated with proton A, while electron 2 is associated with proton B. For simplicity we assume that both $\psi_A(1)$ and $\psi_B(2)$ are ground-state hydrogen wavefunctions. Of course, the electrons are indistinguishable so that an equally appropriate two-electron wavefunction is

$$\psi(1, 2) = \psi_A(2)\psi_B(1) \tag{10.20b}$$

which describes a state of the same energy as the state of Eq. (10.20*a*). Therefore, any linear combination of the wavefunctions given by Eqs. (10.20*a*) and (10.20*b*) can be used to describe the molecule.

A different choice would be to use products of single-particle wavefunctions of the form

$$\psi_A(1)\psi_A(2) \quad \text{or} \quad \psi_B(1)\psi_B(2) \tag{10.21}$$

Such terms describe the physical situation where both electrons are associated with one of the protons, either A or B. This configuration does not occur often for H_2 because it implies that at large interatomic distances R, the H_2 molecule would dissociate into a H^- ion and a proton, which is contrary to the experimental facts. At short distances, however, we cannot exclude the possibility of the presence of terms of the form given by Eq. (10.21). For instance, this could occur when the two protons are very close to each other ($R \ll a_0$). Then the molecule resembles a helium atom and both electrons are simultaneously associated with both protons. It is also possible to use a combination of product functions such as given by both Eqs. (10.20*a*) and (10.20*b*) and Eqs. (10.21).

In recent years great progress has been made in the determination of the wavefunctions of complex molecular systems. A detailed discussion of these methods is beyond our scope here, and we will therefore restrict ourselves to the simplest approach. This is the method of Heitler and London first published in 1927, also known as the *valence bond* method. They used properly symmetrized linear combinations of wavefunctions of the type given by Eqs. (10.20*a*) and (10.20*b*) and were able to show that two hydrogen atoms do bind into a molecule. As we will see, the binding is due to an exchange term similar to the one we encountered in the study of the helium atom. Thus, molecular binding is a purely quantum-mechanical effect.

Heitler–London Method for the H_2 Molecule

We treat the two protons A and B as fixed and separated by a distance R, as shown in Fig. 10.10. The energy of the system will strongly depend on R, and the equilibrium separation R_0 is the value of R at which the total energy of the molecule is a minimum. The two electrons are labeled 1 and 2, and the coordinate convention is as indicated in the figure. The Hamiltonian for this system is given by

$$\hat{H}=\left(\frac{\hat{p}_1^2}{2m_e}-\frac{e^2}{4\pi\varepsilon_0}\frac{1}{r_{1A}}\right)+\left(\frac{\hat{p}_2^2}{2m_e}-\frac{e^2}{4\pi\varepsilon_0}\frac{1}{r_{2B}}\right)+\frac{e^2}{4\pi\varepsilon_0}\left(\frac{1}{R}-\frac{1}{r_{1B}}-\frac{1}{r_{2A}}+\frac{1}{r_{12}}\right) \tag{10.22}$$

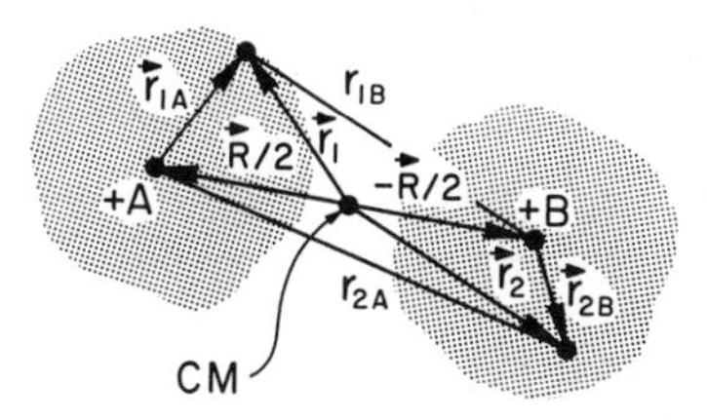

FIGURE 10.10. Schematic representation of a diatomic molecule and the coordinate system used to describe it.

For the two-electron wavefunction we use linear combinations of the form given by Eqs. (10.20*a*) and (10.20*b*). The spins of the two electrons can combine into a triplet or a singlet total-spin state just as for the helium atom. Since the overall wavefunction must be antisymmetric we must combine a symmetric space function $u^{(s)}(1, 2)$ with the singlet spin function and combine an antisymmetric space function $u^{(a)}(1, 2)$ with the triplet spin function. We write

$$\psi(1, 2) = [\psi_A(1)\psi_B(2) - \psi_A(2)\psi_B(1)] \tag{10.23a}$$

where

$$\psi(1, 2) = \begin{cases} u^{(s)}(1, 2)\chi^{(S=0)} \\ u^{(a)}(1, 2)\chi^{(S=1)} \end{cases} \tag{10.23b}$$

Here $\chi^{(S=0)}$ and $\chi^{(S=1)}$ are the singlet and triplet spin wavefunctions of the two electrons [as in Eqs. (10.11)], whereas $u^{(s)}(1, 2)$ and $u^{(a)}(1, 2)$ are symmetric and antisymmetric space functions, namely,

$$\begin{aligned} u^{(s)}(1, 2) &= [u_A(\mathbf{r}_1)u_B(\mathbf{r}_2) + u_A(\mathbf{r}_2)u_B(\mathbf{r}_1)] \\ u^{(a)}(1, 2) &= [u_A(\mathbf{r}_1)u_B(\mathbf{r}_2) - u_A(\mathbf{r}_2)u_B(\mathbf{r}_1)] \end{aligned} \tag{10.24}$$

$u_A(\mathbf{r}_1)$, $u_B(\mathbf{r}_1)$, etc., are hydrogen wavefunctions with respect to protons A and B, and $\mathbf{r}_1$ is the coordinate of electron 1 from some convenient origin. We have chosen the origin at the center of mass of the molecule.

We are interested in the ground state of the H_2 molecule and, therefore, we will use for both $u_A(\mathbf{r})$ and $u_B(\mathbf{r})$ ground-state wavefunctions of the hydrogen atom. These are labeled by the indices $n = 1$, $l = 0$, and $m = 0$. We write

$$u_A(\mathbf{r}_1) = \frac{1}{(\pi a_0^3)^{1/2}} e^{-r_{1A}/a_0} = \frac{1}{(\pi a_0^3)^{1/2}} e^{-|\mathbf{r}_1 - \mathbf{R}/2|/a_0} \tag{10.25a}$$

$$u_B(\mathbf{r}_1) = \frac{1}{(\pi a_0^3)^{1/2}} e^{-r_{1B}/a_0} = \frac{1}{(\pi a_0^3)^{1/2}} e^{-|\mathbf{r}_1 + \mathbf{R}/2|/a_0} \tag{10.25b}$$

and correspondingly for $u_A(\mathbf{r}_2)$ and $u_B(\mathbf{r}_2)$. The wavefunctions of Eqs. (10.25*a*) and (10.25*b*) correctly describe the electron "clouds" sketched in Fig. 10.10, but they are *not* identical to one another. Only when $R \to 0$ do they become identical, in which case we say that $u_A(\mathbf{r}_1)$ and $u_B(\mathbf{r}_1)$ overlap completely. Then they describe the same state of electron 1. When $R \to \infty$, there is no overlap and the two wavefunctions describe entirely different states of electron 1, in spite of having the same quantum numbers. Therefore, even for the ground state, $u^{(a)}(1,2)$ does not vanish unless $R=0$.

We now substitute Eqs. (10.25a) and (10.25b) into $u^{(s)}(1,2)$ and $u^{(a)}(1,2)$ as given by Eqs. (10.24). A consequence of the different space dependence of Eq. (10.25a) from Eq. (10.25b) is that the functions $u^{(s)}(1,2)$ and $u^{(a)}(1,2)$ are no longer normalized to unity. To obtain the normalization constant we perform the integration

$$\begin{aligned}\int |u(1,2)|^2\, d^3r_1\, d^3r_2 \\ &= \int [|u_A(\mathbf{r}_1)|^2\,|u_B(\mathbf{r}_2)|^2 + |u_A(\mathbf{r}_2)|^2\,|u_B(\mathbf{r}_1)|^2]\, d^3r_1\, d^3r_2 \\ &\pm 2\,\mathrm{Re} \int [u_A^*(\mathbf{r}_1)u_B(\mathbf{r}_1)u_A(\mathbf{r}_2)u_B^*(\mathbf{r}_2)]\, d^3r_1\, d^3r_2 \\ &= 2\{1 \pm \mathrm{Re}\,\{S^2(R)\}\}\end{aligned} \qquad (10.26a)$$

where $S(R)$ is the *overlap integral*

$$S(R) = \int u_A^*(\mathbf{r}_1)u_B(\mathbf{r}_1)\, d^3r_1 \qquad (10.26b)$$

Introducing the wavefunctions of Eqs. (10.25a) and (10.25b) we find

$$S(R) = \left(1 + \frac{R}{a_0} + \frac{R^2}{3a_0^2}\right)e^{-R/a_0} \qquad (10.26c)$$

$S(R)$ reduces to zero when $R \to \infty$, in which case the normalization is $1/2^{1/2}$—as expected for noninteracting particles. When $R \to 0$, $S(R)$ reduces to 1, in which case the normalization for $u^{(s)}(1,2)$ is 1/2, whereas $u^{(a)}(1,2)$ vanishes. This is as expected since when $R \to 0$ the two terms in Eqs. (10.24) are equal. Thus, the Heitler–London wavefunction for the H_2 molecule is given by

$$u_{\pm}(1,2) = \frac{1}{\{2 \pm 2\,\mathrm{Re}\,[S^2(R)]\}^{1/2}}[u_A(\mathbf{r}_1)u_B(\mathbf{r}_2) \pm u_A(\mathbf{r}_2)u_B(\mathbf{r}_1)] \qquad (10.27)$$

Here $u_+(1,2)$ is coupled with the spin singlet state, while $u_-(1,2)$ must be coupled with the spin triplet state. We now use these wavefunctions as zero-order approximations and evaluate the expectation value of the Hamiltonian [Eq. (10.22)] as a function of the interatomic distance R.

Since the $S=0$ and $S=1$ spin functions are orthogonal, a stationary state of the molecule is described by the space function $u_+(1,2)$

or $u_-(1,2)$, but not by a combination of both. Thus the expectation value of the ground-state energy is given by

$$\langle E\rangle_\pm = \int u_\pm^*(1,2)\hat{H}u_\pm(1,2)\, d^3r_1\, d^3r_2 \tag{10.28a}$$

Recall that $u_\pm(1,2)$ are eigenstates of the single-particle Hamiltonian such as those given in the first two sets of parentheses in Eq. (10.22). Let us then designate the eigenvalues of the single particle Hamiltonians by E_1 and E_2 which is the energy of the free hydrogen atoms. For the ground state $E_1 = E_2 = -\text{Ry}$, and therefore

$$\langle E\rangle_\pm = -2\,\text{Ry} + \frac{e^2}{4\pi\varepsilon_0}\int u_\pm^*(1,2)\left(\frac{1}{R} + \frac{1}{r_{12}} - \frac{1}{r_{1B}} - \frac{1}{r_{2A}}\right)u_\pm(1,2)\, d^3r_1\, d^3r_2 \tag{10.28b}$$

At this point we define the energy $E_\pm$ of the system to be zero when the molecule is dissociated so that

$$E_\pm = \langle E\rangle_\pm + 2\,\text{Ry} \tag{10.29}$$

Finally, expanding $u_\pm(1,2)$ we obtain the complete expression for the ground-state energy of the H_2 molecule

$$\begin{aligned}E_\pm(R) &= \frac{e^2}{4\pi\varepsilon_0}\frac{1}{R}\\ &+ \frac{e^2}{4\pi\varepsilon_0[1\pm \text{Re}\,(S^2)]}\int |u_A(\mathbf{r}_1)|^2\,|u_B(\mathbf{r}_2)|^2\left(\frac{1}{r_{12}} - \frac{1}{r_{1B}} - \frac{1}{r_{2A}}\right)d^3r_1\, d^3r_2\\ &\pm \left\{\frac{e^2}{4\pi\varepsilon_0[1\pm \text{Re}\,(S^2)]}\text{Re}\int u_A^*(\mathbf{r}_1)u_B(\mathbf{r}_1)u_A(\mathbf{r}_2)u_B^*(\mathbf{r}_2)\right.\\ &\times \left.\left(\frac{1}{r_{12}} - \frac{1}{r_{1B}} - \frac{1}{r_{2A}}\right)d^3r_1\, d^3r_2\right\}\end{aligned} \tag{10.30}$$

We note that Eq. (10.30) contains two integrals. We designate the first integral by J and call it the *direct* integral. It represents the mutual electrostatic attraction between one electron cloud and the proton of the other atom ($1/r_{1B}$ and $1/r_{1A}$) as well as the electrostatic repulsion between the two electrons ($1/r_{12}$). The second integral is designated by K and is known as the *exchange* integral. It is due to the exchange mechanism and depends on the overlap of the electron wavefunctions. K must vanish as $R \to \infty$—it has its maxi-

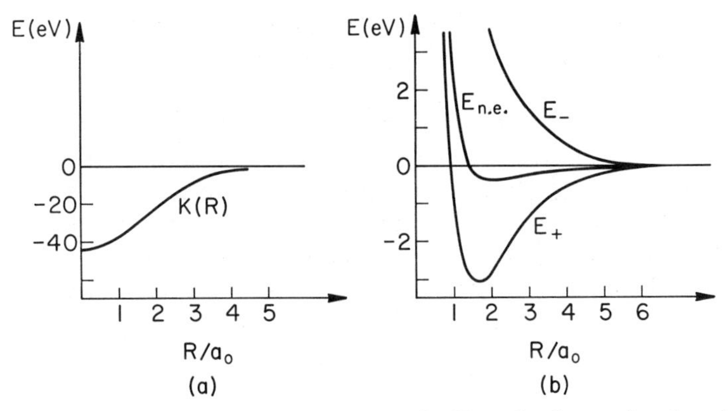

FIGURE 10.11. (a) The exchange integral for the H_2 molecule as a function of the interatomic distance (given in dimensionless units R/a_0). (b) The potential energy for the H_2 molecule: the spin singlet state corresponding to the curve E_+ is bound, whereas the spin triplet state E_- is not bound. The curve $E_{\text{n.e.}}$ represents the calculated potential energy when exchange is not included.

mum value as $R \to 0$—and it is *negative*, as shown in Fig. 10.11(a). Adding the three terms in Eq. (10.30) we obtain the curves labeled by E_+ and E_- in Fig. 10.11(b). Because $K < 0$, E_+ is always lower than E_-. The divergence at $R \to 0$ is due to the e^2/R term. The curve labeled $E_{\text{n.e.}}$ is obtained if we ignore the exchange contribution in which case only a very weak binding results. The $u_-(1, 2)$ state is unbound, a result in agreement with our intuition since it is coupled to a symmetric spin function. This implies that the two electrons have parallel spins, and by the exclusion principle they cannot be found at the same point of space. Thus, their wavefunctions have only small overlap.

The numerical values for the interatomic distance R_0, the dissociation energy, and the vibrational frequency as obtained by the perturbation calculation we described, are given in Table 10.2. Also included in the table are the measured values for these observables for the hydrogen molecule as well as results from a variational calculation. More accurate variational calculations can be brought into excellent agreement with experiment and indicate that the binding of the H_2 molecule is mainly due to the valence bond ($\sim 3/4$) but contains also a small part due to ionic binding ($\sim 1/4$). With such procedures the properties of the excited states of molecules and their physical parameters can be calculated. This is the field of

Table 10.2. Parameters of the H_2 Molecule

	Perturbation theory	Variational calculation[a]	Experimental values[b]
R_0 (Å)	0.80	0.76	0.74
Dissociation energy (eV)	3.14	3.76	4.48
Vibrational frequency (cm^{-1})	4800	4900	4320

[a] Varying $Z_{\rm eff}$; minimum at $Z_{\rm eff} = 1.166$.
[b] American Institute of Physics Handbook.

quantum chemistry that is highly successful, thus providing one more undisputable confirmation of the principles of quantum mechanics.

The H_2 Molecule as a Two-Level System

The hydrogen molecule can be treated in quite general terms as a two-state system. It is instructive to examine this approach since it is completely equivalent to our analysis of the ammonia molecule in Section 4.4. We clearly see that the binding of the molecule is due to the possibility of exchanging an electron between the two atoms. As we know, this is a consequence of the requirement that the overall wavefunctions have definite properties under particle exchange.

As in our recent analysis, let us assume that the molecule consists of two distinct hydrogen atoms whose nuclei are labeled by A and B. We define state $|\mathrm{I}\rangle$ of the system to be the configuration in which electron 1 is associated with A and electron 2 with B. State $|\mathrm{II}\rangle$ then corresponds to the reverse configuration, where electron 2 is associated with A and electron 1 with B. When the two atoms are at large distances, states $|\mathrm{I}\rangle$ and $|\mathrm{II}\rangle$ are stationary states and are, of course, degenerate. Their energy is $2E_0$, where E_0 is the ground-state energy of one atom. In this representation the Hamiltonian matrix is diagonal and of the form

$$\hat{H}(R \to \infty) = \begin{array}{c|cc} & |\mathrm{I}\rangle & |\mathrm{II}\rangle \\ \hline |\mathrm{I}\rangle & 2E_0 & 0 \\ |\mathrm{II}\rangle & 0 & 2E_0 \end{array} \qquad (10.31a)$$

Next, let the interatomic distance R be of the order of the "size" of the atoms. The electron wavefunctions will overlap, and therefore there is a finite probability that electrons 1 and 2 will interchange their roles. This gives rise to an off-diagonal element $K(R)$ that connects states $|\mathrm{I}\rangle$ and $|\mathrm{II}\rangle$. Furthermore, each state acquires additional potential energy e^2/R because of the electrostatic repulsion of the two nuclei and potential energy $J(R)$ due to the electrostatic repulsion (attraction) between the two electrons (the electron and the opposite nucleus). Therefore, the Hamiltonian matrix, when R is finite and of the order of a_0, takes the form

$$\hat{H}(R) = \begin{array}{c|cc} & |\mathrm{I}\rangle & |\mathrm{II}\rangle \\ \hline |\mathrm{I}\rangle & 2E_0 + e^2/R + J(R) & K(R) \\ |\mathrm{II}\rangle & K(R) & 2E_0 + e^2/R + J(R) \end{array} \tag{10.31b}$$

In the limit $R \to \infty$, the terms $K(R)$ and $J(R)$ tend to zero so that Eq. (10.31*b*) reduces to Eq. (10.31*a*).

From the structure of the Hamiltonian of Eq. (10.31*b*) it is clear that states $|\mathrm{I}\rangle$ and $|\mathrm{II}\rangle$ are *not* stationary states of the system. To obtain the stationary states we must diagonalize the matrix. The eigenstates in this new representation are the linear combinations [see Eqs. (4.53)]

$$\begin{aligned} |+\rangle &= \frac{1}{2^{1/2}}[|\mathrm{I}\rangle + |\mathrm{II}\rangle] \\ |-\rangle &= \frac{1}{2^{1/2}}[|\mathrm{I}\rangle - |\mathrm{II}\rangle] \end{aligned} \tag{10.32a}$$

and the Hamiltonian matrix referred to the new basis states is

$$\hat{H}'(R) = \begin{array}{c|cc} & |+\rangle & |-\rangle \\ \hline |+\rangle & 2E_0 + e^2/R + J(R) + K(R) & 0 \\ |-\rangle & 0 & 2E_0 + e^2/R + J(R) - K(R) \end{array} \tag{10.32b}$$

The energy of the states $|+\rangle$ and $|-\rangle$ is given by the diagonal elements of the above matrix, and it differs between the two states by the exchange term $K(R)$. This result is exactly the same as obtained in Eq. (10.30), and one needs to know the details of the wavefunction in order to evaluate the integrals $J(R)$ and $K(R)$.

However, the general features of the solution are immediately evident from this simple analysis.

10.6. SYSTEMS WITH IDENTICAL BOSONS

We now turn to an example of the effects of exchange symmetry for a system of identical bosons. The electromagnetic radiation contained in a cavity is described by such a system. We can think of this system as a *photon gas*, and we will show how Planck's radiation law follows from the requirement that the overall photon wavefunction must be symmetric under exchange.

Consider first a system of m identical noninteracting bosons enclosed in a finite volume. Since the particles are noninteracting the state of the system can be represented as a direct product of single-particle states

$$|\phi_m\rangle = |1\rangle\,|2\rangle \cdots |m\rangle \tag{10.33a}$$

where $1, 2, \ldots, m$ labels the individual bosons. Furthermore, we assume that all particles are in the same state α. Equation (10.33a) is symmetric under any permutation of the particles as required for an assembly of bosons, and is properly normalized. We will now try to build up this state starting from the vacuum state by introducing into the volume one boson at a time. We designate the vacuum state by $|\phi_0\rangle$, so that the amplitude $\langle\phi_1|\,\hat{H}_{\text{int}}\,|\phi_0\rangle$ indicates the transition from the vacuum to the state where one boson has been introduced in the volume. Here $\hat{H}_{\text{int}}$ is the Hamiltonian operator that causes the transition

$$\langle\phi_1|\,\hat{H}_{\text{int}}\,|\phi_0\rangle = w \tag{10.33b}$$

The probability for this transition to occur is $p_1 = |w|^2$. Furthermore, if we designate the boson introduced in the volume by the label "1," we can identify the state $|\phi_1\rangle$ of the system, with the single-particle state $|1\rangle$; namely,

$$|\phi_1\rangle = |1\rangle \tag{10.33c}$$

Next, we introduce the second boson so as to realize the state $|\phi_2\rangle$

of the system. In terms of single-particle states, $|\phi_2\rangle$ can be represented by

$$|\phi_2\rangle = |1\rangle\,|2\rangle \tag{10.34a}$$

Since bosons 1 and 2 are indistinguishable we cannot tell whether 1 was introduced first, followed by 2, or whether 2 was introduced first, followed by 1. Thus, we must symmetrize $|\phi_2\rangle$

$$|\phi_2\rangle = \frac{1}{2^{1/2}}[|1\rangle\,|2\rangle + |2\rangle\,|1\rangle] \tag{10.34b}$$

Physically this means that the state $|\phi_2\rangle$ can be reached from the vacuum by two indistinguishable processes. Therefore, the amplitude† for introducing two bosons into the volume is

$$\langle\phi_2|\,\hat{H}^2_{\text{int}}\,|\phi_0\rangle = \frac{1}{2^{1/2}}[\langle 2\,|\hat{H}_{\text{int}}\,|\phi_0\rangle\langle 1|\,\hat{H}_{\text{int}}\,|\phi_0\rangle + \langle 1|\,\hat{H}_{\text{int}}\,|\phi_0\rangle\langle 2|\,\hat{H}_{\text{int}}\,|\phi_0\rangle] \tag{10.34c}$$

Since the bosons are not interacting, the amplitudes $\langle 1|\,\hat{H}_{\text{int}}\,|\phi_0\rangle$ and $\langle 2|\,\hat{H}_{\text{int}}\,|\phi_0\rangle$ are as given by Eq. (10.33b), so that the probability for reaching the two-boson state is

$$P_2(\text{Bose}) = |\langle\phi_2|\,\hat{H}^2_{\text{int}}\,|\phi_0\rangle|^2 = \tfrac{1}{2}\,|w^2 + w^2|^2 = 2(|w|^2)^2 \tag{10.35a}$$

If the particles were distinguishable, as in classical physics, the probability would be the average of the probabilities of the various possibilities of obtaining a two-boson state. That is

$$\tfrac{1}{2}\{|\langle 2|\,\hat{H}_{\text{int}}\,|\phi_0\rangle\langle 1|\,\hat{H}_{\text{int}}\,|\phi_0\rangle|^2 + |\langle 1|\,\hat{H}_{\text{int}}\,|\phi_0\rangle\langle 2|\,\hat{H}_{\text{int}}\,|\phi_0\rangle|^2\}$$

or

$$P_2(\text{distinguishable}) = \tfrac{1}{2}\{|w\cdot w|^2 + |w\cdot w|^2\} = (|w|^2)^2 \tag{10.35b}$$

This last result is what one would expect intuitively. If $|w|^2$ is the probability for introducing one particle, then the probability for finding two particles in the same volume is $(|w|^2)^2$. This argument is

† The factor of $1/2^{1/2}$ in Eq. (10.34*b*) is the correct one because the states $|1\rangle\,|2\rangle$ and $|2\rangle\,|1\rangle$ are orthogonal and different from one another, even though they both lead to indistinguishable physical systems. Note also that since $|\phi_2\rangle$ can be reached in two different ways, one would need to divide the final probability by two, if we had not normalized Eqs. (10.34*b*) and (10.34*c*).

not true in quantum mechanics. We have seen that the probability of finding two identical bosons inside the volume is twice as large as the probability of finding two distinguishable particles. Consequently, identical bosons tend to occupy the same state more often compared to distinguishable particles: they "bunch together."† The opposite is true for identical fermions that cannot occupy the same state: they are "spread apart."

Next we introduce three bosons into our volume. There are 3! permutations of 3 particles, and therefore 6 indistinguishable processes that can lead from $|\phi_0\rangle$ to the state $|\phi_3\rangle$. As before, we must add the amplitudes corresponding to these processes and then square them. Taking account of the normalization [see also Eqs. (10.5a) and (10.8a)] we find

$$P_3(\text{Bose}) = \frac{1}{3!}|(3!)www|^2 = (3!)(|w|^2)^3 \tag{10.35c}$$

It is clear how to generalize these conclusions to n identical bosons

$$\begin{aligned} P_{n+1} &= [(n+1)!](|w|^2)^{n+1} \\ P_n &= (n!)(|w|^2)^n \\ &\text{etc.} \end{aligned} \tag{10.36}$$

If the system is already in a state containing n identical bosons, then the probability for introducing one additional identical boson is given by the ratio of P_{n+1} to P_n

$$P_{n\to n+1}(\text{Bose}) = \frac{P_{n+1}(\text{Bose})}{P_n(\text{Bose})} = (n+1)\,|w|^2$$

Therefore, the amplitude for this process is

$$\langle n+1|\,\hat{H}_{\text{int}}\,|n\rangle = (n+1)^{1/2}w \tag{10.37a}$$

Here we have chosen the amplitude to be real. The amplitude for removing one boson from a state containing $n+1$ identical bosons can be obtained by recalling that interchange of initial and final states results in complex conjugation of the amplitude, so that from Eq. (10.37a) we obtain

$$\langle n|\,\hat{H}_{\text{int}}\,|n+1\rangle = (n+1)^{1/2}w$$

† This effect has been clearly observed in the correlation in time of photons arriving at a detector from a coherent source. See, for instance, G. A. Rebka and R. V. Pound, *Nature*, **180**, 1035 (1957).

or equivalently

$$\langle n-1|\,\hat{H}_{\text{int}}\,|n\rangle = n^{1/2}w \qquad (10.37b)$$

Equations (10.37a) and (10.37b) are a direct consequence of the symmetry under exchange for a system of identical bosons. We recognize that these equations exhibit exactly the same factors as the matrix elements of the raising and lowering operators of the one-dimensional s.h.o. [Eqs. (9.16b)]. Therefore, a system of identical bosons obeys the same algebra and can be described by the same representations as the one-dimensional s.h.o. The nth state of the oscillator corresponds to the state $|\phi_n\rangle$ containing n bosons, the ground state to $|\phi_0\rangle$, etc. It should now be clear why the raising operator $\hat{a}^\dagger$ is referred to as the creation operator: it increases the occupation number n to $n+1$; that is, an identical boson is created. Correspondingly, the lowering operator $\hat{a}$ is referred to as the destruction operator. It should not be surprising that in classical mechanics an electromagnetic field in a cavity was also described by an assembly of simple harmonic oscillators.

Planck's Radiation Law

We discussed Planck's law in Section 5.4 where we used it to derive the rate for spontaneous transitions of a two-level system that could interact with the electromagnetic field. Here we will derive the radiation law as a consequence of the properties of an assembly of bosons as given by Eqs. (10.37a) and (10.37b). As in Section 5.4 we will examine the radiation when it is in equilibrium in a cavity at temperature T, and we will make use of the Boltzmann distribution. However, the properties of the atoms in the cavity walls play no role in the derivation.

Let the volume of the cavity be V and the temperature T. We assume that the atoms in the walls of the cavity† can absorb and emit photons of angular frequency ω when they undergo transitions from a state $|a\rangle$ to a state $|b\rangle$, or vice versa. Here $E_b - E_a = \hbar\omega$ as shown in Fig. 10.12. We let $R_u(R_d)$ be the rate of absorption

† The walls of the cavity absorb all the radiation incident on it and do not reflect any. In practice, an enclosed cavity is a blackbody. A small hole is provided so that one can observe the radiation contained in the cavity.

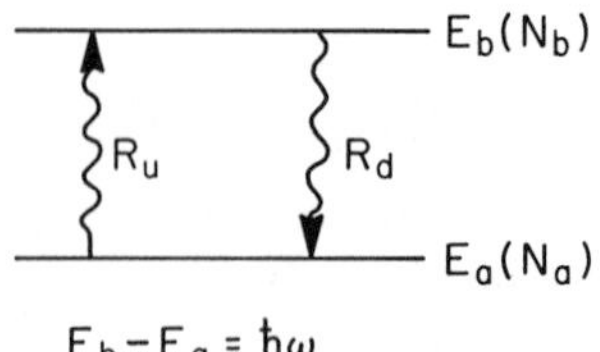

FIGURE 10.12. An atom with levels at energy E_b and E_a where $E_b = E_a + \hbar\omega$ can undergo transitions between these levels. The sketch implies an ensemble where N_b atoms are in the upper level and N_a atoms in the lower level. The rate for transitions *up* is designated by R_u and for transitions *down* by R_d.

(emission) of photons. At equilibrium the two rates must be equal

$$R_u = R_d$$

The rate $R_u(R_d)$ is given by the number of atoms in the initial state $N_a(N_b)$ times the probability for a transition per unit time

$$R_u = N_a P_{a\to b} = R_d = N_b P_{b\to a} \tag{10.38a}$$

At a given temperature T the number of atoms in the states $|a\rangle$ and $|b\rangle$ is determined by the Boltzmann distribution

$$N_a = N_0 e^{-E_a/kT}$$

$$N_b = N_0 e^{-E_b/kT}$$

where N_0 is a normalizing factor. Thus

$$\frac{N_a}{N_b} = e^{-(E_a - E_b)/kT} = e^{\hbar\omega/kT} \tag{10.38b}$$

In the transitions $a \to b$ the number of photons in the cavity decreases by one: thus the transition probability is the square of the amplitude of Eq. (10.37*b*)

$$P_{a\to b} = |\langle \bar{n} - 1|\, \hat{H}_{\text{int}}\, |\bar{n}\rangle|^2 = \bar{n}\, |w|^2 \tag{10.39a}$$

In the transition $b \to a$ the number of photons in the cavity increases by one; thus the transition probability is the square of the amplitude of Eq. (10.37*a*)

$$P_{b\to a} = |\langle \bar{n} + 1|\, \hat{H}_{\text{int}}\, |\bar{n}\rangle|^2 = (\bar{n} + 1)\, |w|^2 \tag{10.39b}$$

In Eqs. (10.39*a*)–(10.39*c*) we use $\bar{n}$ to indicate the average number of photons of angular frequency ω contained in the cavity. It follows that

$$\frac{P_{b\to a}}{P_{a\to b}}=\frac{\bar{n}+1}{\bar{n}} \tag{10.39c}$$

Introducing Eqs. (10.38*b*) and (10.39*a*)–(10.39*b*) in Eq. (10.38*a*) we can solve for $\bar{n}$

$$1+\frac{1}{\bar{n}}=e^{\hbar\omega/kT}$$

or

$$\bar{n}=\frac{1}{e^{\hbar\omega/kT}-1} \tag{10.40a}$$

The energy carried by the $\bar{n}$ photons of angular frequency ω is $E(\omega)=\bar{n}\hbar\omega$, so that†

$$E(\omega)=\frac{\hbar\omega}{e^{\hbar\omega/kT}-1} \tag{10.40b}$$

The above result gives the distribution of the energy in the black-body spectrum at discrete energies $\hbar\omega$. In practice, the observed distribution of energies, or frequencies ω, is *continuous* because a very large number of discrete frequencies can exist in a cavity whose dimensions are large as compared to the wavelength.‡

To convert Eq. (10.40*b*) to a continuous distribution in frequency we must multiply it by the number of discrete states contained in an interval of frequency $d\omega$ at the frequency ω. This is the familiar factor of the density of final states which we derived and used previously. The number of states dN in the phase-space volume Vd^3p is [see Eqs. (5.21)]

$$dN=\frac{V}{(2\pi\hbar)^3}d^3p$$

and therefore

$$\frac{dN}{d(\hbar\omega)}=\frac{dN}{dE}=\frac{V}{(2\pi\hbar)^3}\frac{p^2}{v}d\Omega \tag{10.41a}$$

† Here we neglect the zero-point energy of the field since $\bar{n}\gg 1$.

‡ Note that the energy spectrum of the radiation is discrete (quantized), whereas the energy states of the atoms are continuous since they correspond to their thermal motion. It is for this reason that the Boltzmann distribution was assumed in arriving at Eq. (10.38*b*).

For photons $p = \hbar\omega/c$, $v = c$, and we integrate over the angles to replace $d\Omega$ by 4π. We must also take account of the polarization of the photons.

At each frequency ω there are two photon states corresponding to their possible transverse polarization; thus Eq. (10.41a) must be multiplied by an additional factor of 2. Therefore,

$$\frac{dN}{d\omega} = \frac{V}{\pi^2\hbar^2}\frac{(\hbar\omega)^2}{c^3} \tag{10.41b}$$

Multiplying Eq. (10.40b) by the density of final states gives the continuous distribution of energies in the interval $d\omega$ at the frequency ω. Finally, we divide by the volume to obtain the distribution of the *energy density* $u(\omega)$, that is, the energy density in the frequency interval $d\omega$.

$$\boxed{u(\omega) = \frac{1}{V}\frac{dE(\omega)}{d\omega} = \frac{1}{\pi^2 c^3\hbar^2}\frac{(\hbar\omega)^3}{e^{\hbar\omega/kT} - 1}} \tag{10.42}$$

This result is Planck's law, which correctly describes the experimental observations. Planck derived the blackbody radiation law from thermodynamic arguments by postulating that the energy at a particular frequency ω could take only the discrete values $\hbar\omega$. This was the origin of the quantum idea, even though Planck himself was extremely troubled by the evolving discrepancy with the classical concepts. It was only five years later, when Einstein pointed out the explanation of the photoelectric effect in terms of Planck's energy quanta that the necessity of a complete revision of the classical approach became evident.

Comparison with Experiment

In comparing Planck's law with experimental data, Eq. (10.42) is written in terms of the intensity of radiation at the frequency ω in $d\omega$. The intensity $I(\omega)\,d\omega$ is the flux of energy in the interval $d\omega$, i.e., energy crossing unit area per unit time. If follows that

$$I(\omega) = vu(\omega) = cu(\omega)$$

and therefore

$$I(\omega)\,d\omega = \frac{\omega^2\,d\omega}{\pi^2c^2}\frac{\hbar\omega}{e^{\hbar\omega/kT}-1} \tag{10.43a}$$

For the limiting cases $\hbar\omega \ll kT$ and $\hbar\omega \gg kT$, Eq. (10.43a) reduces to

$$I(\omega)\,d\omega = kT\frac{\omega^2\,d\omega}{\pi^2c^2} \qquad \hbar\omega \ll kT$$

$$I(\omega)\,d\omega = \frac{\hbar\omega^3\,d\omega}{\pi^2c^2}e^{-\hbar\omega/kT} \qquad \hbar\omega \gg kT$$

The first limit is known as the Rayleigh–Jeans law and can be derived from classical arguments. In terms of the frequency ν

$$I(\nu)\,d\nu = \frac{8\pi h\nu^3}{c^3}\,d\nu\frac{1}{e^{h\nu/kT}-1} \tag{10.43b}$$

If one observes the intensity of radiation escaping from a small hole in the cavity a further geometric factor of $\frac{1}{4}$ must be included in the above results.

The intensity of the radiation per unit frequency has a maximum when the derivative of Eq. (10.43b) is zero or

$$\left(1-\frac{1}{3}\frac{h\nu}{kT}\right)e^{h\nu/kT}-1=0$$

with the solution

$$\frac{h\nu_{\max}}{kT} = \text{const} = 2.831 \tag{10.43c}$$

Namely, $\nu_{\max}/T$ is a constant. As the temperature increases, the frequency where the spectrum is maximum also increases. This result, known as Wien's displacement law, can be deduced from classical thermodynamic arguments from which, however, one can not predict the value of the constant in Eq. (10.43c).

Experimental data for the distribution of the radiation are shown in Fig. 10.13. Here the intensity per unit wavelength $I(\lambda)$ is plotted versus wavelength in μm. There is perfect agreement with the

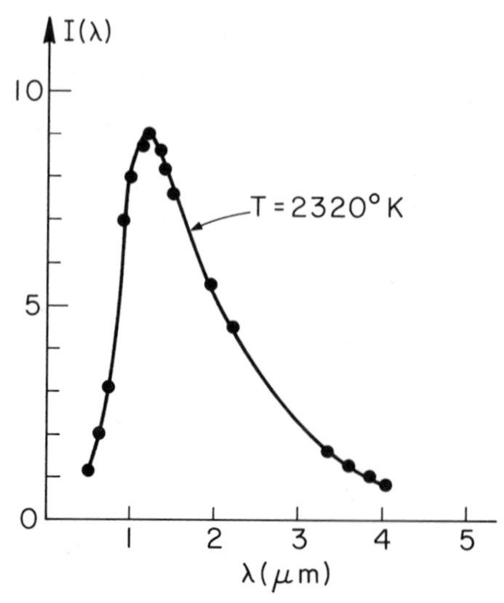

FIGURE 10.13. Distribution of the radiation emitted by a blackbody heated to $T = 2320$ K. The curve is the prediction of Planck's radiation law.

predicted distribution

$$I(\lambda)\, d\lambda = \frac{8\pi ch}{\lambda^5}\, d\lambda \frac{1}{e^{hc/\lambda kT} - 1}$$

We note in passing that since the visible spectrum lies approximately between $0.4\ \mu\text{m} < \lambda < 0.7\ \mu\text{m}$, incandescent lamps which operate with filaments of $T \sim 2{,}000°\text{C}$ are highly inefficient devices.

A more recent measurement of data relating to the radiation law is shown in Fig. 10.14. Here we give the observed spectrum of the cosmic background radiation first discovered by Penzias and Wilson in 1965. It is believed that this radiation was produced at the creation of the universe and is reaching us now as the universe expands. In the *big bang theory* the spectrum of the background radiation must be that of a blackbody, and the data are in close agreement as shown by the solid curve in the figure. The peak of the spectrum is at $(1/c)\nu_{\text{max}} = 5.8\ \text{cm}^{-1}$ which, according to Eq. (10.43*c*), yields $T = 2.96$ K for the apparent temperature. Taking account of the Doppler shift due to the expansion of the universe, this corresponds to $T \simeq 4{,}500$ K at a time approximately 500,000

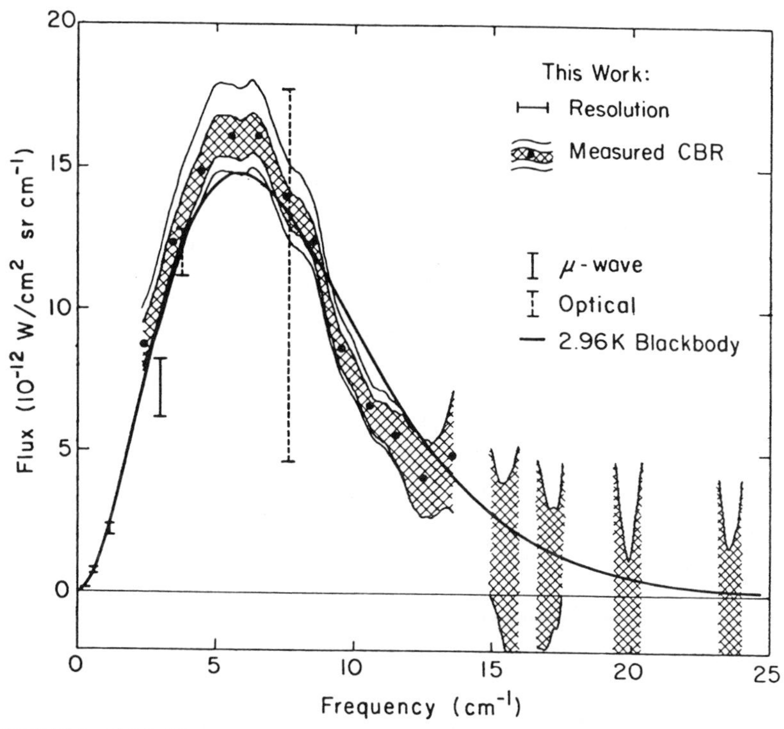

FIGURE 10.14. The measured spectrum of the cosmic background radiation is shown by the shaded areas. The best fit to these data is given by the solid line, which is Planck's blackbody radiation spectrum for $T = 2.96$ K. [From D. P. Woody and P. L. Richards, *Phys. Rev. Lett.* **42,** 925 (1979)].

years after the creation of the universe. This is the temperature at which the early universe became transparent to electromagnetic radiation, according to the current models of nucleosynthesis.

Problems

PROBLEM 1

Show that if the Pauli raising and lowering matrices $\hat{\sigma}_+$ and $\hat{\sigma}_-$ are interpreted as destruction ($\hat{b}=\hat{\sigma}_+$) and creation ($\hat{b}^\dagger=\hat{\sigma}_-$) operators they obey the anticommutation relation

$$\{\hat{b}, \hat{b}^\dagger\} = \hat{b}\hat{b}^\dagger + \hat{b}^\dagger\hat{b} = 1$$

Next, show that you can have *only* zero or one particle (of the type created by $\hat{b}^\dagger$) in any state. This is, of course, the Pauli principle.

PROBLEM 2

Consider three identical particles bound in a three-dimensional s.h.o. potential.

(*a*) Write the $|\text{ket}\rangle$ corresponding to the lowest energy as a product of the $|\text{ket}\rangle$'s of each particle when they have *spin* 1/2.

(*b*) Write the $|\text{ket}\rangle$ of same energy as in (*a*) when the particles have *spin 0*.

PROBLEM 3

The three nucleons in ^3_2He have spin angular momentum 1/2 each and (practically) orbital angular momentum zero. Let

$$|m_1, m_2, m_3\rangle$$

represent the state with m_1, m_2, and m_3 as the z-components of the angular momentum of the nucleons 1, 2, and 3, respectively. We suppress the quantum numbers j_1, j_2, and j_3, since they all take value 1/2. If ${}_2^3$He had a total angular momentum 3/2 (which it actually does not).

(*a*) Write the state $|J=3/2, M=3/2\rangle$ corresponding to the nucleus having total angular momentum 3/2 and z-component 3/2 in terms of $|m_1, m_2, m_3\rangle$ states.

(*b*) From the result of (*a*) obtain the state corresponding to angular momentum $J=3/2$ and $M=1/2$ as linear combinations of states $|m_1, m_2, m_3\rangle$.

PROBLEM 4

(*a*) Give the possible values of the total angular momentum in the configurations implied by the spectroscopic notation

$$^1S, {}^3S, {}^3P, {}^2D, {}^4D$$

(*b*) Determine the electron configuration in the ground state of oxygen (O), chlorine (Cl), and iron (Fe).

Note: Compare your answers with the established configurations and understand the differences, if any.

PROBLEM 5

Make an approximate plot of the low-lying energy levels of the mercury atom ($Z=80$). To proceed, consider the valence electrons and see what states are available when the coupling is of the $L-S$ form (this is, in fact, the case). Show also the spectrum if the coupling was of the $J-J$ form. Label all states by spectroscopic notation.

PROBLEM 6

A block of copper is bombarded by an electron beam accelerated through a potential of 40 kV. Make a plot as a function of energy of the characteristic x-ray lines that are emitted. Label appropriately. If you can discuss their relative intensity, do so. For this last part you will need some reference beyond the text.

PROBLEM 7

Calculate the potential energy due to the interaction of the magnetic moments (spin–spin interaction) of the two electrons in the helium atom. Assume some reasonable average distance (you can use the result of the calculation of the term $e^2/|\mathbf{r}_1 - \mathbf{r}_2|$). The magnetic moments are known. Compare this with the potential energy due to the electrostatic forces.

PROBLEM 8

Make a plot to scale for the Raman spectrum of H_2 and D_2 in their ground state, give the relative intensities of the lines. You should make a reasonable assumption about the rotational constants of these two molecules. You can assume that the interatomic spacing is $R = 1$ Å.

In plotting the line intensities, include the fact that the states are populated according to a Boltzmann distribution at room temperature.

PROBLEM 9

Starting from the Hamiltonian of Eq. (10.22) calculate the weak attractive force between two neutral hydrogen atoms that are separated by a large distance. This is the van der Waals force.

PROBLEM 10

Tritium is an isotope of hydrogen where the nucleus ${}^3_1H$, consists of a proton and two neutrons. (*a*) By how much does the energy of the ground state of the tritium atom differ from the hydrogen atom ground state? Is it larger or smaller?

The $^3_1$H nucleus can decay into a $^3_2$He nucleus (two protons and one neutron) with the emission of an electron and an antineutrino. (*b*) What would be the energy of the ground state of the *ionized* $^3_2$He atom? (that is when only one electron is bound to the $^3_2$He nucleus). What is the wavefunction? (*c*) The nucleus of an $^3_1$H atom decays *instantaneously* into a $^3_2$He nucleus. Find the probability that the $^3_2$He ion will be in its ground state.

Note: For part (*c*) explain your procedure, set up appropriate equations, and only then perform the necessary integral.

PROBLEM 11

Consider a two-electron atom in a state with $L=2$, and let the total spin of the electrons be $S=1$. (*a*) Give the possible J-values. (*b*) Make a plot of the fine structure of the $L=2$ states [in units of the radial matrix element $\mathcal{M}=(e^2/4\pi\varepsilon_0)(\hbar^2/2m^2c^2)\langle(1/r)(dV/dr)\rangle_{n,l}$; do *not* evaluate $\mathcal{M}$]. (*c*) Define the center of gravity of the fine-structure multiplet by assigning to each J-level a weight equal to the number of M-states corresponding to it. Show that the center of gravity corresponds to the energy of the $L=2$ state as obtained without the consideration of spin–orbit coupling.

PROBLEM 12

Positronium is a bound system of an electron (e^-) and a positron (e^+) where the binding arises from the Coulomb force. Treat the problem in the center-of-mass system and find

(*a*) the ground state energy in eV;
(*b*) the possible values of the total spin;
(*c*) Make a plot of all the $n=1$ and $n=2$ states of the system including the spin–orbit interaction (ignore the spin–spin interaction). Be quantitative.
(*d*) Evaluate the probability that the electron and positron are within a relative distance $r=0.01a_0$ (a_0 = Bohr radius) when the system is in the ground state.

(*e*) From part (*d*) try to estimate the lifetime of positronium if annihilation $e^+ + e^- \rightarrow \gamma + \gamma$ occurs whenever the electron and positron are within a relative distance $\Delta r = \hbar/mc$. You can use a semiclassical argument here.

PROBLEM 13

The π^--meson is an elementary particle with spin $S = 0$. It can be bound to a deuteron ($S = 1$) to form a meso-deuterium atom. The π^--meson is captured from an atomic S-state leading to the reaction

$$\pi^- + d \rightarrow n + n$$

(*a*) Making use of the conservation of total angular momentum and of the spin-statistics theorem establish the (relative) angular momentum of the two neutrons.
(*b*) From (*a*) above show that in order to have parity conservation in the above reaction we must assign a negative (intrinsic) parity to the π^- meson.

Note: The (intrinsic) parity of the neutron and proton is taken to be positive by convention.

Chapter 11

SCATTERING: PART I

The phenomenon of scattering is quite important in the study of classical systems. But when the size of systems become so small that we cannot directly see them, or when the forces involved are so short-ranged that we cannot directly measure them, scattering becomes an indispensable probe. Consider, for example, the structure of atoms: the fact that atoms consist of a positively charged nucleus surrounded by a cloud of electrons was understood by scattering alpha particles off thin layers of matter. Similar experiments indicate that the electron is a point particle up to very small distances. The proton, on the other hand, exhibits a bound-state structure made up of yet smaller constituents named quarks. It is probably true that scattering is the primary method of investigation of high-energy phenomena. Scattering is also important in the study of transport phenomena such as conductivity or the specific heat of bulk matter since these properties are affected by the scattering of free electrons and phonons (the vibrational quanta) inside the material.

The simplest form of scattering we can think of is that of a beam of particles—with well-defined momenta—from a fixed center of force. Classically, the trajectories of the particles change due to the force experienced, and the deflection of the trajectories from the initial direction defines the scattering angle. Even though a scattering experiment may consist of measuring the trajectories of individual particles, the result is expressed in terms of the *differential cross section,* which is determined from the number of particles scattered into a solid angle. This statistical measurement is quite useful in the sense that it provides information on the shape of the potential, the nature of the force, and so on.

In classical mechanics, given the initial conditions, we can theoretically calculate the trajectory exactly. In quantum mechanics, on the other hand, the notion of a trajectory does not exist, and one must solve the Schrödinger equation for the scattering problem. In Chapters 8 and 9 we studied bound systems; these are described by the stationary states of the Hamiltonian, subject to the boundary condition that the probability amplitude vanish as the separation between the two particles tends to infinity. The same Hamiltonian, as in the case of bound systems, can also give scattering wavefunctions. These obey the boundary condition that at infinite separation they must reduce to free-particle wave functions.† Once the scattering solutions are known it is a simple matter to obtain the differential cross section.

In this chapter we first define various relevant quantities such as the differential cross section. For most potentials, the scattering equations take a much simpler form in the center-of-mass frame. However, since the measurements are done in the laboratory, we indicate how observables are transformed from the center-of-mass frame to the laboratory frame. We also discuss the boundary conditions on the scattering solutions. The scattering equation is then solved by the method of partial-wave analysis, which is quite useful for relatively low-energy scattering. We derive the optical theorem by this method and apply the method of partial-wave analysis to various processes of physical interest. The effects of spin are treated briefly in a qualitative way. In Chapter 12 we will discuss solutions to the scattering equation by other methods and then apply them to other processes of physical interest, namely, scattering phenomena at high energies.

11.1. DIFFERENTIAL AND TOTAL CROSS SECTION

In a typical scattering experiment the number of particles involved is quite large so that even classically it is impractical to measure the details of each individual trajectory. What one measures is the initial velocity of the particles (they are usually prepared to be monoenergetic) and their final velocity. That is, if the direction of the incident beam is taken to be at $\theta = 0$, then one measures the number of particles that scatter into a solid angle $d\Omega$ at (θ, ϕ).

† This is exactly true for short-range potentials but also remains a good approximation for long-range potentials such as the Coulomb potential.

Hence, if N is the number of incident particles per unit area per unit time and ΔN of them scatter by an angle (θ, ϕ) into the solid angle $\Delta\Omega$ in unit time, then the differential cross section is defined to be†

$$\sigma(\theta, \phi) = \frac{d\sigma}{d\Omega}(\theta, \phi) = \frac{1}{N}\frac{\Delta N}{\Delta\Omega} \tag{11.1a}$$

in the limit $\Delta\Omega \to 0$. The differential cross section is a useful quantity since it probes the shape of the potential or the nature of the force—the larger the differential cross section, the stronger the force. In most scattering phenomena, however, the potentials are spherically symmetric. Consequently, physical observables cannot have an azimuthal dependence (no ϕ dependence). One can in such cases define the differential cross section for scattering into a ring around the beam axis by integrating out the azimuthal angle, i.e.,

$$\sigma(\theta) = \int d\phi\sigma(\theta, \phi) = 2\pi\sigma(\theta, \phi) \tag{11.1b}$$

The ring is at angle θ and has width $d\theta$ and radius $r \sin\theta$ as shown in Fig. 11.1. From Eqs. (11.1a) and (11.1b) we can also define the total cross section as

$$\sigma_T = \sigma_{\text{tot}} = \int \sigma(\theta, \phi)\, d\Omega = \int_0^{\pi} \sigma(\theta) \sin\theta\, d\theta \tag{11.1c}$$

The total cross section measures the total number of scattered particles (i.e., at all angles) per unit time when one particle is incident per unit area in unit time.

It is easy to see from the definition of Eq. (11.1a) that both the differential and the total cross sections have the dimensions of an area. Geometrically, one can think of the total cross section as the transverse area subtended by the scatterer. Thus, classically, if we have a potential with range a and there is azimuthal symmetry, then the part of the incident beam that would suffer any scattering is that which is confined to an area πa^2 perpendicular to the beam direction. The total cross section in this case, is simply πa^2. This simple

† The differential cross section should be properly designated by $d\sigma/d\Omega$ since it is defined as the ratio of two infinitesimal quantities. However, for simplicity in writing equations we will adopt the more condensed notation $\sigma(\theta, \phi)$. The reader should keep in mind, however, the important difference between an integrated cross section such as σ_T and a differential cross section $\sigma(\theta, \phi)$, in spite of the notational similarity.

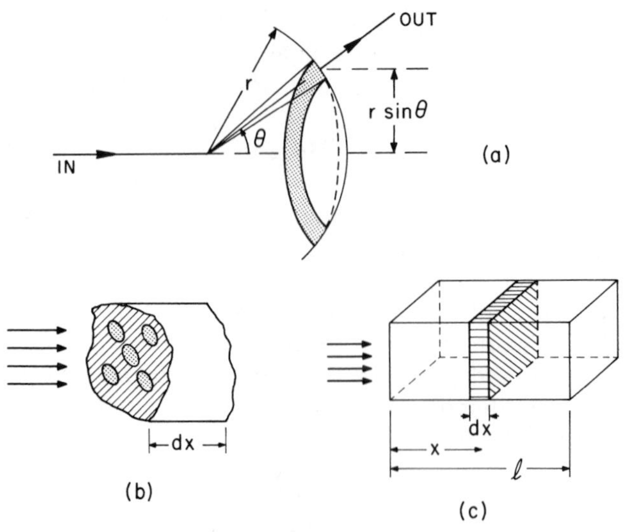

FIGURE 11.1. (a) Scattering of an incident particle through an angle θ. The area of the ring perpendicular to the scattered trajectory is $2\pi r \sin\theta \, d\theta$. (b) The target in a scattering experiment consists of many individual scatterers that can be thought of as each subtending an area equal to the total cross section. (c) A target of finite length where the element of thickness dx at a distance x is indicated by the shading.

derivation fails in quantum mechanics, as we will show later, and this is primarily due to the wave nature of particles. The unit in which scattering cross sections are measured in nuclear and high-energy physics is called a *barn* and corresponds to

$$1 \text{ barn} = 10^{-24} \text{ cm}^2$$

Most frequently in scattering experiments a beam of particles is incident on a fixed target of finite thickness. Let us derive the appropriate expression for the total cross section in such a case. Each target particle acts as a scattering source. Let the shaded area around each target particle [Fig. 11.1(b)] represent the total cross section. If the incident particle strikes the shaded area it will scatter, otherwise it will not interact.† If the target is of infinitesimal

† This picture fails for long-range potentials for which $\sigma_T \to \infty$. In practice, however, σ_T can always be treated as finite.

thickness dx and has a density of ρ particles per unit volume, the fraction of the transverse area that is shaded is

$$\frac{dA}{A} = \sigma_T \rho \, dx \tag{11.2a}$$

Therefore, the probability that an incident particle will scatter in the interval dx is

$$P(x)\, dx = \frac{dA}{A} = \sigma_T \rho \, dx \tag{11.2b}$$

We define N_0 as the number of particles incident on the target and $N(x)$ as the number of incoming particles at a distance x from the edge of the target as shown in Fig. 11.1(c). Whenever a particle scatters, $N(x)$ decreases such that $dN_s(x) = -dN(x)$, where $dN_s(x)$ represents the number of scattered particles at the distance x (in the interval dx). Furthermore, we know that

$$\begin{aligned} dN(x) &= -N(x)P(x)\, dx \\ &= -N(x)\sigma_T \rho \, dx \end{aligned}$$

or

$$N(x) = N_0 e^{-\sigma_T \rho x} \tag{11.3a}$$

It is clear that the number of particles scattered from a target of thickness l is

$$\begin{aligned} N_s(l) &= N_0 - N(l) = N_0 - N_0 e^{-\sigma_T \rho l} \\ &= N_0(1 - e^{-\sigma_T \rho l}) \end{aligned}$$

If $\sigma_T \rho l \ll 1$, then we can write

$$N_s(l) \simeq N_0 \sigma_T \rho l \tag{11.3b}$$

so that the total cross section in this case is given by

$$\sigma_T = \frac{N_s(l)}{N_0 \rho l} \tag{11.3c}$$

The above results are obtained with the assumption that the transverse dimensions of the beam are smaller than those of the target; this is true in most experimental arrangements.

In defining the cross section we have implicitly used the concept of flux of incoming and outgoing particles. This concept was first introduced in Section 6.7 and we recall that the flux $\mathbf{j}$ gives the number of particles crossing the unit area normal to their direction of motion per unit time. If $\mathbf{v}$ is the velocity of the particles and ρ their density, then

$$\mathbf{j} = \rho \mathbf{v} \tag{11.4a}$$

Furthermore, if the amplitude for the incoming particles is normalized to one per unit volume, then the flux coincides with the probability current density and is given by

$$\mathbf{j} = -\frac{i\hbar}{2m}(\psi^* \nabla\psi - (\nabla\psi^*)\psi] \tag{11.4b}$$

This measures the probability that a particle will cross a unit area normal to the direction of the current in unit time.

Transformation from the Center-of-Mass Frame to the Laboratory Frame

We have already shown in Section 8.2 that when the potentials involved are central,† the motion can be decomposed into that of the center of mass and an effective reduced system. The motion of the center of mass is not subject to any forces and therefore in the center-of-mass frame the Hamiltonian under study becomes much simpler. For example, for two spinless particles, the number of degrees of motion (freedom) reduces by one half. Therefore, calculation of quantities of physical interest in the center-of-mass frame results in great simplification.

In fact, in studying the bound-state solutions we neglected the center-of-mass motion completely since it does not affect the properties of the stationary states. In the case of scattering, however, although it may be simpler to calculate observable quantities in the center-of-mass frame, the actual observations are made in the laboratory. Thus, we must know how to transform various quantities such as the scattering angle and differential cross section back into the laboratory frame. It is worth pointing out here that the total

† We do not consider noncentral potentials which arise in the presence of spin-dependent forces; but see Section 11.5.

cross section σ_T is the same in both frames since it represents the total probability that an incident particle will scatter; hence it must be independent of the choice of the reference frame.

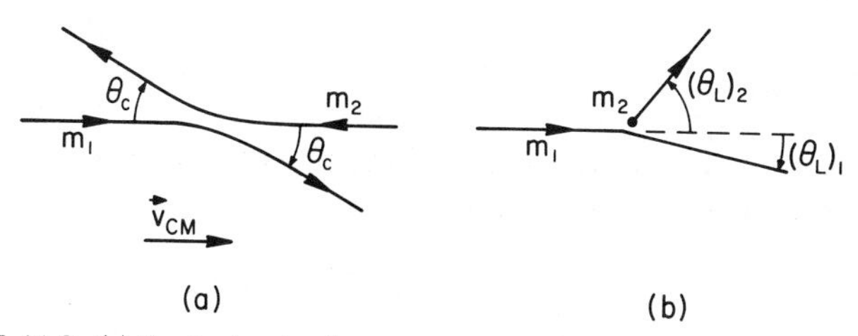

FIGURE 11.2. (a) Scattering in the center-of-mass system. (b) The same scattering process as observed in the laboratory.

Let us consider the scattering of two particles with masses m_1 and m_2. Let m_2 be stationary in the laboratory frame and let m_1 be moving along the Z-axis with a velocity† $v_1 \ll c$. This is the usual case in experiments except for colliding-beam arrangements. Furthermore, let us use the subscripts C and L to denote various quantities in the center-of-mass and laboratory frames, respectively, as shown in Fig. 11.2. For this system, therefore,

$$\mu = \text{reduced mass} = \frac{m_1 m_2}{m_1 + m_2} \tag{11.5a}$$

and, by definition, the velocity of the center of mass is

$$\mathbf{v}_{CM} = \frac{m_1 \mathbf{v}_1}{m_1 + m_2} = \frac{m_1 (\mathbf{v}_1)_L}{m_1 + m_2} \tag{11.5b}$$

In the center-of-mass frame, therefore, the particles have velocities given by

$$(\mathbf{v}_1)_C = (\mathbf{v}_1)_L - \mathbf{v}_{CM} = \frac{m_2 (\mathbf{v}_1)_L}{m_1 + m_2}$$

$$(\mathbf{v}_2)_C = (\mathbf{v}_2)_L - \mathbf{v}_{CM} = -\frac{m_1 (\mathbf{v}_1)_L}{m_1 + m_2} \tag{11.5c}$$

† We will derive and use nonrelativistic transformation equations. When v_1 approaches the velocity of light c, one must use relativistic kinematics, which add to the complexity of the equations but are otherwise equivalent to the nonrelativistic case.

Thus, from Eq. (11.5*c*) it is clear that in the center-of-mass frame the two particles move toward each other with equal and opposite momenta as shown in Fig. 11.2(a). After the scattering they must again move with equal and opposite momenta. Thus, in the center-of-mass frame the two particles scatter by equal and opposite angles. The scattering is not so simple in the laboratory frame and is shown in Fig. 11.2(b).

If we denote by primes the velocities after scattering and remember that the scattering is elastic, Eq. (11.5*c*) gives

$$(\mathbf{v}_1')_C = (\mathbf{v}_1')_L - \mathbf{v}_{CM}$$

We express this equation in terms of its components along the beam direction

$$v_{1C} \cos \theta_C = v_{1L} \cos(\theta_L)_1 - v_{CM}$$

or

$$v_{1L} \cos(\theta_L)_1 = v_{CM} + v_{1C} \cos \theta_C \tag{11.6a}$$

and transverse to the beam direction

$$v_{1C} \sin \theta_C = v_{1L} \sin(\theta_L)_1 \tag{11.6b}$$

The ratio of the two expressions above gives

$$\tan(\theta_L)_1 = \frac{\sin \theta_C}{\cos \theta_C + v_{CM}/v_{1C}} = \frac{\sin \theta_C}{\cos \theta_C + \gamma} \tag{11.7a}$$

where

$$\gamma = \frac{v_{CM}}{v_{1C}} = \frac{m_1}{m_2} \tag{11.7b}$$

which follows from Eqs. (11.5*b*) and (11.5*c*). Furthermore, it can be shown easily that

$$(\theta_L)_2 = \tfrac{1}{2}(\pi - \theta_C) \tag{11.7c}$$

Equation (11.7*a*) holds for both elastic as well as inelastic scatterings. In the case of inelastic scattering the particle number is not necessarily conserved and as a consequence rest-mass energy (internal energy) is converted into kinetic energy of particles. Suppose we are considering the following inelastic scattering process

$$m_1 + m_2 \rightarrow m_3 + m_4$$

The amount of mass converted to kinetic energy of the final particles is known as the Q-value of the reaction and is designated by Q. It then holds that

$$\gamma = \left(\frac{m_1 m_3}{m_2 m_4} \frac{E}{E+Q}\right)^{1/2} \tag{11.7d}$$

where E is the total initial energy of the two particles in the center-of-mass frame. Clearly, for elastic scattering, i.e., $Q=0$, $m_1=m_3$, and $m_2=m_4$, the above expression reduces to Eq. (11.7b).

The relation between the differential cross section in the center-of-mass frame and the laboratory frame can be obtained from the simple observation that the same particles that go into the solid angle $d\Omega_C$ at θ_C in the center-of-mass frame go into the solid angle $d\Omega_L$ at θ_L in the laboratory frame. That is,†

$$\sigma_L(\theta_L) \sin\theta_L \, d\theta_L = \sigma_C(\theta_C) \sin\theta_C \, d\theta_C$$

or

$$\sigma_L(\theta_L) = \sigma_C(\theta_C) \left| \frac{d\cos\theta_C}{d\cos\theta_L} \right| \tag{11.8a}$$

The function $|d\cos\theta_C/d\cos\theta_L|$ is the Jacobian of the transformation and can be evaluated from Eqs. (11.7a) and (11.7b). The relation for the cross sections turns out to be [see Problem 11.5]

$$\begin{aligned} \sigma_L(\theta_1) &= \sigma_C(\theta_C) \frac{[1+2\gamma\cos\theta_C+\gamma^2]^{3/2}}{|1+\gamma\cos\theta_C|} \\ &= \sigma_C(\theta_C) \frac{\{[1-\gamma^2\sin^2(\theta_L)_1]^{1/2}+\gamma\cos(\theta_L)_1\}^2}{[1-\gamma^2\sin^2(\theta_L)_1]^{1/2}} \\ \sigma_L(\theta_2) &= \sigma_C(\theta_C) 4\cos(\theta_L)_2 \end{aligned} \tag{11.8b}$$

Using the last of Eqs. (11.8b) and Eq. (11.7c) it is a simple matter to verify that the total cross section in the laboratory frame is the same as in the center of mass. That is,

$$(\sigma_T)_L = (\sigma_T)_C$$

† Here we are assuming rotational symmetry, and hence the differential cross section does not depend on ϕ.

For the special case of equal mass particles, i.e., $m_1 = m_2$ (therefore, $\gamma = 1$) the transformation equations reduce to

$$(\theta_L)_1 = \frac{\theta_C}{2}$$

$$(\theta_L)_1 + (\theta_L)_2 = \frac{\pi}{2} \tag{11.8c}$$

$$\sigma_L(\theta_{1,2}) = \sigma_C(\theta_C) 4 \cos(\theta_L)_{1,2}$$

When the target particle is much more massive than the beam particle, $\gamma \rightarrow 0$ and the laboratory and center-of-mass frames are equivalent.

Equipped with these kinematic relations we are now ready to solve the scattering equation. Of course, as we pointed out earlier it is much simpler to deal with the theoretical equation in the center-of-mass frame since there the problem can be thought of as the scattering of a particle with mass μ from a fixed potential. We do so in the next section.

11.2. SOLUTIONS OF THE SCATTERING EQUATION: THE METHOD OF PARTIAL WAVES

In the center-of-mass frame, scattering is governed by the same Hamiltonian as for the bound states of the system. Namely,

$$\hat{H} = -\frac{\hbar^2}{2\mu}\nabla^2 + U(|\mathbf{r}|) \tag{11.9a}$$

where μ is the reduced mass of the system. The stationary states of the Hamiltonian take the same form as for the bound states

$$\psi(\mathbf{r}, t) = \psi(\mathbf{r}) e^{-iEt/\hbar} \tag{11.9b}$$

In this case, however, the energy of the system E is positive and, as we shall see, has a continuous spectrum. Furthermore, if we are considering elastic scattering, the energy E does not change, and hence we need only study the space part of the wavefunction that satisfies the equation

$$\hat{H}\psi(\mathbf{r}) = E\psi(\mathbf{r})$$

or

$$\left[\nabla^2 + k^2 - \frac{2\mu}{\hbar^2} U(|\mathbf{r}|)\right]\psi(\mathbf{r}) = 0, \qquad k^2 = \frac{2\mu E}{\hbar^2} \tag{11.9c}$$

Before solving Eq. (11.9c) let us establish the boundary conditions that the wave function must satisfy. First, if the potential vanishes for infinite separation and if we assume that initially the particle was far from the scattering center, the solutions before the scattering ($t \to -\infty$) must satisfy

$$(\nabla^2 + k^2)\psi_{\text{in}}(\mathbf{r}) = 0 \qquad \begin{cases} r \to \infty \\ t \to -\infty \end{cases} \tag{11.10a}$$

that is, they must be essentially described by free-particle solutions. If the particle is incident along the Z-axis we can write the incident wavefunction as a plane wave

$$\psi_{\text{in}}(\mathbf{r}) = e^{ikz} \tag{11.10b}$$

In the scattering region, of course, the wavefunction must depend on the form of the potential and must satisfy

$$\left[\nabla^2 + k^2 - \frac{2\mu}{\hbar^2} U(|\mathbf{r}|)\right]\psi(\mathbf{r}) = 0 \tag{11.11a}$$

After scattering, however, the wavefunction must again have a free-particle form far from the scattering source, and thus must satisfy

$$(\nabla^2 + k^2)\psi_f(\mathbf{r}) = 0 \qquad \begin{cases} r \to \infty \\ t \to +\infty \end{cases} \tag{11.11b}$$

It is easy to visualize this sequence of events (see Fig. 11.3). Initially a plane wave is incident along the Z-axis. If we assume that the scattering potential is spherically symmetric, then at very far distances from the scattering source we will have spherically outgoing scattered waves. Thus, the form of the total wavefunction at large distances from the scattering center (after the scattering has occurred) is given by†

$$\psi_f(\mathbf{r}) \xrightarrow[r\to\infty]{} \psi_{\text{in}}(\mathbf{r}) + \psi_{\text{sc}}(\mathbf{r}) = e^{ikz} + f(\theta, \phi)\frac{e^{ikr}}{r} \tag{11.12}$$

† Of course, we are assuming that the potential is centered at the origin.

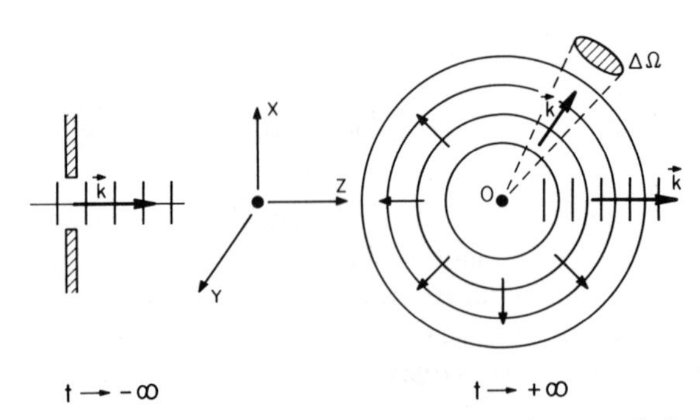

FIGURE 11.3. Diagrammatic sketch of the asymptotic states before and after scattering from a fixed potential.

Here $f(\theta, \phi)$ measures the angular distribution of the scattered wave and is known as the scattering amplitude. The wavefunctions will be normalized to one particle per unit volume, and it can be easily shown that $\psi_{sc}(\mathbf{r})$ also satisfies the free-particle equation. The factor $1/r$ in $\psi_{sc}(\mathbf{r})$ is necessary to insure that the probability of scattered particles crossing a solid angle $d\Omega$ does not depend on the radial distance r.

Let us now calculate the flux associated with the incident as well as the scattered wavefunctions for large distances.

$$\mathbf{j}_{\text{in}} = -\frac{i\hbar}{2\mu}(\psi_{\text{in}}^* \nabla \psi_{\text{in}} - (\nabla \psi_{\text{in}}^*)\psi)$$

$$= \frac{\hbar k}{\mu}\mathbf{n}_z \tag{11.13a}$$

Similarly, the flux of the scattered wave is defined as

$$\mathbf{j}_{\text{sc}} = -\frac{i\hbar}{2\mu}(\psi_{\text{sc}}^* \nabla \psi_{\text{sc}} - (\nabla \psi_{\text{sc}}^*)\psi_{\text{sc}})$$

Using spherical coordinates (See Appendix 8) the radial component of the flux at large distances becomes

$$(\mathbf{j}_{\text{sc}})_r = -\frac{i\hbar}{2\mu}\left[\psi_{\text{sc}}^* \frac{\partial}{\partial r}\psi_{\text{sc}} - \left(\frac{\partial}{\partial r}\psi_{\text{sc}}^*\right)\psi_{\text{sc}}\right]$$

or

$$(\mathbf{j}_{\text{sc}})_r \xrightarrow[r\to\infty]{} \frac{\hbar k}{\mu}\frac{|f(\theta, \phi)|^2}{r^2} + \mathrm{O}\left(\frac{1}{r^3}\right) \tag{11.13b}$$

Equation (11.13b) shows that the probability of particles scattering across an area dS of a sphere of radius r is (in the limit $r \to \infty$)

$$\mathbf{j}_{sc} \cdot d\mathbf{S} = \frac{\hbar k}{\mu} |f(\theta, \phi)|^2 \frac{dS}{r^2}$$

$$= \frac{\hbar k}{\mu} |f(\theta, \phi)|^2 \, d\Omega \tag{11.13c}$$

where $dS/r^2 = d\Omega$ is the solid angle subtended at the origin by the area element.

From Eqs. (11.13a) and (11.13c) we immediately see that the differential cross section is given by

$$\sigma(\theta, \phi) = |f(\theta, \phi)|^2 \tag{11.14}$$

This is the fundamental relation between the observable differential cross section and the quantum-mechanical scattering amplitude. It is similar to the familiar relation between the probability amplitude $\psi(x) = \langle x \mid \psi \rangle$ and the observable probability of finding a particle at x; hence the function $f(\theta, \phi)$ is referred to as an amplitude.

It is evident now that solving the scattering equation amounts to finding the scattering amplitude. There are several ways of doing this and we will discuss here the method of partial waves. This method was first introduced by Lord Rayleigh in connection with the scattering of sound waves and was later adapted to quantum mechanics. The method consists of writing the wavefunction as a superposition of various angular momentum components. This point needs clarification and can be understood as follows: Classically, a particle with well-defined momentum p and impact parameter b has a fixed angular momentum bp. In quantum mechanics, on the other hand, if we choose the particle to have well-defined momentum, its position becomes completely uncertain. Consequently, it no longer has a unique angular momentum, and hence the wavefunction can be written as a superposition of various angular momentum components.

In this method, therefore, the scattering amplitide is expressed as a sum over all partial-wave contributions, and, in principle, the result is exact. However, the method can be useful only if a few angular momenta contribute dominantly to the scattering. If the range of the potential is a_0 and the momentum of the particles is $\mathbf{p}$,

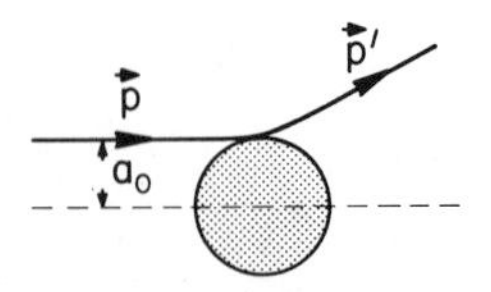

FIGURE 11.4. The relation between the maximum angular momentum wave and the range of the potential is simply $l_{max} = a_0 p$.

then only those angular momentum components would contribute that satisfy (see Fig. 11.4)

$$\hbar[l(l+1)]^{1/2} \simeq \hbar l \ll a_0 |\mathbf{p}|$$

or

$$l \ll a_0 |\mathbf{k}| \tag{11.15}$$

The method of partial-wave analysis is therefore useful for low-energy scattering.

To obtain the scattering amplitude we proceed in three steps as follows: First, we find the free-particle eigenfunctions expressed in spherical coordinates, so that they correspond to specific states of angular momentum. We usually refer to these eigenfunctions as *spherical waves*; they form a complete set in which any wavefunction can be expanded. The second step consists of expanding the incident particle wavefunction $\psi_{in}(\mathbf{r})$ given by Eq. (11.10*b*) as a plane wave, into angular momentum eigenstates. The algebraic equations leading to the desired result in this step appear forbidding but in reality are simple, especially since we are interested only in the asymptotic form $r \to \infty$. The result is given in Eq. (11.20*a*), and some of the intermediate steps are separated from the main text so that they can be omitted on a first reading. Last, we similarly expand the final wavefunction $\psi_f(\mathbf{r})$ given by Eq. (11.12). It is then possible to identify the scattering amplitude $f(\theta, \phi)$ with a sum of the coefficients of the expansion.

Free-Particle Solutions in Spherical Coordinates

Since the scattering phenomena mostly involve rotationally invariant potentials, we follow the derivation of Section 8.2 and decompose the total wavefunction into angular momentum eigenstates [see Eqs.

(8.14)]

$$\psi_{l,m}(r,\theta,\phi) = N_l R_l(r) Y_{l,m}(\theta,\phi) = N_l \frac{u_l(r)}{r} Y_{l,m}(\theta,\phi) \tag{11.16a}$$

Here, $Y_{l,m}(\theta, \phi)$ are the spherical harmonics used previously and discussed in Appendix 9. It is sufficient to recall that they are related to the associated Legendre polynomials through

$$Y_{l,m}(\theta,\phi) = \varepsilon\left[\frac{2l+1}{4\pi}\frac{(l-|m|)!}{(l+|m|)!}\right]^{1/2} P_{l,m}(\cos\theta)e^{im\phi} \tag{11.16b}$$

where $\varepsilon = (-1)^m$ for $m > 0$, and 1 otherwise. Furthermore, if $\psi_{l,m}(r, \theta, \phi)$ satisfies the free-particle equation, i.e.,

$$(\nabla^2 + k^2)\psi_{l,m}(r,\theta,\phi) = 0$$

then it follows (see Section 8.3) that the radial functions satisfy

$$\frac{d^2u_l}{dr^2} + \left[k^2 - \frac{l(l+1)}{r^2}\right]u_l(r) = 0 \tag{11.17a}$$

with the boundary condition

$$u_l(r) \to 0 \quad \text{as} \quad r \to 0 \tag{11.17b}$$

If we define the new variable

$$\rho = kr = \left(\frac{2\mu E}{\hbar^2}\right)^{1/2} r$$

then Eq. (11.17a) becomes

$$\frac{d^2u_l}{d\rho^2} + \left[1 - \frac{l(l+1)}{\rho^2}\right]u_l(\rho) = 0 \tag{11.17c}$$

The solutions to this equation are related to the spherical Bessel functions $j_l(\rho)$, and $n_l(\rho)$, such that (see Appendix 12)

$$u_l(\rho) = \rho j_l(\rho) \qquad \text{or} \qquad u_l(\rho) = \rho n_l(\rho)$$

The most general solution is

$$u_l(\rho) = u_l(kr) = kr[a_1 j_l(kr) + a_2 n_l(kr)] \tag{11.18a}$$

From the limiting form of these functions as $\rho \to 0$ (see Appendix 12)

$$j_l(\rho) \propto \rho^l, \qquad \rho \to 0$$
$$n_l(\rho) \propto \frac{1}{\rho^{l+1}}, \qquad \rho \to 0 \tag{11.18b}$$

we see that the boundary condition of Eq. (11.17*b*) requires that

$$a_2 = 0$$

The solution to the radial equation becomes

$$u_l(kr) = a_1 kr j_l(kr)$$

and hence the free-particle solution in spherical coordinates is given by

$$\begin{aligned}\psi_{l,m}(r, \theta, \phi) &= N_l R_l(r) Y_{l,m}(\theta, \phi) \\ &= \tilde{N}_l j_l(kr) Y_{l,m}(\theta, \phi)\end{aligned} \tag{11.19a}$$

The normalization constant $\tilde{N}_l$ is obtained from the orthonormality relations

$$\int d\Omega\, Y^*_{l,m}(\theta, \phi) Y_{l,m}(\theta, \phi) = 1$$
$$\int r^2\, dr j_l(kr) j_l(k'r) = \frac{\pi}{2k^2} \delta(k - k') \tag{11.19b}$$

The normalization constant is

$$\tilde{N}_l = \left(\frac{2k^2}{\pi}\right)^{1/2}$$

Here we have chosen the phase to be real. Therefore the normalized free-particle solutions in spherical coordinates are given by

$$\psi_{l,m}(r, \theta, \phi) = \left(\frac{2k^2}{\pi}\right)^{1/2} j_l(kr) Y_{l,m}(\theta, \phi) \tag{11.19c}$$

Expansion of a Plane Wave into Spherical Waves

The above wavefunctions form a complete basis and any wavefunction can be expressed in terms of them. In particular, let us expand the incident plane wave in terms of the angular momentum components. We know that

$$\psi_{\text{in}} = e^{ikz} = e^{ikr\cos\theta}$$

This does not depend on the azimuthal angle ϕ, which simply reflects the fact that the wave has no angular momentum along the Z-axis. Therefore, its expansion in the spherical basis would only involve $m = 0$ components. In the limit $r \to \infty$ one obtains

$$e^{ikz} = \sum_{l=0}^{\infty} \frac{2l+1}{kr} i^l \sin\left(kr - \frac{l\pi}{2}\right) P_l(\cos\theta), \qquad r \to \infty \tag{11.20a}$$

where $P_l(\cos\theta)$ are the Legendre polynomials.

The expansion of Eq. (11.20a) is obtained as follows. We use Eq. (11.19c) to write

$$\begin{aligned}
\psi_{\text{in}} = e^{ikz} = e^{ikr\cos\theta} &= \sum_{l=0}^{\infty} a_l \psi_{l,0}(kr, \theta, \phi) \\
&= \sum_{l=0}^{\infty} a_l \left(\frac{2k^2}{\pi}\right)^{1/2} j_l(kr) Y_{l,0}(\theta, \phi) \\
&= \left(\frac{2k^2}{\pi}\right)^{1/2} \sum_{l=0}^{\infty} a_l \left(\frac{2l+1}{4\pi}\right)^{1/2} j_l(kr) P_l(\cos\theta)
\end{aligned} \tag{11.20b}$$

The expansion coefficients a_l are obtained from the orthogonality relation of the Legendre functions and the integral representation of the spherical Bessel functions.

$$\begin{aligned}
&\int_0^{\pi} P_l(\cos\theta) P_{l'}(\cos\theta) \sin\theta \, d\theta = \frac{2}{2l+1} \delta_{ll'} \\
&\frac{1}{2i^l} \int_0^{\pi} e^{ix\cos\theta} P_l(\cos\theta) \sin\theta \, d\theta = j_l(x)
\end{aligned} \tag{11.20c}$$

Multiplying Eq. (11.20b) by $P_l(\cos\theta)$ and integrating over all solid angles we obtain

$$\begin{aligned}
\int d\Omega P_l(\cos\theta) e^{ikr\cos\theta} = &\left(\frac{2k^2}{\pi}\right)^{1/2} \sum_{l'=0}^{\infty} a_{l'} \left(\frac{2l'+1}{4\pi}\right)^{1/2} j_{l'}(kr) \\
&\times \int d\Omega P_l(\cos\theta) P_{l'}(\cos\theta)
\end{aligned}$$

If we use the relations given in Eq. (11.20*c*) we obtain

$$2\pi 2i^l j_l(kr) = \left(\frac{2k^2}{\pi}\right)^{1/2} a_l \left(\frac{2l+1}{4\pi}\right)^{1/2} j_l(kr) 2\pi \frac{2}{2l+1}$$

or

$$a_l = 2\pi i^l \left(\frac{2l+1}{2k^2}\right)^{1/2} \tag{11.20d}$$

The incident wave in the spherical basis becomes

$$\psi_{\text{in}} = e^{ikz} = \sum_{l=0}^{\infty} 2\pi i^l \left(\frac{2l+1}{2k^2}\right)^{1/2} \left(\frac{2k^2}{\pi}\right)^{1/2} \left(\frac{2l+1}{4\pi}\right)^{1/2} j_l(kr) P_l(\cos\theta)$$

$$= \sum_{l=0}^{\infty} (2l+1) i^l j_l(kr) P_l(\cos\theta) \tag{11.21a}$$

We also note that (see Appendix 12)

$$j_l(kr) \xrightarrow[kr\to\infty]{} \frac{1}{kr} \sin\left(kr - \frac{l\pi}{2}\right)$$

and hence for large distances

$$\psi_{\text{in}} \xrightarrow[r\to\infty]{} \sum_{l=0}^{\infty} \frac{2l+1}{kr} i^l \sin\left(kr - \frac{l\pi}{2}\right) P_l(\cos\theta) \tag{11.21b}$$

This confirms our earlier assertion that a plane wave is a superposition of an infinite number of spherical waves of various angular momenta. Furthermore, it contains both incoming as well as outgoing waves.

Expansion of the Scattering Amplitude

If the potential $U(|\mathbf{r}|)$ falls off faster than $1/r^2$, then for large distances the final wavefunction would also be a free-particle solution. This is because the centrifugal barrier in this case would dominate over the potential-energy term. However, the final wavefunction at large distances after scattering would undergo a phase change relative to the incident wave. This can be seen intuitively as follows: If the scattering potential is attractive the particle will be accelerated, and consequently the wavelength would be shorter near the scattering source. On the other hand, if the potential is repulsive, then the particle would be decelerated and thus would have a longer wavelength in the scattering region. In either case, when the particle emerges from the scattering region, its phase would be different from the case when there is no scattering. Thus, the final wavefunction for large distances must have the form

$$\psi_f \xrightarrow[r\to\infty]{} \sum_{l=0}^{\infty} A_l \frac{2l+1}{kr} i^l \sin\left(kr - \frac{l\pi}{2} + \delta_l\right) P_l(\cos\theta) \tag{11.21d}$$

where δ_l is the phase shift that the lth partial wave suffers. It is positive if the potential is attractive and negative if the potential is repulsive. The constant A_l is determined from the observation that

$$\psi_f \xrightarrow[r\to\infty]{} e^{ikz} + f(\theta, \phi)\frac{e^{ikr}}{r}$$

Subtracting the expansion of Eq. (11.20a) from that of ψ_f as given by Eq. (11.21d) we can write

$$f(\theta, \phi)\frac{e^{ikr}}{r} = \sum_{l=0}^{\infty} A_l \frac{2l+1}{kr} i^l \sin\left(kr - \frac{l\pi}{2} + \delta_l\right) P_l(\cos\theta) - \sum_{l=0}^{\infty} \frac{2l+1}{kr} i^l \sin\left(kr - \frac{l\pi}{2}\right) P_l(\cos\theta) \tag{11.22a}$$

Since the lhs of Eq. (11.22a) contains no term in e^{-ikr} (incoming waves†) the same must be true of the rhs. By writing

$$\sin\left(kr - \frac{l\pi}{2} + \delta_l\right) = \frac{1}{2i}\left(e^{i(kr - l\pi/2 + \delta_l)} - e^{-i(kr - l\pi/2 + \delta_l)}\right)$$

we see that the above condition can be satisfied only if $A_l = e^{i\delta_l}$. In that case the scattered wave takes the form

$$f(\theta, \phi)\frac{e^{ikr}}{r} = \frac{e^{ikr}}{r} \sum_{l=0}^{\infty} \frac{2l+1}{2ik}(e^{2i\delta_l} - 1) P_l(\cos\theta) \tag{11.22c}$$

This gives an expression for the scattering amplitude in terms of

† Given the time dependence of Eq. (11.9b) and that $E > 0$, clearly a term of the form e^{-ikr} contributes incoming spherical waves $e^{-i(kr+Et)}$.

the phase shifts

$$f(\theta, \phi) = \sum_{l=0}^{\infty} \frac{2l+1}{2ik}(e^{2i\delta_l} - 1)P_l(\cos\theta)$$
$$= \sum_{l=0}^{\infty} \frac{2l+1}{k} e^{i\delta_l} \sin\delta_l P_l(\cos\theta)$$
$$= \sum f_l(\theta, \phi) \tag{11.23a}$$

where $f_l(\theta, \phi)$ can be thought of as the scattering amplitude for the lth partial wave. The differential cross section is given by [see Eq. (11.14)]

$$\sigma(\theta, \phi) = |f(\theta, \phi)|^2 = \frac{1}{k^2}\left|\sum_{l=0}^{\infty}(2l+1)e^{i\delta_l}\sin\delta_l P_l(\cos\theta)\right|^2 \tag{11.23b}$$

The total cross section becomes

$$\sigma_T = \int d\Omega\, |f(\theta, \phi)|^2$$
$$= \sum_{l=0}^{\infty}\sum_{l'=0}^{\infty} \frac{(2l+1)}{k}\frac{(2l'+1)}{k} e^{i(\delta_l - \delta_{l'})}\sin\delta_l \sin\delta_{l'}$$
$$\times \int d\Omega P_l(\cos\theta)P_{l'}(\cos\theta)$$
$$= \sum_{l=0}^{\infty}\sum_{l'=0}^{\infty} \frac{(2l+1)}{k}\frac{(2l'+1)}{k} e^{i(\delta_l - \delta_{l'})}\sin\delta_l \sin\delta_{l'} 2\pi\frac{2}{2l+1}\delta_{ll'}$$

$$\boxed{\sigma_T = \frac{4\pi}{k^2}\sum_{l=0}^{\infty}(2l+1)\sin^2\delta_l} \tag{11.23c}$$

We see that if we know the phase shifts for every partial wave we know everything about the scattering. And it is clear that the method is useful only if the series can be approximated by a few terms, i.e., if only the first few angular momentum contributions are significant. We still have to determine the relevant phase shifts depending on the particular form of the potential, and this we will do in the next section.

The Optical Theorem

It is interesting to note that if we calculate the forward scattering amplitude from Eq. (11.23a) we obtain

$$f(\theta=0)=\sum_{l=0}^{\infty}\frac{2l+1}{k}e^{i\delta_l}\sin\delta_l P_l(1)$$

$$=\sum_{l=0}^{\infty}\frac{2l+1}{k}e^{i\delta_l}\sin\delta_l$$

It follows that

$$\mathrm{Im}[f(\theta=0)]=\sum_{l=0}^{\infty}\frac{2l+1}{k}\sin^2\delta_l=\frac{k}{4\pi}\sigma_T$$

or

$$\sigma_T=\frac{4\pi}{k}\mathrm{Im}[f(\theta=0)] \tag{11.24}$$

Equation (11.24) represents an important result that relates the total scattering cross section to the imaginary part of the forward amplitude and is known as the *optical theorem*. Even though we derived it using the method of partial waves, it is true in general and has a natural physical interpretation. When scattering occurs, part of the energy carried by the incoming wave is radiated into all angles. This energy must be removed from the incident wave. Consequently the energy flowing in the forward direction is reduced and this modifies the scattering amplitude in that direction, i.e., at $\theta=0$. We will make use of the optical theorem when we discuss applications of elastic and inelastic scattering.

11.3. APPLICATIONS OF PARTIAL-WAVE ANALYSIS

In the previous section we developed the formalism of the partial-wave expansion but gave no indication how the phase shifts can be determined from the form of the scattering potential. We will do so now by considering some simple potentials and restricting ourselves to the first few partial waves.

Scattering from a Delta Potential

Let the potential be of the form

$$U(r) = \gamma\delta(r-a)$$

as shown in Fig. 11.5. Here γ measures the strength of the potential. Furthermore, we consider extremely low energy particles so that only $l=0$ components contribute to the scattering, that is, only the S-waves suffer appreciable scattering.

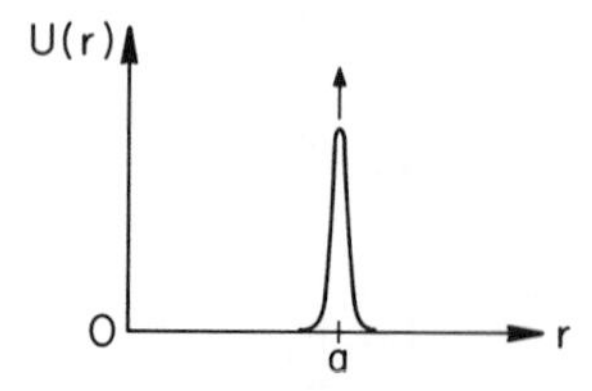

FIGURE 11.5. Representation of the radial dependence of a delta function shell potential.

For $r<a$, the radial equation is given by

$$\frac{d^2u}{dr^2} + k^2u = 0 \tag{11.25a}$$

where $k^2 = 2\mu E/\hbar^2$ and the solution that vanishes at the origin has the form

$$u(r) = A \sin kr \tag{11.25b}$$

For $r>a$, the radial equation again is

$$\frac{d^2u}{dr^2} + k^2u = 0 \tag{11.26a}$$

However, since we expect the wave to undergo a phase change the solution for $r>a$ is given by

$$u(r) = B \sin(kr + \delta_0) \tag{11.26b}$$

Here δ_0 is the phase change in the wave.

The solutions must match at the boundary and this gives the

condition

$$A \sin ka = B \sin(ka + \delta_0) \tag{11.27a}$$

Now, the radial equation for all values of r including $r = a$ is given by

$$\frac{d^2u}{dr^2} + k^2u = \frac{2\mu}{\hbar^2}\gamma\delta(r-a)u \tag{11.27b}$$

Integrating the above equation between $a-\varepsilon$ and $a+\varepsilon$ we have

$$\lim_{\varepsilon\to 0}\int_{a-\varepsilon}^{a+\varepsilon} dr\left(\frac{d^2u}{dr^2} + k^2u\right) = \lim_{\varepsilon\to 0}\int_{a-\varepsilon}^{a-\varepsilon} dr\frac{2\mu}{\hbar^2}\gamma\delta(r-a)u$$

or,

$$\lim_{\varepsilon\to 0}\left.\frac{du}{dr}\right|_{a+\varepsilon} - \left.\frac{du}{dr}\right|_{a-\varepsilon} = \frac{2\mu\gamma}{\hbar^2}u(a)$$

Using the expressions from Eqs. (11.25b) and (11.26b) we have

$$kB\cos(ka+\delta_0) - kA\cos ka = \frac{2\mu\gamma}{\hbar^2}B\sin(ka+\delta_0)$$

or,

$$kB\cos(ka+\delta_0) - \frac{2\mu\gamma}{\hbar^2}B\sin(ka+\delta_0) = kA\cos ka \tag{11.27c}$$

Dividing Eq. (11.27c) by Eq. (11.27a) we obtain

$$k\cot(ka+\delta_0) - \frac{2\mu\gamma}{\hbar^2} = k\cot ka$$

or

$$k\frac{\cot ka\cot\delta_0 - 1}{\cot ka + \cot\delta_0} = \frac{2\mu\gamma}{\hbar^2} + k\cot ka$$

which we simplify to

$$\cot\delta_0 = -\cot ka - \frac{k\hbar^2}{2\mu\gamma}\csc^2 ka$$

Since we have assumed the particles to have low energy, i.e., $ka \ll 1$, we can expand the trigonometric functions, and this gives

$$\cot\delta_0 = -\frac{1 + 2\mu\gamma a/\hbar^2}{(2\mu\gamma/\hbar^2)ka^2} \tag{11.28a}$$

It is clear that for $l=0$ (S-wave scattering), the differential cross section is isotropic. Using Eq. (11.23c) the total cross section can be written as

$$\sigma_T = \frac{4\pi}{k^2}\sin^2\delta_0 = \frac{4\pi}{k^2}\frac{1}{\cot^2\delta_0+1}$$

Introducing the result of Eq. (11.28a) and expanding in the small quantity $ka \ll 1$, we obtain

$$\sigma_T \simeq 4\pi a^2\left(\frac{2\mu\gamma a/\hbar^2}{1+2\mu\gamma a/\hbar^2}\right)^2 \tag{11.28b}$$

Note that for low-energy scattering the scattering cross section is independent of the energy.

Square-Well Potential

As a second example we consider the spherical square-well potential [see Fig. 11.6(a)] in three dimensions given by

$$U(r) = \begin{cases} -U_0 & \text{for } r<a \\ 0 & \text{for } r>a \end{cases} \tag{11.29a}$$

This potential approximates the strong force between the neutron and the proton, and as we saw in Section 9.7 can give rise to a bound state in that system. For the present, however, let us consider only a shallow potential that does not allow for any bound states to exist. We again consider low energy particles such that only $l=0$ waves scatter. Of course, for the energy, $E>0$.

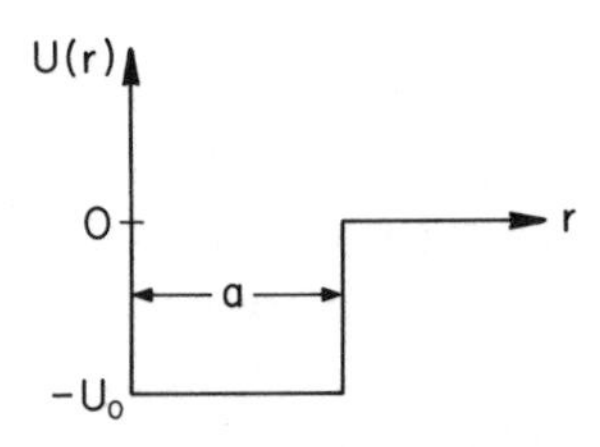

FIGURE 11.6. Representation of the radial dependence of a square-well potential.

We proceed in a manner exactly analogous to the discussion in Section 9.6. The radial equation for $r<a$ is given by

$$\frac{d^2u}{dr^2}+(k^2+k_0^2)u=0 \tag{11.29b}$$

where $k^2=2\mu E/\hbar^2$ and $k_0^2=2\mu U_0/\hbar^2$. The solution to the above equation, which vanishes at the origin, is given by

$$u(r)=A\sin k_1r \quad \text{where} \quad k_1^2=k^2+k_0^2 \tag{11.29c}$$

The radial equation for $r>a$ is the free-particle equation

$$\frac{d^2u}{dr^2}+k^2u=0 \tag{11.30a}$$

where the phase-shifted form of the solution is

$$u(r)=B\sin(kr+\delta_0) \tag{11.30b}$$

Furthermore, the solutions and their derivatives have to be continuous at the boundary, which implies the relations

$$A\sin k_1a=B\sin(ka+\delta_0) \tag{11.31a}$$

$$k_1A\cos k_1a=kB\cos(ka+\delta_0) \tag{11.31b}$$

Dividing Eq. (11.31a) by Eq. (11.31b) We obtain

$$\frac{1}{k}\tan(ka+\delta_0)=\frac{1}{k_1}\tan k_1a \tag{11.31c}$$

or,

$$\frac{1}{k}\frac{\tan ka+\tan\delta_0}{1-\tan ka\tan\delta_0}=\frac{1}{k_1}\tan k_1a$$

or

$$\tan\delta_0(k_1+k\tan ka\tan k_1a)=k\tan k_1a-k_1\tan ka \tag{11.31d}$$

The particles are of low energy so that $ka\ll 1$. If we further assume that the potential is shallow, i.e., $k_1a\simeq k_0a$ is also small, then the above expression reduces to

$$\tan\delta_0=\frac{k\tan k_1a-k_1ka}{k_1}$$

or

$$\delta_0 = k\left(\frac{\tan k_1 a - k_1 a}{k_1}\right) \tag{11.32a}$$

Again, since we are considering only S-wave scattering, the cross section is isotropic and the total cross section is given by

$$\begin{aligned}\sigma_T &= \frac{4\pi}{k^2}\sin^2\delta_0 = \frac{4\pi}{k^2}\delta_0^2 \\ &= 4\pi\left(\frac{\tan k_1 a - k_1 a}{k_1}\right)^2 \\ &= 4\pi a^2\left(\frac{\tan k_1 a - k_1 a}{k_1 a}\right)^2 \\ &= 4\pi a^2\left(\frac{\tan k_0 a - k_0 a}{k_0 a}\right)^2 \end{aligned} \tag{11.32b}$$

In deriving Eq. (11.32b) we assumed that since $ka \ll 1$, $k_1 a \simeq k_0 a$, and also that the potential is shallow enough so that no bound states are possible. The cross section is different if the potential well is deep enough, and this case will be discussed later. Note here that the cross section again does not depend on the energy of the particles

The Hard Sphere

The low-energy scattering cross section from a repulsive square-well potential can be obtained simply from Eq. (11.32b). Note that the potential in this case has the form

$$U(\mathbf{r}) = \begin{cases} U_0, & r < a \\ 0, & r > a \end{cases}$$

The difference between an attractive and a repulsive potential amounts to replacing k_0 by $i\kappa_0$

$$k_0 \rightarrow i\kappa_0$$

For a repulsive potential the low-energy scattering cross section becomes [see Eq. (11.32*b*)]

$$\sigma_T = 4\pi a^2 \left(\frac{\tan(i\kappa_0 a) - i\kappa_0 a}{i\kappa_0 a}\right)^2$$

$$= 4\pi a^2 \left(\frac{\tanh(\kappa_0 a) - \kappa_0 a}{\kappa_0 a}\right)^2 \tag{11.33}$$

The cross section for low-energy scattering from a hard sphere is obtained simply by noting that in that case

$$\kappa_0 = \left(\frac{2\mu U_0}{\hbar^2}\right)^{1/2} \to \infty \quad \text{since} \quad U_0 \to \infty$$

Hence the cross section becomes

$$\sigma_T = 4\pi a^2 \tag{11.34a}$$

Again, the cross section does not depend on the energy for low-energy scattering and equals four times the classical value of πa^2. We would expect to obtain the classical result in the limit of high-energy scattering. But even in that case the cross section has the value $\sigma_T(k \to \infty) = 2\pi a^2$, as can be seen from the following nonrigorous argument.†

The largest l value that contributes to scattering is given by $l_{\max} = ak$. Therefore,

$$\sigma_T(k \to \infty) = \sum_{l=0}^{l_{\max}=ak} \frac{4\pi}{k^2}(2l+1)\sin^2\delta_l \tag{11.34b}$$

If we assume that the phase shifts are completely random, then we can replace $\sin^2\delta_l$ by its average value $\langle \sin^2\delta_l \rangle = 1/2$. Performing the summation in Eq. (11.34*b*) we obtain

$$\sigma_T(k \to \infty) = \frac{4\pi}{k^2} \sum_{l=0}^{ak} \tfrac{1}{2}(2l+1)$$

$$= \frac{4\pi}{k^2} \tfrac{1}{2}(ka+1)^2 \simeq 2\pi a^2 \tag{11.34c}$$

The discrepancy with the classical geometrical value in both cases

† For a rigorous proof see, for example, L. I. Schiff, *Quantum Mechanics*, Third Edition, p. 125. McGraw Hill, New York, 1968.

arises because of the wave nature of particles. There is diffraction around the edges of the sphere which leads to constructive interference and thus to an increased cross section.

11.4. LOW-ENERGY n–p SCATTERING

We have already discussed the deuteron, which is a bound state of the neutron–proton system (see Section 9.7). The scattering of neutrons from protons is determined by the same Hamiltonian as for the bound-state case except that now $E>0$, and hence the boundary conditions are different. Scattering of low-energy neutrons reveals more information about the interaction between neutrons and protons than can be obtained from the study of the bound state alone. In fact, neutron scattering is well suited for studying the short-range nuclear force because of the absence in this case of Coulomb forces.

First, we note that if we are considering incident energies of the order of 10 MeV and if we assume that the nuclear force has a range of about 1.4 F, then

$$l_{\max} \simeq ak = a\left(\frac{2\mu E}{\hbar^2}\right)^{1/2}$$

$$= a\left[\frac{2\mu c^2 E}{(\hbar c)^2}\right]^{1/2} = a\left[\frac{2(0.5\times 10^3\ \text{MeV})\times 10\ \text{MeV}}{(200\ \text{MeV-F})^2}\right]^{1/2} = 0.7$$

where we use $\mu \sim m_p/2 \sim 0.5\times 10^3\ \text{MeV}$ and $\hbar c \sim 200\ \text{MeV-F}$. Therefore, we need consider only S-wave scattering ($l=0$). However, as we have seen earlier, bound states of the n–p system can also occur for $l=0$ with almost zero energy. Since the low-energy neutrons can, in principle, resonate with this state, the scattering in this case has to be treated more carefully.

Let us recapitulate quickly the conclusions we reached from the study of the bound state of the system. The energy of the bound state is $-E_0 = W$, with $W \simeq 0$. Thus, for $r<a$ the radial equation for

the bound state is

$$\frac{d^2u_b}{dr^2}+\frac{2\mu}{\hbar^2}(-W+U_0)u_b=0$$

or

$$\frac{d^2u_b}{dr^2}+k_0^2u_b=0, \qquad W \ll U_0$$

where as before $k_0^2=2\mu U_0/\hbar^2$, and the solution that vanishes at the origin is

$$u_b(r)=A\sin k_0 r \tag{11.35a}$$

For $r>a$ the radial equation is

$$\frac{d^2u_b}{dr^2}-\frac{2\mu W}{\hbar^2}u_b=0$$

or

$$\frac{d^2u_b}{dr^2}-\beta^2u_b=0$$

where $\beta^2=2\mu W/\hbar^2$, and the solution that vanishes at $r\to\infty$ has the form

$$u_b(r)=Be^{-\beta r} \tag{11.35b}$$

The solutions and their derivatives have to be continuous at the boundary that leads to the relation [see Eq. (9.54c), where we used α instead of k_0]

$$k_0\cot k_0a=-\beta \tag{11.36}$$

or

$$k_0\tan\left(\frac{\pi}{2}-k_0a\right)=-\beta$$

or, expanding the tangent,

$$k_0a=\frac{\pi}{2}+\frac{\beta}{k_0} \tag{11.37}$$

Here we have assumed that the value of k_0a is very close to $\pi/2$, the value for which a bound state exists. Thus, we see that the existence of a bound state constrains the value of k_0 to that given by Eq. (11.37).

On the other hand, for low-energy scattering the relation that must be satisfied according to Eq. (11.31c) is

$$\frac{1}{k}\tan(ka+\delta_0)=\frac{1}{k_1}\tan k_1 a \tag{11.38a}$$

where $k_1^2=k_0^2+k^2$, and for $ka \ll 1$ we can write $k_1 \simeq k_0$. Equation (11.38a) then becomes

$$k \tan k_0 a = k_0 \tan(ka+\delta_0)$$

which in view of Eq. (11.37) can be written in the form

$$k \tan\left(\frac{\pi}{2}+\frac{\beta}{k_0}\right)=k_0\frac{\tan ka+\tan\delta_0}{1-\tan ka \tan\delta_0}$$

Furthermore, approximating $\tan ka \simeq ka$, we obtain

$$-k\cot\frac{\beta}{k_0}=k_0\frac{ka+\tan\delta_0}{1-ka\tan\delta_0}$$

Since $W \ll U_0$, it follows that $\beta/k_0 \ll 1$, and we approximate $\cot(\beta/k_0)\simeq k_0/\beta$. Thus

$$-k\frac{k_0}{\beta}=k_0\frac{ka+\tan\delta_0}{1-ka\tan\delta_0}$$

This expression can be solved for $\tan\delta_0$ to yield

$$\cot\delta_0=\frac{1}{\tan\delta_0}=-\frac{\beta-k^2a}{k(\beta a+1)}$$

or,

$$k\cot\delta=-\frac{\beta}{\beta a+1}+\frac{a}{\beta a+1}k^2 \tag{11.38b}$$

If we further assume $\beta a \ll 1$ and $ka \ll 1$, the above result is simplified to

$$k\cot\delta_0 \simeq -\beta$$

It can be shown that for any form of the potential the S-wave scattering phase shift can be expressed by an expansion of the form

$$k\cot\delta_0=\frac{1}{a_0}+\tfrac{1}{2}r_0k^2+O(k^4) \tag{11.39a}$$

Such expressions are known as *effective range expansions*. Here the constant a_0 is called the Fermi *scattering length*, and r_0 is known as the effective range. Comparing Eqs. (11.38*b*) and (11.39*a*) and assuming $\beta a \ll 1$ and $ka \ll 1$, we obtain

$$\frac{1}{a_0} \simeq -\beta \tag{11.39b}$$

The scattering length and its relation to the phase shift δ_0 of the wavefunction has a geometrical interpretation which is indicated in Fig. 11.7. The case of the deuteron corresponds to that in part (d) of the figure since a bound state can be formed. However, because the state is so weakly bound, $a_0 \sim -1/\beta \sim -4.3$ F is much longer than the range of the potential.

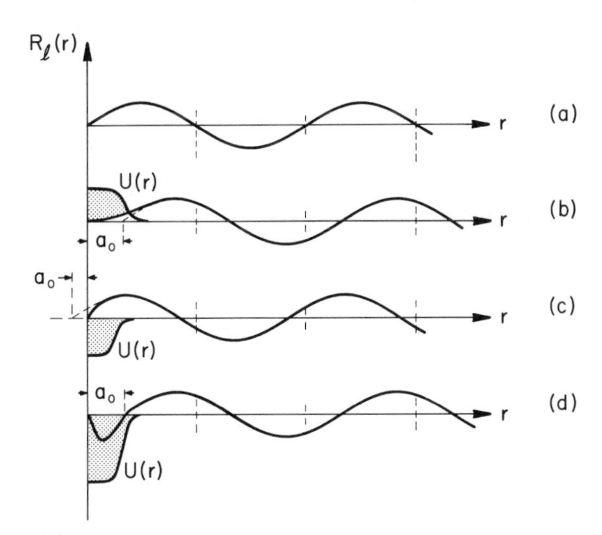

FIGURE 11.7. Relation between the scattering length a_0 and phase shift δ_0. The curves shown represent the radial wavefunction for the outgoing wave for an incident wavenumber k. (a) No scattering, $a_0 = 0$. (b) Repulsive potential, $a_0 < 0$. (c) Weak attractive potential, $a_0 > 0$. (d) Strong attractive potential, $a_0 < 0$.

The scattering cross section can be expressed completely in terms of the scattering length

$$\sigma_T(k \to 0) = \frac{4\pi}{k^2} \sin^2 \delta_0 = \frac{4\pi}{k^2} \frac{1}{\cot^2 \delta_0 + 1}$$

$$= \frac{4\pi}{k^2} \frac{1}{(1/ka_0)^2 + 1}$$

or

$$\sigma_T(k \to 0) = \frac{4\pi a_0^2}{1 + k^2 a_0^2} \tag{11.40}$$

We note here two points: First, the scattering is isotropic. However, the scattering cross section in this case depends on the energy. Second, for $ka \ll 1$, i.e., $E \to 0$, the total cross section is equal to $4\pi a_0^2$. This is an enhanced value for scattering since the scattering length a_0 for the n–p system is quite large. The physical way to understand this result is that if the well has a bound-state level whose energy is close to zero, then the scattered particles have a tendency to get bound to the well. However, they cannot really form a bound state since their energy is not negative. Rather, they tend to interact much more strongly which leads to the enhancement in the scattering cross section.

In the case of n–p scattering we can deduce the value of the scattering length—without any knowledge of the potential—simply from the study of the bound-state system. We know that $1/\beta$ measures the spatial extent of the deuteron wavefunction. From Eq. (9.61c) we are given that $1/\beta = 4.31$ F and if we use $a = 1.2$ F, we obtain for the n–p scattering length [see Eq. (11.38b)] the value

$$\frac{1}{a_0} = -\frac{\beta}{1 + \beta a}$$

or

$$a_0 = -5.5 \text{ F} \tag{11.41a}$$

Therefore, the n–p scattering cross section at low energies follows from Eq. (11.40) to be

$$\sigma_T(k \to 0) = 4\pi a_0^2 = 3.8 \times 10^{-24} \text{ cm}^2 \qquad \text{(predicted)}$$

The actual data on low-energy n–p scattering are shown in Fig. 11.8 and do indeed exhibit the energy dependence predicted by Eq. (11.40). However, the limiting value of σ_T is

$$\sigma_T(k \to 0) = 20.36 \times 10^{-24} \text{ cm}^2 \qquad \text{(observed)}$$

which is in complete disagreement with the prediction.

The reason for this discrepancy is that in the deuteron, the neutron and proton are always in a $S = 1$ (*triplet*) *state*; namely, the neutron and the proton have their spins aligned. However, in the

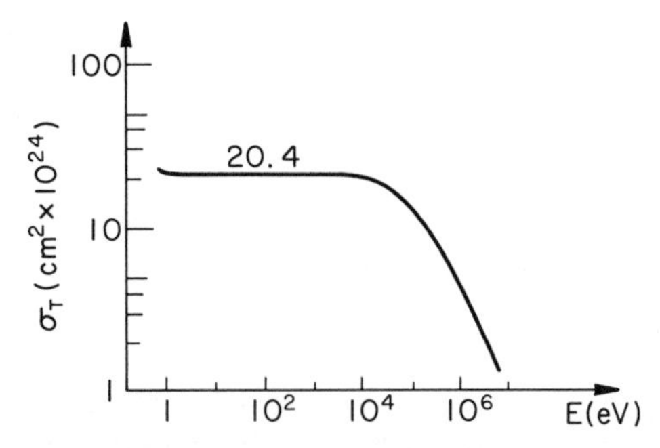

FIGURE 11.8. Cross section for the scattering of neutrons from protons for laboratory energies from 1 eV to 10 MeV. Note that the incident energy is plotted on a logarithmic scale so as to expand the region where $ka_0 \ll 1$.

scattering process, the neutron and the proton can be either in the $S=1$ *or* $S=0$ state. We conclude that the excess cross section is due to scattering in the $S=0$ (singlet) state. The cross section averaged over the singlet and the triplet states is given by

$$\sigma_{\text{av}} = \tfrac{1}{4}\sigma_s + \tfrac{3}{4}\sigma_t \tag{11.41b}$$

where the factors $\frac{1}{4}$ and $\frac{3}{4}$ reflect the *statistical weight* $(2S+1)$ of the singlet and triplet states, respectively. By this we mean that there is only one $(M=0)$ state for $S=0$, whereas for $S=1$ there are three $(M=+1, 0, -1)$ states. Accepting this explanation we find that

$$\sigma_s \simeq 68 \times 10^{-24} \text{ cm}^2$$

which translates into a scattering length for the singlet state of the order of

$$|(a_0)_s| \simeq 23 \text{ F}$$

We see that the n–p force depends on the orientation of the spins, an effect that has been verified by the scattering of polarized neutrons. From such experiments it is found further that $(a_0)_s$ is positive. In other words, there is no bound state in the spin singlet configuration. The large scattering length, however, implies that the system is "almost" bound: If the scattering was due to the presence of a bound state in the $S=0$ configuration the binding energy would be

$$W \simeq \frac{(\hbar c)^2}{2\mu c^2} \frac{1}{|(a_0)_s|^2} \sim 70 \text{ keV}$$

This condition is sometimes interpreted as indicating the existence of a *virtual* excited state for the deuteron.

11.5. ANGULAR DISTRIBUTION

The applications we have discussed so far have been to low-energy scattering, where only S-waves have to be considered. As we have seen, S-wave scattering is isotropic. In this section we qualitatively discuss and show examples of angular distributions observed when higher partial waves contribute to the scattering amplitude. In general, however, when only one partial wave, l, contributes to the scattering, the differential cross section will have an angular distribution proportional to the absolute square of the lth Legendre polynomial

$$\sigma(\theta, \phi) \propto |P_l(\cos\theta)|^2$$

In practice it often happens that at a particular energy one partial wave is dominant, i.e., one of the phase shifts is large ($\delta_l \to \pi/2$), whereas all other phase shifts remain small.

As an illustration, the differential cross section $\sigma_C(\theta, \phi)$ for n–p scattering at neutron laboratory energies $E = 14.1$ MeV and $E = 92$ MeV is shown in Fig. 11.9. The center-of-mass differential cross section is plotted versus the cosine of the center-of-mass scattering angle. At the low energy of $E = 14.1$ MeV the cross section is practically isotropic, indicating that only the $l = 0$ wave contributes. At $E = 92$ MeV the angular distribution has a strong component proportional to

$$\sigma_C(\theta) \propto |P_1(\cos\theta)|^2 = \cos^2\theta$$

indicating the contribution of the $l = 1$ wave. The n–p scattering, however, is complicated by the fact that both the neutron and proton have spin 1/2. In the preceding section we saw that at low energies n–p scattering in the $S = 0$ state is quite different from scattering in the $S = 1$ state, even though the orbital angular momentum was the same in both cases and restricted to $l = 0$ waves.

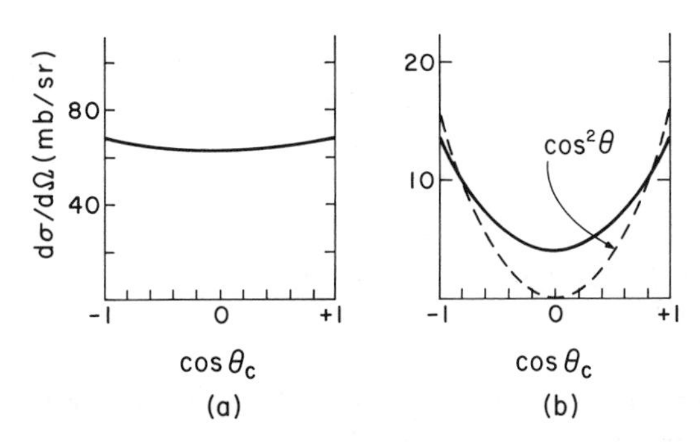

FIGURE 11.9. The center-of-mass differential cross section for n–p scattering plotted in mb/sr (1 barn $= 10^{-24}\,\text{cm}^2$) versus $\cos(\theta_C)$. (a) For laboratory energy $E = 14.1$ MeV. (b) For $E = 92$ MeV. The solid curve represents the data and the dashed curve is proportional to $\cos^2(\theta_C)$.

Therefore, it would be interesting to examine the scattering of two spin-zero particles, for instance of two π-mesons. This is difficult in practice because while beams of $\pi^{\pm}$-mesons are available in the laboratory, it is not possible to construct a target consisting of π-mesons. Nevertheless, $\pi^+\pi^-$ scattering can be observed indirectly by studying inelastic reactions in which a π-meson is produced, as in the reaction

$$\pi^- p \to \pi^+ \pi^- n$$

With the appropriate choice of the production angle of the π^+ we can interpret this reaction as shown in Fig. 11.10. The proton emits a π^+ and becomes a neutron which emerges from the scattering region. The momentum of the virtual π^+ is determined by requiring that the momentum balances† at the production vertex

$$\mathbf{p}_p = \mathbf{p}_n + \mathbf{p}_{\text{“}\pi^+\text{”}}$$

The virtual π^+ then scatters from the incident π^-, and the two pions emerge as real particles which are observed at $r \to \infty$. The scattering angle θ is determined in the $\pi^+\pi^-$ center-of-mass system (not in the

† Energy cannot be balanced at the n–p vertex, but this is allowed as long as $\Delta E\, \Delta t \sim \hbar$. Here ΔE is the energy imbalance, and Δt is the time interval over which the interaction takes place; Δt can be very small!

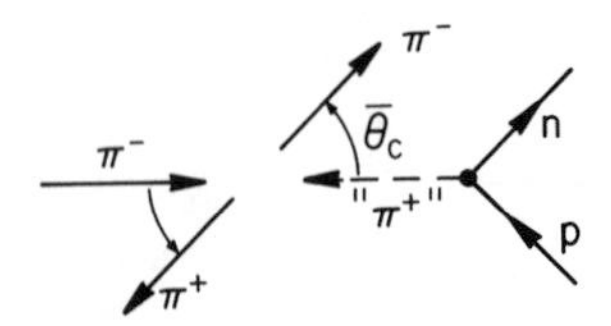

FIGURE 11.10. $\pi^-\pi^+$ scattering in the reaction $\pi^- p \to \pi^-\pi^+ n$. The angle $\bar{\theta}_C$ is measured in the $\pi^-\pi^+$ center-of-mass system.

$\pi^- p$ center of mass), and the Z-axis is chosen along the incident π direction (transformed into the $\pi^+\pi^-$ center of mass).

The resulting angular distributions at $(\pi^-\pi^+)$ center-of-mass energies $E_{\pi^+\pi^-} = 750$ MeV and $E_{\pi^+\pi^-} = 1270$ MeV are shown in Fig. 11.11. At the higher energy the scattering is dominated by the $l = 2$ wave and

$$\sigma_C(\theta) \propto |P_2(\cos\theta)|^2 = \tfrac{1}{4}(9\cos^4\theta - 6\cos^2\theta + 1)$$

At the lower energy the scattering is dominated by the $l = 1$ wave and

$$\sigma_C(\theta) \propto |P_1(\cos\theta)|^2 = \cos^2\theta$$

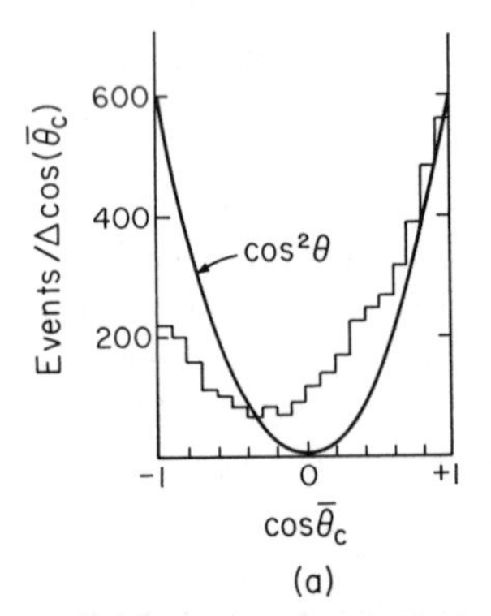

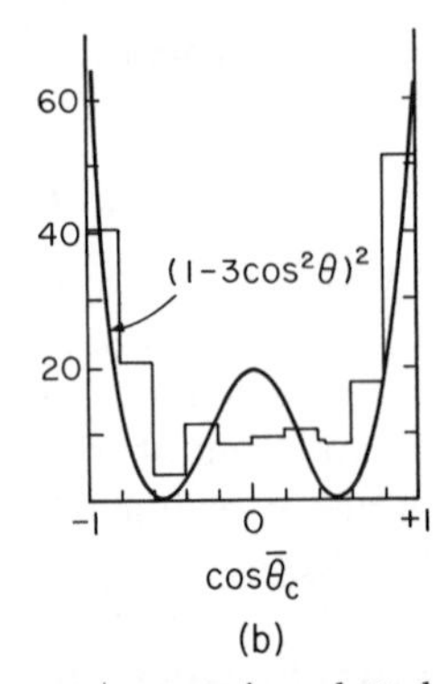

FIGURE 11.11. Angular distribution for $\pi^-\pi^+$ scattering plotted versus cos $\bar{\theta}_C$, where $\bar{\theta}_C$ is the scattering angle in the $\pi^-\pi^+$ center-of-mass system. The differential cross section is plotted in the form of events in a given interval of cos $\bar{\theta}_C$. (a) At a center-of-mass energy $E = 750$ MeV as obtained from the reaction $\pi^+ n \to \pi^+\pi^- p$ at a π^+ incident energy of 7 GeV. The solid curve is $|P_1(\cos\theta)|^2$. Note the presence of interference with the $l = 0$ wave. [From A. Engler *et al.*, *Phys. Rev.* D **10,** 2070 (1974)]. (b) $E = 1270$ MeV as obtained from the reaction $\pi^- p \to \pi^+\pi^- n$ at incident π^- energy of 8 GeV. The solid curve is $|P_2(\cos\theta)|^2$. [From H. Yuta, University of Pennsylvania, Ph.D thesis, 1966].

In this case there is a considerable discrepancy between the prediction and the data due to the presence of other partial waves that *interfere* with the dominant wave. In fact, if we assume the presence of $l=0$ and $l=1$ waves, the scattering amplitude is

$$f(\theta, \phi) \propto A + B \cos \theta$$

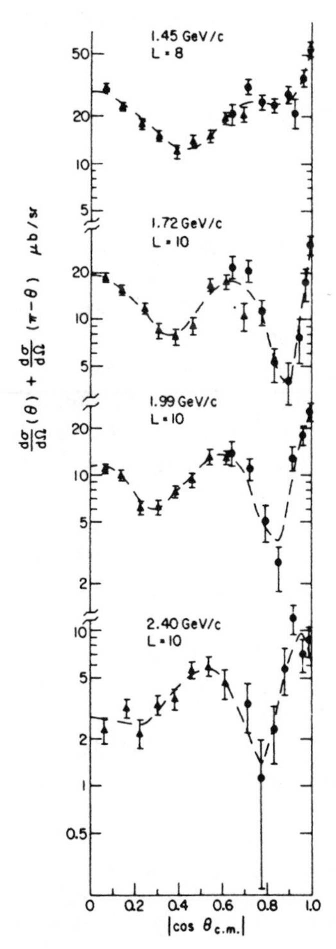

FIGURE 11.12. Angular distribution for $\bar{p}p \to \pi^+\pi^-$ plotted as a function of the π^- angle in the $\bar{p}p$ center-of-mass system for different energies. [From H. Nicholson *et al.*, *Phys. Rev. Lett.* **23**, 603 (1969)].

and therefore

$$\sigma_C(\theta, \phi) \propto [|A|^2 + 2\,\mathrm{Re}(AB^*)\cos\theta + |B|^2\cos^2\theta]$$

which gives rise to the observed asymmetry† in $\cos\theta$.

As partial waves with higher l-values contribute to the scattering, the angular distributions become correspondingly more complex. As an illustration, in Fig. 11.12 are shown angular distribution of π mesons produced in proton–antiproton annihilations in the reaction

$$\bar{p}p \rightarrow \pi^+\pi^-$$

Note how the angular distributions change rapidly with incident energy.

Qualitative Discussion of the Effects of Spin

So far we have ignored the fact that in actual scattering experiments both the incident beam and the target particles have spin. The method of partial waves remains applicable in the presence of spin but must be modified to take spin into account. We will not discuss the modifications in detail but restrict ourselves to a qualitative discussion of π^+p scattering; namely, the scattering of a spin-0 particle from a spin-1/2 target.

As before, we choose the Z-axis along the incident particle direction. Therefore, the initial state always‡ has $m_l = 0$. Since both the angular momentum and its z-component must be conserved, we conclude that for a central potential *in the absence of spin*, $m_l = 0$ in the final state as well. We made use of this fact in Section 11.2 by noting that only $m = 0$ eigenstates contribute to the scattering amplitude. Consider next the case where the target has spin $s = 1/2$ and therefore m_s can equal $+1/2$ or $-1/2$. To simplify the discussion we examine the case of scattering in the $l = 1$ partial wave and let $m_s = +1/2$, as shown in Fig. 11.13.

---

† The fact that the angular distribution of Fig. 11.11(a) can be fitted by using powers no greater than $\cos^2\theta$ indicates that the highest contributing partial wave has $l = 1$. It is interesting that $\sigma(\theta, \phi)$ for the $\pi^+\pi^0$ system at the same center-of-mass energy is completely symmetric. This is because for $\pi^+\pi^0$ the $l = 0$ wave cannot interfere with the $l = 1$ wave due to isotopic spin considerations.

‡ From a classical analogy the orbital angular momentum vector must be normal to the scattering plane; thus the projection of $\mathbf{L}$ onto the incident beam direction is zero. This corresponds to the $m_l = 0$ state.

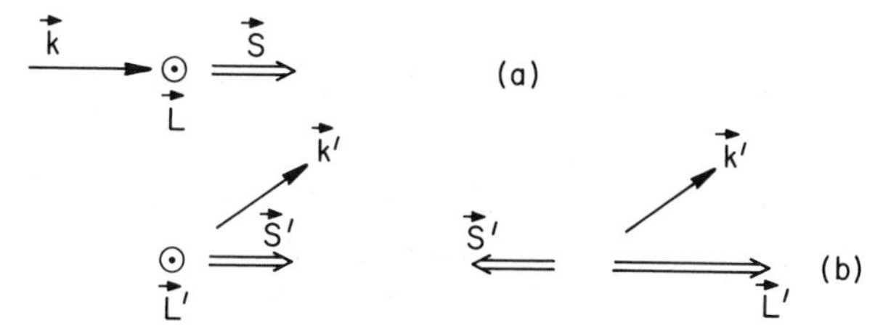

FIGURE 11.13. Vector model representation of the scattering of a spin-zero particle from a target particle with spin 1/2. (a) The initial state. (b) The two possibilities for the final state represent no-flip and spin-flip scattering.

From the rules of composition of angular momentum we know that the total angular momentum can be either $J = 1/2$ or $J = 3/2$. Let us choose $J = 3/2$ in which case the *initial state* is characterized by the quantum numbers

$$l = 1 \qquad s = 1/2 \qquad J = 3/2$$
$$m_l = 0 \qquad m_s = +1/2 \qquad M = m_l + m_s = +1/2$$

For the *final state*, J and M must be the same as in the initial state because of angular momentum conservation. m_l and m_s, however, need not be conserved independently.† There are two possible configurations for the final state

$$(1) \quad l = 1 \qquad s = 1/2 \qquad J = 3/2$$
$$m_l = 0 \qquad m_s = +1/2 \qquad M = +1/2$$
$$(2) \quad l = 1 \qquad s = 1/2 \qquad J = 3/2$$
$$m_l = +1 \qquad m_s = -1/2 \qquad M = +1/2$$

The two possible final states represent very different processes. In case (1) the orientation of the target's spin remains unchanged and the scattering amplitude $g(\theta)$ is the same as for spinless particles [Eq. (11.23a)]; This process is called *direct* or *no-flip* scattering. For case (2) the target spin is flipped and we speak of *spin-flip* scattering. The spin-flip amplitude $h(\theta, \phi)$ is characterized by $Y_l^1(\theta, \phi)$. Thus, it can give rise to a ϕ-dependence, provided the target is polarized transverse to the beam direction. Note that the spin-flip amplitude always vanishes at $\theta = 0$ and $\theta = \pi$ because it is

† Since the outgoing particle is no longer moving along, the Z-axis the projection of the angular momentum onto the Z-axis can be different from zero.

proportional to sin θ. It is also important to recognize that $g(\theta)$ and $h(\theta, \phi)$ are orthogonal to each other because they correspond to *different spin eigenstates* (orientations). Thus, we have

$$f(\theta, \phi) = g(\theta) + h(\theta, \phi)$$

and

$$\sigma(\theta, \phi) = |f(\theta)|^2 = |g(\theta)|^2 + |h(\theta, \phi)|^2 \tag{11.42}$$

When the target is unpolarized the probability for finding the particle in either the $m_s = +1/2$ or the $m_s = -1/2$ states is 1/2. We must average over these two configurations of the initial state. As a result the total cross section [Eq. (11.23c)] takes the form

$$\sigma_T = \frac{4\pi}{k^2} \frac{1}{(2s+1)} \sum_{j=1/2}^{j_{\max}} (2j+1) \sin^2 \delta_j \tag{11.43}$$

These ideas can be verified in $\pi^+ p$ scattering at a laboratory energy $E_\pi = 200$ MeV, where the scattering is dominated by the $J = 3/2, l = 1$ wave. One finds for the amplitudes

$$\begin{aligned} g(\theta) &= \frac{1}{k} e^{i\delta_{3/2}} (\sin \delta_{3/2}) 2 \cos \theta \qquad \text{direct} \\ h(\theta, \phi) &= \frac{1}{k} e^{i\delta_{3/2}} (\sin \delta_{3/2}) \sin \theta e^{i\phi} \qquad \text{spin-flip} \end{aligned} \tag{11.44a}$$

and thus the differential cross section is given by

$$\sigma(\theta, \phi) = |g(\theta)|^2 + |h(\theta, \phi)|^2 = \frac{1}{k^2} \sin^2 \delta_{3/2} (1 + 3 \cos^2 \theta) \tag{11.44b}$$

The angular distribution for the data is shown in Fig. 11.14(a) and agrees very closely with the prediction for scattering in a pure $l = 1, j = 3/2$ state. The total cross section is given by

$$\sigma_T = \int \sigma(\theta, \phi)\, d\Omega = \frac{2\pi}{k^2} \sin^2 \delta_{3/2} \int (1 + 3 \cos^2 \theta)\, d \cos \theta = \frac{8\pi}{k^2} \sin^2 \delta_{3/2}$$

Since $\sin^2 \delta_{3/2}$ cannot exceed unity we predict that

$$\sigma_T \leqslant \frac{8\pi}{k^2}$$

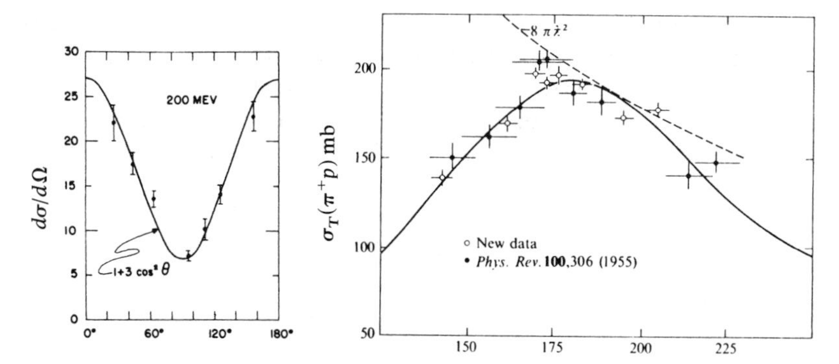

FIGURE 11.14. Scattering of π^+ mesons from protons. (a) Angular distribution at $E_\pi = 200$ MeV; (b) The total cross section as a function of laboratory energy of the incident π^+. [(a) From Mukhin *et al. Soviet Physics JETP* **4,** 237 (1957), (b) From S. J. Lindenbaum and L. C. L. Yuan, *Phys. Rev.* **111,** 1380 (1958)].

This upper bound is shown by the dashed curve in Fig. 11.14(b) and we note that the inequality is verified. At a laboratory energy $E_\pi \sim 200$ MeV, $\sin \delta_{3/2}$ reaches its maximal value

$$\sin \delta_{3/2} = 1 \quad \text{or} \quad \delta_{3/2} = \pi/2$$

If, for a particular energy, the lth phase shift becomes 90° we say that the system has a *resonance* in the lth partial wave at that energy. This is obviously the case for the $\pi^+ p$ system at $E_\pi \sim$ 200 MeV.

11.6. CONCLUSION

We have introduced the relevant ideas for the study of scattering theory. The Schrödinger equation for this phenomenon was set up and we discussed the appropriate boundary conditions. We solved the scattering equation by the method of partial waves and expressed the scattering amplitude in terms of the phase shifts that various wave components undergo. As applications of this method we considered the low-energy scattering of particles from a delta potential, the square-well potential, and a hard sphere. The physical example of low-energy n–p scattering was analyzed and compared with the experimental data. Finally, we discussed the angular distribution of the differential cross sections in general, and the effects of spin qualitatively.

Problems

PROBLEM 1

A free particle of energy $E > 0$ moving in one dimension is incident on an attractive potential well of width $2a = \hbar/(2mE)^{1/2}$. Find the *phase shift* of the wavefunction (*a*) when $-U_0 = |E|$; (*b*) when $-U_0 = 3\,|E|$. Discuss the physical consequences of case (*b*).

PROBLEM 2

A particle is scattered by a potential at sufficiently low energy so that the phase shifts $\delta_l = 0$ for $l > 1$. Assume rotational symmetry.

(*a*) Show that the differential scattering cross section has the form

$$\sigma(\theta, \phi) = A + B\cos\theta + C\cos^2\theta$$

and determine A, B, and C in terms of phase shifts.

(*b*) Determine the total cross section in terms of A, B, and C.

(*c*) Assume that the differential cross section is known for $\theta = 90°(\sigma = \alpha^2)$; $\theta = 180°$ $(\sigma = \beta^2)$, and $\theta = 45°$ $(\sigma = \gamma^2)$. Determine $\sigma(\theta, \phi)$ for $\theta = 0°$ in terms of α, β, and γ.

(*d*) Obtain the imaginary part of the forward scattering amplitude in terms of α, β, and γ.

PROBLEM 3

What must U_0a^2 be for a three-dimensional square-well potential (attractive) so that the scattering cross section is zero at zero bombarding energy. (This is the Ramsauer–Townsend effect)

PROBLEM 4

Consider a spherically symmetric potential of the form

$$U(r) = \begin{cases} -\dfrac{\hbar^2}{2m} K_1^2, & 0 \leqslant r < R_1 \\ \dfrac{\hbar^2}{2m} K_0^2, & R_1 \leqslant r < R_0 \\ 0, & R_0 \leqslant r \end{cases}$$

Calculate the phase shift δ_0 (for $l=0$) for the cases $K_0R_0=4$, $K_1R_1=1.5$, and $R_1=(1/2)R_0$.

PROBLEM 5

Obtain the relation between the differential cross section $\sigma(\theta)$ in the laboratory and center-of-mass frames. That is, derive Eqs. (11.8*b*).

Chapter 12

SCATTERING: PART II

We introduced the formalism for scattering and studied low-energy scattering from various potentials in Chapter 11. In all those cases we assumed that the target particle was different from the incident particle. In $n-p$ scattering, for example, the neutron and proton are distinguishable. The formulae for scattering when the particles involved are indistinguishable, however, need modification, and we discuss this in the first section of this chapter. We then consider the energy dependence of the scattering amplitude and discuss the pnenomenon of resonance in general.

The scattering equation can be solved in many different ways. The partial-wave analysis, which was developed in the last chapter, is not well suited for the study of high-energy scattering since the number of waves that contribute to the scattering increases. The method of Lippman and Schwinger, more commonly known as the integral solution to the scattering problem, becomes useful in this case. We discuss the method of obtaining such solutions. These solutions take a strikingly simple form in an approximation method known as the Born approximation. We examine the validity of this approximation and give physical examples of it. Finally, we consider qualitatively the case of inelastic scattering and show how the elastic scattering results must be modified in order to accommodate inelastic processes.

12.1. SCATTERING OF IDENTICAL PARTICLES

Classically we can follow the trajectory of each particle, so that dealing with a system of identical particles does not pose a special problem. In quantum mechanics, however, the notion of a trajectory does not exist. Therefore, if we are studying the scattering of identical particles, it is impossible to distinguish the target particle from the incident particle in the final state. As we have seen in Section 10.1 the wavefunction of a system containing identical particles has to be symmetrized or antisymmetrized depending on whether the system consists of identical bosons or identical fermions.

We shall handle the symmetry properties of the system of identical particles in the following way. Recall that the total wavefunction of the two-particle system is given by

$$\psi_{\text{tot}}(\mathbf{r}_1, \mathbf{r}_2) = \psi_C(\mathbf{R})\psi(\mathbf{r}) \tag{12.1}$$

where

$$\mathbf{R} = \tfrac{1}{2}(\mathbf{r}_1 + \mathbf{r}_2)$$

is the center-of-mass coordinate [see Eq. (8.3a) with $m_1 = m_2$], and

$$\mathbf{r} = \mathbf{r}_1 - \mathbf{r}_2$$

is the relative coordinate. It is clear that under an exchange of the two particles

$$\mathbf{r}_1 \leftrightarrow \mathbf{r}_2$$

and therefore

$$\mathbf{R} \rightarrow \mathbf{R} \qquad \mathbf{r} \rightarrow -\mathbf{r} \tag{12.2a}$$

Clearly, the center-of-mass wavefunction remains invariant under the exchange of particles. That is

$$\psi_C(\mathbf{R}) \rightarrow \psi_C(\mathbf{R}) \tag{12.2b}$$

The symmetry properties of the system are completely determined by the symmetry of the wavefunction for the reduced system. As a simple example, we consider the scattering of two identical spinless bosons. In this case the total wavefunction has to be symmetric under the exchange of the particles

$$\psi(\mathbf{r}) \rightarrow \psi(-\mathbf{r}) = \psi(\mathbf{r}) \tag{12.3a}$$

This means that the final-state wavefunction must be symmetrized. From Fig. 12.1 we recognize that $\mathbf{r} \rightarrow -\mathbf{r}$ implies

$$\begin{aligned} r &\rightarrow r \\ \theta &\rightarrow \pi - \theta \\ \phi &\rightarrow \pi + \phi \end{aligned} \tag{12.3b}$$

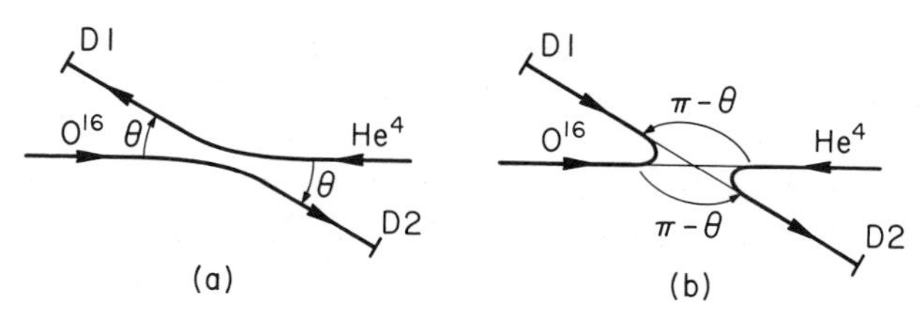

FIGURE 12.1. Scattering of two distinguishable particles in their center-of-mass system. (a) He^4 is detected in D1. (b) He^4 is detected in D2. If the two particles were identical the two processes (a) and (b) could not be distinguished.

The scattered wavefunction [see Eq. (11.12)] in this case takes the form

$$\psi_{\text{sc}}^{(\mathbf{r})} \xrightarrow[r\to\infty]{} f(\theta, \phi)\frac{e^{ikr}}{r} + f(\pi-\theta, \pi+\phi)\frac{e^{ikr}}{r}$$

$$= [f(\theta, \phi) + f(\pi-\theta, \pi+\phi)]\frac{e^{ikr}}{r} \tag{12.3c}$$

The scattering amplitude, therefore is symmetric and is given by

$$f_{\text{sym}}(\theta, \phi) = [f(\theta, \phi) + f(\pi-\theta, \pi+\phi)] \tag{12.4a}$$

The differential cross section is obtained from the absolute square of the scattering amplitude

$$\begin{aligned}\sigma_C(\theta, \phi) &= |f_{\text{sym}}(\theta, \phi)|^2 \\ &= |f(\theta, \phi)|^2 + |f(\pi-\theta, \pi+\phi)|^2 \\ &\quad + 2\,\text{Re}[f^*(\theta, \phi)f(\pi-\theta, \pi+\phi)]\end{aligned} \tag{12.4b}$$

The first two terms in the above equation are what we would obtain if the two particles were indistinguishable, but with the provision that the cross sections as measured for particle 1 at θ, ϕ and particle 2 at $(\pi-\theta)$, $(\pi+\phi)$ are added. This point can be clarified with the help of Fig. 12.1 where both processes (a) and (b) will contribute at detector D1 if the particles are identical. The cross term represents the quantum-mechanical interference that manifests itself in any system containing identical particles.

This is observed in the low-energy scattering of C^{12} nuclei from C^{12} as shown in Fig. 12.2. The C^{12} nuclei have spin zero and are therefore bosons. At low energies the scattering is due to the Coulomb force between the two nuclei for which the scattering

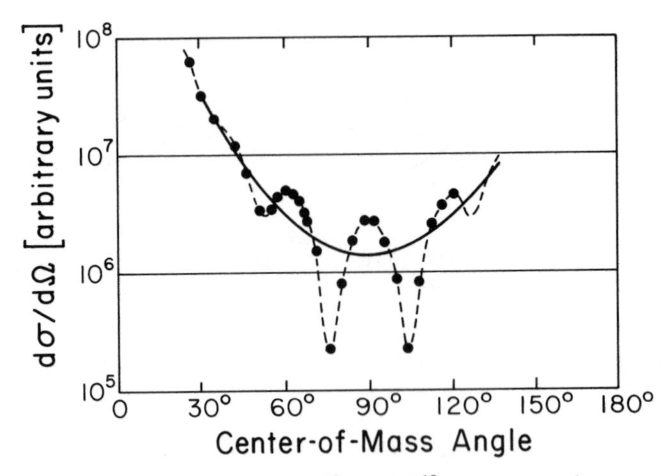

FIGURE 12.2. Elastic scattering of C^{12} from C^{12} at a center-of-mass energy of 5 MeV. [From A. D. Bromley *et al.*, *Phys. Rev. Lett.* **4,** 365 (1960).] The dashed curve includes the effects of interference whereas the solid curve does not.

amplitude is

$$f(\theta, \phi) = \frac{Ze^2}{4\pi\varepsilon_0} \frac{\mu}{2p_{CM}^2} \frac{1}{\sin^2 \theta/2} \tag{12.5a}$$

This result was obtained by time-dependent perturbation theory methods in Chapter 5 [see Eq. (5.38)]. Introducing the above amplitude into Eq. (12.4*b*) gives the differential cross section for this process to be

$$\sigma_C(\theta, \phi) = \left(\frac{Ze^2}{4\pi\varepsilon_0}\right)^2 \frac{\mu^2}{4p_{CM}^4} \left(\frac{1}{\sin^4 \theta/2} + \frac{1}{\cos^4 \theta/2} + \frac{2}{\sin^2 \theta/2 \cos^2 \theta/2}\right) \tag{12.5b}$$

where θ is the center-of-mass scattering angle.

In Fig. 12.2 the solid curve represents the differential cross section calculated as if the two particles were distinguishable; that is, it does not include the interference term of Eq. (12.5*b*). The dashed curve incorporates the effect of symmetrization and agrees completely with the experimental data. Note that the observed cross section is exactly symmetric with respect to 90° since it is *impossible* *to distinguish* between scattering in the forward or backward directions.

As a second example, we consider the scattering of two identical fermions, say electrons. Each of them has a spin angular momentum equal to 1/2. Thus, the two-electron system can have total spin angular momentum equal to 1 or 0. We refer to these values of the total spin by saying that the particles are in the triplet or singlet spin state.

The wavefunction in the present case can be taken as a product of a spin part and a space part. Because we are dealing with identical fermions, the total wavefunction has to be antisymmetric under the exchange of the two particles. Let us assume that the electrons are in the triplet state. In the spin space the triplet state is symmetric and, therefore, for the total wavefunction to be antisymmetric, the space part of the wavefunction for the triplet state must be antisymmetric. That is,

$$\psi_{\text{triplet}}(\mathbf{r}) \rightarrow \psi_{\text{triplet}}(-\mathbf{r}) = -\psi_{\text{triplet}}(\mathbf{r}) \qquad (12.6a)$$

Following the ideas that led to Eq. (12.3c) we obtain an antisymmetric scattering amplitude

$$f_{\text{triplet}}(\theta, \phi) = [f(\theta, \phi) - f(\pi - \theta, \pi + \phi)] \qquad (12.6b)$$

Therefore, the differential cross section for the triplet state becomes

$$\begin{aligned} \sigma_{\text{triplet}}(\theta, \phi) &= |f_{\text{triplet}}(\theta, \phi)|^2 \\ &= |f(\theta, \phi)|^2 + |f(\pi - \theta, \pi + \phi)|^2 \\ &\quad - 2[\operatorname{Re} f^*(\theta, \phi) f(\pi - \theta, \pi + \phi)] \qquad (12.6c) \end{aligned}$$

On the other hand, if the electrons are in the singlet state, then their spins are antiparallel and the wavefunction is antisymmetric in the spin space. Correspondingly, the space part of the wavefunction must be symmetric under the exchange of particles. Thus

$$\psi_{\text{singlet}}(\mathbf{r}) \rightarrow \psi_{\text{singlet}}(-\mathbf{r}) = \psi_{\text{singlet}}(\mathbf{r}) \qquad (12.7a)$$

This determines the form of the scattering amplitude to be symmetric. That is,

$$f_{\text{singlet}}(\theta, \phi) = [f(\theta, \phi) + f(\pi - \theta, \pi + \phi)] \qquad (12.7b)$$

Therefore, the differential cross section is given by

$$\begin{aligned}\sigma_{\text{singlet}}(\theta, \phi) &= |f_{\text{singlet}}(\theta, \phi)|^2 \\ &= |f(\theta, \phi)|^2 + |f(\pi-\theta, \pi+\phi)|^2 \\ &\quad + 2\,\text{Re}[f^*(\theta, \phi) f(\pi-\theta, \pi+\phi)] \quad (12.7c)\end{aligned}$$

In many scattering experiments, however, unpolarized particles are used. In that case the two fermions can be in a triplet or singlet state. Consequently, one defines a spin-averaged cross section. In the case of two electrons there are four final states available, out of which three belong to the triplet state and one to the singlet state. Thus, the spin-averaged differential cross section is given by

$$\begin{aligned}\sigma_{\text{av}}(\theta, \phi) &= \tfrac{1}{4}[3\sigma_{\text{triplet}}(\theta, \phi) + \sigma_{\text{singlet}}(\theta, \phi)] \\ &= |f(\theta, \phi)|^2 + |f(\pi-\theta, \pi+\phi)|^2 - \text{Re}[f^*(\theta, \phi) f(\pi-\theta, \pi+\phi)] \\ &\qquad (12.8)\end{aligned}$$

These predictions are confirmed by scattering polarized beams of electrons off polarized targets. A further check using unpolarized beams is possible by measuring the scattering at $\theta = \pi/2$ in the center of mass. Then, of course,

$$f(\theta) = f(\pi - \theta)$$

and our analysis indicates that

$$\sigma_{\text{triplet}}(\pi/2) = 0 \qquad (12.9a)$$

or

$$\sigma_{\text{av}}(\pi/2) = |f(\pi/2)|^2 \qquad (12.9b)$$

where $f(\pi/2)$ is the scattering amplitude for the scattering of distinguishable particles.

12.2. ENERGY DEPENDENCE AND RESONANCE SCATTERING

We have already seen in the case of low-energy $n-p$ scattering that the scattering amplitude, and hence the cross section, varies as a function of energy. A further example is shown in Fig. 12.3 which

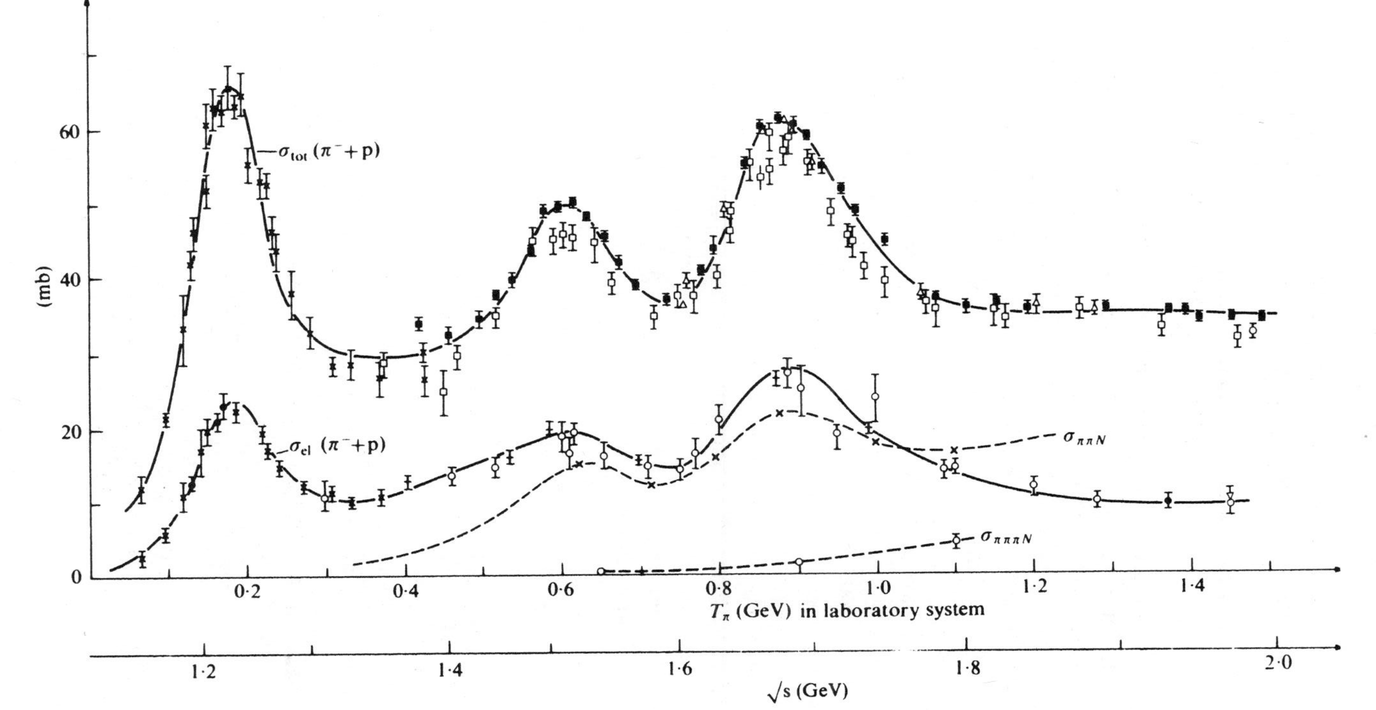

FIGURE 12.3. $\sigma_{tot}(\pi^- + p)$, $\sigma_{Cl}(\pi^- + p)$ (in mb) plotted against the laboratory kinetic energy. [The data is from a compilation by Focacci and Giacomelli, CERN 66-18 (1966) by permission]. $\sigma_{\pi\pi N}$ is also shown. A. H. Rosenfeld and P. Soding *Pion–Nucleon Scattering*, Wiley-Interscience, New York (1968)].

gives the total cross section and the elastic cross section for $\pi^- p$ scattering up to an incident energy of 1.4 GeV. The difference between σ_T and σ_{elastic} is the inelastic cross section $\sigma_{\text{inelastic}}$. A collision is inelastic when the *internal* structure of the colliding particles changes. For instance, the processes

$$\pi^- p \to \pi^\circ n$$

$$\pi^- p \to \pi^+ \pi^- \pi^- p$$

represent possible inelastic final states in $\pi^- p$ collisions.

In Fig. 12.3 we note that at particular energies the cross section exhibits peaks. From the expression of σ_T in the partial-wave analysis [see Eq. (11.23*c*)] we recognize that if the peak occurs in the lth partial wave it implies that $\sin \delta_l$ tends to its maximum value at that energy, namely, $\delta_l = \pi/2, 3\pi/2, \ldots, (2n+1)\pi/2$, etc. In such a case we say that the scattering amplitude has a resonance at that particular energy.

We see that the phase shift as well as the scattering cross section become functions of energy, as already seen in the case of $n-p$ scattering. To fix our ideas about the resonance phenomenon more clearly let us look at only the S-wave scattering from an attractive square-well potential. As we can see from Eq. (11.31*d*), if we consider only low-energy scattering, i.e., $ka \ll 1$, we obtain

$$\tan \delta_0 = \frac{ka(\tan k_1 a/k_1 a - 1)}{1 + k^2 a^2(\tan k_1 a/k_1 a)} \tag{12.10a}$$

where $k_1 = (k^2 + 2\mu U_0/\hbar^2)^{1/2}$. From the above expression we immediately derive that

$$\sin^2 \delta_0 = \frac{\tan^2 \delta_0}{1 + \tan^2 \delta_0} = \frac{k^2 a^2(\tan k_1 a/k_1 a - 1)^2}{(1 + k^2 a^2)[1 + k^2 a^2(\tan^2 k_1 a/k_1^2 a^2)]}$$

Since we assume $ka \ll 1$, the denominator can be simplified, and

$$\sin^2 \delta_0 \simeq \frac{k^2 a^2(\tan k_1 a/k_1 a - 1)^2}{1 + k^2 a^2(\tan^2 k_1 a/k_1^2 a^2)} \tag{12.10b}$$

As long as $k_1 a \neq (2n+1)\pi/2$

$$\sin^2 \delta_0 \simeq (ka)^2 \tag{12.11a}$$

That is, the cross section is small, since

$$\sigma_T = \frac{4\pi}{k^2}\sin^2\delta_0$$
$$= 4\pi a^2 \tag{12.11b}$$

However, when $k_1 a \to (2n+1)\pi/2$, $\sin^2\delta_0 \to 1$ and the cross section becomes

$$\sigma_T = \frac{4\pi}{k^2} = \frac{1}{k^2a^2}4\pi a^2 \tag{12.11c}$$

This is much larger than the nonresonant cross section since we have assumed $ka \ll 1$. Therefore, the cross section when plotted against $k_1 a$ will exhibit peaks whenever $k_1 a$ equals an odd multiple of $\pi/2$, as shown in Fig. 12.4.

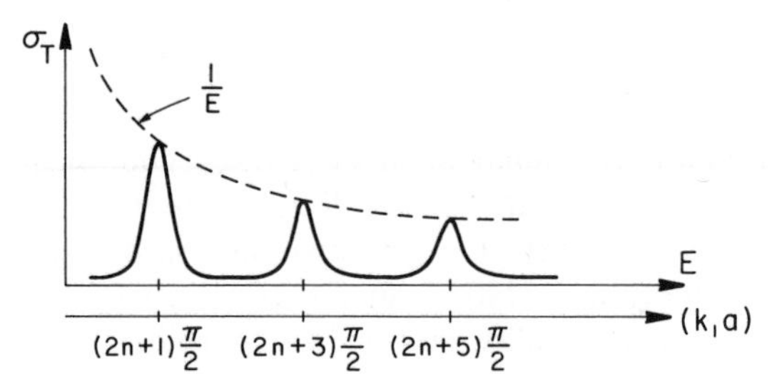

FIGURE 12.4. Resonance peaks in the scattering from an attractive square well in the limit $ka \ll 1$.

Note here that

$$k_1 = \left(k^2 + \frac{2\mu U_0}{\hbar^2}\right)^{1/2} = \frac{(2\mu U_0)^{1/2}}{\hbar}\left(1 + \frac{\hbar^2 k^2}{2\mu U_0}\right)^{1/2}$$
$$\simeq \frac{(2\mu U_0)^{1/2}}{\hbar} + \frac{(2\mu U_0)^{1/2}}{\hbar}\frac{1}{2}\frac{E}{U_0} = \frac{(2\mu U_0)^{1/2}}{\hbar} + \frac{1}{\hbar}\left(\frac{\mu}{2U_0}\right)^{1/2}E \tag{12.12}$$

where we have used $E = \hbar^2 k^2/2\mu$. It is clear from Eq. (12.12) that the abscissa of the graph can be labeled by the energy of the incident particles, which is the origin of the energy dependence of the scattering amplitude and thus of the cross section as well.

If the potential well is strong enough so that the phase shift δ_0 is a multiple of π, then $\sigma_T = 0$, and a phenomenon opposite to resonance occurs. The potential behaves as if it were completely transparent to the particles and there is no scattering at all. From Eq. (12.10b) we see that for $ka \ll 1$ this occurs if $k_1 a = \pi, 2\pi, \ldots, n\pi$. This effect is actually observed in nature. When electrons with energy $E \sim 0.7$ eV are incident on noble gas atoms, they can move through the medium with practically no scattering. This is known as the *Ramsauer–Townsend effect* and implies that for the case at hand the depth and width of the potential are such that one-half wavelength of the interior wavefunction fits almost exactly in the well. The Ramsauer–Townsend effect cannot occur for a repulsive potential since in that case it is not possible to satisfy simultaneously the conditions $ka \ll 1$ and $k_1 a = \pi$.

The Breit–Wigner Formula

We now examine the behavior of the cross section near a resonance in general. We will find that the system can be described in terms of quasi-stationary states such as those we discussed in Section 5.7. We assume that the phase shift δ_l in the lth partial wave reaches the value $\pi/2$ at the particular energy E_0. Therefore,

$$\delta_l(E_0) = \frac{\pi}{2}, \qquad \cos\delta_l(E_0) = 0, \qquad \sin\delta_l(E_0) = 1 \quad (12.13a)$$

If $E \simeq E_0$, then we can make a Taylor expansion around E_0 so that

$$\sin\delta_l(E) \simeq \sin\delta_l(E_0) + \left[\cos\delta_l(E)\frac{d\delta_l(E)}{dE}\right]_{E=E_0}(E-E_0) = 1 \qquad (12.13b)$$

and

$$\cos\delta_l(E) \simeq \cos\delta_l(E_0) - \left[\sin\delta_l(E)\frac{d\delta_l(E)}{dE}\right]_{E=E_0}(E-E_0)$$
$$= -\frac{d\delta_l(E)}{dE}\bigg|_{E-E_0}(E-E_0) = -\frac{2}{\Gamma}(E-E_0) \qquad (12.13c)$$

Here we have defined the rate of change of the phase shift with respect to energy near the resonance as

$$\left.\frac{d\delta_l(E)}{dE}\right|_{E=E_0}=\frac{2}{\Gamma}$$

with Γ a real constant. The scattering amplitude for the lth partial wave is

$$f_l(\theta, E)=\frac{2l+1}{k}\, e^{i\delta_l(E)} \sin \delta_l(E) P_l(\cos \theta) \tag{12.14a}$$

Apart from the slow variation of $1/k$, it is clear that the energy dependence of the scattering amplitude is given by

$$\begin{aligned} f_l(E) &= e^{i\delta_l(E)} \sin \delta_l(E) \\ &= \frac{\sin \delta_l(E)}{\cos \delta_l(E) - i \sin \delta_l(E)} \end{aligned} \tag{12.14b}$$

If we use the relations of Eqs. (12.13*b*) and (12.13*c*) the amplitude near the resonance becomes

$$f_l(E) \simeq \frac{1}{-(2/\Gamma)(E-E_0)-i} = -\frac{\Gamma/2}{(E-E_0)+i\Gamma/2} \tag{12.15a}$$

Therefore, the energy dependence of the cross section is given by

$$|f_l(E)|^2=\frac{\Gamma^2/4}{(E-E_0)^2+\Gamma^2/4} \tag{12.15b}$$

This result is known as the Breit–Wigner formula and has the typical shape of a resonance curve as shown in Fig. 12.5. Up to normalization factors, this is the identical result to that obtained in Eq. (5.47*c*). It equals unity when $E=E_0$, and Γ represents the full width of the curve at its half maximum. The cross section appears to have a prominent peak when $\Gamma \ll E_0$. The total scattering cross section for the lth partial wave for the scattering of spinless particles near the resonance is given by

$$\begin{aligned} \sigma_{lT} &= \frac{4\pi}{k^2}(2l+1)\,|f_l(E)|^2 \\ &= \frac{4\pi}{k^2}(2l+1)\frac{\Gamma^2/4}{(E-E_0)^2+\Gamma^2/4} \end{aligned} \tag{12.15c}$$

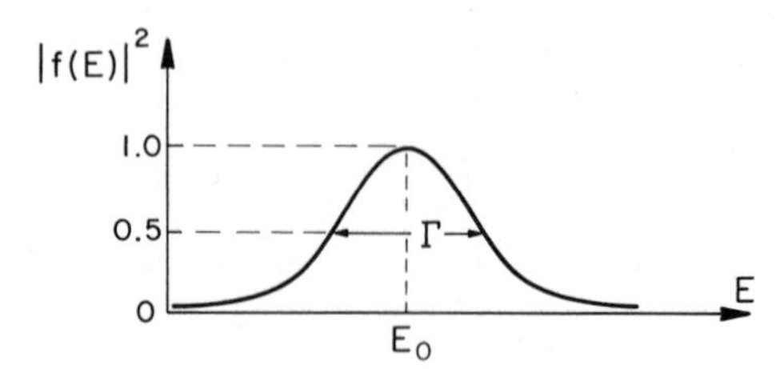

FIGURE 12.5. The Breit–Wigner resonance curve.

We note that when $E \ll E_0$, the amplitude given in Eq. (12.15*a*) is small, positive, and almost real. When $E \gg E_0$, it is small, negative, and almost real. At resonance (i.e., $E = E_0$), $f_l(E)$ is purely imaginary and equal to i. The behavior of the complex amplitude $f_l(E)$ is shown in Fig. 12.6 which is known as an Argand plot. The complex vector $f_l(E)$ begins at the origin and its tip traces out in a *counterclockwise* direction a circle of radius 1/2 centered at the point $+i/2$ in the complex plane†; thus the angle subtended at the center equals 2δ. The requirement of counterclockwise motion is equivalent to the condition $\Gamma > 0$.

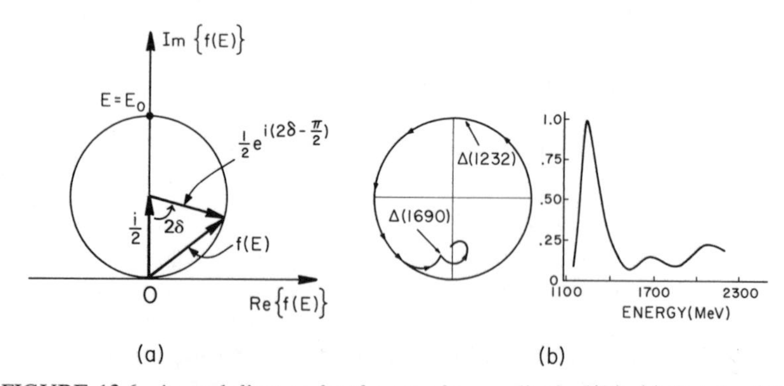

FIGURE 12.6. Argand diagram for the complex amplitude $f(E)$. (a) Details of the construction of the diagram. (b) $f(E)$ and $|f(E)|^2$ for the $j = 3/2$ wave in π^+p scattering (for total spin $S = 3/2$).

† To verify this construction note that

$$f_l(E) = e^{i\delta_l} \sin \delta_l = (i/2)(1 - e^{2i\delta_l}) = i/2 + (1/2)e^{i(2\delta_l - \pi/2)}$$

The complex vector $(1/2)e^{i(2\delta_l - \pi/2)}$ rotates around the center at a rate $d\delta_l(E)/dE$, and when $\delta_l = 0$ it points along the negative imaginary axis. Therefore, $2\delta_l$ must be measured from that axis as shown in the figure. By construction, the vector $f_l(E)$ is the sum of the vectors $i/2$ and $(1/2)e^{i(2\delta_l - \pi/2)}$.

The actual Argand diagram for the $j = 3/2$ wave in $\pi^+ p$ scattering is snown in Fig. 12.6(b). In the presence of inelastic scattering the phase shift becomes complex and $|e^{2i\delta_l}| < 1$. Consequently, $f_l(E)$ moves inside the unit circle. Such behavior is observed in the figure for center-of-mass energies $E > 1.5$ GeV and indicates the threshold for inelastic scattering.

Quasi-Stationary States

The peaks observed in the scattering cross section as a function of the incident energy are more pronounced as the ratio Γ/E_0 becomes smaller. In particular, in the limit $\Gamma \to 0$, the scattering amplitude in the resonating wave becomes sigular at $E = E_0$. This is reminiscent of the discrete energies that characterize the stationary states of a bound system. However, bound states have negative energy and the energy in a scattering process is positive. Thus, these cannot be bound states and we call such states resonances. Furthermore, we note that the energy of the resonances in the scattering system is not sharp but has a width Γ. Therefore, the energy of such a state is uncertain by

$$\Delta E = \Gamma/2 \tag{12.16a}$$

From the uncertainty relation we deduce that the time interval over which we can observe such a system is

$$\Delta E \, \Delta t \simeq \frac{\hbar}{2}$$

or

$$\Delta t \simeq \frac{\hbar/2}{\Delta E} = \frac{\hbar/2}{\Gamma/2} = \frac{\hbar}{\Gamma} = \tau \tag{12.16b}$$

We call $\tau = \hbar/\Gamma$ the *lifetime* of the resonance since it represents the time interval over which two particles stay in the vicinity of each other before receding in opposite directions in the center of mass. It is also clear that such a state cannot be a stationary state if it has a finite lifetime. We have called such states quasi-stationary, and we showed in Section 5.7 that they are eigenstates of the equation of motion with complex eigenvalues.

In Chapter 5 we began from a decaying state and obtained the probability distribution of the energy. In the present case we are given the scattering amplitude $f_l(E)$ which we can consider as an energy wavefunction† $\chi(E)$. Then the wavefunction $\psi_l(t)$ as a function of time is related to $\chi(E)$ through a Fourier transform. The procedure is similar to that followed in Section 5.7, but for completeness we present it here as well. We express $\psi_l(t)$ as follows [see also Eq. (5.47a)]

$$\begin{aligned}\psi_l(t) &= \int_0^\infty f_l(E)e^{-iEt/\hbar}\,dE = -\int_0^\infty \frac{\Gamma/2}{(E-E_0)+i\Gamma/2}e^{-iEt/\hbar}\,dE \\ &= -e^{-i(E_0-i\Gamma/2)t/\hbar}\int_0^\infty \frac{\Gamma/2}{(E-E_0)+i\Gamma/2}e^{-i[(E-E_0)+i\Gamma/2]t/\hbar}\,dE \\ &= -\frac{i\pi\Gamma}{4}e^{-i(E_0-i\Gamma/2)t/\hbar} \qquad (12.17a)\end{aligned}$$

where we have used the fact that

$$\int_{-\infty}^{\infty} dz\frac{e^{i\alpha z}}{z} = i\pi$$

with z-complex. It is clear then that the total wavefunction can be written as

$$\psi_l(\mathbf{r},t) = \psi_l(\mathbf{r})\psi_l(t) = N_t\psi_l(\mathbf{r})e^{-i(E_0-i\Gamma/2)t/\hbar} \qquad (12.17b)$$

Furthermore, remembering that

$$\hat{H}\psi = i\hbar\frac{\partial\psi}{\partial t}$$

we can express the time-independent equation that $\psi_l(\mathbf{r})$ obeys as

$$\hat{H}\psi_l(\mathbf{r}) = (E_0 - i\Gamma/2)\psi_l(\mathbf{r}) \qquad (12.18a)$$

Namely, the energy eigenvalue of the state is

$$E_l = E_0 - i\Gamma/2 \qquad (12.18b)$$

The physical meaning of the complex-energy eigenvalue becomes clear when we examine the probability of finding the system in the

† Here we are neglecting the energy dependence $1/k \sim 1/(E)^{1/2}$ in the scattering amplitude which is not affected by the Fourier transformation.

state ψ_l

$$P_l(t) = |\psi_l(\mathbf{r}, t)|^2 = |\psi_l(\mathbf{r})|^2 e^{-(\Gamma/\hbar)t}$$
$$= |\psi_l(\mathbf{r})|^2 e^{-t/\tau} \quad (12.19)$$

We see that the probability decreases exponentially with a time constant $\tau = \hbar/\Gamma$, where τ is the lifetime of the state. The state is not stationary since the probability of finding the system in that state depends on time. However, for a time interval $t \ll \tau$, the system behaves as if it is in a stationary state with well-defined quantum numbers. Note that the above interpretation depends on Γ being real and positive, which is indeed always the case.

12.3. THE LIPPMAN–SCHWINGER EQUATION

The central idea in the method of Lippman and Schwinger is that rather than analyzing each angular momentum component separately, we try to obtain the scattering amplitude as a whole by solving an integral equation. It is clear that such an approach is useful for studying high-energy scattering processes where a significant number of angular momentum components scatter appreciably.

The time-independent Schrödinger equation for scattering is given by

$$(\nabla^2 + k^2)\psi(\mathbf{r}) = \frac{2\mu}{\hbar^2} U(r)\psi(\mathbf{r}) \quad (12.20)$$

where $k^2 = 2\mu E/\hbar^2$. This equation has a close similarity to the Poisson equation of electrostatics

$$\nabla^2\phi = -4\pi\rho \quad (12.21a)$$

The best way to solve such equations is by Greens method: one defines a Greens function $G(\mathbf{r})$ for the Laplacian through

$$\nabla^2 G(\mathbf{r}) = -4\pi\delta^3(\mathbf{r}) \quad (12.21b)$$

It is known that $G(\mathbf{r}) = 1/|\mathbf{r}|$, from which it follows that

$$\phi(\mathbf{r}) = \int d^3r' G(\mathbf{r} - \mathbf{r}')\rho(\mathbf{r}') = \int d^3r' \frac{\rho(\mathbf{r}')}{|\mathbf{r} - \mathbf{r}'|} \quad (12.21c)$$

Thus we see that the solution of Poisson's equation can be given as an integral of the sources over all space.

We can also apply the above method to the problem of scattering. We have to define a Greens function for the problem. Furthermore, our solution for large spatial distances must have the form

$$\psi(\mathbf{r}) \xrightarrow[r\to\infty]{} e^{i\mathbf{k}\cdot\mathbf{r}} + f(\theta, \phi)\frac{e^{ikr}}{r} \tag{12.22}$$

Let us assume that $G(\mathbf{r}, \mathbf{r}')$ is the Greens function for the scattering problem; that is,

$$(\nabla^2 + k^2)G(\mathbf{r}, \mathbf{r}') = \delta^3(\mathbf{r} - \mathbf{r}') \tag{12.23a}$$

We still have to determine the form of $G(\mathbf{r}, \mathbf{r}')$. But assuming that such an object exists, it is clear that the solution to the Schrödinger equation [see Eq. (12.20)] can be written as

$$\psi(\mathbf{r}) = \frac{2\mu}{\hbar^2}\int d^3r'\, G(\mathbf{r}, \mathbf{r}')U(\mathbf{r}')\psi(\mathbf{r}') \tag{12.23b}$$

This is because

$$\begin{aligned}(\nabla^2 + k^2)\psi(\mathbf{r}) &= \frac{2\mu}{\hbar^2}\int d^3r'\,(\nabla^2 + k^2)G(\mathbf{r}, \mathbf{r}')U(\mathbf{r}')\psi(\mathbf{r}') \\ &= \frac{2\mu}{\hbar^2}\int d^3r'\,\delta^3(\mathbf{r} - \mathbf{r}')U(\mathbf{r}')\psi(\mathbf{r}') \\ &= \frac{2\mu}{\hbar^2}U(\mathbf{r})\psi(\mathbf{r}) \end{aligned} \tag{12.23c}$$

which is nothing other than Eq. (12.20). This shows that as for the Poisson equation, the solution $\psi(\mathbf{r})$ in the present case can also be written as an integral of the source over the whole space, except that here the source depends on the solution $\psi(\mathbf{r})$ itself. Such a solution is known as an integral equation.

The solution in Eq. (12.23b) is not unique in the sense that

$$\psi(\mathbf{r}) = \psi^{(0)}(\mathbf{r}) + \frac{2\mu}{\hbar^2}\int d^3r'\, G(\mathbf{r}, \mathbf{r}')U(\mathbf{r}')\psi(\mathbf{r}') \tag{12.24a}$$

is also a solution of Eq. (12.20) if

$$(\nabla^2 + k^2)\psi^{(0)}(\mathbf{r}) = 0 \tag{12.24b}$$

This simply reflects the fact that we can always add any homogeneous solution of the differential equation to the total solution. This nonuniqueness is, however, fixed from the physical requirement that when $U=0$, i.e., when there is no scattering, the solution must have the form

$$\psi(\mathbf{r}) \xrightarrow[U=0]{} e^{i\mathbf{k}\cdot\mathbf{r}} \tag{12.25a}$$

This fixes the solution given in Eq. (12.24a) as

$$\begin{aligned}\psi(\mathbf{r}) &= e^{i\mathbf{k}\cdot\mathbf{r}} + \frac{2\mu}{\hbar^2}\int d^3r' G(\mathbf{r},\mathbf{r}')U(\mathbf{r}')\psi(\mathbf{r}') \\ &= e^{i\mathbf{k}\cdot\mathbf{r}} + \psi_{\text{sc}}(\mathbf{r}) \xrightarrow[r\to\text{large}]{} e^{i\mathbf{k}\cdot\mathbf{r}} + f(\theta,\phi)\frac{e^{ikr}}{r}\end{aligned} \tag{12.25b}$$

By comparison we see that the scattered wave is given by

$$\psi_{\text{sc}}(\mathbf{r}) = \frac{2\mu}{\hbar^2}\int d^3r' G(\mathbf{r},\mathbf{r}')U(\mathbf{r}')\psi(\mathbf{r}') \tag{12.25c}$$

The integral solution to the Schrödinger equation given in Eq. (12.25b) is known as the Lippman–Schwinger equation. However, to be able to make use of this solution we must know the form of the Greens function.

The Greens Function for the Scattering Problem

We define the Greens function through

$$(\nabla^2 + k^2)G(\mathbf{r},\mathbf{r}') = \delta^3(\mathbf{r}-\mathbf{r}') \tag{12.26a}$$

From translational invariance, it is clear that $G(\mathbf{r},\mathbf{r}')$ must have the form

$$G(\mathbf{r},\mathbf{r}') = G(\mathbf{r}-\mathbf{r}') \tag{12.26b}$$

Let us introduce the Fourier transforms

$$\left.\begin{aligned}G(\mathbf{r}-\mathbf{r}') &= \int d^3q\tilde{G}(\mathbf{q})e^{i\mathbf{q}\cdot(\mathbf{r}-\mathbf{r}')} \\ \delta^3(\mathbf{r}-\mathbf{r}') &= \frac{1}{(2\pi)^3}\int d^3q e^{i\mathbf{q}\cdot(\mathbf{r}-\mathbf{r}')}\end{aligned}\right\} \tag{12.26c}$$

If we substitute these relations into Eq. (12.26a) the differential equation reduces to an algebraic equation for $\tilde{G}(q)$, namely,

$$(-q^2+k^2)\tilde{G}(\mathbf{q}) = \frac{1}{(2\pi)^3}$$

or

$$\tilde{G}(\mathbf{q}) = -\frac{1}{(2\pi)^3}\frac{1}{q^2-k^2} \tag{12.27}$$

The Greens function is obtained by noting that

$$\begin{aligned} G(\mathbf{r}-\mathbf{r}') &= \int d^3q\tilde{G}(\mathbf{q})e^{i\mathbf{q}\cdot(\mathbf{r}-\mathbf{r}')} \\ &= -\frac{1}{(2\pi)^3}\int d^3q\frac{1}{q^2-k^2}e^{i\mathbf{q}\cdot(\mathbf{r}-\mathbf{r}')} \end{aligned} \tag{12.28a}$$

We define the variable

$$\mathbf{r}-\mathbf{r}'=\mathbf{R}$$

so that

$$\begin{aligned} G(\mathbf{R}) &= -\frac{1}{(2\pi)^3}\int d^3q\frac{1}{q^2-k^2}e^{i\mathbf{q}\cdot\mathbf{R}} \\ &= -\frac{1}{(2\pi)^3}\int q^2\,dq\sin\theta\,d\theta\,d\phi\frac{1}{q^2-k^2}e^{iqR\cos\theta} \\ &= -\frac{1}{(2\pi)^2 iR}\int_{-\infty}^{\infty}dq\frac{qe^{iqR}}{q^2-k^2} \end{aligned} \tag{12.28b}$$

The integrand has simple poles at $q=\pm k$ as shown in Fig. 12.7(a). Thus, the integral can be evaluated using Cauchy's residue theorem, but to do that we must specify the contour of the integration.

First, note that the contour of integration has to be closed in the upper half-plane because only then will the exponential be damped. There are then three distinct ways of choosing the contour along the real axis. If we choose the contour along the real axis, as in shown in part (b), then only the residue of the pole at $q=-k$ will contribute. Consequently,

$$\begin{aligned} G(\mathbf{r}-\mathbf{r}') = G(\mathbf{R}) &= (2\pi i)\lim_{q\to -k}\left[-\frac{1}{(2\pi)^2 iR}\frac{(q+k)qe^{iqR}}{q^2-k^2}\right] \\ &= -\frac{1}{2\pi R}\frac{(-k)e^{-ikr}}{-2k} = -\frac{1}{4\pi R}e^{-ikR} = -\frac{1}{4\pi\,|\mathbf{r}-\mathbf{r}'|}e^{-ik\,|\mathbf{r}-\mathbf{r}'|} \end{aligned} \tag{12.29a}$$

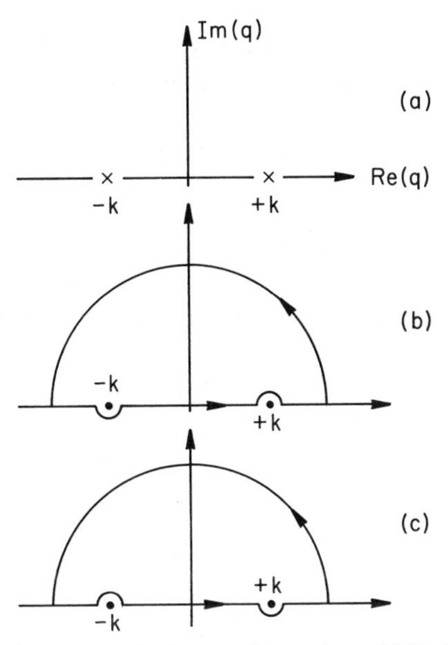

FIGURE 12.7. Integration contour in the complex q-plane. (a) Position of the poles of the Greens function. (b) Contour yielding an incoming wave solution. (c) Contour for outgoing wave solution.

On the other hand, if we had chosen the contour along the real axis, as is shown in part (c), then only the residue of the pole at $q = k$ would contribute, and hence the Greens function would have the form

$$G(\mathbf{r}-\mathbf{r}') = G(\mathbf{R}) = (2\pi i) \lim_{q\to k} \left[-\frac{1}{(2\pi)^2 iR} \frac{(q-k)qe^{iqR}}{q^2-k^2} \right]$$

$$= -\frac{1}{2\pi R}\frac{ke^{ikR}}{2k} = -\frac{1}{4\pi R}e^{ikR} = -\frac{1}{4\pi\,|\mathbf{r}-\mathbf{r}'|}e^{ik\,|\mathbf{r}-\mathbf{r}'|} \qquad (12.29b)$$

Finally, one could have chosen the principal value in evaluating the integral and would have obtained yet another form for the Greens function. These are known, respectively, as the advanced, retarded, and stationary Greens functions.

For our purpose, however, we recall that the Greens function is the scattered wavefunction at $\mathbf{r}$ for a delta source at $\mathbf{r}'$. Furthermore, since we want the scattered wavefunction to be outgoing we must choose the contour of integration such that the Greens function has an outgoing form†. Thus, we see that the proper boundary condition is imposed by choosing the contour shown in part (c) and the Greens function has the form

$$\boxed{G(\mathbf{r}-\mathbf{r}') = -\frac{e^{ik|\mathbf{r}-\mathbf{r}'|}}{4\pi|\mathbf{r}-\mathbf{r}'|}} \tag{12.30a}$$

Thus, the solution to the scattering problem is given by

$$\begin{aligned}\psi(\mathbf{r}) &= e^{i\mathbf{k}\cdot\mathbf{r}} + \frac{2\mu}{\hbar^2}\int d^3r' G(\mathbf{r}-\mathbf{r}')U(\mathbf{r}')\psi(\mathbf{r}') \\ &= e^{i\mathbf{k}\cdot\mathbf{r}} - \frac{\mu}{2\pi\hbar^2}\int d^3r' \frac{e^{ik|\mathbf{r}-\mathbf{r}'|}}{|\mathbf{r}-\mathbf{r}'|}U(\mathbf{r}')\psi(\mathbf{r}')\end{aligned} \tag{12.30b}$$

2.4. BORN APPROXIMATION

Although we have obtained the form of the Greens function, Eq. (12.30*b*) is not easy to solve. In fact, it is not always possible to solve an integral equation exactly. One can, however, solve it iteratively. Let us take the general solution of the scattering problem and introduce the notation

$$\psi(\mathbf{r}) = \psi^{(0)}(\mathbf{r}) + \frac{2\mu}{\hbar^2}\int d^3r' G(\mathbf{r}-\mathbf{r}')U(\mathbf{r}')\psi(\mathbf{r}')$$

or

$$\psi = \psi^{(0)} + \frac{2\mu}{\hbar^2} GU\psi \tag{12.31}$$

Here we have represented $e^{i\mathbf{k}\cdot\mathbf{r}}$ by $\psi^{(0)}$. If we now substitute the lowest-order solution to the scattering equation on the rhs of Eq. (12.31), we obtain to first order

$$\psi^{(1)} = \psi^{(0)} + \frac{2\mu}{\hbar^2} GU\psi^{(0)} \tag{12.32a}$$

† See footnote on p. 495.

Written out in full, this is

$$\psi^{(1)}(\mathbf{r}) = \psi^{(0)}(\mathbf{r}) + \frac{2\mu}{\hbar^2}\int d^3r' G(\mathbf{r}-\mathbf{r}')U(\mathbf{r}')\psi^{(0)}(\mathbf{r}') \qquad (12.32b)$$

If we substitute the first-order solution on the rhs of Eq. (12.31) we obtain to second-order

$$\psi^{(2)} = \psi^{(0)} + \frac{2\mu}{\hbar^2} GU\psi^{(0)} + \left(\frac{2\mu}{\hbar^2}\right)^2 GUGU\psi^{(0)} \qquad (12.33a)$$

This, written out in detail, stands for

$$\psi^{(2)}(\mathbf{r}) = \psi^{(0)}(\mathbf{r}) + \frac{2\mu}{\hbar^2}\int d^3r' G(\mathbf{r}-\mathbf{r}')U(\mathbf{r}')\psi^{(0)}(\mathbf{r}')$$

$$+\left(\frac{2\mu}{\hbar^2}\right)^2 \int d^3r'\, d^3r'' G(\mathbf{r}-\mathbf{r}')U(\mathbf{r}')G(\mathbf{r}'-\mathbf{r}'')U(\mathbf{r}'')\psi^{(0)}(\mathbf{r}'') \qquad (12.33b)$$

We can iterate to as many orders as we wish. This expansion is known as the Born approximation. It is quite useful if the strength of the potential is such that only few terms in the expansion are to be kept. Note also that the iterative solutions are of the same form as the perturbative solutions for the wavefunction resulting from the time-independent perturbation theory introduced in Section 4.2. Here the scattering potential plays the role of the perturbation.

We restrict ourselves now to the *first* Born approximation so that

$$\psi(\mathbf{r}) = \psi^{(0)}(\mathbf{r}) + \frac{2\mu}{\hbar^2}\int d^3r' G(\mathbf{r}-\mathbf{r}')U(\mathbf{r}')\psi^{(0)}(\mathbf{r}')$$

$$= e^{i\mathbf{k}\cdot\mathbf{r}} - \frac{\mu}{2\pi\hbar^2}\int d^3r' \frac{e^{ik|\mathbf{r}-\mathbf{r}'|}}{|\mathbf{r}-\mathbf{r}'|} U(\mathbf{r}')e^{i\mathbf{k}\cdot\mathbf{r}'} \qquad (12.34a)$$

The scattered wave in this approximation is given by

$$\psi_{\text{sc}}(\mathbf{r}) = -\frac{\mu}{2\pi\hbar^2}\int d^3r' \frac{e^{ik|\mathbf{r}-\mathbf{r}'|}}{|\mathbf{r}-\mathbf{r}'|} U(\mathbf{r}')e^{i\mathbf{k}\cdot\mathbf{r}'} \qquad (12.34b)$$

We examine the behavior of the scattered wave as $r \to \infty$. Since the range over which the potential is nonzero is finite, $r' \ll r$, and hence we can expand

$$|\mathbf{r}-\mathbf{r}'|=(r^2+r'^2-2\mathbf{r}\cdot\mathbf{r}')^{1/2}$$

$$=r\left(1+\frac{r'^2}{r^2}-\frac{2\mathbf{r}\cdot\mathbf{r}'}{r^2}\right)^{1/2}\simeq r\left(1-\frac{\mathbf{r}\cdot\mathbf{r}'}{r^2}\right) \qquad (12.35a)$$

We see that for large r we can write

$$\frac{1}{|\mathbf{r}-\mathbf{r}'|}\simeq\frac{1}{r} \qquad (12.35b)$$

Furthermore,

$$e^{ik|\mathbf{r}-\mathbf{r}'|}\simeq e^{ikr(1-\mathbf{r}\cdot\mathbf{r}'/r^2)}=e^{ikr-i\mathbf{k}_f\cdot\mathbf{r}'} \qquad (12.35c)$$

where we have used the notation $\mathbf{k}_f=k\mathbf{r}/r$ for the momentum of the outgoing wave. For elastic scattering $\mathbf{k}_f$ has the same magnitude as the momentum of the incident wave $\mathbf{k}_i$ but is along the direction of the outgoing particle.

The scattered wave for large distances takes the form

$$\psi_{sc}(r)\xrightarrow[r\to\infty]{}-\frac{\mu}{2\pi\hbar^2}\int d^3r'\,\frac{e^{ikr-i\mathbf{k}_f\cdot\mathbf{r}'}}{r}\,U(\mathbf{r}')e^{i\mathbf{k}_i\cdot\mathbf{r}'}$$

$$=-\frac{\mu}{2\pi\hbar^2}\frac{e^{ikr}}{r}\int d^3r'e^{i(\mathbf{k}_i-\mathbf{k}_f)\cdot\mathbf{r}'}U(\mathbf{r}')$$

$$=f(\theta,\phi)\frac{e^{ikr}}{r} \qquad (12.36a)$$

Here we have used $\mathbf{k}_i$ to represent the momentum of the incident wave. Thus, in the first Born approximation the scattering amplitude is given by

$$\boxed{f(\theta,\phi)=f(\mathbf{k}_i,\mathbf{k}_f)=-\frac{\mu}{2\pi\hbar^2}\int d^3r'e^{i(\mathbf{k}_i-\mathbf{k}_f)\cdot\mathbf{r}'}U(\mathbf{r}')} \qquad (12.36b)$$

Note that the scattering amplitude in the Born approximation is proportional to the Fourier transform of the potential with respect to the momentum transfer $\mathbf{q}=\mathbf{k}_i-\mathbf{k}_f$; the angular dependence of the amplitude is contained in the factor $e^{i\mathbf{q}\cdot\mathbf{r}'}$.

Validity of the Born Approximation

The wavefunction for the scattering problem is given by [see Eq. (12.30*b*)]

$$\begin{aligned}\psi(\mathbf{r}) &= e^{i\mathbf{k}\cdot\mathbf{r}} + \psi_{sc}(\mathbf{r}) \\ &= e^{i\mathbf{k}\cdot\mathbf{r}} - \frac{\mu}{2\pi\hbar^2}\int d^3r'\,\frac{e^{ik|\mathbf{r}-\mathbf{r}'|}}{|\mathbf{r}-\mathbf{r}'|}\,U(\mathbf{r}')\psi(\mathbf{r}')\end{aligned} \tag{12.37a}$$

In the first Born approximation we replace the wavefunction under the integral sign by the incident wave. Thus

$$\psi_B(\mathbf{r}) = e^{i\mathbf{k}\cdot\mathbf{r}} - \frac{\mu}{2\pi\hbar^2}\int d^3r'\,\frac{e^{ik|\mathbf{r}-\mathbf{r}'|}}{|\mathbf{r}-\mathbf{r}'|}\,U(\mathbf{r}')e^{i\mathbf{k}\cdot\mathbf{r}'} \tag{12.37b}$$

We expect this approximation to be a good approximation if in the range of the potential

$$|\psi_{sc}(\mathbf{r})| \ll |e^{i\mathbf{k}\cdot\mathbf{r}}| = 1 \tag{12.38a}$$

Since the influence of the potential is the strongest at the origin, if

$$|\psi_{sc}(0)| \ll 1 \tag{12.38b}$$

then this approximation should be reliable.

Let us assume that the potential is spherically symmetric. That is, $U(\mathbf{r}') = U(r')$. Then the Born approximation is valid if [from Eq. (12.38*b*)]

$$\left|\frac{\mu}{2\pi\hbar^2}\int d^3r'\,\frac{e^{ikr'}}{r'}\,U(r')e^{i\mathbf{k}\cdot\mathbf{r}'}\right| \ll 1$$

or

$$\left|\frac{\mu}{2\pi\hbar^2}2\pi\int r'^2\,dr'\,\sin\theta'\,d\theta'\,\frac{e^{ikr'}}{r'}\,U(r')e^{ikr'\cos\theta'}\right| \ll 1$$

or

$$\frac{2\mu}{\hbar^2 k}\left|\int_0^\infty dr' e^{ikr'}U(r')\sin kr'\right| \ll 1 \tag{12.39a}$$

This is the condition for validity of the Born approximation. We note that at low energies $kr' \to 0$. Therefore, $\sin kr' \simeq kr'$ and $e^{ikr'} \simeq 1$, so that the condition for validity becomes

$$\frac{2\mu}{\hbar^2}\left|\int_0^\infty dr'r'U(r')\right| \ll 1 \tag{12.39b}$$

If the potential has a height U_0 and range r_0, then the Born approximation is valid at low energies only if

$$\frac{\mu\,|U_0|\,r_0^2}{\hbar^2} \ll 1 \tag{12.40a}$$

On the other hand, at high energies $kr \to \infty$, the exponential in Eq. (12.39a) oscillates rapidly and picks up contributions only from $r' \leqslant 1/k$, so that the condition for validity at high energies becomes

$$\frac{2\mu}{\hbar^2 k}\frac{|U_0|}{2k} \ll 1$$

or

$$\frac{\mu\,|U_0|\,r_0^2}{\hbar^2} \ll (kr_0)^2 \tag{12.40b}$$

This shows that if the Born approximation is valid at low energies it is also valid at high energies; the converse, however, is not true. Equations (12.40a) and (12.40b) are equivalent to the condition of validity given in Section 5.5 [p. 204].

Coulomb Scattering Revisited

We obtained the solution to the problem of the scattering of a charged particle in a Coulomb potential in Section 5.5. There we used time-dependent perturbation theory and commented that the result agreed with the classical calculation. We now show that the same result is obtained in the first Born approximation. Of course, some of the mathematical steps are identical in the two calculations, but it is important to point out the differences in the two methods and thus gain further insight into this important problem.

The Coulomb potential, as we well know, is a long-range potential. In fact, its range is infinite, and as a result the conventional phase-shift analysis does not work. The reason is not too hard to see physically. In the usual phase-shift analysis we assume that the incident wave represents free particles at infinite separation and hence can be described by a plane wave. But it is clear that because the Coulomb potential has an infinite range, the incident particles feel the Coulomb force even at infinite separations and thus cannot be represented by a plane wave. Furthermore, the long range also has the consequence that even at low energies quite a few partial waves suffer appreciable scattering so that partial-wave analysis is not the right method to deal with Coulomb scattering.

It is also clear that even at low energies the Born approximation for Coulomb scattering is not valid. Given that $U(r) = Ze^2/4\pi\varepsilon_0 r$, the condition for validity given by Eq. (12.39b) is not fulfilled since

$$\int_0^\infty dr\, rU(r) = \int_0^\infty dr\, r\frac{Ze^2}{4\pi\varepsilon_0 r} \to \infty \tag{12.41a}$$

We recognize that this problem arises because of the long-range nature of the Coulomb potential. In practice, however, the Coulomb potential is screened, so we can get around the difficulty of Eq. (12.41a) by modifying the potential to be†

$$U(r) = \frac{Ze^2}{4\pi\varepsilon_0 r}e^{-mr} \tag{12.41b}$$

The parameter m that we introduced defines the inverse range of the potential. It is clear now that the Born approximation would be valid if

$$\frac{2\mu}{\hbar^2}\left|\int_0^\infty dr\, rU(r)\right| = \frac{2\mu}{\hbar^2}\left|\int_0^\infty dr\, r\frac{Ze^2}{4\pi\varepsilon_0 r}e^{-mr}\right|$$

$$= \frac{2\mu Ze^2}{\hbar^2 m(4\pi\varepsilon_0)} \ll 1$$

This condition can be met by an appropriate choice of the parameter m. Thus, we can do all scattering calculations with the modified potential given in Eq. (12.41b) and in the end take the limit $m \to 0$ to obtain the result for Coulomb scattering.

† This is known as the *Yukawa potential* and was postulated to explain the force between nucleons arising from the exchange of mesons of mass m.

The Born amplitude for this process is

$$f_B(\theta, \phi) = f_B(\mathbf{k}_i, \mathbf{k}_f) = -\frac{\mu}{2\pi\hbar^2}\int d^3r' e^{(i/\hbar)\mathbf{q}\cdot\mathbf{r}'} U(\mathbf{r}')$$
$$= -\frac{\mu}{2\pi\hbar^2}\int d^3r' e^{(i/\hbar)\mathbf{q}\cdot\mathbf{r}'} \frac{Ze^2}{4\pi\varepsilon_0 r'} e^{-mr'} \tag{12.42}$$

Here $\mathbf{q} = \hbar(\mathbf{k}_i - \mathbf{k}_f)$ so that for elastic scattering

$$q^2 = \hbar^2(\mathbf{k}_i - \mathbf{k}_f)^2 = \hbar^2(k_i^2 + k_f^2 - 2\mathbf{k}_i \cdot \mathbf{k}_f)$$
$$= \hbar^2(k^2 + k^2 - 2k^2 \cos\theta) = 4p^2 \sin^2\frac{\theta}{2}$$

or

$$q = 2p \sin\frac{\theta}{2} \tag{12.43}$$

which is in agreement with our previous definition given in Eq. (5.34*b*); θ is the center-of-mass scattering angle. The integral in Eq. (12.42) is similar to that performed in Eq. (5.35*b*) and yields

$$f_B(\theta, \phi) = -\frac{\mu}{2\pi\hbar^2}\frac{Ze^2}{4\pi\varepsilon_0}\int r'^2\, dr' \sin\theta'\, d\theta'\, d\phi' e^{(i/\hbar)qr'\cos\theta'} \frac{e^{-mr}}{r'}$$
$$= -\frac{Ze^2\mu}{2\pi\hbar(4\pi\varepsilon_0)} 2\pi \int_0^\infty r'\, dr' e^{-mr'} \frac{1}{iqr'}\left(e^{(i/\hbar)qr'} - e^{-(i/\hbar)qr'}\right)$$
$$= -\frac{Ze^2\mu}{i\hbar q(4\pi\varepsilon_0)}\int_0^\infty dr'\left(e^{-[m-(i/\hbar)q]r'} - e^{-[m+(i/\hbar)q]r'}\right)$$
$$= -\frac{Ze^2\mu}{i\hbar q(4\pi\varepsilon_0)}\left[\frac{1}{m-(i/\hbar)q} - \frac{1}{m+(i/\hbar)q}\right]$$
$$= -\frac{Ze^2\mu}{i\hbar^2 q(4\pi\varepsilon_0)}\frac{2iq}{m^2+q^2/\hbar^2}$$
$$= -\frac{2\mu}{\hbar^2}\frac{1}{m^2+q^2/\hbar^2}\frac{Ze^2}{4\pi\varepsilon_0} \tag{12.44a}$$

It suffices now to let $m \to 0$, and thus the differential cross section is

$$\sigma_B(\theta, \phi) = |f_B(\theta, \phi)|^2 = \left(\frac{Ze^2}{4\pi\varepsilon_0}\right)^2 \left(\frac{2\mu}{q^2}\right)^2$$
$$= \left(\frac{Ze^2}{4\pi\varepsilon_0}\right)^2 \frac{\mu^2}{4p^4}\frac{1}{\sin^4\theta/2} \tag{12.45}$$

where we have used Eq. (12.43). This is in exact agreement with Eq. (5.38).

Note that in the present calculation we did not have to introduce specific wavefunctions for the initial and final particles, even though we have done so implicitly by using the first Born approximation. Thus, it is not surprising that the two results agree. On the other hand, the agreement with the classical result—as already mentioned—is one of the accidental features associated with the Coulomb potential.

12.5. INELASTIC SCATTERING

We have seen in Section 11.2 that the final wavefunction in a scattering process can be written as [see Eq. (11.22*a*)]

$$\psi_f(\mathbf{r}) = \psi_{\text{in}}(\mathbf{r}) + \psi_{\text{sc}}(\mathbf{r})$$

$$\begin{aligned}\psi_f(r) \xrightarrow[r\to\infty]{} & \sum_{l=0}^{\infty} \frac{2l+1}{2ikr}\, i^l\left(e^{2i\delta_l} e^{i(kr-l\pi/2)} - e^{-i(kr-l\pi/2)}\right) P_l(\cos\theta) \\ = & \sum_{l=0}^{\infty} \frac{2l+1}{2ikr}\left[e^{2i\delta_l} e^{ikr} + (-1)^{l+1} e^{-ikr}\right] P_l(\cos\theta) \quad (12.46a)\end{aligned}$$

Here the phase shifts δ_l are all real since the potential is real. Furthermore, if we define

$$S_l = e^{2i\delta_l} \tag{12.46b}$$

then the wavefunction for large distances can be written as

$$\psi_f(\mathbf{r}) \xrightarrow[r\to\infty]{} \sum_{l=0}^{\infty} \frac{2l+1}{2ikr}\left[S_l e^{ikr} + (-1)^{l+1} e^{-ikr}\right] P_l(\cos\theta) \tag{12.46c}$$

It is clear from Eq. (12.46*b*) that since the phase shift is real

$$|S_l| = 1 \tag{12.47}$$

Furthermore, from Eq. (12.46*c*) we see that this implies that the normalization of the outgoing wave is the same as for the incoming wave. This, of course, physically implies that the total number of particles that come in is equal to the total number of particles going out. This can be seen in detail as follows: The radial flux at large distances is given by

$$j_r = -\frac{i\hbar}{2\mu}\left[\psi_f^*\left(\frac{\partial}{\partial r}\psi_f\right) - \left(\frac{\partial}{\partial r}\psi_f^*\right)\psi_f\right]$$

$$j_r \xrightarrow[r\to\infty]{} = \frac{\hbar k}{\mu}\sum_{l,l'}\frac{2l+1}{2ikr}\frac{2l'+1}{2ikr}[S_l^* S_{l'} + (-1)^{l+l'+1}]P_l(\cos\theta)P_{l'}(\cos\theta) \tag{12.48a}$$

Consequently, the flux of probability out of a sphere of large radius R is given by

$$\int_R \mathbf{j}\cdot d\mathbf{S} = \int j_r R^2 \sin\theta\, d\theta\, d\phi = 2\pi R^2 \int_0^\pi j_r \sin\theta\, d\theta \tag{12.48b}$$

If we use the expression for j_r from Eq. (12.48a) and the orthonormality relations for the Legendre polynomials we obtain the expression for the net flux out of a large sphere as

$$\int_R \mathbf{j}\cdot d\mathbf{S} = \frac{\pi\hbar}{\mu k}\sum_{l=0}^{\infty}(2l+1)(|S_l|^2 - 1) \tag{12.48c}$$

If the phase shifts are real so that $|S_l|^2 = 1$, then the net flux out of the sphere is zero. This is a statement of conservation of probability. It says that the number of particles that enter into the interaction region is the same as the number of particles that exit; this is the case for elastic scattering.

However, there occur in nature processes where the internal structure of the system changes. Such processes are known as inelastic scattering processes. For example, if we scatter neutrons off a complex nucleus, the neutron may scatter elastically. It may also scatter by raising the nucleus to an excited state or be absorbed by the nucleus. This means that the net flux out of a sphere in the presence of such processes (i.e., inelastic scattering) need not vanish anymore. In fact, it should be negative since we are losing a fraction of the incoming beam to other processes. This, therefore, implies from Eq. (12.48c) that in the presence of inelastic scattering.

$$|S_l|^2 < 1 \tag{12.49a}$$

In this case we can no longer write

$$S_l = e^{2i\delta_l}$$

with δ_l real. But if we let

$$\delta_l \to \delta_l + i\eta_l \tag{12.49b}$$

with δ_l and η_l real we obtain

$$S_l = e^{-2\eta_l} e^{2i\delta_l} \tag{12.49c}$$

and this expression can describe processes where the particle flux is not conserved. It also shows that in the presence of inelastic scattering the phase shifts become complex. To understand the meaning of all this let us go back to the Schrödinger equation assuming the potential $U(r)$ to be real

$$i\hbar \frac{\partial \psi}{\partial t} = \left(-\frac{\hbar^2}{2\mu} \nabla^2 + U \right) \psi \tag{12.50a}$$

$$-i\hbar \frac{\partial \psi^*}{\partial t} = \left(-\frac{\hbar^2}{2\mu} \nabla^2 + U \right) \psi^* \tag{12.50b}$$

Multiplying Eq. (12.50a) by ψ^* from the left and Eq. (12.50b) by ψ from the right and subtracting the second from the first we obtain the following

$$i\hbar \left(\psi^* \frac{\partial \psi}{\partial t} + \frac{\partial \psi^*}{\partial t} \psi \right) = \psi^* \left(-\frac{\hbar^2}{2\mu} \nabla^2 + U \right) \psi - \left(-\frac{\hbar^2}{2\mu} \nabla^2 + U \right) \psi^* \psi$$

or,

$$\begin{aligned} i\hbar \frac{\partial}{\partial t} (\psi^* \psi) &= -\frac{\hbar^2}{2\mu} [\psi^* \nabla^2 \psi - (\nabla^2 \psi^*) \psi] \\ &= -\frac{\hbar^2}{2\mu} \boldsymbol{\nabla} \cdot [\psi^* \boldsymbol{\nabla} \psi - (\boldsymbol{\nabla} \psi^*) \psi] \end{aligned}$$

Namely

$$\frac{\partial}{\partial t} (\psi^* \psi) = \frac{i\hbar}{2\mu} \boldsymbol{\nabla} \cdot [\psi^* \boldsymbol{\nabla} \psi - (\boldsymbol{\nabla} \psi^*) \psi]$$

which we recognize as the continuity equation

$$\frac{\partial}{\partial t} P(r, t) = -\boldsymbol{\nabla} \cdot \mathbf{j} \tag{12.51a}$$

Integrating as before over the volume of a large sphere we have

$$\int_\Omega d^3r \frac{\partial}{\partial t} P(r, t) = \int_\Omega d^3r(-\nabla \cdot \mathbf{j}) = -\int_s d\mathbf{S} \cdot \mathbf{j}$$

or,

$$\frac{\partial}{\partial t} \int_\Omega d^3r P(r, t) = -\int_s d\mathbf{S} \cdot \mathbf{j} \tag{12.51b}$$

This tells us that if the flux out of a closed surface is zero, then the particles are in a stationary state and the total probability of finding them in the enclosed volume does not change with time. That is, there are no sources or sinks of particles. This result was derived by assuming that the potential is real. Let us now allow for a complex potential. The continuity equation then becomes

$$i\hbar \frac{\partial}{\partial t}(\psi^* \psi) = -\frac{\hbar^2}{2\mu} \nabla \cdot (\psi^* \nabla \psi - \nabla \psi^* \psi) + (U - U^*)\psi^* \psi \tag{12.52a}$$

Furthermore, if we write

$$U = U_R - iU_I \tag{12.52b}$$

with U_R and U_I real, then the continuity equation becomes

$$i\hbar \frac{\partial}{\partial t} P(\mathbf{r}, t) = -\frac{\hbar^2}{2\mu} \nabla \cdot (\psi^* \nabla \psi - \nabla \psi^* \psi) - 2iU_I P(\mathbf{r}, t)$$

or,

$$\frac{\partial}{\partial t} P(\mathbf{r}, \mathbf{t}) = -\nabla \cdot \mathbf{j} - \frac{2}{\hbar} U_I P(\mathbf{r}, t)$$

and finally

$$\frac{\partial}{\partial t} P(\mathbf{r}, t) + \nabla \cdot \mathbf{j} = -\frac{2}{\hbar} U_I P(\mathbf{r}, t) \tag{12.52c}$$

Note here that two distinct cases may arise. If the net flux out of a closed surface vanishes, then

$$\nabla \cdot \mathbf{j} = 0$$

and

$$\frac{\partial}{\partial t} P(\mathbf{r}, t) = -\frac{2}{\hbar} U_I P(\mathbf{r}, t)$$

or,

$$P(\mathbf{r}, t) = e^{-2U_I t/\hbar} \tag{12.53a}$$

That is, the probability of finding particles in the enclosed volume changes with time. Therefore, particles are no longer in stationary states. It is clear from Eq. (12.53*a*) that if $U_I > 0$ the potential acts as a sink, whereas if $U_I < 0$ the potential behaves like a source of particles.

In scattering theory, however, we assume the wavefunctions to be stationary states. This leads to the second possibility, namely,

$$\frac{\partial}{\partial t} P(\mathbf{r}, t) = 0$$

and hence

$$\nabla \cdot \mathbf{j} = -\frac{2}{\hbar} U_I P(\mathbf{r}, t) \tag{12.53b}$$

Integrating over a large volume, we obtain

$$\int_\Omega d^3r \nabla \cdot \mathbf{j} = -\frac{2}{\hbar} \int_\Omega d^3r U_I P(\mathbf{r}, t)$$

or,

$$\int_S \mathbf{j} \cdot d\mathbf{S} = -\frac{2}{\hbar} \int_\Omega d^3r U_I \, |\psi|^2 \tag{12.53c}$$

This means that the inelastic processes such as absorption in scattering can be described by introducing complex potentials which in turn lead to complex phase shifts and result in a nonzero flux out of a closed surface. Furthermore, the lhs of Eq. (12.53*c*) simply measures the flux removed from the incident beam. Hence

$$\sigma_{\text{absorption}} = \sigma_{\text{inelastic}} = -\frac{\mu}{\hbar k} \int_S \mathbf{j} \cdot d\mathbf{S}$$

$$= \frac{\pi}{k^2} \sum_{l=0}^{\infty} (2l+1)(1-|S_l|^2) \tag{12.54}$$

Here we have used the expression of Eq. (12.48*c*) for the net flux out of a closed sphere. From the definition of the scattering amplitude

$$f(\theta)=\sum_{l=0}^{\infty}\frac{2l+1}{2ik}(e^{2i\delta_l}-1)P_l(\cos\theta)$$

$$=\sum_{l=0}^{\infty}\frac{2l+1}{2ik}(S_l-1)P_l(\cos\theta) \qquad (12.55a)$$

we obtain the total cross section for *elastic* scattering as

$$\sigma_{\text{elastic}}=\int \sin\theta\, d\theta\, d\phi\, |f(\theta)|^2$$

$$=\frac{\pi}{k^2}\sum_{l=0}^{\infty}(2l+1)\,|S_l-1|^2 \qquad (12.55b)$$

The total cross section, which is the sum of the elastic and inelastic scattering, is given by

$$\sigma_{\text{tot}}=\sigma_{\text{elastic}}+\sigma_{\text{inelastic}}$$

$$=\frac{\pi}{k^2}\sum_{l=0}^{\infty}(2l-1)(|S_l-1|^2+1-|S_l|^2)$$

$$=\frac{2\pi}{k^2}\sum_{l=0}^{\infty}(2l+1)[1-\text{Re}(S_l)] \qquad (12.56a)$$

This reduces to the familiar expression for total elastic scattering when the phase shifts are real [Eq. (11.23*c*)]. It is easy to see from Eqs. (12.55*a*) and (12.56*a*) that

$$\sigma_{\text{tot}}=\frac{4\pi}{k}\text{Im}[f(\theta=0)] \qquad (12.56b)$$

That is, the optical theorem remains valid even in the presence of inelastic scattering.

An example of inelastic scattering is shown in Fig. 12.3 where the inelastic cross section $\sigma_{\text{inelastic}}$ is given by the difference of the two curves which represent σ_T and σ_{elastic}, respectively. A further example of inelastic scattering is shown in Fig. 12.6(b) where the phase shift δ_l is plotted in the Argand diagram. Clearly, for energies $E>1500$ MeV, the magnitude of S_l is $|S_l|<1$, indicating that the phase shift in that particular wave has become complex.

From Eqs. (12.54), (12.55*b*), and (12.56*a*) we see that when

$S_l = 1$, there is no scattering whatsoever in the lth wave. When $S_l = 0$, there is complete absorption in that wave and we have

$$\sigma_{\text{elastic}} = \sigma_{\text{inelastic}} = \tfrac{1}{2}\sigma_{\text{tot}} = \frac{\pi}{k^2}\sum_{l=0}^{\infty}(2l+1) \tag{12.57a}$$

If the absorbing potential has a range a, then in the limit of very high energies $l_{\max} = ka$. Using this in Eq. (12.57a) we obtain

$$\begin{aligned} \sigma_{\text{elastic}} &= \sigma_{\text{inelastic}} = \pi a^2 \\ \sigma_{\text{tot}} &= 2\pi a^2 \end{aligned} \tag{12.57b}$$

This is referred to as scattering from a *black disk*, and the elastic scattering is called *shadow scattering*. Note the close analogy of the terminology with the *diffraction* of light from an opaque object.

We see that whenever inelastic scattering takes place it is always accompanied by elastic scattering. However, the converse is not true since if $|S_l| = 1$, $\sigma_{\text{inelastic}} = 0$ but $\sigma_{\text{elastic}} \propto \sin^2\delta_l$. Furthermore, the angular distribution in the reaction channels for inelastic scattering is determined by the dominant partial wave, just as it is for elastic channels. The same correspondence holds for the energy dependence of the reaction amplitudes. For example, a resonance can occur simultaneously in several reaction channels and will always be reflected by a resonance in the elastic channel. In that case each resonating channel contributes a partial width Γ_r, the total width of the resonance being given by

$$\Gamma = \sum_r \Gamma_r \tag{12.58}$$

An interesting example is neutrino inelastic scattering

$$\nu_\mu + n \to p + \mu^- + X$$

Here X stands for combinations of any other particles. The observed energy dependence for this process [with $(E_\nu)_{\text{lab}}$ in GeV] is

$$\sigma_{\text{tot}} = (0.65 \times 10^{-38})(E_\nu)_{\text{lab}} \qquad (\text{cm}^2) \tag{12.59a}$$

Neutrinos interact only weakly and we believe that the weak interaction is of *very* short range. For practical purposes we treat this as a point interaction and hence only $l = 0$ waves can contribute.

Thus, theoretically,

$$\sigma_{\text{tot}} < \frac{4\pi}{k^2} = \frac{4\pi}{c^2(\hbar k)^2}(\hbar c)^2 = \frac{4\pi}{E_{CM}^2}(\hbar c)^2 \tag{12.59b}$$

Here we have assumed that we are looking at very high energy scattering so that $\hbar kc = E_{CM}$ (i.e., the neutron mass has been neglected). We can show for the present case that $E_{CM}^2 = 2m_n c^2 (E_\nu)_{\text{lab}}$ so that combining Eqs. (12.59*a*) and (12.59*b*) we obtain the bound

$$\sigma_{\text{tot}} = (0.65 \times 10^{-38}\ \text{cm}^2)\frac{E_{CM}^2}{2m_n c^2} < \frac{4\pi}{E_{CM}^2}(\hbar c)^2$$

or,

$$E_{CM}^4 < 1.6 \times 10^{12}\ (\text{GeV})^4$$

or,

$$(E_{CM})_{\text{lim}} < 10^3\ \text{GeV} \tag{12.59c}$$

We conclude that σ_{tot} for neutrino scattering cannot continue to depend linearly on energy when $E \sim 10^3$ GeV. At such energies partial waves with $l \geqslant 1$ must participate in the scattering. If this is true the range of the weak interactions must be finite and greater than

$$a \geqslant \frac{1}{k_{\text{lim}}} = \frac{\hbar c}{c\hbar k_{\text{lim}}} = \frac{\hbar c}{(E_{CM})_{\text{lim}}} = \frac{0.2\ \text{GeV}-\text{F}}{10^3\ \text{GeV}}$$
$$= 2 \times 10^{-4}\ \text{F}$$

In fact, recent well-established theoretical models predicted the range of the weak interactions to be about 10 times above this limit, and this has been brilliantly confirmed by the experimental discovery of the $W^{\pm}$ and Z^0 particles.

12.6. CONCLUSION

We have discussed how to incorporate the indistinguishability of particles into the scattering formalism. Resonance scattering and the energy dependence of the scattering amplitude have also been discussed in detail. The integral equation method for solving the scattering problem is quite useful. We have discussed such solutions with emphasis on the Greens function techniques. Furthermore, we explained the Born approximation and discussed its validity. It was then applied to the case of Coulomb scattering. Finally, we also discussed the modifications necessary in the scattering formalism in the presence of inelastic scattering.

Problems

PROBLEM 1

Consider the scattering of two electrons in the center-of-mass system. Assume (1) *no spin flip*; (2) unpolarized beam; (3) unpolarized target; (4) no polarization measurement in the detectors. Find the probability for scattering through an angle θ as given by Eqs. (12.6) and (12.7) of the text. Do this by direct counting of all spin states and by more general arguments using eigenstates of the total spin operator. Show that you obtain the same result in both cases.

PROBLEM 2

(*a*) The range of nuclear forces is of the order of 1 F so that the relevant Yukawa potential [see Eq. (12.42)] has $m = 1\ \text{F}^{-1}$. Use this value to plot the differential cross section for Yukawa and Coulomb potentials as a function of q^2 for a particle of incident momentum $p = 10$ GeV, can you tell the two apart?

(*b*) Discuss the physical assumptions implicit in the Yukawa theory and its relation to the mass of the π-meson.

PROBLEM 3

Derive the Greens functions for one-dimensional scattering:

$$\frac{d^2G}{dx^2}(x, x') + k^2 G(x, x') = \delta(x - x')$$

with the property that

$$G(x, x') \sim \begin{cases} f_+ e^{ik(x-x')} & x - x' \to \infty \\ f_- e^{-ik(x-x')} & x - x' \to -\infty \end{cases}$$

PROBLEM 4

Consider a potential of Gaussian shape

$$U(r) = Ae^{-r^2/2R_0^2}$$

where R_0 can be called the root mean square radius. The data in the figure [from Barbiellini *et al.*, *Phys. Lett.* **39B,** 663 (1972)] give

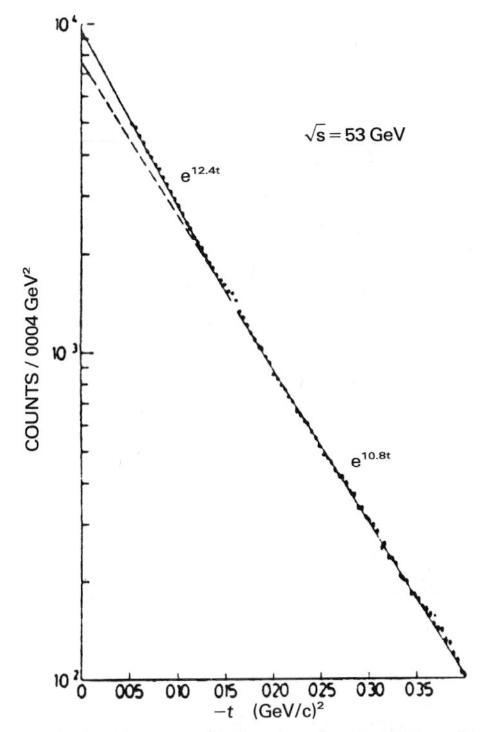

Differential cross section for p–p elastic scattering at center-of-mass energy squared $s - 2800$ (GeV)^2.

the differential cross section for $p-p$ elastic scattering in terms of the momentum transfer

$$-t \simeq q^2 = 4p^2 \sin^2\frac{\theta}{2}$$

$$\frac{d\sigma}{dt} = \frac{\pi}{p^2}\frac{d\sigma}{d\Omega}$$

at an equivalent incident momentum $p \simeq 1500$ GeV (1 GeV $=$ 10^9 eV).

(*a*) Obtain the differential cross section $d\sigma/d\Omega$ for the Gaussian potential.

(*b*) From the data of the figure estimate the root mean square radius R_0.

PROBLEM 5

Let $U(x, y, z) = 0$ except for the region

$$x^2 < \left(\frac{a}{2}\right)^2, \quad y^2 < \left(\frac{a}{2}\right)^2, \quad z^2 < \left(\frac{a}{2}\right)^2$$

which is a cube of side length a. The strength of the potential is V_0 inside this cube. A plane wave e^{ikz} is incident on the cube. What is the differential cross section as a function of the angles θ and ϕ in the Born approximation?

PROBLEM 6

Consider a spherically symmetric repulsive potential

$$U(r) = \frac{\alpha}{r^2}, \qquad \alpha > 0$$

Use the Born approximation to calculate the angular dependence as well as the energy dependence of the differential cross section.

PROBLEM 7

Use the Born approximation to determine the total scattering cross section for particles of low energy in a potential of the form

$$U(r)=\frac{\alpha}{r^4}, \qquad \alpha>0$$

Appendix 1

The Fourier Transform

Consider a function $f(x)$ defined in the interval $-\infty \leqslant x \leqslant \infty$. If we can express $f(x)$ in the form

$$f(x) = \frac{1}{(2\pi)^{1/2}} \int_{-\infty}^{\infty} dk e^{ikx} g(k) \tag{A1.1}$$

then the function $g(k)$ is said to be the Fourier transform of $f(x)$. We see that the Fourier transform can be thought of formally as the limit of a Fourier series when the period tends to infinity. Furthermore, we note that the relation of Eq. (A1.1) is invertible. That is, multiplying both sides of Eq. (A1.1) by an exponential and integrating over x we obtain

$$\frac{1}{(2\pi)^{1/2}} \int_{-\infty}^{\infty} dx e^{-ikx} f(x) = \frac{1}{(2\pi)^{1/2}} \int dx e^{-ikx} \frac{1}{(2\pi)^{1/2}} \int dx' e^{ik'x} g(k')$$

$$= \int dk' g(k') \frac{1}{2\pi} \int dx e^{-i(k-k')x}$$

We recognize [see Eq. (A2.7)] that

$$\frac{1}{2\pi} \int dx e^{-i(k-k')x} = \delta(k-k') \tag{A1.2a}$$

where $\delta(k-k')$ is the Dirac delta function. Using this relation and the properties of the delta function, we obtain

$$\frac{1}{(2\pi)^{1/2}} \int dx e^{-ikx} f(x) = \int dk' g(k') \delta(k-k') = g(k)$$

or

$$g(k)=\frac{1}{(2\pi)^{1/2}}\int dxe^{-ikx}f(x) \tag{A1.2b}$$

That is, if $g(k)$ is the Fourier transform of $f(x)$, then $f(x)$ is the inverse Fourier transform of $g(k)$. Note that the factor of $1/2\pi$ is used differently by different authors. We have defined it so that the Fourier transform and its inverse are symmetrical. The advantage of this definition is that if a function is normalized, then so is its Fourier transform.

To see this note that if

$$\int dxf^*(x)f(x)=1$$

Then

$$\begin{aligned}\int dkg^*(k)g(k)&=\int dk\frac{1}{(2\pi)^{1/2}}\int dxf^*(x)e^{ikx}\frac{1}{(2\pi)^{1/2}}\int dx'e^{-ikx'}f(x')\\&=\int dx\,dx'f^*(x)f(x')\frac{1}{2\pi}\int dke^{-ik(x'-x)}\\&=\int dx\,dx'f^*(x)f(x')\delta(x-x')\\&=\int dxf^*(x)f(x)=1\end{aligned} \tag{A1.3}$$

Examples of Fourier Transforms

(i) Let $f(x)=\delta(x)$. Clearly in this case

$$g(k)=\frac{1}{(2\pi)^{1/2}}\int dxe^{-ikx}\delta(x)=\frac{1}{(2\pi)^{1/2}} \tag{A1.4}$$

In other words, the Dirac delta function in coordinate space, has a constant value of $1/(2\pi)^{1/2}$ in momentum space. Conversely, we can show that a momentum delta function has a constant value $1/(2\pi)^{1/2}$ in coordinate space.

(ii) Let $f(x)=e^{-\alpha^2x^2/2}$. This is a Gaussian peaked at the origin and with a width proportional to $1/\alpha$. For this function

$$g(k)=\frac{1}{(2\pi)^{1/2}}\int dxe^{-ikx}e^{-\alpha^2x^2/2}$$

$$=\frac{1}{(2\pi)^{1/2}}\int dxe^{-1/2(\alpha^2x^2+2ikx-k^2/\alpha^2+k^2/\alpha^2)}$$

$$=\frac{1}{(2\pi)^{1/2}}\int dxe^{-1/2(\alpha x+ik/\alpha)^2}e^{-k^2/2\alpha^2}$$

$$=\frac{1}{(2\pi)^{1/2}}e^{-k^2/2\alpha^2}\int dxe^{-1/2(\alpha x+ik/\alpha)^2}$$

$$=\frac{1}{(2\pi)^{1/2}}e^{-k^2/2\alpha^2}\frac{1}{\alpha}\int dze^{-z^2/2}$$

The integral (where we have set $z=\alpha x+ik/\alpha$) can be evaluated by contour integration and is well known to yield $(2\pi)^{1/2}$. Thus

$$g(k)=\frac{1}{\alpha}e^{-k^2/2\alpha^2} \tag{A1.5}$$

This shows that the Fourier transform of a Gaussian is again a Gaussian but with an inverse width. That is, in this case the width of the function is proportional to α, so if we have a very narrow Gaussian function in coordinate space, then it will be spread out in momentum space, and vice versa.

(iii) Note that a Fourier transform takes a function from configuration space to momentum space, and vice versa. In general, it can be defined for any pair of conjugate variables. This is useful because sometimes working in the conjugate space can simplify a problem. We now show this with the following example. Let

$$g(k)=\frac{1}{(2\pi)^{1/2}}\int dxe^{-ikx}f(x)$$

If $G(k)$ is the Fourier transform for df/dx, then

$$G(k)=\frac{1}{(2\pi)^{1/2}}\int dxe^{-ikx}\frac{df(x)}{dx}$$

$$=\frac{1}{(2\pi)^{1/2}}\int dx\left\{\frac{d}{dx}[e^{-ikx}f(x)]+ike^{-ikx}f(x)\right\}$$

$$=\frac{1}{(2\pi)^{1/2}}e^{-ikx}f(x)|_{-\infty}^{\infty}+ik\frac{1}{(2\pi)^{1/2}}\int dxe^{-ikx}f(x)$$

If the function $f(x)$ is regular, then the product $e^{-ikx}f(x)$ evaluated at infinity vanishes (because of the rapid oscillation of e^{-ikx}). Thus, the first term drops out and we obtain

$$G(k)=ikg(k) \tag{A1.6}$$

This shows that the differential operator d/dx behaves like a multiplicative factor in momentum space. In general, differential operators in one space behave as multiplicative factors in the conjugate space. Hence, the study of differential equations (such as the Schrödinger equation) in coordinate space reduces to a study of algebraic equations in momentum space.

(iv) We conclude by listing some useful relations between Fourier transforms. If $g(k)$ is the Fourier transform of $f(x)$, then the transform of the functions listed below is given in terms of $g(k)$ as indicated

Function	Fourier transform
$f(x)$	$g(k)$
$f(x+a)$	$e^{ika}g(k)$
$e^{i\mu x}f(x)$	$g(k-\mu)$
$f^*(x)$	$g^*(-k)$
$f(-x)$	$g(-k)$

Appendix 2

The Dirac Delta Function

A. PROPERTIES

The Dirac delta function is a generalized function defined by the relation

$$\int_{-\infty}^{\infty} dx\, \delta(x-x_0) f(x) = f(x_0) \tag{A2.1}$$

The integral picks out only the first term in the Taylor expansion of the function $f(x)$ around x_0, and this relation must hold for any function. Suppose we now choose an arbitrary function that is nonzero everywhere except at the point x_0 where it vanishes:

$$\begin{aligned} f(x) &= 0 && \text{at } x = x_0 \\ &= \text{nonzero} && \text{everywhere also} \end{aligned} \tag{A2.2a}$$

Clearly in this case Eq. (A2.1) gives

$$\int dx\, \delta(x-x_0) f(x) = 0 \tag{A2.2b}$$

and since this relation must be true for any arbitrary form of $f(x)$ outside of the point x_0, we conclude that

$$\delta(x-x_0) = 0 \quad \text{if} \quad x \neq x_0$$

It is clear from Eq. (A2.1) also that†

$$\delta(x-x_0) = \infty \quad \text{at} \quad x = x_0$$

† Note that the Riemann–Lebesgue definition of integration gives

$$\int dx f(x) g(x) = \lim_{a\to 0} \left\{ a \sum_i f(x_i) g(x_i) \right\}$$

where the path along which integration is performed is divided into equal parts of length a.

Thus, we see that

$$\delta(x - x_0) = 0 \quad \text{if} \quad x \neq x_0$$
$$= \infty \quad \text{if} \quad x = x_0 \qquad \text{(A2.3)}$$

The delta function is normalized to unity as can be seen from the defining relation Eq. (A2.1) if we choose $f(x) = 1$. That is

$$\int dx\, \delta(x - x_0) = 1 \qquad \text{(A2.4)}$$

All the above results show that the delta function cannot be thought of as a function in the usual sense. However, we can think of it as a limit of a sequence of regular functions, as we will see when we discuss the representations of the delta function. We often depict the delta function by a curve as shown in Fig. A2.1, where the width is supposed to tend to zero and the peak to infinity, keeping the area under the curve finite.

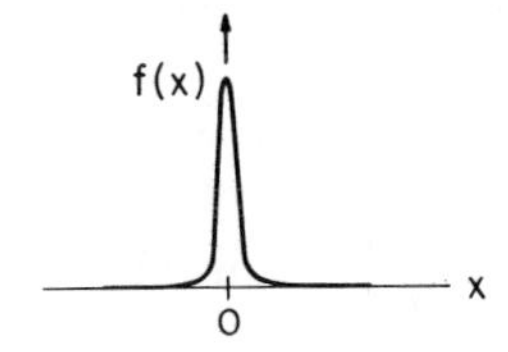

FIGURE A2.1. Schematic representation of the Dirac delta function $\delta(x)$.

Let us derive some of the properties of the delta function without assuming any particular representation for it.

(i) $\int dx\, \delta(x) f(x) = f(0)$ (definition of the delta function)

(ii) $\delta^*(x) = \delta(x)$ i.e., it is real.

To see this note that since (i) must be true for any function, if we choose a real function we have

$$[\int dx\, \delta(x) f(x)]^* = f^*(0) = f(0)$$

or

$$\int dx\, \delta^*(x) f(x) = f(0) = \int dx\, \delta(x) f(x)$$

Comparison gives $\delta^*(x) = \delta(x)$

(iii) $\delta(x) = \delta(-x)$ i.e., the delta function is an even function

To see this we again take the defining relation and use an even function $f(x)$. In that case

$$\int dx\, \delta(x) f(x) = f(0)$$

If we make the change of variable

$$x \to -x$$

then the lhs becomes

$$\int dx\, \delta(-x) f(-x) = \int dx\, \delta(-x) f(x)$$

The value of the integral, however, does not change under a change of variable, so we conclude that

$$\int dx\, \delta(-x) f(x) = f(0) = \int dx\, \delta(x) f(x)$$

This shows that $\delta(-x) = \delta(x)$

(iv) $\delta(ax) = \dfrac{1}{a}\, \delta(x) \qquad \text{for } a > 0$

From the normalization condition we see that

$$\int dx\, \delta(ax) = \int \frac{1}{a}\, d(ax)\, \delta(ax)$$

$$= \frac{1}{a} \int dy\, \delta(y) = \frac{1}{a}, \qquad y = ax$$

Comparison with $\int dx\, \delta(x) = 1$, shows then that

$$\delta(ax) = \frac{1}{a}\, \delta(x)$$

We list below some further properties of the delta function without actually proving them. They are, however, relatively simple to prove.

(v) $\int dx\, \delta'(x) f(x) = -f'(0), \qquad \text{where } \delta'(x) = \dfrac{d}{dx}\, \delta(x)$

(vi) $\delta'(-x) = -\delta'(x)$

(vii) $x\, \delta(x) = 0$

(viii) $\delta(x^2 - a^2) = \dfrac{1}{2a} [\delta(x-a) + \delta(x+a)] \qquad \text{for } a > 0$

(ix) $f(x)\, \delta(x-a) = f(a)\, \delta(x-a)$

(x) $\int dx\, \delta(x-b)\, \delta(a-x) = \delta(a-b)$

B. REPRESENTATIONS

As we noted earlier, the delta function can be thought of as a limit of a sequence of regular functions. We now show this by giving some simple representations of the delta function.

(1) A representation of the delta function is given by

$$\delta(x) = \lim_{g\to\infty} \frac{1}{\pi} \frac{\sin gx}{x} \tag{A2.5}$$

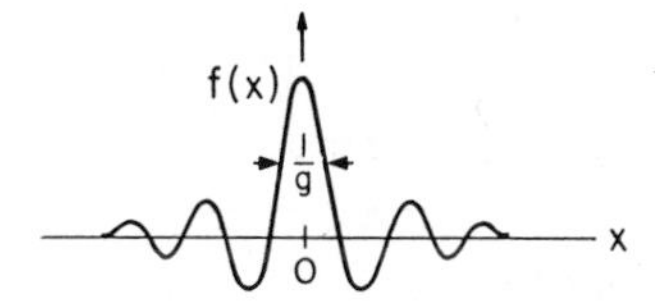

FIGURE A2.2. The "diffraction amplitude" $[\sin(gx)]/x$ which is a representation of the δ-function in the limit $g \to \infty$.

Pictorially, this function for any particular value of g looks like a diffraction amplitude with a width proportional to $1/g$; it is shown in Fig. A2.2. For every g the function is regular. As we increase the value of g the function peaks more strongly at $x = 0$, and hence in the limit $g \to \infty$ it behaves like the delta function. That it has the properties of the delta function can be seen by evaluating the integral

$$I = \lim_{g\to\infty} \int_{-\varepsilon}^{\varepsilon} dx \frac{1}{\pi} \frac{\sin gx}{x} f(x) = \lim_{g\to\infty} \frac{1}{\pi i} \int_{-\varepsilon}^{\varepsilon} dx \frac{e^{igx}}{x} f(x)$$

$$= \lim_{g\to\infty} \frac{1}{\pi i} \int_{-\varepsilon}^{\varepsilon} \frac{d(gx)}{g} \frac{e^{igx}}{x} f\left(\frac{1}{g} gx\right)$$

Defining $y = gx$, the integral can be written as

$$I = \lim_{g\to\infty} \frac{1}{\pi i} \int_{-g\varepsilon}^{g\varepsilon} dy \frac{e^{iy}}{y} f\left(\frac{y}{g}\right) = \lim_{g\to\infty} \frac{1}{\pi i} \int_{-\infty}^{\infty} dy \frac{e^{iy}}{y} f\left(\frac{y}{g}\right)$$

This integral can be evaluated by using the method of residues. There is a pole at $y = 0$, and the residue of the integrand at the pole is $f(0)$. Therefore the value of the integral is

$$I = \lim_{g\to\infty} \int_{-t}^{t} dx \frac{1}{\pi} \frac{\sin gx}{x} f(x) = \frac{1}{\pi i} \pi i f(0) = f(0) \qquad \text{(A2.6)}$$

Equation (A2.6) is nothing other than the definition of the delta function, so we can identify

$$\delta(x) = \lim_{g\to\infty} \frac{1}{\pi} \frac{\sin gx}{x}$$

(2) It follows from the above that an alternate representation of the delta function is

$$\boxed{\delta(x) = \frac{1}{2\pi} \int_{-\infty}^{\infty} dk e^{ikx}} \qquad \text{(A2.7)}$$

This can be seen as follows.

$$\frac{1}{\pi} \frac{\sin gx}{x} = \frac{1}{2\pi} \int_{-g}^{g} dk e^{ikx}$$

Therefore

$$\delta(x) = \lim_{g\to\infty} \frac{1}{\pi} \frac{\sin gx}{x} = \lim_{g\to\infty} \frac{1}{2\pi} \int_{-g}^{g} dk e^{ikx}$$

$$= \frac{1}{2\pi} \int_{-\infty}^{\infty} dk e^{ikx}$$

From the definition of the Fourier transform, we immediately see that the delta function is simply the Fourier transform of the constant $1/(2\pi)^{1/2}$.

(3) Another useful representation of the delta function is

$$\delta(x) = \lim_{\alpha\to\infty} \left(\frac{\alpha}{\pi}\right)^{1/2} e^{-\alpha x^2} \qquad \text{(A2.8)}$$

This is a normalized Gaussian function of standard deviation $1/(2\alpha)^{1/2}$ which tends to zero as $\alpha \to \infty$. Its height is proportional to $\alpha^{1/2}$ and it peaks at $x = 0$. Mathematically, the equivalence of this function to the delta function is established by showing that

$$\lim_{\alpha\to\infty} \left[\int dx \left(\frac{\alpha}{\pi}\right)^{1/2} e^{-\alpha x^2} f(x) \right] = f(0)$$

Other representations that are encountered in various applications are

(4)
$$\delta(x) = \lim_{\varepsilon \to 0} \frac{1}{\pi} \frac{\varepsilon}{x^2 + \varepsilon^2} \qquad \text{(A2.9)}$$

(5)
$$\delta(x) = \frac{d}{dx} \theta(x) \qquad \text{(A2.10)}$$

where $\theta(x)$ is the step function which is defined as a generalized function through

$$\begin{aligned} \theta(x) &= 0, \qquad x \leqslant 0 \\ &= 1, \qquad x \geqslant 0 \end{aligned}$$

The step function $\theta(x)$ is shown in Fig. (A2.3).

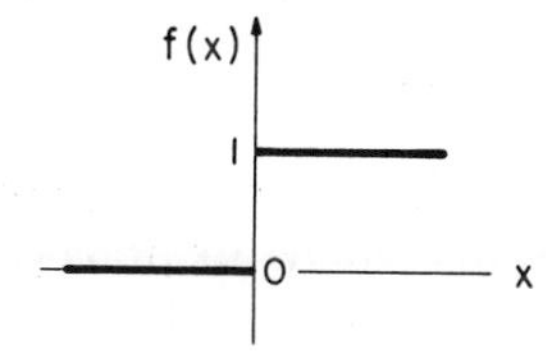

FIGURE A2.3. The step-function $\theta(x)$.

Appendix 3

Normalization of Plane Waves and Wave Packets

In this appendix we discuss the question of normalization of wavefunctions. Let us begin with plane waves defined by

$$\phi_k(x, t) = Ae^{-i(\omega t - kx)} \tag{A3.1a}$$

where A is a constant. We have noted that such a function is not normalizable because

$$\int_{-\infty}^{\infty} dx\phi_k^*(x, t)\phi_k(x, t) = \int_{-\infty}^{\infty} dx\, |A|^2 \to \infty \tag{A3.1b}$$

As explained in the text, a plane wave represents a particle whose position is completely uncertain. In practice, however, we always deal with particles that are localized to some degree. Furthermore, the form of the plane wave is simple enough so that we can circumvent the difficulty of normalization in the following way.

A. NORMALIZATION TO DIRAC DELTA FUNCTION

The form of the plane waves allows us to normalize them to the Dirac delta function. From the definition of Eq. (A3.1a) we see that

$$\begin{aligned} \int_{-\infty}^{\infty} dx\phi_k^*(x, t)\phi_{k'}(x, t) &= \int_{\infty}^{\infty} dx\, |A|^2\, e^{i(\omega-\omega')t} e^{-i(k-k')x} \\ &= |A|^2\, e^{i(\omega-\omega')t} \int_{-\infty}^{\infty} dx e^{-i(k-k')x} \\ &= |A|^2\, e^{i(\omega-\omega')t} 2\pi\, \delta(k-k') \\ &= 2\pi\, |A|^2\, \delta(k-k') \end{aligned} \tag{A3.2a}$$

Here we have used the definition of the delta function [see Eq. (A2.7)] as well as the fact that when $k = k'$, then $\omega = \omega'$. Thus, if we require that

$$\int_{-\infty}^{\infty} dx \phi_k^*(x, t) \phi_{k'}(x, t) = \delta(k - k') \tag{A3.2b}$$

the normalized amplitude of the plane waves will be determined to be

$$A = A^* = \frac{1}{(2\pi)^{1/2}}$$

or

$$\boxed{\phi_k(x, t) = \frac{1}{(2\pi)^{1/2}} e^{-i(\omega t - kx)}} \tag{A3.2c}$$

Note that we have chosen the phase of the amplitude to be real. Furthermore, from the properties of the delta function we recognize that this normalization is divergent at $k = k'$. Therefore, the expectation value of operators, or the average value of observables, has to be defined carefully as

$$\langle Q \rangle = \frac{\int_{-\infty}^{\infty} dx \phi_k^*(x, t) Q \phi_k(x, t)}{\int_{-\infty}^{\infty} dx \phi_k^*(x, t) \phi_k(x, t)} \tag{A3.3}$$

We emphasize here that this has to be calculated carefully because in this case both the numerator and the denominator diverge.

B. BOX NORMALIZATION

An alternate, simpler, and more frequently used normalization for the plane waves is the box normalization. Here we assume that the physical space (in this case one-dimensional) extends from $-L$ to $+L$, where L is large. Clearly then we can normalize the plane waves in this box by demanding that

$$\int_{-L}^{L} dx \phi_k^*(x, t) \phi_k(x, t) = 1$$

or,

$$\int_{-L}^{L} dx\,|A|^2 = |A|^2\,2L = 1$$

leading to

$$A = A^* = \frac{1}{(2L)^{1/2}} \tag{A3.4a}$$

Therefore the plane waves can be normalized inside the box in which case we write

$$\boxed{\phi_k(x, t) = \frac{1}{(2L)^{1/2}}\, e^{-i(\omega t - kx)}} \tag{A3.4b}$$

Note that this form of the wavefunction is properly normalized for every value of the length L. However, orthogonality of the wavefunctions requires that we use *periodic boundary conditions* when using box normalization. This can be seen as follows

$$\int_{-L}^{L} dx \phi_k^*(x, t)\phi_{k'}(x, t) = \frac{1}{2L}\int_{-L}^{L} dx e^{i(\omega-\omega')t} e^{-i(k-k')x}$$

$$= e^{i(\omega-\omega')t}\frac{\sin(k-k')L}{(k-k')L} \tag{A3.5a}$$

If $k \neq k'$, this will be zero only if $(k-k')L = l\pi$, where l is an integer. This condition is satisfied by

$$\begin{aligned} kL &= n\pi \\ k'L &= m\pi \qquad (n, m \text{ integers}) \end{aligned} \tag{A3.5b}$$

Clearly, with this condition,

$$\begin{aligned} \phi_k(L, t) &= e^{-i(\omega t - kL)} = e^{-i(\omega t + kL)} e^{2ikL} \\ &= \phi_k(-L, t) e^{2in\pi} \end{aligned}$$

or,

$$\phi_k(L, t) = \phi_k(-L, t) \tag{A3.5c}$$

In the case of box normalization, the average value of observables can be written simply as

$$\langle Q \rangle = \int_{-L}^{L} dx \phi_k^*(x, t) Q \phi_k(x, t) \tag{A3.6}$$

Furthermore, since the box length $2L$ is an artifact of our method of calculation, physical observables must be independent of L. Therefore, at the end of a calculation we can take the limit of the box having infinite length to recover the entire physical space.

C. NORMALIZATION OF WAVE PACKETS

Wave packets, as we have already explained in the text, are a superposition of plane wave states of different momentum. They are sufficiently localized in space so that there is no difficulty in normalizing them. This should be contrasted with the plane waves that extend over the entire physical space, which is the sole cause of their nonnormalizability. Given any wave packet, however, we can always normalize it by multiplying by an appropriate constant. We now show this for the case of the square-wave packet. That is, we set

$$\begin{aligned} A(k) &= 1 \quad \text{for} \quad k_1 \leq k \leq k_2 \\ &= 0 \quad \text{for} \quad k < k_1 \quad \text{and} \quad k > k_2 \end{aligned} \tag{A3.7a}$$

The momentum space wavefunction is defined [see Eq. (1.59)] through

$$\tilde{\phi}(k, t) = N_k e^{-i\omega t} A(k) \tag{A3.7b}$$

Here N_k is the normalization constant to be determined. We impose the normalization condition

$$\int_{-\infty}^{\infty} dk \tilde{\phi}^*(k, t) \tilde{\phi}(k, t) = 1$$

so that

$$|N_k|^2 \int_{-\infty}^{\infty} dk \, |A(k)|^2 = 1 \tag{A3.8a}$$

From the form of $A(k)$ in Eq. (A3.7a) we see that the integral on the lhs reduces to

$$|N_k|^2(k_2-k_1)=1$$

or,

$$N_k=N_k^*=\frac{1}{(\Delta k)^{1/2}} \qquad (\Delta k=k_2-k_1) \tag{A3.8b}$$

Thus, the normalized momentum wavefunction has the form

$$\tilde{\phi}(k,t)=\begin{cases}\dfrac{1}{(\Delta k)^{1/2}}e^{-i\omega t} & k_1\leq k\leq k_2\\ =0 & k<k_1 \text{ and } k>k_2\end{cases} \tag{A3.8c}$$

The coordinate space wavefunction can be obtained by taking the inverse Fourier transform of $\tilde{\phi}(k,t)$:

$$\phi(x,t)=\frac{1}{(2\pi)^{1/2}}\int_{-\infty}^{\infty}dk\tilde{\phi}(k,t)e^{ikx}=\frac{1}{(2\pi\,\Delta k)^{1/2}}\int_{k_1}^{k_2}dke^{-i(\omega t-kx)} \tag{A3.9a}$$

Expanding $\omega(k)$ around the central value $k_0=(k_1+k_2)/2$ and keeping only the linear terms, i.e., writing

$$\omega(k)=\omega(k_0)+(k-k_0)\left.\frac{d\omega}{dk}\right|_{k_0}=\omega_0+(k-k_0)v_g \tag{A3.9b}$$

the coordinate space wavefunction becomes

$$\begin{aligned}\phi(x,t)&=\frac{1}{(2\pi\,\Delta k)^{1/2}}\int_{k_1}^{k_2}dke^{-i[\omega_0t+(k-k_0)v_gt-kx]}\\&=\frac{1}{(2\pi\,\Delta k)^{1/2}}e^{-i(\omega_0t-k_0x)}\int_{k_1}^{k_2}dke^{-i(k-k_0)(v_gt-x)}\\&=\frac{1}{(2\pi\,\Delta k)^{1/2}}e^{-i(\omega_0t-k_0x)}\int_{-\Delta k/2}^{\Delta k/2}dze^{-iz(v_gt-x)}, \qquad (z=k-k_0)\\&=\frac{1}{(2\pi\,\Delta k)^{1/2}}e^{-i(\omega_0t-k_0x)}\frac{\sin[\Delta k(v_gt-x)/2]}{(v_gt-x)/2}\end{aligned} \tag{A3.9c}$$

which is in agreement with the form of Eq. (1.61b). Furthermore, remembering that normalization is maintained under a Fourier transformation [see Eq. (A1.3)], we recognize that the wavefunction $\phi(x, t)$ must be normalized.

Conversely, we could have started with the coordinate wavefunction calculated in the text [see Eq. (1.61*b*)].

$$G(x, t) = e^{-i(\omega_0 t - k_0 x)} \frac{\sin \Delta k/2(v_g t - x)}{(v_g t - x)/2} \tag{A3.10}$$

which as can be readily checked is not normalized to unity. To define a normalized wavefunction, we let

$$\begin{aligned} \phi(x, t) &= AG(x, t) \\ &= Ae^{-i(\omega_0 t - k_0 x)} \frac{\sin \Delta k/2(v_g t - x)}{(v_g t - x)/2} \end{aligned} \tag{A3.11}$$

where A is the normalization constant to be determined. The normalization condition requires

$$\int dx \phi^*(x, t)\phi(x, t) = 1$$

or,

$$|A|^2 \int_{-\infty}^{\infty} dx \left[\frac{\sin \Delta k/2(v_g t - x)}{(v_g t - x)/2}\right]^2 = 1$$

namely

$$|A|^2\, 2\, \Delta k \int_{-\infty}^{\infty} dy \frac{\sin^2 y}{y^2} = 1 \tag{A3.12a}$$

Here we have defined $y = \Delta k/2(v_g t - x)$. The integral can be evaluated by the method of residues and has the value π. The normalization condition therefore constrains the constant A to take the value such that

$$|A|^2\, 2\, \Delta k \pi = 1$$

or

$$A = A^* = \frac{1}{(2\pi\, \Delta k)^{1/2}} \tag{A3.12b}$$

Here again we have chosen the phase of the constant to be real. The normalized wavefunction in the coordinate space is then given by

$$\boxed{\phi(x, t) = \frac{1}{(2\pi\,\Delta k)^{1/2}}\, e^{-i(\omega_0 t - k_0 x)}\, \frac{\sin \Delta k/2(v_g t - x)}{(v_g t - x)/2}} \tag{A3.12c}$$

which is in exact agreement with the result of Eq. (A3.9*c*).

It is straightforward to show that the Fourier transform of $\phi(x, t)$ as given by Eq. (A3.12*c*) reproduces the normalized wave packet in momentum space $\tilde{\phi}(k, t)$ as given by Eq. (A3.8*c*). This is to be expected and we do not give the calculation here. One essentially retraces the steps that led to Eq. (A3.9*c*) and to do so one must perform a contour integral. The integral yields zero when $k < k_1$ or $k > k_2$, while it is finite and constant when $k_1 \leqslant k \leqslant k_2$.

Similar considerations apply to all forms of wave packets. In addition to the square-wave packet which we have analyzed here, the Gaussian packet is also often used. [For more details the reader can consult L. I. Shiff, '*Quantum Mechanics*', Third Edition, p. 60. McGraw Hill, N.Y. 1968].

Appendix 4

Review of Matrix Algebra

A. Definitions

A properly arranged array of $n \times m$ numbers is called a matrix. We will restrict ourselves to *square matrices* and will represent an n-dimensional square matrix by writing

$$\begin{bmatrix} a_{11} & a_{12} & a_{13} & \cdots & a_{1n} \\ a_{21} & a_{22} & a_{23} & \cdots & a_{2n} \\ \vdots & \vdots & \vdots & \vdots & \vdots \\ a_{n1} & a_{n2} & a_{n3} & \cdots & a_{nn} \end{bmatrix} \tag{A4.1}$$

The number a_{ij} is the *matrix element* belonging to the *i-row* and *j-column* of the array. We will consider matrices with elements that can be *complex numbers*. In this appendix matrices will be indicated by using a caret or an open font symbol. For example,

$$\hat{A}, \hat{\sigma}, \mathbb{1}$$

indicate matrices. The symbol $\mathbb{1}$ is reserved for the unit matrix of any dimensionality

$$\begin{bmatrix} 1 & 0 & 0 & \cdots & 0 \\ 0 & 1 & 0 & \cdots & 0 \\ 0 & 0 & 1 & \cdots & 0 \\ \vdots & \vdots & \vdots & \vdots & \vdots \\ 0 & 0 & 0 & \cdots & 1 \end{bmatrix}$$

The matrix elements of $\mathbb{1}$ are given by the Kronecker symbol

$$\delta_{ij} \tag{A4.2}$$

which equals 1 if $i = j$, and zero otherwise.

The sum of two matrices

$$\hat{A} + \hat{B} = \hat{C} \tag{A4.3a}$$

is obtained by the addition of the corresponding elements

$$C_{ij} = A_{ij} + B_{ij} \tag{A4.3b}$$

and similarly for subtraction. A matrix can be multiplied by a scalar (number)

$$\hat{D} = \lambda \hat{A} \tag{A4.4a}$$

by multiplying every element of the matrix by λ

$$D_{ij} = \lambda A_{ij} \tag{A4.4b}$$

The associative and distributive properties of addition hold with respect to multiplication by scalars.

B. MATRIX MULTIPLICATION

The product of two matrices $\hat{A}$ and $\hat{B}$ is another matrix $\hat{D}$

$$\hat{A} \cdot \hat{B} = \hat{D} \tag{A4.5a}$$

where the matrix elements of $\hat{D}$ are

$$D_{ij} = \sum_k A_{ik} B_{kj} \tag{A4.5b}$$

A useful mnemonic for the definition of Eq. (A4.5*b*) is to recall that to form D_{ij} we take the i-row of matrix $\hat{A}$ and rotate it so as to juxtapose it to the j-column of matrix $\hat{B}$. We then multiply the pairs of numbers and add them up. This operation is depicted in Fig. A4.1 through the motion of the thumb and index of the left hand. It is clear from the above that square matrices can be multiplied only if they have the same dimensionality.

As an example consider the matrices

$$\hat{A} = \begin{pmatrix} 2 & 3 & 5 \\ -1 & 0 & 3 \\ 4 & -7 & 9 \end{pmatrix} \qquad \hat{B} = \begin{pmatrix} 4 & 3 & 1 \\ 2 & 2 & 6 \\ 1 & 5 & 4 \end{pmatrix}$$

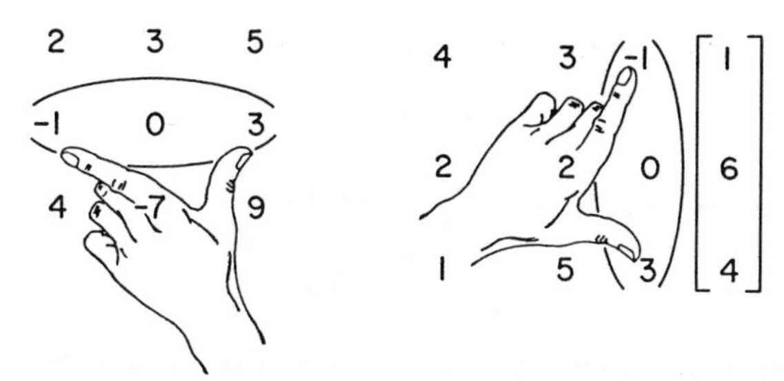

FIGURE A4.1. Forming the D_{23} element of the product matrix $\hat{D} = \hat{A} \cdot \hat{B}$. Place the thumb and index of the left hand to span the *second* row of matrix A and rotate it to bring it adjacent to the *third* column of matrix B. Then multiply the corresponding elements and add: $D_{23} = (-1)\times 1 + 0\times 6 + 3\times 4 = 11$.

Their product is

$$\hat{D} = \hat{A} \cdot \hat{B} = \begin{pmatrix} 19 & 37 & 40 \\ -1 & 12 & 11 \\ 11 & 43 & -2 \end{pmatrix}$$

Matrix multiplication is *not* commutative. In general,

$$\hat{B} \cdot \hat{A} = \hat{F} \neq \hat{D} = \hat{A} \cdot \hat{B} \tag{A4.6a}$$

Clearly, the elements of $\hat{F}$ are given by

$$F_{ij} = \sum_{l} B_{il} A_{lj} \tag{A4.6b}$$

That $\hat{F} \neq \hat{D}$ is easily checked. Our previous example yields

$$\hat{F} = \hat{B} \cdot \hat{A} = \begin{pmatrix} 9 & 5 & 38 \\ 26 & -36 & 70 \\ 13 & -25 & 56 \end{pmatrix}$$

which differs from the matrix $\hat{D}$.

We can also define the multiplication of a column vector by a matrix: the result is a new column vector. Let the vector $\mathbf{x}$ be defined by the components $(x_1, x_2, \ldots, x_n)$. Then

$$\hat{A} \cdot \mathbf{x} = \mathbf{y} \tag{A4.7a}$$

where the components of $\mathbf{y}$ are given by

$$y_i = \sum_{k} A_{ik} x_k \tag{A4.7b}$$

Similarly a row vector $\boldsymbol{\xi}$ can premultiply a matrix $\hat{B}$ to yield a new row vector $\boldsymbol{\eta}$

$$\boldsymbol{\xi} \cdot \hat{B} = \boldsymbol{\eta} \tag{A4.8a}$$

where the components of $\boldsymbol{\eta}$ are given by

$$\eta_j = \sum_l \eta_l B_{lj} \tag{A4.8b}$$

These operations can be visualized in terms of our mnemonic rule as shown in the following examples

$$\mathbf{y} = \hat{A} \cdot \mathbf{x} = \begin{pmatrix} 2 & 3 & 5 \\ -1 & 0 & 3 \\ 4 & -7 & 9 \end{pmatrix} \begin{pmatrix} 2 \\ 3 \\ -2 \end{pmatrix} = \begin{pmatrix} 3 \\ -8 \\ -31 \end{pmatrix}$$

$$\boldsymbol{\eta} = \boldsymbol{\xi} \cdot \hat{B} = \begin{pmatrix} 2 & -2 & 1 \end{pmatrix} \begin{pmatrix} 4 & 3 & 1 \\ 2 & 2 & 6 \\ 1 & 5 & 4 \end{pmatrix} = \begin{pmatrix} 5 & 7 & -6 \end{pmatrix}$$

C. DETERMINANTS AND INVERSE MATRICES

We know that for every vector we can form its magnitude $|\mathbf{x}| = (\sum_i x_i^2)^{1/2}$, which is a scalar quantity. Analogous scalar quantities for a matrix are its *trace* and its *determinant.* The trace of a matrix is the sum of its diagonal elements

$$\text{Tr } \hat{A} = \sum_i A_{ii} \tag{A4.9}$$

For instance, for the matrices $\hat{A}$ and $\hat{B}$ used in the previous numerical examples

$$\text{Tr } \hat{A} = 11 \quad \text{Tr } \hat{B} = 10$$

The determinant of a matrix $\hat{A}$ is the scalar resulting from the sum of all possible products of the elements of the matrix, but where one and only *one* element from *each row* and *each column* is involved. Each product is assigned a plus or minus sign according to the following rule: we join *every pair* of elements involved in a product by a line segment; if the total number of segments sloping upward to the right is odd, the product gets a minus sign. Otherwise it gets a plus sign.

This definition seems cumbersome and can be better understood for matrices of low dimensionality. Consider a 2×2 matrix $\hat{G}$. The determinant is obtained by forming the two products along the diagonals as shown

$$\hat{G}=\begin{pmatrix} a & b \\ c & d \end{pmatrix} \qquad \det \hat{G}=\begin{vmatrix} a & b \\ c & d \end{vmatrix}$$

(−) (+)

$$\det \hat{G}=ad-cb=g_{11}g_{22}-g_{21}g_{12}$$

The method of the diagonals can be extended to a 3×3 matrix (but *not* to matrices of dimension $n>3$). Given a matrix $\hat{K}$

$$\hat{K}=\begin{pmatrix} a & b & c \\ d & e & f \\ g & h & i \end{pmatrix} \qquad \det \hat{K}=\left|\begin{matrix} a_{+} & b_{+} & c_{+} \\ d & e & f \\ g^{-} & h^{-} & i^{-} \end{matrix}\right|\begin{matrix} a & b \\ d & e \\ g & h \end{matrix}$$

its determinant is defined as

$$\begin{aligned}\det \hat{K} &= aei+bfg+cdh-gec-hfa-idb \\ &= K_{11}K_{22}K_{33}+K_{12}K_{23}K_{31}+K_{13}K_{21}K_{32}-K_{31}K_{22}K_{13} \\ &\quad -K_{32}K_{23}K_{11}-K_{33}K_{21}K_{12}\end{aligned}$$

Next we consider an arbitrary square matrix $\hat{A}$ and delete the j-row and i-column containing the element A_{ij}. The determinant of the remaining matrix is called the *minor* of A_{ij} and we indicate it by M_{ij}. The cofactor of the element A_{ij} is

$$C_{ij}=(-1)^{i+j}M_{ij} \tag{A4.10}$$

In terms of the cofactors the determinant of $\hat{A}$ can be written as

$$\det \hat{A}=\sum_{j=1}^{n} A_{ij}C_{ij} \tag{A4.11a}$$

where i can be any one row of the matrix, or equivalently

$$\det \hat{A} = \sum_{i=1}^{n} A_{ij} C_{ij} \tag{A4.11b}$$

where j is any one column of the matrix. Note that Eqs. (A4.11a) and (A4.11b) do *not* represent matrix multiplications. As an application let us obtain the determinant of the matrix $\hat{K}$ used in the previous example by expanding along the first row

$$\begin{aligned}\det \hat{K} &= aM_{11} - bM_{12} + cM_{13} \\ &= a(ei - hf) - b(di - gf) + c(dh - eg)\end{aligned}$$

This result is the same as obtained previously.

We note some of the properties of the determinant:

1. If all elements of any row or column are zero the determinant is zero.
2. If two rows or two columns are interchanged the determinant changes sign.
3. If two rows or two columns are equal to each other or in constant ratio the determinant is zero.
4. The determinant of the product of two matrices equals the product of the determinants

$$\det(\hat{A} \cdot \hat{B}) = (\det \hat{A})(\det \hat{B}) \tag{A4.12}$$

Given a matrix $\hat{A}$ with $\det \hat{A} \neq 0$ we can find its *inverse* matrix $\hat{A}^{-1}$ such that

$$\hat{A} \cdot \hat{A}^{-1} = \mathbb{1} \tag{A4.13a}$$

The elements of $\hat{A}^{-1}$ are given by the cofactor of the *transposed* element of $\hat{A}$ divided by the determinant of $\hat{A}$

$$(A^{-1})_{ij} = \frac{1}{\det \hat{A}} C_{ji} \tag{A4.13b}$$

To show the validity of this equation note that

$$\sum_i A_{ki}(A^{-1})_{ij} = \frac{1}{\det \hat{A}} \sum_i A_{ki} C_{ji} \tag{A4.13c}$$

If $k = j$ in the above equation the sum on the rhs equals $\det \hat{A}$, and thus Eq. (A4.13c) equals 1. If $k \neq j$ the sum on the rhs represents the determinant of a matrix with two identical rows; this equals zero. Thus

$$\sum_i A_{ki}(A^{-1})_{ij} = \delta_{kj}$$

as demanded by Eq. (A4.13*a*).

As an example consider the matrix

$$\hat{A} = \begin{pmatrix} 1 & 1 & -3 \\ 1 & 0 & 1 \\ -1 & 2 & 1 \end{pmatrix}$$

and let us find its inverse. The determinant has the value

$$\det \hat{A} = -10$$

The cofactors of A are obtained from Eq. (A4.10) $C_{ij} = (-)^{i+j}M_{ij}$, so that

$$\hat{C} = \begin{pmatrix} -2 & -2 & 2 \\ -7 & -2 & -3 \\ 1 & -4 & -1 \end{pmatrix}$$

and therefore the inverse matrix is (remember to transpose and that $\det \hat{A} = -10$)

$$\hat{A}^{-1} = \frac{1}{10}\begin{pmatrix} 2 & 7 & -1 \\ 2 & 2 & 4 \\ -2 & 3 & 1 \end{pmatrix}$$

The reader should check by straightforward multiplication that indeed $\hat{A} \cdot \hat{A}^{-1} = \mathbb{1}$.

D. SPECIAL MATRICES

We have already encountered some special matrices:

(1) The unit matrix δ_{ij}.
(2) The inverse matrix.

We now define a matrix to be:

(3) *Diagonal* if

$$a_{ij} = \lambda_i, \qquad \text{when } i = j$$
$$= 0, \qquad \text{otherwise}$$

(4) *Symmetric* if

$$A_{ij} = A_{ji}$$

(5) *Antisymmetric* if

$$A_{ij} = -A_{ji}$$

Clearly, all the diagonal elements of an antisymmetric matrix vanish.

(6) The *transpose* of a matrix $\hat{A}$ is the matrix $\hat{A}^T$, which has elements

$$(A^T)_{ij} = A_{ji}$$

(7) A matrix with real elements is *orthogonal* if

$$\hat{A}^T = \hat{A}^{-1}$$

Namely,

$$\sum_j (A^T)_{ij} A_{jk} = \sum_j A_{ji} A_{jk} = \delta_{ik}$$

(8) For a matrix with complex elements the *adjoint* matrix $\hat{A}^\dagger$ is defined as the transposed complex conjugate of $\hat{A}$

$$(A^\dagger)_{ij} = A^*_{ji}$$

(9) A matrix is *hermitian* if it is equal to its adjoint

$$\hat{H}^\dagger = \hat{H} \qquad \text{(hermitian matrix)}$$

A hermitian matrix with real elements is a symmetric matrix.

(10) A matrix is *unitary* if its inverse is equal to the adjoint matrix

$$\hat{U}^\dagger = \hat{U}^{-1} \qquad \text{(unitary matrix)}$$

A unitary matrix with real elements is an orthogonal matrix.

E. DIAGONALIZATION OF A HERMITIAN MATRIX

Two matrices $\hat{A}$ and $\hat{A}'$ are said to be *equivalent* if they can be connected by a similarity (also referred to as unitary) transformation

$$\hat{A}' = \hat{S} \cdot \hat{A} \cdot \hat{S}^{-1} \tag{A4.14}$$

where $\hat{S}$ is a unitary matrix. Similarity transformations leave the trace and the determinant of a matrix invariant.

Every hermitian matrix is equivalent to a diagonal matrix. That is, given $\hat{H}$ where $\hat{H} = \hat{H}^{\dagger}$, we can form

$$\hat{H}' = \hat{S} \cdot \hat{H} \cdot \hat{S}^{-1} \tag{A4.15a}$$

where

$$(H')_{ij} = \lambda_i \, \delta_{ij} \tag{A4.15b}$$

The diagonal elements λ_i are called the *eigenvalues* of $\hat{H}'$, they are also the eigenvalues of $\hat{H}$ and of all equivalent matrices. All equivalent hermitian matrices have the same eigenvalues.

For every eigenvalue λ_α there corresponds a vector $\mathbf{x}^{(\alpha)}$ such that

$$\hat{H} \cdot \mathbf{x}^{(\alpha)} = \lambda_\alpha \mathbf{x}^{(\alpha)} \tag{A4.16a}$$

If the matrix $\hat{H}$ acting on $\mathbf{x}^{(\alpha)}$ reproduces $\mathbf{x}^{(\alpha)}$ multiplied by a number λ_α, the vector $\mathbf{x}^{(\alpha)}$ is called the *eigenvector* of $\hat{H}$ corresponding to the αth eigenvalue.

If we act with the matrix $\hat{S}$ on Eq. (A4.16a) and insert $\hat{S}^{-1} \cdot \hat{S} = \mathbb{1}$ we obtain

$$\hat{S} \cdot \hat{H} \cdot \mathbf{x}^{(\alpha)} = (\hat{S} \cdot \hat{H} \cdot \hat{S}^{-1}) \cdot (\hat{S} \cdot \mathbf{x}^{(\alpha)}) = \lambda_a (\hat{S} \cdot \mathbf{x}^{(\alpha)})$$

or

$$\hat{H}' \cdot \mathbf{y}^{(\alpha)} = \lambda_\alpha \mathbf{y}^{(\alpha)} \tag{A4.16b}$$

with

$$\mathbf{y}^{(\alpha)} = S \cdot \mathbf{x}^{(\alpha)}$$

That is, the components of the eigenvector depend on the representation in which the equivalent matrices $\hat{H}$ are expressed.

It is always possible to find the eigenvalues λ_i without knowing the similarity transformation matrix $\hat{S}$ that makes $\hat{H}$ diagonal. To show this consider Eq. (A4.16a) which represents n-linear equations

$$\sum_j H_{ij} x_j^{(\alpha)} = \lambda_\alpha x_i^{(\alpha)}, \qquad i = 1, 2, \ldots, n$$

We rewrite

$$\lambda_\alpha x_i^{(\alpha)} = \sum_j \lambda_\alpha \, \delta_{ij} x_j^{(\alpha)}$$

and therefore

$$\sum_j H_{ij} x_j^{(\alpha)} = \sum_j \lambda_\alpha \, \delta_{ij} x_j^{(\alpha)}$$

or

$$\sum_j (H_{ij} - \lambda_\alpha \, \delta_{ij}) x_j^{(\alpha)} = 0 \tag{A4.17a}$$

Equation (A4.17a) represents a set of n homogeneous linear algebraic equations in the n unknowns $x_j^{(\alpha)}$ ($j = 1, 2, \ldots, n$). They can have a solution only if the determinant of the coefficients equals zero. It suffices to set

$$\det(H_{ij} - \lambda_\alpha \, \delta_{ij}) = 0 \tag{A4.17b}$$

to determine the eigenvalues λ_α. In fact, Eq. (A4.17b) is a polynomial of the nth degree in λ_α and therefore has n roots. Each root gives one of the eigenvalues λ_α; some roots may be degenerate or zero.

As an example, consider the matrix

$$\hat{H} = \begin{pmatrix} 0 & 1 & 0 \\ 1 & 0 & 1 \\ 0 & 1 & 0 \end{pmatrix} \tag{A4.18a}$$

Its eigenvalues are found by setting

$$\det \begin{vmatrix} -\lambda & 1 & 0 \\ 1 & -\lambda & 1 \\ 0 & 1 & -\lambda \end{vmatrix} = -\lambda^3 + 2\lambda = 0$$

Thus

$$\lambda(\lambda^2 - 2) = 0$$

or

$$\lambda_1 = 0 \qquad \lambda_2 = \sqrt{2} \qquad \lambda_3 = -\sqrt{2}$$

The diagonal matrix equivalent to $\hat{H}$ is written as

$$\hat{H}' = \begin{pmatrix} \sqrt{2} & 0 & 0 \\ 0 & 0 & 0 \\ 0 & 0 & -\sqrt{2} \end{pmatrix} \tag{A4.18b}$$

The order in which we introduce the eigenvalues when writing $\hat{H}'$ is arbitrary, but it affects the form of the matrix $\hat{S}$ that connects $\hat{H}$ and $\hat{H}'$.

F. FINDING THE EIGENVECTORS

In the diagonal representation the eigenvectors $\mathbf{y}^{(\alpha)}$ take a trivial form: they can be represented by column vectors with a 1 entry in the α-slot and zeros everywhere else. What we are really interested in is to find the eigenvectors $\mathbf{x}^{(\alpha)}$ in the original representation of $\hat{H}$.

We begin with Eq. (A4.16*a*) and note that

$$(\hat{H} - \lambda_a \mathbb{1}) \cdot \mathbf{x}^{(\alpha)} \equiv \hat{E}^{(\alpha)} \cdot \mathbf{x}^{(\alpha)} = 0 \tag{A4.19a}$$

corresponds to the n linear equations

$$\sum_j E_{ij}^{(\alpha)} x_j^{(\alpha)} = 0, \qquad i = 1, 2, \ldots, n \tag{A4.19b}$$

We know that the determinant of $\hat{E}^{(\alpha)}$ is zero by definition [see Eq. (A4.17*b*)]. Thus

$$\det \hat{E}^{(\alpha)} = \sum_j E_{ij}^{(\alpha)} C_{ij} = 0 \tag{A4.19c}$$

Comparing Eqs. (A4.19*b*) and (A4.19*c*) it is clear that the components $x_j^{(\alpha)}$ are given by the cofactors of any one *row* of $E_{ij}^{(\alpha)}$. This definition is valid up to a constant which can be determined by demanding an appropriate normalization of the eigenvectors. Thus

$$x_j^{(\alpha)} = \text{cofactor of } E_{ij}^{(\alpha)} \tag{A4.20}$$

As an example let us find the eigenvectors for the matrix $\hat{H}$ given by Eq. (A4.18*a*). For $\lambda_\alpha = \sqrt{2}$

$$\hat{E}^{(\alpha)} = \begin{pmatrix} -\sqrt{2} & 1 & 0 \\ 1 & -\sqrt{2} & 1 \\ 0 & 1 & -\sqrt{2} \end{pmatrix} \qquad \begin{aligned} C_{11} &= 1 \\ C_{12} &= -(-\sqrt{2}) \\ C_{13} &= 1 \end{aligned}$$

For $\lambda_\beta = 0$

$$\hat{E}^{(\beta)} = \begin{pmatrix} 0 & 1 & 0 \\ 1 & 0 & 1 \\ 0 & 1 & 0 \end{pmatrix} \qquad \begin{aligned} C_{11} &= -1 \\ C_{12} &= -(0) \\ C_{13} &= 1 \end{aligned}$$

For $\lambda_\gamma = -\sqrt{2}$

$$\hat{E}^{(\gamma)} = \begin{pmatrix} \sqrt{2} & 1 & 0 \\ 1 & \sqrt{2} & 1 \\ 0 & 1 & \sqrt{2} \end{pmatrix} \qquad \begin{aligned} C_{11} &= 1 \\ C_{12} &= -\sqrt{2} \\ C_{13} &= 1 \end{aligned}$$

We normalize the eigenvectors and note that the matrix S can be constructed by writing the eigenvectors as row vectors in proper order

$$\hat{S} = \begin{pmatrix} x_1^{(\alpha)} & x_2^{(\alpha)} & \cdots & x_n^{(\alpha)} \\ x_1^{(\beta)} & x_2^{(\beta)} & \cdots & x_n^{(\beta)} \\ \vdots & \vdots & \vdots & \vdots \\ x_1^{(\nu)} & x_2^{(\nu)} & \cdots & x_n^{(\nu)} \end{pmatrix}$$

This follows because we want

$$\hat{S} \cdot \mathbf{x}^{\alpha} = \mathbf{y}^{(\alpha)}$$

with

$$\mathrm{y}^{(\alpha)} = \begin{pmatrix} 1 \\ 0 \\ 0 \end{pmatrix}, \qquad \mathrm{y}^{(\beta)} = \begin{pmatrix} 0 \\ 1 \\ 0 \end{pmatrix}, \qquad \text{etc.}$$

Note also that

$$\mathbf{x}^{(\alpha)} \cdot \mathbf{x}^{(\beta)} = \delta_{\alpha\beta}$$

Thus, for our example,

$$\mathbf{x}^{(\alpha)} = \frac{1}{2}\begin{pmatrix} 1 \\ \sqrt{2} \\ 1 \end{pmatrix} \qquad \mathbf{x}^{(\beta)} = \frac{1}{2}\begin{pmatrix} -\sqrt{2} \\ 0 \\ \sqrt{2} \end{pmatrix} \qquad \mathbf{x}^{(\gamma)} = \frac{1}{2}\begin{pmatrix} 1 \\ -\sqrt{2} \\ 1 \end{pmatrix}$$

and

$$S = \frac{1}{2}\begin{pmatrix} 1 & \sqrt{2} & 1 \\ -\sqrt{2} & 0 & \sqrt{2} \\ 1 & -\sqrt{2} & 1 \end{pmatrix} \tag{A4.21}$$

The unitary transformation matrix $\hat{S}$ is identical to the matrix used in Section 2.3. The reader should check by direct multiplication that indeed

$$\hat{H}' = \hat{S} \cdot \hat{H} \cdot \hat{S}^{-1}$$

for the matrices given by Eqs. (A4.18) and (A4.21).

Appendix 5

Quantum Conditions and the Poisson Brackets

The quantum conditions were introduced in Eqs. (6.9*b*) and can be taken as the defining axioms for quantum mechanics. However, as also alluded to in the text, they have a very close relation to classical physics. If one accepts the precept that the classical domain is an approximation to quantum physics in the limit of large quantum numbers (i.e., the correspondence principle), then it is helpful to use classical equations as a guide to obtain quantum relations. This is clearly demonstrated below where we show how the quantum-mechanical equivalent of the Poisson brackets lead to the quantum conditions.

For two functions A and B of the canonical variables, the Poisson bracket is defined as

$$\{A, B\} = \sum_i \left(\frac{\partial A}{\partial x_i} \frac{\partial B}{\partial p_i} - \frac{\partial A}{\partial p_i} \frac{\partial B}{\partial x_i} \right) \tag{A5.1a}$$

and satisfies formal conditions such as

$$\begin{aligned} \{A, B\} &= -\{B, A\} \\ \{A, C\} &= 0 \end{aligned} \tag{A5.1b}$$

where C is a constant. Furthermore,

$$\begin{aligned} \{A_1 + A_2, B\} &= \{A_1, B\} + \{A_2, B\} \\ \{A, B_1 + B_2\} &= \{A, B_1\} + \{A, B_2\} \\ \{A_1 A_2, B\} &= A_1\{A_2, B\} + \{A_1, B\}A_2 \\ \{A, B_1 B_2\} &= \{A, B_1\}B_2 + B_1\{A, B_2\} \end{aligned} \tag{A5.1c}$$

We now introduce a quantum Poisson bracket that satisfies all the conditions of Eqs. (A5.1*a*)–(A5.1*c*). However, we must now maintain the order of the operators $\hat{A}$ and $\hat{B}$ as given in the above relations because quantum-mechanical operators in general do not commute. We calculate the following quantum Poisson bracket in two different ways:

$$\begin{aligned}\{A_1A_2, B_1B_2\}_Q &= \{A_1A_2, B_1\}_Q B_2 + B_1\{A_1A_2, B_2\}_Q \\ &= A_1\{A_2, B_1\}_Q B_2 + \{A_1, B_1\}_Q A_2 B_2 \\ &\quad + B_1 A_1\{A_2, B_2\}_Q + B_1\{A_1, B_2\}_Q A_2 \end{aligned} \tag{A5.2a}$$

But also

$$\begin{aligned}\{A_1A_2, B_1B_2\}_Q &= A_1\{A_2, B_1B_2\}_Q + \{A_1, B_1B_2\}_Q A_2 \\ &= A_1B_1\{A_2, B_2\}_Q + A_1\{A_2, B_1\}_Q B_2 \\ &\quad + B_1\{A_1, B_2\}_Q A_2 + \{A_1, B_1\}_Q B_2 A_2 \end{aligned} \tag{A5.2b}$$

Equating the relations in Eqs. (A5.2*a*) and (A5.2*b*) we have

$$(A_1B_1 - B_1A_1)\{A_2, B_2\}_Q = \{A_1, B_1\}_Q(A_2B_2 - B_2A_2) \tag{A5.2c}$$

This condition can be satisfied if

$$\begin{aligned}[\hat{A}_1, \hat{B}_1] &\equiv (\hat{A}_1\hat{B}_1 - \hat{B}_1\hat{A}_1) = i\hbar\{A_1, B_1\}_Q \\ [\hat{A}_2, \hat{B}_2] &\equiv (\hat{A}_2\hat{B}_2 - \hat{B}_2\hat{A}_2) = i\hbar\{A_2, B_2\}_Q \end{aligned} \tag{A5.3}$$

Here $\hbar$ is a constant of proportionality experimentally determined to be Planck's constant (divided by 2π). From Eq. (A5.3) it is a simple matter to derive the quantum conditions, as in Eqs. (6.9).

Appendix 6

Composition of Angular Momenta

In Section 7.6 we discussed in a qualitative way and by example that a system with two constituents that are in eigenstates of the angular momentum operators can be represented by an eigenstate of the total angular momentum. Here we generalize this discussion to arbitrary values of the angular momentum and indicate how these two representations of the system are related.

We begin by constructing an operator for the total angular momentum as defined in Eq. (7.60*a*)

$$\hat{\mathbf{J}} = \hat{\mathbf{J}}^{(1)} + \hat{\mathbf{J}}^{(2)} \tag{A6.1}$$

$\hat{\mathbf{J}}$ obeys the commutation relations for angular momentum†

$$\hat{\mathbf{J}} \times \hat{\mathbf{J}} = i\hbar\hat{\mathbf{J}} \tag{A6.2}$$

From the triplet of operators $\hat{\mathbf{J}}$ we construct the corresponding operators

$$\begin{aligned} \hat{J}_z &= \hat{J}_z^{(1)} + \hat{J}_z^{(2)} \\ \hat{J}_+ &= \hat{J}_+^{(1)} + \hat{J}_+^{(2)} \\ \hat{J}_- &= \hat{J}_-^{(1)} + \hat{J}_-^{(2)} \end{aligned} \tag{A6.3}$$

where $\hat{J}_\pm$ is defined in the usual way as in Eqs. (7.37*a*). The operator for the square of the total angular momentum is

$$\begin{aligned} \hat{\mathbf{J}}^2 &= [\hat{\mathbf{J}}^{(1)} + \hat{\mathbf{J}}^{(2)}]^2 = (\hat{\mathbf{J}}^{(1)})^2 + (\hat{\mathbf{J}}^{(2)})^2 + 2\hat{\mathbf{J}}^{(1)} \cdot \hat{\mathbf{J}}^{(2)} \\ &= (\hat{\mathbf{J}}^{(1)})^2 + (\hat{\mathbf{J}}^{(2)})^2 + \hat{J}_+^{(1)}\hat{J}_-^{(2)} + \hat{J}_-^{(1)}\hat{J}_+^{(2)} + 2\hat{J}_z^{(1)}\hat{J}_z^{(2)} \end{aligned} \tag{A6.4}$$

This expansion follows directly from the defining properties of Eqs. (A6.3).

The eigenstates of $\hat{\mathbf{J}}^2$, $\hat{J}_z$ will be indicated by the bra $|J, M\rangle$, where the indices J, M refer to the eigenvalues of the state in this rep-

† It can be shown in general that if $\hat{\mathbf{J}}^{(1)}$, $\hat{\mathbf{J}}^{(2)}$ are angular momentum operators, then $\hat{\mathbf{J}} = \alpha\hat{\mathbf{J}}^{(1)} + \beta\hat{\mathbf{J}}^{(1)}$ is also an angular momentum operator if and only if $\alpha, \beta = 0$ or 1.

resentation. Correspondingly, $|j_1 m_1\rangle$ and $|j_2 m_2\rangle$ are the eigenstates of the operators $(\hat{\mathbf{J}}^{(1)})^2, \hat{J}_z^{(1)}$ and $(\hat{\mathbf{J}}^{(2)})^2, \hat{J}_z^{(2)}$. It should be clear that a system admitting the two operators $\hat{\mathbf{J}}^{(1)}$ and $\hat{\mathbf{J}}^{(2)}$ can be represented in a basis of *product* states labeled by the four indices $(j_1, m_1; j_2, m_2)$. We indicate these basis states by

$$|j_1, j_2; m_1, m_2\rangle \equiv |j_1 m_1\rangle\, |j_2 m_2\rangle \tag{A6.5}$$

The system can, however, also be represented by the indices J and M of the total angular momentum together with the two indices j_1 and j_2 referring to the angular momentum of the constituent parts of the system. We indicate the basis states in this representation by

$$|j_1, j_2; J, M\rangle \tag{A6.6}$$

We emphasize that Eqs. (A6.5) and (A6.6) express the basis states for two different representations of the angular momentum of a quantum system. Thus, they must be related by a linear transformation, just as in Eq. (2.22*b*). The matrix elements of this transformation are known as the Clebsch–Gordan coefficients. The quantum numbers (indices) of the two representations are not completely free but are related: M is given by

$$M = m_1 + m_2 \tag{A6.7}$$

which follows directly from the first of Eqs. (A6.3). For the possible values of J we have

$$J = (j_1 + j_2), (j_1 + j_2 - 1), \ldots, |j_1 - j_2| \tag{A6.8}$$

as already indicated in Eq. (7.63*a*). That the values of J must be bounded between $(j_1 + j_2)$ and $|j_1 - j_2|$ can be seen from Eq. (A6.4). That all integers values between these limits are possible requires a more systematic study using the properties of the raising and lowering operators of Eqs. (A6.3). We indicate below how this is done.

If we form all possible product states, the state with $m_1 = j_1$ and $m_2 = j_2$ corresponds to the *stretched* case of Figs. 7.2 and 7.3. This state is by necessity the same as the $J = M = j_1 + j_2$ state in the representation of the total angular momentum

$$|j_1, j_2; J_{\max}, M = J\rangle = |j_1, m_1 = j_1\rangle\, |j_2, m_2 = j_2\rangle \tag{A6.9}$$

We can then operate on this stretched state with $\hat{J}_-$ to obtain all the states with $J = J_{max}$; there will be $(2J_{max}+1)$ such states. Next we examine the product states where

$$m_1 + m_2 = M = j_1 + j_2 - 1$$

There are two such states corresponding to $m_1 = j_1$, $m_2 = j_2 - 1$ or $m_1 = j_1 - 1$, $m_2 = j_2$. Out of these we form linear combinations. One of the linear combinations will be the state belonging to J_{max} and $M = J_{max} - 1$ already found previously. The orthogonal combination will be the state with $J = J_{max} - 1$ and $M = J_{max} - 1$. We then apply the lowering operator to this latter state to find all states with $J = J_{max} - 1$. This procedure is continued until all states are exhausted and leads to the fact that J can take values only between $|j_1 - j_2| \leqslant J \leqslant (j_1 + j_2)$.

The two representations [see Eqs. (A6.5) and (A6.6)] will be related by a unitary transformation. Taking into account the constraints imposed by Eqs. (A6.7) and (A6.8) we can write this linear transformation as

$$|j_1, j_2; J, M\rangle = \sum_m C_m \, |j_1, j_2; m, M-m\rangle \tag{A6.10}$$

The coefficients C_m are known as the Clebsch–Gordan coefficients and can be at least symbolically expressed by forming the inner product of Eq. (A6.10) (closing it from the left) with the basis states of the product space representation [i.e., as given by Eq. (A6.5)]. These states are, of course, orthonormal. Thus, we write Eq. (A6.10) in the form

$$|j_1, j_2; J, M\rangle = \sum_m |j_1, j_2; m, M-m\rangle\langle j_1, j_2; m, M-m \, |j_1, j_2; J, M\rangle \tag{A6.11}$$

The inverse expansion, from the total angular momentum representation to the product space eigenstates, is written as

$$|j_1, j_2; m_1 m_2\rangle = \sum_J |j_1, j_2; J, M\rangle\langle j_1, j_2; J, M \, |j_1, j_2; m_1, m_2\rangle \qquad M = m_1 + m_2 \tag{A6.12}$$

where the sum over J is restricted as in Eq. (A6.8).

The Clebsch–Gordan coefficients are found by constructing the appropriate eigenstates as outlined in the preceding paragraphs and by using the properties of the lowering operators given in Eqs. (7.50). These coefficients are tabulated in many books† and reference manuals. The Clebsch–Gordan coefficients for the simple case of addition of two spin-1/2 angular momenta can be obtained from Eqs. (7.62). The coefficients in this case are 1 and $\pm 1/\sqrt{2}$, as the reader can verify. In our present notation we write Eqs. (7.62) as

$$
\begin{aligned}
&|1/2, 1/2; J=1, M=1\rangle = |1/2, 1/2; 1/2, 1/2\rangle \\
&|1/2, 1/2; J=1, M=0\rangle \\
&\quad = \frac{1}{2^{1/2}}\{|1/2, 1/2; 1/2, -1/2\rangle + |1/2, 1/2; -1/2, 1/2\rangle\} \\
&|1/2, 1/2; J=1, M=-1\rangle = |1/2, 1/2; -1/2, -1/2\rangle \\
&|1/2, 1/2; J=0, M=0\rangle \\
&\quad = \frac{1}{2^{1/2}}\{|1/2, 1/2; 1/2, -1/2\rangle - |1/2, 1/2; -1/2, 1/2\rangle\}
\end{aligned}
\tag{A6.13}
$$

† See, for instance, the classic text by E. U. Condon and G. H. Shortley, *The Theory of Atomic Spectra*, Cambridge University Press (1957).

Appendix 7

Matrix Elements of Vector Operators

As pointed out in Chapter 7, since many physical systems have spherical symmetry their stationary states are labeled by the eigenstates of angular momentum. These eigenstates form a complete orthonormal set and if we use the notation $|l, m\rangle$ for an eigenstate of orbital angular momentum it must satisfy

$$\langle l_1, m_1 \mid l_2, m_2\rangle = \delta_{l_1,l_2}\,\delta_{m_1,m_2} \tag{A7.1}$$

If an operator $\hat{Q}$ has specific transformation properties under rotations, then the matrix elements of $\hat{Q}$ between eigenstates of angular momentum are related to one another, and often vanish. These results are obtained as a direct consequence of the rotational symmetry which affects both the operator and the eigenstates in a well-defined way.

Let us first consider an operator $\hat{Q}$ that is invariant under rotations, that is,

$$\hat{Q}' = \hat{R}(\theta, \phi)\hat{Q}\hat{R}^{-1}(\theta, \phi) = \hat{Q} \tag{A7.2}$$

where $\hat{R}(\theta, \phi)$ represents an arbitrary rotation. Operators such as $\hat{Q}$ are called *scalar* operators, examples of which are $|r| = (x^2+y^2+z^2)^{1/2}$, or ∇^2, etc. Consider then an arbitrary state that is also an eigenstate of angular momentum, e.g., $|n, l, m\rangle$. Here the index n stands for the quantum numbers of the energy operator and other constants of the motion. When $\hat{Q}$ acts on that state it *cannot* change the angular momentum (since $\hat{Q}$ is rotationally invariant), even though it may affect the index n. In view of the orthogonality of Eq. (A7.1) it will always hold that

$$\langle n_1, l_1, m_1|\, \hat{Q}\, |n_2, l_2, m_2\rangle = 0 \qquad \text{if } l_1 \neq l_2 \quad \text{or} \quad m_1 \neq m_2 \tag{A7.3}$$

Relations such as that indicated by Eq. (A7.3) are known as selection rules. In this case we see that a scalar operator can have a matrix element (connect) only between states with the same value of l and m.

Next we consider an operator such as $\hat{x}$, the position operator along the X-axis. This operator is not invariant under rotations and we refer to the triplet $\hat{x}$, $\hat{y}$, $\hat{z}$ as a *vector* operator, designating it by $\hat{\mathbf{x}}$ or $\hat{\mathbf{r}}$. To obtain the matrix elements of vector operators we will use the coordinate representation† in which the eigenstates of angular momentum can be represented by the spherical harmonics [See Eq. (7.57)]. In this representation the matrix element of $\hat{x}$ can be calculated by integrating over the unit sphere

$$\langle l_1, m_1|\hat{x}|l_2, m_2\rangle = \int Y^*_{l_1,m_1}(\theta, \phi)xY_{l_2,m_2}(\theta, \phi)\, d(\cos\theta)\, d\phi \tag{A7.4}$$

The transformation properties of x under rotations can be found by expressing x in terms of spherical harmonics. By inverting Eqs. (7.59*b*) of the text we find

$$\begin{aligned} x &= r\sin\theta\cos\phi = -\left(\frac{2\pi}{3}\right)^{1/2}(Y_{1,1} - Y_{1,-1})r \\ y &= r\sin\theta\sin\phi = i\left(\frac{2\pi}{3}\right)^{1/2}(Y_{1,1} + Y_{1,-1})r \\ z &= r\cos\theta = \left(\frac{4\pi}{3}\right)^{1/2} Y_{1,0}r \end{aligned} \tag{A7.5}$$

Here $r = (x^2 + y^2 + z^2)^{1/2}$ is a scalar operator and will not affect the integral in Eq. (A7.4). Using the first of Eqs. (A7.5) the angular part of the integral takes the form

$$-\left(\frac{2\pi}{3}\right)^{1/2} r\int (Y^*_{l_1,m_1}Y_{1,1}Y_{l_2m_2})\, d(\cos\theta)\, d\phi + \left(\frac{2\pi}{3}\right)^{1/2} r\int (Y^*_{l_1m_1}Y_{1,-1}Y_{l_2,m_2})\, d(\cos\theta)\, d\phi \tag{A7.6}$$

It is clear from Eqs. (7.57) that the integration over $d\phi$ will give

† We restrict this discussion to orbital angular momentum in order to be able to work in the coordinate representation. Similar results are valid in general.

zero unless

$$\begin{aligned} m_1 &= m_2 + 1 \qquad \text{for the first term} \\ m_1 &= m_2 - 1 \qquad \text{for the second term} \end{aligned} \tag{A7.7}$$

Furthermore, from the properties of the spherical harmonics, the integral over $d(\cos\theta)$ vanishes unless

$$l_1 = l_2 \pm 1 \tag{A7.8}$$

Equations (A7.7) and (A7.8) are the selection rules for the matrix elements of the $\hat{x}$ operator. Even though we derived them in the coordinate representation, they are matrix elements of an operator and, therefore, are generally valid. We summarize the selection rules for the three components of the vector operator $\hat{\mathbf{r}}$ as follows:

$$\begin{aligned} \langle l_1, m_1|\,(x+iy)\,|l_2, m_2\rangle &= 0 \qquad \text{unless} \begin{cases} (l_1 - l_2) = \pm 1 \\ m_1 = m_2 + 1 \end{cases} \\ \langle l_1, m_1|\,(x-iy)\,|l_2, m_2\rangle &= 0 \qquad \text{unless} \begin{cases} (l_1 - l_2) = \pm 1 \\ m_1 = m_2 - 1 \end{cases} \\ \langle l_1, m_1|\,z\,|l_2, m_2\rangle &= 0 \qquad \text{unless} \begin{cases} (l_1 - l_2) = \pm 1 \\ m_1 = m_2 \end{cases} \end{aligned} \tag{A7.9}$$

These selection rules apply to any operator that transforms under rotations as the operator $\hat{\mathbf{r}}$. Such operators are called vector operators.

The structure of Eq. (A7.6) suggests that when a vector operator acts on an eigenstate of angular momentum the result is some linear combination of angular momentum eigenstates. These states appear to be those that can be obtained by composing the initial angular momentum state with an angular momentum $|l, m\rangle$, where l and m refer to the transformation properties of the operator. This observation is generally true and, therefore, from Appendix 6 we would expect that when a vector operator—for instance $\hat{z}$, for which $l = 1$, $m = 0$—acts on the state $|l_2, m_2\rangle$ it would produce the states

$$|l_2 + 1, m_2\rangle, \qquad |l_2, m_2\rangle, \qquad \text{and} \quad |l_2 - 1, m_2\rangle \tag{A7.10}$$

However, the state $|l_2, m_2\rangle$ seems to be missing according to Eq. (A7.8). This must indeed be so because the vector operator $\hat{z}$ has negative parity: it changes sign under inversion of the coordinates. Thus, the states resulting from $\hat{z}\,|l_2, m_2\rangle$ must have the opposite parity from that of the state $|l_2, m_2\rangle$. Therefore, the matrix element of $\hat{z}$ vanishes when $\Delta l = 0$ and this is indicated in Eqs. (A7.9).

These selection rules can be extended to the case when the eigenstates refer to spin or total angular momentum states, but then one must exercise more care with the parity argument. Similar selection rules exist for operators of higher rank—for instance for tensor operators such as $\hat{x}^2$ or $\hat{x}\hat{y}$—and for all symmetry transformations in general. These results are known as the Wigner–Eckart theorem.

Appendix 8

Operators in Spherical Coordinates

A. IMPORTANT NOTE

In this appendix we will use carets to indicate the unit vectors along the coordinate axes. Thus

$$\begin{aligned} \hat{x} &\equiv \mathbf{n}_x & \qquad & \hat{r} \equiv \mathbf{n}_r \\ \hat{y} &\equiv \mathbf{n}_y & \text{and} \qquad & \hat{\theta} \equiv \mathbf{n}_\theta \\ \hat{z} &\equiv \mathbf{n}_z & & \hat{\phi} \equiv \mathbf{n}_\phi \end{aligned}$$

This departure from the convention adopted throughout the main text has been deemed necessary in order to simplify the notation that otherwise would be less transparent. Furthermore, we will *not* use carets to indicate quantum-mechanical operators in this appendix, it being understood that $\mathbf{L}$, L_x, L_y, L_z, and L^2 are the quantum-mechanical angular momentum operators.

B. COORDINATE TRANSFORMATIONS

The cartesian coordinates of a point (x, y, z) can be expressed in terms of the spherical coordinates (r, θ, ϕ) as

$$\begin{aligned} x &= r \sin\theta \cos\phi \\ y &= r \sin\theta \sin\phi \\ z &= r \cos\theta \end{aligned} \tag{A8.1}$$

The spherical coordinates form an orthonormal basis satisfying the relations

$$\hat{r}\cdot\hat{\theta}=\hat{r}\cdot\hat{\phi}=\hat{\theta}\cdot\hat{\phi}=0; \qquad \hat{r}\cdot\hat{r}=\hat{\theta}\cdot\hat{\theta}=\hat{\phi}\cdot\hat{\phi}=1$$
$$\hat{r}\times\hat{\theta}=\hat{\phi}, \qquad \hat{r}\times\hat{\phi}=-\hat{\theta}, \qquad \hat{\theta}\times\hat{\phi}=\hat{r} \tag{A8.2}$$

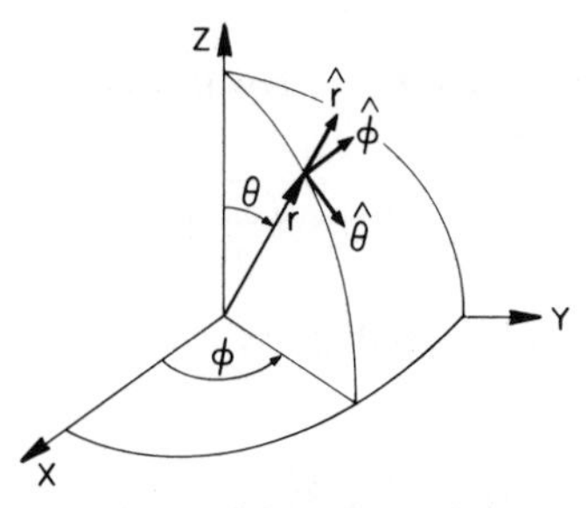

FIGURE A8.1. Spherical coordinates and the corresponding (position dependent) unit vectors, $\hat{r}$, $\hat{\theta}$, $\hat{\phi}$.

As can be seen from Fig. A8.1 these unit vectors can be written in terms of the unit vectors of the cartesian coordinates as

$$\begin{aligned}\hat{r}&=\hat{x}\sin\theta\cos\phi+\hat{y}\sin\theta\sin\phi+\hat{z}\cos\theta\\ \hat{\theta}&=\hat{x}\cos\theta\cos\phi+\hat{y}\cos\theta\sin\phi-\hat{z}\sin\theta\\ \hat{\phi}&=-\hat{x}\sin\phi+\hat{y}\cos\phi\end{aligned} \tag{A8.3a}$$

We can also invert these relations and write

$$\begin{aligned}\hat{x}&=\hat{r}\sin\theta\cos\phi+\hat{\theta}\cos\theta\cos\phi-\hat{\phi}\sin\phi\\ \hat{y}&=\hat{r}\sin\theta\sin\phi+\hat{\theta}\cos\theta\sin\phi+\hat{\phi}\cos\phi\\ \hat{z}&=\hat{r}\cos\theta-\hat{\theta}\sin\theta\end{aligned} \tag{A8.3b}$$

We are now in a position to express the gradient operator

$$\nabla=\hat{x}\frac{\partial}{\partial x}+\hat{y}\frac{\partial}{\partial y}+\hat{z}\frac{\partial}{\partial z}$$

in spherical coordinates. To do so we use the chain rule of differentiation

$$\frac{\partial}{\partial r}=\frac{\partial x}{\partial r}\frac{\partial}{\partial x}+\frac{\partial y}{\partial r}\frac{\partial}{\partial y}+\frac{\partial z}{\partial r}\frac{\partial}{\partial z}, \qquad \text{etc.}$$

and the relations given by Eqs. (A8.3b). Thus

$$\begin{aligned}\nabla &= \hat{x}\frac{\partial}{\partial x}+\hat{y}\frac{\partial}{\partial y}+\hat{z}\frac{\partial}{\partial z}\\ &= \hat{r}\left(\sin\theta\cos\phi\frac{\partial}{\partial x}+\sin\theta\sin\phi\frac{\partial}{\partial y}+\cos\theta\frac{\partial}{\partial z}\right)\\ &\quad+\hat{\theta}\left(\cos\theta\cos\phi\frac{\partial}{\partial x}+\cos\theta\sin\phi\frac{\partial}{\partial y}-\sin\theta\frac{\partial}{\partial z}\right)\\ &\quad+\hat{\phi}\left(-\sin\phi\frac{\partial}{\partial x}+\cos\phi\frac{\partial}{\partial y}\right)\end{aligned}$$

Furthermore, using the partial derivatives of Eqs. (A8.1) we can write the above result as

$$\begin{aligned}\nabla &= \hat{r}\left(\frac{\partial x}{\partial r}\frac{\partial}{\partial x}+\frac{\partial y}{\partial r}\frac{\partial}{\partial y}+\frac{\partial z}{\partial r}\frac{\partial}{\partial z}\right)\\ &\quad+\hat{\theta}\left(\frac{1}{r}\frac{\partial x}{\partial\theta}\frac{\partial}{\partial x}+\frac{1}{r}\frac{\partial y}{\partial\theta}\frac{\partial}{\partial y}+\frac{1}{r}\frac{\partial z}{\partial\theta}\frac{\partial}{\partial z}\right)\\ &\quad+\hat{\phi}\left(\frac{1}{r\sin\theta}\frac{\partial x}{\partial\phi}\frac{\partial}{\partial x}+\frac{1}{r\sin\theta}\frac{\partial y}{\partial\phi}\frac{\partial}{\partial y}\right)\end{aligned}$$

The chain rule of differentiation allows us to combine the above into

$$\boxed{\nabla=\hat{r}\frac{\partial}{\partial r}+\hat{\theta}\frac{1}{r}\frac{\partial}{\partial\theta}+\hat{\phi}\frac{1}{r\sin\theta}\frac{\partial}{\partial\phi}} \tag{A8.4}$$

This is the expression given in Eq. (8.12a) of the text.

To obtain the Laplacian we need to know how the unit vectors $\hat{r}$, $\hat{\theta}$, and $\hat{\phi}$ vary when the coordinates change. This variation can be obtained by taking the partial derivatives of Eqs. (A8.3a) with respect to the coordinates. We have

$$\begin{array}{lll}\dfrac{\partial\hat{r}}{\partial r}=0 & \dfrac{\partial\hat{r}}{\partial\theta}=\hat{\theta} & \dfrac{\partial\hat{r}}{\partial\phi}=\hat{\phi}\sin\theta\\ \dfrac{\partial\hat{\theta}}{\partial r}=0 & \dfrac{\partial\hat{\theta}}{\partial\theta}=-\hat{r} & \dfrac{\partial\hat{\theta}}{\partial\phi}=\hat{\phi}\cos\theta\\ \dfrac{\partial\hat{\phi}}{\partial r}=0 & \dfrac{\partial\hat{\phi}}{\partial\theta}=0 & \dfrac{\partial\hat{\phi}}{\partial\phi}=-(\hat{r}\sin\theta+\hat{\theta}\cos\theta)\end{array} \tag{A8.5}$$

The above relations indicate that unlike the cartesian coordinates the unit vectors of the spherical coordinates vary from point to point. This can be seen from considering the definition of the unit vectors on the sphere (see Fig. A8.1). Of course, the relations expressed by Eqs. (A8.5) are completely analogous to the more familiar relations for polar coordinates on a flat surface.

Using Eq. (A8.4) and the above results we can evaluate

$$\nabla^2 = \boldsymbol{\nabla} \cdot \boldsymbol{\nabla} = \left(\hat{r}\frac{\partial}{\partial r} + \frac{\hat{\theta}}{r}\frac{\partial}{\partial \theta} + \frac{\hat{\phi}}{r \sin \theta}\frac{\partial}{\partial \phi}\right) \cdot \left(\hat{r}\frac{\partial}{\partial r} + \frac{\hat{\theta}}{r}\frac{\partial}{\partial \theta} + \frac{\hat{\phi}}{r \sin \theta}\frac{\partial}{\partial \phi}\right)$$

$$= \left(\hat{r}\frac{\partial}{\partial r}\right) \cdot \left(\hat{r}\frac{\partial}{\partial r} + \frac{\hat{\theta}}{r}\frac{\partial}{\partial \theta} + \frac{\hat{\phi}}{r \sin \theta}\frac{\partial}{\partial \phi}\right)$$

$$+ \left(\hat{\theta}\frac{1}{r}\frac{\partial}{\partial \theta}\right) \cdot \left(\hat{r}\frac{\partial}{\partial r} + \frac{\hat{\theta}}{r}\frac{\partial}{\partial \theta} + \frac{\hat{\phi}}{r \sin \theta}\frac{\partial}{\partial \phi}\right)$$

$$+ \left(\hat{\phi}\frac{1}{r \sin \theta}\frac{\partial}{\partial \phi}\right) \cdot \left(\hat{r}\frac{\partial}{\partial r} + \frac{\hat{\theta}}{r}\frac{\partial}{\partial \theta} + \frac{\hat{\phi}}{r \sin \theta}\frac{\partial}{\partial \phi}\right)$$

If we remember that the coordinates are independent variables, i.e., $\partial r/\partial \theta = 0$, $\partial \phi/\partial \theta = 0$, etc., we can perform the indicated differentiations using Eqs. (A8.5). The resulting algebraic expression can be simplified using Eqs. (A8.2) to yield the following expression for the Laplacian

$$\nabla^2 = \hat{r} \cdot \left(\hat{r}\frac{\partial^2}{\partial r^2}\right) + \frac{\hat{\theta}}{r} \cdot \left(\hat{\theta}\frac{\partial}{\partial r} + \frac{\hat{\theta}}{r}\frac{\partial^2}{\partial \theta^2}\right)$$

$$+ \frac{\hat{\phi}}{r \sin \theta} \cdot \left(\hat{\phi} \sin \theta \frac{\partial}{\partial r} + \frac{\hat{\phi} \cos \theta}{r}\frac{\partial}{\partial \theta} + \frac{\hat{\phi}}{r \sin \theta}\frac{\partial^2}{\partial \phi^2}\right)$$

$$= \frac{\partial^2}{\partial r^2} + \frac{2}{r}\frac{\partial}{\partial r} + \frac{1}{r^2}\frac{\partial^2}{\partial \theta^2} + \frac{\cot \theta}{r^2}\frac{\partial}{\partial \theta} + \frac{1}{r^2 \sin^2 \theta}\frac{\partial^2}{\partial \phi^2}$$

or

$$\boxed{\nabla^2 = \frac{1}{r^2}\left[\frac{\partial}{\partial r}\left(r^2\frac{\partial}{\partial r}\right) + \frac{1}{\sin \theta}\frac{\partial}{\partial \theta}\left(\sin \theta \frac{\partial}{\partial \theta}\right) + \frac{1}{\sin^2 \theta}\frac{\partial^2}{\partial \phi^2}\right]} \quad \text{(A8.6)}$$

C. ANGULAR MOMENTUM OPERATORS

The form of the angular momentum operators in spherical coordinates [given in Eqs. (7.56)] can be obtained in the following way.

$$
\begin{aligned}
\mathbf{L} &= \mathbf{r} \times \mathbf{p} = r[\hat{r} \times (-i\hbar \boldsymbol{\nabla})] \\
&= (-i\hbar r)\hat{r} \times \left(\hat{r}\frac{\partial}{\partial r} + \frac{\hat{\theta}}{r}\frac{\partial}{\partial \theta} + \frac{\hat{\phi}}{r \sin\theta}\frac{\partial}{\partial \phi} \right) \\
&= -i\hbar r\left(\frac{\hat{\phi}}{r}\frac{\partial}{\partial \theta} - \frac{\hat{\theta}}{r\sin\theta}\frac{\partial}{\partial \phi} \right) \\
&= -i\hbar\left(\hat{\phi}\frac{\partial}{\partial \theta} - \frac{\hat{\theta}}{\sin\theta}\frac{\partial}{\partial \phi} \right)
\end{aligned} \tag{A8.7}
$$

To obtain the cartesian components of $\mathbf{L}$ we simply project onto the axes

$$
\begin{aligned}
L_x &= \hat{x} \cdot \mathbf{L} = \hat{x} \cdot (-i\hbar)\left(\hat{\phi}\frac{\partial}{\partial \theta} - \frac{\hat{\theta}}{\sin\theta}\frac{\partial}{\partial \phi} \right) \\
&= i\hbar\left(-\sin\phi\frac{\partial}{\partial \theta} - \frac{\cos\theta\cos\phi}{\sin\theta}\frac{\partial}{\partial \phi} \right) \\
&= i\hbar\left(\sin\phi\frac{\partial}{\partial \theta} + \cot\theta\cos\phi\frac{\partial}{\partial \phi} \right) \\
L_y &= \hat{y} \cdot \mathbf{L} = \hat{y} \cdot (-i\hbar)\left(\hat{\phi}\frac{\partial}{\partial \theta} - \frac{\hat{\theta}}{\sin\theta}\frac{\partial}{\partial \phi} \right) \\
&= i\hbar\left(-\cos\phi\frac{\partial}{\partial \theta} + \cot\theta\sin\phi\frac{\partial}{\partial \phi} \right) \\
L_z &= \hat{z} \cdot \mathbf{L} = \hat{z} \cdot (-i\hbar)\left(\hat{\phi}\frac{\partial}{\partial \theta} - \frac{\hat{\theta}}{\sin\theta}\frac{\partial}{\partial \phi} \right)
\end{aligned}
$$

or

$$
\boxed{L_z = -i\hbar\frac{\partial}{\partial \phi}} \tag{A8.8}
$$

Furthermore, we can immediately write

$$L_+ = L_x + iL_y = \hbar e^{i\phi}\left(\frac{\partial}{\partial\theta} + i\cot\theta\frac{\partial}{\partial\phi}\right)$$
$$L_- = L_x - iL_y = \hbar e^{-i\phi}\left(-\frac{\partial}{\partial\theta} + i\cot\theta\frac{\partial}{\partial\phi}\right) \tag{A8.9}$$

We obtain L^2 by evaluating $\mathbf{L}\cdot\mathbf{L}$ or $L_x^2+L_y^2+L_z^2$. For instance,

$$L^2 = \mathbf{L}\cdot\mathbf{L} = -i\hbar\left(\hat{\phi}\frac{\partial}{\partial\theta} - \frac{\hat{\theta}}{\sin\theta}\frac{\partial}{\partial\phi}\right)\cdot(-i\hbar)\left(\hat{\phi}\frac{\partial}{\partial\theta} - \frac{\hat{\theta}}{\sin\theta}\frac{\partial}{\partial\phi}\right)$$
$$= -\hbar^2\left\{\hat{\phi}\cdot\left[\frac{\partial}{\partial\theta}\left(\hat{\phi}\frac{\partial}{\partial\theta} - \frac{\hat{\theta}}{\sin\theta}\frac{\partial}{\partial\phi}\right)\right]\right.$$
$$\left. - \left(\frac{\hat{\theta}}{\sin\theta}\frac{\partial}{\partial\phi}\right)\cdot\left(\hat{\phi}\frac{\partial}{\partial\theta} - \frac{\hat{\theta}}{\sin\theta}\frac{\partial}{\partial\phi}\right)\right\}$$

The expression is reduced by using Eqs. (A8.5) and after some algebra leads to

$$L^2 = -\hbar^2\left[\frac{1}{\sin\theta}\frac{\partial}{\partial\theta}\left(\sin\theta\frac{\partial}{\partial\theta}\right) + \frac{1}{\sin^2\theta}\frac{\partial^2}{\partial\phi^2}\right] \tag{A8.10}$$

Comparing Eq. (A8.10) with Eq. (A8.6) we see that we can also write the Laplacian

$$\nabla^2 = \frac{1}{r^2}\frac{\partial}{\partial r}\left(r^2\frac{\partial}{\partial r}\right) - \frac{1}{\hbar^2 r^2}L^2 \tag{A8.11}$$

as given in Eq. (8.12*b*).

Appendix 9

Legendre Polynomials and Spherical Harmonics

A. THE DIFFERENTIAL EQUATION

As we know from Section 7.5, and as is shown in detail in Appendix 8 [Eqs. (A8.8) and (A8.10)], the angular momentum operators $\hat{L}_z$ and $\hat{L}^2$ can be expressed in the coordinate representation in spherical coordinates by

$$\hat{L}_z = -i\hbar \frac{\partial}{\partial \phi}$$
$$\hat{L}^2 = -\hbar^2 \left[\frac{1}{\sin\theta} \frac{\partial}{\partial \theta} \left(\sin\theta \frac{\partial}{\partial \theta} \right) + \frac{1}{\sin^2\theta} \frac{\partial^2}{\partial \phi^2} \right] \tag{A9.1a}$$

We also know from the study of angular momentum [see Eqs. (7.49)] that the eigenvalues of the above operators are

$$\hbar m \quad \text{and} \quad \hbar^2 l(l+1) \tag{A9.1b}$$

respectively. Here both m and l take integer values, l is positive, and $-l \leqslant m \leqslant l$. Note that we exclude 1/2 integer values, which in general are allowed, from the commutation properties of angular momentum. Furthermore, $\hat{L}_z$ and $\hat{L}^2$ commute and hence can have simultaneous eigenstates.

We now try to construct the simultaneous eigenstates of $\hat{L}^2$ and $\hat{L}_z$ in the coordinate representation. These will be labeled by the eigenvalues m and l, and will be separable in the coordinates θ and ϕ. Hence we can write

$$\psi_{l,m}(\theta, \phi) = P(\theta)Q(\phi) \tag{A9.2}$$

Since the $\psi_{l,m}(\theta, \phi)$ have to be eigenstates of $\hat{L}_z$ with eigenvalue $\hbar m$, we see that

$$\hat{L}_z \psi_{l,m}(\theta, \phi) = \hbar m \psi_{l,m}(\theta, \phi)$$
$$-i\hbar \frac{dQ}{d\phi} = \hbar m Q(\phi)$$

or

$$Q_m(\phi) = e^{im\phi} \tag{A9.3}$$

This determines the ϕ-dependence of the solutions uniquely and shows that the functions are periodic in ϕ with a period 2π since m is an integer. This should necessarily be so because otherwise $Q_m(\phi)$ would not be single-valued.

The functions $\psi_{l,m}(\theta, \phi)$ are also eigenstates of $\hat{L}^2$ with eigenvalues $\hbar^2 l(l+1)$. Thus, we have

$$\hat{L}^2 \psi_{l,m}(\theta, \phi) = \hbar^2 l(l+1)\psi_{l,m}(\theta, \phi)$$

or

$$-\hbar^2\left(\frac{1}{\sin\theta}\frac{\partial}{\partial\theta}\sin\theta\frac{\partial}{\partial\theta} + \frac{1}{\sin^2\theta}\frac{\partial^2}{\partial\phi^2}\right)P_{l,m}(\theta)Q_m(\phi)$$
$$= \hbar^2 l(l+1)P_{l,m}(\theta)Q_m(\phi)$$

or

$$-\hbar^2 e^{im\phi}\left(\frac{1}{\sin\theta}\frac{d}{d\theta}\sin\theta\frac{d}{d\theta} + l(l+1) - \frac{m^2}{\sin^2\theta}\right)P_{l,m}(\theta) = 0 \tag{A9.4a}$$

Multiplying the above equation by $(-e^{-im\phi}/\hbar^2)$ and changing variables to

$$x = \cos\theta$$

the equation becomes

$$(1-x^2)\frac{d^2}{dx^2}P_{l,m}(x) - 2x\frac{dP_{l,m}(x)}{dx} + \left[l(l+1) - \frac{m^2}{1-x^2}\right]P_{l,m}(x) = 0 \tag{A9.4b}$$

Here we must remember that $-1 \leqslant x \leqslant 1$ and that $l \geqslant |m|$. Equation (A9.4b) can also be written as

$$\boxed{\frac{d}{dx}\left[(1-x^2)\frac{dP_{l,m}}{dx}\right] + \left[l(l+1) - \frac{m^2}{1-x^2}\right]P_{l,m} = 0} \tag{A9.4c}$$

The solutions of Eq. (A9.4b) or Eq. (A9.4c) are known as the *associated Legendre polynomials*. When $m = 0$, then the equation simplifies to

$$\frac{d}{dx}\left[(1-x^2)\frac{dP_l}{dx}\right] + l(l+1)P_l(x) = 0$$

or

$$(1-x^2)\frac{d^2P_l}{dx^2}-2x\frac{dP_l}{dx}+l(l+1)P_l(x)=0 \tag{A9.5}$$

This is Legendre's equation and its solutions $P_l(x)$ are called the Legendre polynomials.

B. LEGENDRE POLYNOMIALS

The solutions of Eq. (A9.5) are generated by the function

$$T(x, s)=(1-2xs+s^2)^{-1/2}=\sum_{l=0}^{\infty} P_l(x)s^l \tag{A9.6a}$$

That is, the Legendre polynomials $P_l(x)$ can be obtained from $T(x, s)$ by taking the lth derivative with respect to the variable s and evaluating it at $s=0$. In other words

$$\left.\frac{1}{l!}\frac{\partial^l T(x, s)}{\partial s^l}\right|_{s=0}=P_l(x) \tag{A9.6b}$$

That the functions $P_l(x)$ defined by Eq. (A9.6b) satisfy the Legendre equation can be simply seen from the structure and symmetry relations of the generating function $T(x, s)$. A particular representation of the Legendre polynomials is given by the Rodrigues' formula

$$P_l(x)=\frac{1}{2^l l!}\frac{d^l}{dx^l}(x^2-1)^l \tag{A9.6c}$$

This shows that the $P_l(x)$ are polynomials of order l. The first few Legendre polynomials can be obtained from the above relation and are

$$\begin{aligned}
P_0(x)&=1\\
P_1(x)&=x\\
P_2(x)&=\tfrac{1}{2}(3x^2-1)\\
P_3(x)&=\frac{x}{2}(5x^2-3), \qquad \text{etc.}
\end{aligned} \tag{A9.7}$$

The orthonormality relations for the Legendre polynomials are given by

$$\int_{-1}^{+1} dx P_l(x) P_{l'}(x) = \frac{2}{2l+1} \delta_{ll'} \tag{A9.8}$$

For the first few Legendre polynomials listed in Eq. (A9.7) the above relation can be verified by direct evaluation.

The solutions of Eq. (A9.4c), the associated Legendre polynomials, can be obtained from the Legendre polynomials through

$$P_{l,m}(x) = (1-x^2)^{|m|/2} \frac{d^{|m|}}{dx^{|m|}} P_l(x) \tag{A9.9}$$

That the $P_{l,m}(x)$ satisfy the differential equation can be seen by evaluating

$$\frac{d}{dx}\left[(1-x^2)\frac{dP_{lm}}{dx}\right] = \frac{|m|^2}{(1-x^2)} P_{l,m}(x) + (1-x^2)^{|m|/2} \frac{d^{|m|}}{dx^{|m|}} \left\{\frac{d}{dx}\left[(1-x^2)\frac{dP_l}{dx}\right]\right\} \tag{A9.10}$$

where the definition of Eq. (A9.9) was used. Furthermore, the second term on the rhs simplifies upon using Eq. (A9.5) so that we have

$$\begin{aligned} \frac{d}{dx}\left[(1-x^2)\frac{dP_{l,m}}{dx}\right] &= \frac{m^2}{1-x^2} P_{l,m}(x) + (1-x^2)^{|m|/2} \frac{d^{|m|}}{dx^{|m|}} [-l(l+1)P_l(x)] \\ &= \frac{m^2}{1-x^2} P_{l,m}(x) - l(l+1)P_{l,m}(x) \end{aligned}$$

which is the desired result [Eq. (A9.4c)].

This shows that the associated Legendre polynomials can be constructed from the Legendre polynomials themselves. As can be seen from Eq. (A9.9) a generating function can be defined for the associated Legendre polynomials as follows

$$\begin{aligned} T_m(x, s) &= (1-x^2)^{|m|/2} \frac{\partial^{|m|}}{\partial x^{|m|}} T(x, s) \\ &= \sum P_{l,m}(x) s^l \end{aligned} \tag{A9.11a}$$

so that

$$P_{l,m}(x) = \frac{1}{l!} \frac{\partial^l}{\partial s^l} T_m(x, s)|_{s=0} \tag{A9.11b}$$

Similarly, the Rodigues formula for the associated Legendre polynomials generalizes to

$$P_{l,m}(x) = \frac{1}{2^l l!} (1 - x^2)^{|m|/2} \frac{d^{l+|m|}}{dx^{l+|m|}} (x^2 - 1)^l \tag{A9.12}$$

The first few polynomials can be obtained from the above relation and are

$$\begin{aligned}
P_{0,0}(x) &= P_0(x) = 1 \\
P_{1,0}(x) &= P_1(x) = x \\
P_{1,1}(x) &= (1 - x^2)^{1/2} = P_{1,-1}(x) \\
P_{2,0}(x) &= P_2(x) = \tfrac{1}{2}(3x^2 - 1) \\
P_{2,1}(x) &= 3x(1 - x^2)^{1/2} = P_{2,-1}(x) \\
P_{2,2}(x) &= 3(1 - x^2) = P_{2,-2}(x)
\end{aligned} \tag{A9.13}$$

The orthogonality relation for the associated Legendre polynomials is given by

$$\int_{-1}^{+1} dx P_{l,m}(x) P_{l',m}(x) = \frac{2}{2l+1} \frac{(l + |m|)!}{(l - |m|)!} \delta_{ll'} \tag{A9.14}$$

This can again be simply tested for the first few polynomials listed in Eq. (A9.13).

C. SPHERICAL HARMONICS

We can now write the eigenstates of the angular momentum operators $\hat{L}^2$ and $\hat{L}_z$ in the coordinate representation.

$$\psi_{l,m}(\theta, \phi) = P_{l,m}(\theta) e^{im\phi} \tag{A9.15a}$$

The normalized eigenstates defined as

$$Y_{l,m}(\theta, \phi) = \varepsilon\left[\frac{2l+1}{4\pi}\frac{(l-|m|)!}{(l+|m|)!}\right]^{1/2} P_{l,m}(\cos\theta)e^{im\phi} \tag{A9.15b}$$

are known as the *spherical harmonics*. Here ε is a phase factor defined by convention to be

$$\begin{aligned} \varepsilon &= (-1)^m \quad \text{for} \quad m > 0 \\ &= 1 \qquad\quad \text{for} \quad m \leqslant 0 \end{aligned} \tag{A9.15c}$$

From the normalization condition of Eq. (A9.14) it is clear that the spherical harmonics satisfy the orthogonality relation

$$\begin{aligned}
&\int d\Omega\, Y^*_{l,m}(\theta, \phi) Y_{l',m'}(\theta, \phi) \\
&= \varepsilon\varepsilon'\left[\frac{2l+1}{4\pi}\frac{(l-|m|)!}{(l+|m|)!}\right]^{1/2}\left[\frac{2l'+1}{4\pi}\frac{(l'-|m'|)!}{(l'+|m'|)!}\right]^{1/2} \\
&\quad \times \int_0^\pi \sin\theta\, d\theta P_{l,m}(\cos\theta) P_{l',m'}(\cos\theta) \int_0^{2\pi} d\phi e^{-i(m-m')\phi} \\
&= \varepsilon\varepsilon'\left[\frac{2l+1}{4\pi}\frac{(l-|m|)!}{(l+|m|)!}\right]^{1/2}\left[\frac{2l'+1}{4\pi}\frac{(l'-|m'|)!}{(l'+|m'|)!}\right]^{1/2} \\
&\quad \times 2\pi\,\delta_{m,m'} \int_{-1}^{1} dx P_{l,m}(x) P_{l',m}(x) \\
&= \left[\frac{2l+1}{4\pi}\frac{2l'+1}{4\pi}\frac{(l-|m|)!\,(l'-|m|)!}{(l+|m|)!\,(l'-|m|)!}\right]^{1/2} \\
&\quad \times 2\pi\,\delta_{mm'} \int_{-1}^{+1} dx P_{l,m}(x) P_{l',m}(x) \\
&= \left[\frac{2l+1}{4\pi}\frac{2l'+1}{4\pi}\frac{(l-|m|)!\,(l'-|m|)!}{(l+|m|)!\,(l'+|m|)!}\right]^{1/2} \\
&\quad \times 2\pi\,\delta_{mm'}\frac{2\,\delta_{ll'}}{2l+1}\frac{(l+|m|)!}{(l-|m|)!} = \delta_{l,l'}\delta_{m,m'}
\end{aligned} \tag{A9.16}$$

We list below the first few of the spherical harmonics. These are obtained from their definition in Eq. (A9.15*b*)

$$\begin{aligned}
Y_{0,0}(\theta, \phi) &= \frac{1}{(4\pi)^{1/2}} \\
Y_{1,0}(\theta, \phi) &= \left(\frac{3}{4\pi}\right)^{1/2} \cos\theta \\
Y_{1,\pm 1}(\theta, \phi) &= \mp\left(\frac{3}{8\pi}\right)^{1/2} \sin\theta e^{\pm i\phi} \\
Y_{2,0}(\theta, \phi) &= \left(\frac{5}{16\pi}\right)^{1/2} (3\cos^2\theta - 1) \\
Y_{2,\pm 1}(\theta, \phi) &= \mp\left(\frac{15}{8\pi}\right)^{1/2} \cos\theta \sin\theta e^{\pm i\phi} \\
Y_{2,\pm 2}(\theta, \phi) &= \left(\frac{15}{32\pi}\right)^{1/2} \sin^2\theta e^{\pm 2i\phi}
\end{aligned} \tag{A9.17}$$

We also note the relation

$$Y_{l,-m}(\theta, \phi) = (-1)^m [Y_{l,m}(\theta, \phi)]^* \tag{A9.18}$$

Appendix 10

Special Functions: Laguerre and Hermite Polynomials

A. THE RADIAL EQUATION FOR THE HYDROGEN ATOM

We begin with Eq. (8.15) and replace μ by m_e. We set

$$U(r) = -\frac{e^2}{4\pi\varepsilon_0}\frac{1}{r}$$

and remember that $E_{nl} < 0$. Then the radial part of the time-independent Schrödinger equation takes the form

$$\left[\frac{1}{r^2}\frac{d}{dr}\left(r^2\frac{d}{dr}\right) - \frac{2m_e}{\hbar^2}|E_{nl}| + \frac{2m_e e^2}{\hbar^2(4\pi\varepsilon_0)}\frac{1}{r} - \frac{l(l+1)}{r^2}\right]R_{nl}(r) = 0 \tag{A10.1}$$

With appropriate foresight we introduce the dimensionless variable

$$\rho = \frac{2r}{a_0} = 2\frac{m_e e^2}{\hbar^2(4\pi\varepsilon_0)}r \tag{A10.2}$$

and set

$$|E_{nl}| = \frac{m_e e^4}{2\hbar^2}\frac{1}{(4\pi\varepsilon_0)^2}\frac{1}{n^2} = \text{Ry}\frac{1}{n^2} \tag{A10.3}$$

Note that the definition of ρ introduced here differs by a factor of 2 from that used in Eq. (8.21c), and in place of the parameter $1/\lambda^2$ used in the text we have introduced n^2.

With these substitutions Eq. (A10.1) becomes

$$\left\{\frac{1}{\rho^2}\frac{d}{d\rho}\left(\rho^2\frac{d}{d\rho}\right) + \left[\frac{1}{\rho} - \frac{1}{4n^2} - \frac{l(l+1)}{\rho^2}\right]\right\}R_{nl}(\rho) = 0 \tag{A10.4a}$$

We perform one more transformation by introducing the variable

$$x = \frac{1}{n}\rho \tag{A10.4b}$$

in terms of which the radial equation becomes

$$\left[\frac{1}{x^2}\frac{d}{dx}\left(x^2\frac{d}{dx}\right) + \left(\frac{n}{x} - \frac{1}{4} - \frac{l(l+1)}{x^2}\right)\right]R_{nl}(x) = 0 \tag{A10.5}$$

To satisfy the boundary conditions we set

$$R_{n,l}(x) = F(x)e^{-x/2} \tag{A10.6a}$$

where $F(x)$ is a polynomial in x. Substituting in Eq. (A10.4*b*) the equation for $F(x)$ is found to be

$$\frac{d^2F}{dx^2} + \left(\frac{2}{x} - 1\right)\frac{dF}{dx} + \left[\frac{n-1}{x} - \frac{l(l+1)}{x^2}\right]F = 0 \tag{A10.6b}$$

As we know, $F(x)$ must be a polynomial of finite order and for this to occur n must be an integer. Therefore we write

$$F(x) \equiv x^s L(x) \qquad s \geqslant 0 \tag{A10.7a}$$

Introducing this expression in Eq. (A10.6*b*) gives

$$x^2\frac{d^2L}{dx^2} + x[2(s+1) - x]\frac{dL}{dx} + [x(n-s-1) + s(s+1) - l(l+1)]L = 0 \tag{A10.7b}$$

Since L must be finite at $x = 0$ we conclude that $[s(s+1) - l(l+1)]$ must equal zero. As discussed in the text we must choose the solution $s = l$, in which case the differential equation for $L(x)$ is

$$\boxed{x\frac{d^2L}{dx^2} + [2(l+1) - x]\frac{dL}{dx} + (n-l-1)L = 0} \tag{A10.8}$$

Here n is an integer

$$n \geqslant l + 1$$

B. LAGUERRE POLYNOMIALS

The solutions of Laguerre's equation

$$x\frac{d^2}{dx^2}L_q(x)+(1-x)\frac{d}{dx}L_q(x)+qL_q(x)=0 \qquad \text{(A10.9)}$$

are the Laguerre polynomials $L_q(x)$. They are defined in the range $0 \leqslant x < \infty$ and can be obtained from a generating function

$$T(x,s)=\frac{e^{-xs/(1-s)}}{1-s}=\sum_{q=0}^{\infty}\frac{L_q(x)}{q!}s^q \qquad \text{(A10.10}a\text{)}$$

The polynomials are calculated from Eq. (A10.10) by taking derivatives of both sides with respect to s and then setting $s=0$. Clearly, the qth derivative of the rhs evaluated at $s=0$ is identically equal to $L_q(x)$. That is

$$L_q(x)=\left.\frac{\partial^q T(x,s)}{\partial s^q}\right|_{s=0}$$

A particular representation of the Laguerre polynomials is

$$L_q(x)=e^x\frac{d^q}{dx^q}(x^q e^{-x}) \qquad \text{(A10.10}b\text{)}$$

The *associated Laguerre* polynomials are defined through

$$L_q^p(x)=\frac{d^p}{dx^p}L_q(x) \qquad \text{(A10.11)}$$

By differentiating Eq. (A10.9) p times, we find the differential equation obeyed by the associated Laguerre polynomials

$$x\frac{d^2}{dx^2}L_q^p(x)+(p+1-x)\frac{d}{dx}L_q^p(x)+(q-p)L_q^p=0 \qquad \text{(A10.12)}$$

This equation is identical to Eq. (A10.8) if we identify

$$2(l+1)=p+1 \quad \text{and} \quad n-l-1=q-p$$

namely

$$p=2l+1, \qquad q=n+l \qquad \text{(A10.13)}$$

The functions $L_q^p(x)$ are orthogonal in the indices q, and their normalization is given by

$$\int_0^\infty e^{-x} x^{2l}\left[L_{n+l}^{2l+1}(x)\right]^2 x^2\,dx = \frac{2n[(n+l)!]^3}{(n-l-1)!} \tag{A10.14}$$

Thus, the normalized radial functions for the hydrogen atom can be expressed in closed form through

$$R_{n,l}(r) = -\left(\frac{2r}{na_0}\right)^l \left\{\left(\frac{2}{na_0}\right)^3 \frac{(n-l-1)!}{2n[(n+l)!]^3}\right\}^{1/2} e^{-r/na_0} L_{n+l}^{2l+1}\left(\frac{2r}{na_0}\right) \tag{A10.15}$$

A few of the first polynomials are listed below

$$\begin{array}{lll} L_0(x) = 1 & L_1(x) = 1 - x & L_2(x) = 2 - 4x + x^2 \\ & L_1^1(x) = -1 & L_2^1(x) = -4 + 2x \\ & & L_2^2(x) = 2 \end{array} \tag{A10.16}$$

$$L_3(x) = 6 - 18x + 9x^2 - x^3, \qquad \text{etc.}$$

Note on Convention: We follow the older convention for the associated Laguerre polynomials as can be found in Schiff, Pauling and Wilson, or Fermi. In most recent texts, however, a convention that is consistent with the definition of the confluent hypergeometric function is used. In this case the associated Laguerre polynomials are defined through

$$L_q^p(x) = (-1)^p \frac{d^p}{dx^p} L_{q+p}(x)$$

The radial function is then proportional to $(-1)^{2l+1} L_{n-(l+1)}^{2l+1}(x)$.

C. HERMITE POLYNOMIALS

The ground-state wavefunction for the one-dimensional s.h.o. was given in the coordinate representation by Eq. (9.25c) of the text

$$\phi_0(y) = \left(\frac{\beta^2}{\pi}\right)^{1/4} e^{-y^2/2} \tag{A10.17a}$$

where

$$\beta = \left(\frac{m\omega}{\hbar}\right)^{1/2} \quad \text{and} \quad y = \beta x \tag{A10.17b}$$

The wavefunction for the nth eigenstate is obtained by operating on $\phi_0(y)$ with the raising operator

$$\hat{a}^{\dagger} = \frac{1}{2^{1/2}}\left(y - \frac{d}{dy}\right) \tag{A10.17c}$$

n times. Thus, the functions $\phi_n(y)$ will be polynomials of finite order multiplied by $e^{-y^2/2}$. Therefore, we write

$$\phi_n(y) = A_n H_n(y) e^{-y^2/2} \tag{A10.18}$$

where A_n is a normalization constant, and $H_n(y)$ is a polynomial in y of order n.

To find the differential equation obeyed by the $H_n(y)$ we recall Eq. (9.17) where the number operator is diagonal and has eigenvalue n

$$\hat{a}^{\dagger}\hat{a}\ |\phi_n\rangle = n\ |\phi_n\rangle \tag{A10.19}$$

In terms of the representation of Eqs. (A10.17c) and (A10.18) the above relation can be written as

$$\frac{1}{2}\left(y - \frac{d}{dy}\right)\left(y + \frac{d}{dy}\right) H_n e^{-y^2/2} = nH_n e^{-y^2/2} \tag{A10.20a}$$

Performing the indicated differentiations we obtain the relation

$$\frac{d^2}{dy^2} H_n(y) - 2y\frac{d}{dy} H_n(y) + 2nH_n(y) = 0 \tag{A10.20b}$$

This is Hermite's equation and its solutions are defined in the range $-\infty < y < +\infty$; as we know n is a positive integer.

The Hermite polynomials can be obtained from the generating function

$$T(y, s) = e^{-s^2+2sy} = \sum_{n=0}^{\infty} \frac{H_n(y)}{n!} s^n \tag{A10.21a}$$

or alternately from the relation

$$H_n(y) = (-1)^n e^{y^2} \frac{d^n}{dy^n}(e^{-y^2}) \tag{A10.21b}$$

The Hermite polynomials are orthogonal and obey the normalization condition

$$\int_{-\infty}^{+\infty} e^{-y^2} H_n(y) H_m(y)\, dy = \delta_{nm} 2^n n! \sqrt{\pi}$$

Thus, the normalization constant in Eq. (A10.18) is given by

$$A_n = \left(\frac{\beta^2}{\pi}\right)^{1/4} \frac{1}{(2^n n!)^{1/2}}$$

A few of the first polynomials are listed below

$$H_0(y) = 1 \qquad H_1(y) = 2y$$

$$H_2(y) = 4y^2 - 2 \qquad H_3(y) = 8y^3 - 12y, \qquad \text{etc.}$$

D. ASSOCIATED LAGUERRE POLYNOMIALS OF HALF-INTEGER ORDER

The solutions of the radial equation for the three-dimensional s.h.o. can be expressed in terms of the associated Laguerre polynomials of half-integer order. Using the quantum number k defined by Eq. (9.44b), and in terms of the notation of section 9.4, we can write

$$\phi_{k,l,m}(r, \theta, \phi) \propto r^l L_{k/2}^{l+1/2}(\beta^2 r^2) e^{-(\beta^2/2)r^2} Y_{l,m}(\theta, \phi)$$

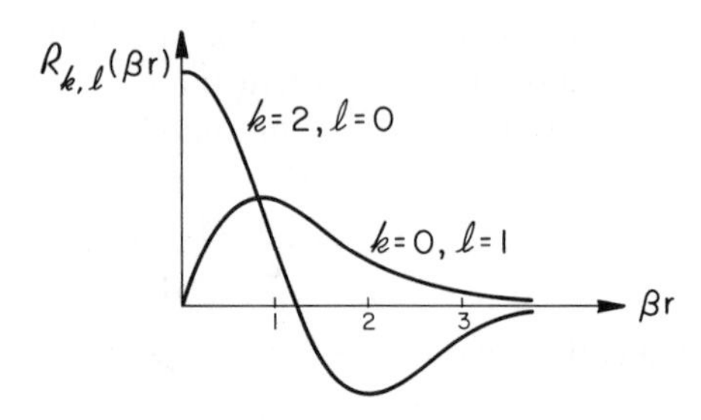

FIGURE A10.1. The lowest two radial eigenfunctions for the 3-dimensional simple harmonic oscillator.

where

$$L_0^a(x) = \Gamma(a+1)$$
$$L_1^a(x) = \Gamma(a+2)[(a+1) - x], \qquad \text{etc.}$$

For instance,

$$L_0^{1/2} = \Gamma(3/2) = \Gamma(1/2) = \sqrt{\pi}$$

The radial wavefunctions for $n = 1(k = 0, l = 1)$ and $n = 2(k = 2, l = 0)$ are shown in Fig. A10.1. Wavefunctions for $k = 0$ are nodeless, for $k = 2$ have 1 node, etc. All $l = 0$ wavefunctions are finite at the origin.

Appendix 11

Expectation Values of $\hat{r}^{-s}$ for Hydrogen

In Chapter 8 we indicated that it is often necessary to evaluate the expectation value of powers of the radial position operator $\hat{r}$ for hydrogenlike wavefunctions. Results for $\hat{r}^s$, where s is a positive or negative integer, were given in Eqs. (8.35) for a few values of s. The expectation values can be obtained by performing the integration over the radial functions and using the properties of the associated Laguerre polynomials $L_{(n+l)}^{(2l+1)}$ [see, Eq. (8.31)]

$$\langle \hat{r}^s \rangle_{n,l} = \left(\frac{na_0}{2}\right)^2 \int_0^\infty \frac{(n-l-1)!}{2n[(n+l)!]^3}\, \rho^{2l+s+2} e^{-\rho} L_{n+l}^{2l+1}(\rho)\, d\rho$$

with

$$\rho = \frac{2r}{na_0}$$

However, these integrals are generally difficult to evaluate and it is therefore instructive to see how we can obtain the result in cases of interest by using special "tricks."

A. EVALUATION OF $\langle 1/\hat{r} \rangle$

We recall that the Hamiltonian for the hydrogen atom is

$$\hat{H}_0 = \frac{\hat{p}^2}{2m_e} - \frac{e^2}{4\pi\varepsilon_0 \hat{r}}$$
$$= \hat{T} + \hat{U} \tag{A11.1}$$

Furthermore, the virial theorem [see, Eq. (6.32b)] applied to the present case ($n = -1$), gives

$$\langle \hat{T} \rangle_{nlm} = -\tfrac{1}{2} \langle \hat{U} \rangle_{nlm}$$

or

$$\langle \hat{T} + \hat{U} \rangle_{nlm} = \tfrac{1}{2} \langle \hat{U} \rangle_{nlm}$$

or

$$E_{nlm} = -\frac{1}{2} \left\langle \frac{e^2}{4\pi\varepsilon_0 \hat{r}} \right\rangle_{nlm}$$

or

$$\left\langle \frac{1}{\hat{r}} \right\rangle_{nlm} = -\frac{8\pi\varepsilon_0}{e^2} E_{nlm} = -\frac{8\pi\varepsilon_0}{e^2} \left(\frac{-e^2}{4\pi\varepsilon_0} \frac{1}{2a_0 n^2} \right)$$

Therefore

$$\boxed{\left\langle \frac{1}{\hat{r}} \right\rangle_{nlm} = \frac{1}{a_0 n^2}} \tag{A11.2}$$

where

$$a_0 = \frac{4\pi\varepsilon_0 \hbar^2}{m_e e^2}$$

is the Bohr radius. This calculation has already been used in a slightly different form in Eqs. (8.37) of the main text.

B. EVALUATION OF $\langle 1/\hat{r}^2 \rangle$

To calculate the expectation value of the operator $1/\hat{r}^2$, we note that if we add to the Hamiltonian a perturbation

$$\hat{H}' = \frac{\lambda}{\hat{r}^2} \tag{A11.3}$$

then the first-order change in the energy is given by

$$\langle \hat{H}' \rangle_{nlm} = \left\langle \frac{\lambda}{\hat{r}^2} \right\rangle_{nlm} = \lambda \left\langle \frac{1}{\hat{r}^2} \right\rangle_{nlm} \tag{A11.4}$$

On the other hand, with this perturbation, the problem can be solved exactly. This is because in this case we can absorb the perturbation into the centrifugal barrier term. We write the Hamiltonian in the coordinate representation as

$$\hat{H}_0 + \hat{H}' = -\frac{\hbar^2}{2m_e}\nabla^2 - \frac{e^2}{4\pi\varepsilon_0 r} + \frac{\lambda}{r^2}$$

$$= -\frac{\hbar^2}{2m_e}\left[\frac{1}{r^2}\frac{\partial}{\partial r}\left(r^2\frac{\partial}{\partial r}\right) - \frac{\hat{L}^2}{\hbar^2 r^2}\right] - \frac{e^2}{4\pi\varepsilon_0 r} + \frac{\lambda}{r^2} \quad \text{(A11.5)}$$

If we now separate out the angular part, then the Hamiltonian for the radial equation is

$$\hat{H}_0 + \hat{H}' = -\frac{\hbar^2}{2m_e}\left[\frac{1}{r^2}\frac{\partial}{\partial r}\left(r^2\frac{\partial}{\partial r}\right) - \frac{l(l+1)}{r^2}\right] - \frac{e^2}{4\pi\varepsilon_0 r} + \frac{\lambda}{r^2}$$

$$= -\frac{\hbar^2}{2m_e}\left\{\frac{1}{r^2}\frac{\partial}{\partial r}\left(r^2\frac{\partial}{\partial r}\right) - \frac{l'(\lambda)[l'(\lambda)+1]}{r^2}\right\} - \frac{e^2}{4\pi\varepsilon_0 r} \quad \text{(A11.6)}$$

where we have introduced the λ-dependent quantity $l'(\lambda)$ defined through

$$l'(\lambda)[l'(\lambda)+1] = l(l+1) + \frac{2m_e\lambda}{\hbar^2} \quad \text{(A11.7)}$$

We recall that the solution of the radial equation for the hydrogen atom results in an energy eigenvalue [see Section 8.3 and in particular the equations leading to Eq. (8.27a); here we use k in place of $s_{\max}$]

$$E_{n,l'} = -\frac{e^2}{4\pi\varepsilon_0}\frac{1}{2a_0}\frac{1}{(k+l'+1)^2} \quad \text{(A11.8)}$$

where k is an integer and $n = (k+l+1)$. Thus, the energy in Eq. (A11.8) is a function of λ, $E_n(\lambda)$.

The next step is to expand $E_n(\lambda)$ in a Taylor series around $\lambda = 0$ and identify the second term with the first-order change in energy given in Eq. (A11.4). We write

$$E_n(\lambda) = E_n(0) + \lambda\left.\frac{dE_n}{d\lambda}\right|_{\lambda=0} + \left.\frac{\lambda^2}{2!}\frac{d^2E_n}{d\lambda^2}\right|_{\lambda=0} + \cdots \quad \text{(A11.9)}$$

It is clear that $E_n(0)$ is the unperturbed value of the energy since when $\lambda = 0$

$$l'(\lambda = 0) = l \tag{A11.10}$$

and in this case the solution reduces to the usual solution for the hydrogen atom. Taking the derivative of Eq. (A11.7) with respect to λ gives

$$[2l'(\lambda) + 1]\frac{dl'(\lambda)}{d\lambda} = \frac{2m_e}{\hbar^2}$$

or

$$\left.\frac{dl'(\lambda)}{d\lambda}\right|_{\lambda=0} = \frac{2m_e}{\hbar^2[2l'(\lambda=0)+1]} = \frac{2m_e}{\hbar^2(2l+1)} \tag{A11.11a}$$

This relation simplifies the calculation of the derivative of the energy because

$$\begin{aligned}
\left.\frac{dE_n(\lambda)}{d\lambda}\right|_{\lambda=0} &= -\frac{e^2}{4\pi\varepsilon_0}\frac{1}{2a_0}\frac{d}{d\lambda}\left.\frac{1}{[k+l'(\lambda)+1]^2}\right|_{\lambda=0} \\
&= -\frac{e^2}{4\pi\varepsilon_0}\frac{1}{2a_0}(-2)\frac{1}{[k+l'(\lambda)+1]^3}\left.\frac{dl'(\lambda)}{d\lambda}\right|_{\lambda=0} \\
&= \frac{e^2}{4\pi\varepsilon_0}\frac{1}{a_0}\frac{1}{(k+l+1)^3}\frac{2m_e}{\hbar^2(2l+1)} \\
&= \frac{m_e e^2}{4\pi\varepsilon_0\hbar^2}\frac{1}{a_0}\frac{1}{n^3}\frac{1}{l+1/2} \\
&= \frac{1}{n^3 a_0^2(l+1/2)}
\end{aligned} \tag{A11.11b}$$

We can now equate the terms linear in λ in Eqs. (A11.4) and (A11.9)

$$\lambda\left\langle\frac{1}{\hat{r}^2}\right\rangle_{nlm} = \lambda\left.\frac{dE_n}{d\lambda}\right|_{\lambda=0} = \lambda\frac{1}{n^3 a_0^2(l+1/2)} \tag{A11.12}$$

to obtain

$$\boxed{\left\langle\frac{1}{\hat{r}^2}\right\rangle_{nlm} = \frac{1}{n^3 a_0^2(l+1/2)}} \tag{A11.13}$$

C. EVALUATION OF $\langle 1/\hat{r}^3 \rangle$

To calculate this expectation value we follow yet another procedure. Recall that the radial momentum operator is defined as

$$\hat{p}_r = -i\hbar\left(\frac{\partial}{\partial r}+\frac{1}{r}\right) \tag{A11.14a}$$

and therefore

$$\hat{p}_r^2 = -\hbar^2\left(\frac{\partial}{\partial r}+\frac{1}{r}\right)\left(\frac{\partial}{\partial r}+\frac{1}{r}\right) = -\hbar^2\frac{1}{r^2}\frac{\partial}{\partial r}\left(r^2\frac{\partial}{\partial r}\right) \tag{A11.14b}$$

This is the familiar expression for the radial part of the Laplacian operator [see Eq. (8.12*b*)]. Thus, the Hamiltonian for the hydrogen atom can be written as

$$\hat{H}_0 = \frac{1}{2m_e}\left(\hat{p}_r^2+\frac{\hat{L}^2}{\hat{r}^2}\right)-\frac{e^2}{4\pi\varepsilon\hat{r}} \tag{A11.14c}$$

Therefore,

$$\begin{aligned}[\hat{H}_0, \hat{p}_r] &= \left[\left\{\frac{1}{2m_e}\left(\hat{p}_r^2+\frac{\hat{L}^2}{\hat{r}^2}\right)-\frac{e^2}{4\pi\varepsilon_0\hat{r}}\right\}, \hat{p}_r\right] \\ &= \frac{\hat{L}^2}{2m_e}\left[\frac{1}{\hat{r}^2}, \hat{p}_r\right]-\frac{e^2}{4\pi\varepsilon_0}\left[\frac{1}{\hat{r}}, \hat{p}_r\right]\end{aligned} \tag{A11.15a}$$

Here we have used the fact that since $\hat{p}_r$ involves only radial coordinates, it commutes with the angular momentum operator $\hat{L}^2$ which involves only the angular coordinates θ and ϕ. The expression in Eq. (A11.15*a*) can be simplified using Eq. (A11.14*a*) to give

$$[\hat{H}_0, \hat{p}_r] = i\hbar\left(\frac{e^2}{4\pi\varepsilon_0\hat{r}^2}-\frac{\hat{L}^2}{m_e\hat{r}^3}\right) \tag{A11.15b}$$

Furthermore, since $\hat{p}_r$ has no explicit time dependence we know that in an energy eigenstate the expectation value of the commutator is zero [see Eq. (6.30)]

$$\langle nlm|\,[\hat{H}_0, \hat{p}_r]\,|nlm\rangle = 0$$

or

$$\frac{e^2}{4\pi\varepsilon_0}\left\langle\frac{1}{\hat{r}^2}\right\rangle_{nlm}-\frac{1}{m_e}\left\langle\frac{\hat{L}^2}{\hat{r}^3}\right\rangle_{nlm}=0$$

or

$$\frac{e^2}{4\pi\varepsilon_0}\left\langle\frac{1}{\hat{r}^2}\right\rangle_{nlm} - \frac{\hbar^2 l(l+1)}{m_e}\left\langle\frac{1}{\hat{r}^3}\right\rangle_{nlm} = 0$$

or

$$\left\langle\frac{1}{\hat{r}^3}\right\rangle_{nlm} = \frac{m_e}{\hbar^2 l(l+1)}\frac{e^2}{4\pi\varepsilon_0}\left\langle\frac{1}{\hat{r}^2}\right\rangle_{nlm}$$

$$= \frac{m_e e^2}{4\pi\varepsilon_0\hbar^2}\frac{1}{l(l+1)}\frac{1}{n^3 a_0^2(l+1/2)}$$

Thus

$$\boxed{\left\langle\frac{1}{\hat{r}^3}\right\rangle_{n,l,m} = \frac{1}{n^3 a_0^3 l(l+1/2)(l+1)}} \qquad \text{(A11.16)}$$

Here we have used the value of $\langle 1/\hat{r}^2\rangle_{nlm}$ from Eq. (A11.13). Although we have derived only the expectation values of operators for the hydrogen atom, the method can be simply extended to hydrogenlike atoms.

Appendix 12

Spherical Bessel Functions

The radial Schrödinger equation for a three-dimensional system subject to a constant potential reduces to

$$\left\{\frac{d^2}{d\rho^2}+\frac{2}{\rho}\frac{d}{d\rho}+\left[1-\frac{l(l+1)}{\rho^2}\right]\right\}R(\rho)=0 \tag{A12.1}$$

Therefore, solutions of the above equation are needed whenever we wish to express free-particle states in spherical coordinates. This is particularly important in scattering theory.

Equation (A12.1) is a Bessel equation and its solutions that are regular at $\rho=0$ are the *spherical Bessel* functions

$$j_l(\rho)=\left(\frac{\pi}{2\rho}\right)^{1/2} J_{l+1/2}(\rho) \tag{A12.2}$$

where $J_{l+1/2}(\rho)$ are the ordinary Bessel functions of half-odd-integer order. That Eq. (A12.2) is a solution of Eq. (A12.1) can be proved by substitution. The first few $j_l(\rho)$ are

$$\begin{aligned} j_0(\rho)&=\frac{\sin\rho}{\rho} \\ j_1(\rho)&=\frac{\sin\rho}{\rho^2}-\frac{\cos\rho}{\rho} \\ j_2(\rho)&=\left(\frac{3}{\rho^3}-\frac{1}{\rho}\right)\sin\rho-\frac{3}{\rho^3}\cos\rho, \qquad \text{etc.} \end{aligned} \tag{A12.3}$$

For $\rho\to 0$ the $j_l(\rho)$ behave as

$$j_l(\rho)\xrightarrow[\rho\to 0]{}\frac{\rho^l}{(2l+1)!!} \quad \text{where} \quad (2l+1)!!=1,3,5,\ldots(2l+1)$$

and for $\rho\to\infty$ as

$$j_l(\rho)\xrightarrow[\rho\to\infty]{}\frac{1}{\rho}\cos[\rho-\tfrac{1}{2}(l+1)\pi]$$

The spherical *Neumann* functions are also solutions of Eq. (A12.1) and are similarly defined as

$$n_l(\rho)=(-1)^{l+1}\left(\frac{\pi}{2\rho}\right)^{1/2}J_{-l-1/2}(\rho)=\left(\frac{\pi}{2\rho}\right)^{1/2}N_{l+1/2}(\rho) \tag{A12.4}$$

The first few $n_l(\rho)$ are

$$\begin{aligned} n_0(\rho)&=-\frac{\cos\rho}{\rho}\\ n_1(\rho)&=-\frac{\cos\rho}{2}-\frac{\sin\rho}{\rho}\\ n_2(\rho)&=-\left(\frac{3}{\rho^3}-\frac{1}{\rho}\right)\cos\rho-\frac{3}{\rho^2}\sin\rho,\qquad \text{etc.}\end{aligned} \tag{A12.5}$$

For $\rho\to 0$ the $n_l(\rho)$ behave as

$$n_l(\rho)\xrightarrow[\rho\to 0]{}-\frac{(2l-1)!!}{\rho^{l+1}}$$

and clearly diverge at the origin. For $\rho\to\infty$ we have

$$n_l(\rho)\xrightarrow[\rho\to\infty]{}\frac{1}{\rho}\sin[\rho-\tfrac{1}{2}(l+1)\pi]$$

The asymptotic expansions as $\rho\to\infty$ are good approximations when $\rho\geqslant 1/2l(l+1)$. [For the properties of these functions see, for instance, L. I. Schiff, *Quantum Mechanics*, Third Edition McGraw-Hill, New York (1968), p. 85.]

It is convenient in scattering problems, when ρ does not extend to the origin, to form linear combinations of the $j_l(\rho)$ and $n_l(\rho)$ functions. These combinations are known as the *spherical Hankel* functions of the first and second kind and are defined through

$$\begin{aligned} h_l^{(1)}(\rho)&=j_l(\rho)+in_l(\rho)\\ h_l^{(2)}(\rho)&=j_l(\rho)-in_l(\rho)=[h_l^{(1)}(\rho)]^*\end{aligned} \tag{A12.6}$$

The first few $h_l^{(1)}(\rho)$ are

$$
\begin{aligned}
h_0^{(1)}(\rho) &= -\frac{i}{\rho}\, e^{i\rho} \\
h_1^{(1)}(\rho) &= -\left(\frac{1}{\rho}+\frac{i}{\rho^2}\right)e^{i\rho} \\
h_2^{(1)}(\rho) &= \left(\frac{i}{\rho}-\frac{3}{\rho^2}-\frac{3i}{\rho^3}\right)e^{i\rho}, \qquad \text{etc.}
\end{aligned}
\tag{A12.7}
$$

For $\rho \to \infty$ the asymptotic expansions are

$$h_l^{(1)}(\rho) \xrightarrow[\rho\to\infty]{} \frac{1}{\rho}\, e^{i[\rho-(l+1)\pi/2]}$$

$$h_l^{(2)}(\rho) \xrightarrow[\rho\to\infty]{} \frac{1}{\rho}\, e^{-i[\rho-(l+1)\pi/2]}$$

that is, they behave as outgoing and incoming spherical waves, respectively.

For certain problems it is necessary to know the zeros of $j_l(\rho)$. We designate these roots by

$$\rho_{n,l} = \pi\beta_{n,l}$$

The values of $\beta_{n,l}$ such that $j_l(\pi\beta_{n,l}) = 0$ are given in the table below for $n = 1$–5 and the first few values of l.

	$n=1$	$n=2$	$n=3$	$n=4$	$n=5$
$l=0$	1.0000	2.0000	3.0000	4.0000	5.0000
$l=1$	1.4303	2.1590	3.4709	4.4775	5.4816
$l=2$	1.8346	2.8950	3.9226	4.9385	5.9189

Appendix 13

Evaluation of Overlap Integrals

In the analysis of the problem of the helium atom in Section 10.4 we needed to evaluate the integral of Eq. (10.17a) which was of the form

$$\Delta E = \frac{e^2}{4\pi\varepsilon_0} \int |u(\mathbf{r}_1)|^2 \frac{1}{|\mathbf{r}_1 - \mathbf{r}_2|} |u(\mathbf{r}_2)|^2 \, d^3r_1 \, d^3r_2 \qquad \text{(A13.1)}$$

with

$$u(\mathbf{r}) = \frac{1}{\pi^{1/2}} \left(\frac{Z}{a_0}\right)^{3/2} e^{-rZ/a_0} \qquad \text{(A13.2)}$$

Then

$$\Delta E = \frac{e^2}{4\pi\varepsilon_0} \left[\frac{1}{\pi}\left(\frac{Z}{a_0}\right)^3\right]^2 \int e^{-2r_1Z/a_0} \frac{1}{|\mathbf{r}_1 - \mathbf{r}_2|} e^{-2r_2Z/a_0} \, d^3r_1 \, d^3r_2 \qquad \text{(A13.3)}$$

It is convenient to express this integral by a two-step process

$$\Delta E = e\frac{1}{\pi}\left(\frac{Z}{a_0}\right)^3 \int e^{-2r_1Z/a_0} U(r_1) \, d^3r_1 \qquad \text{(A13.4}a\text{)}$$

with

$$U(r_1) = \frac{e}{4\pi\varepsilon_0} \frac{1}{\pi}\left(\frac{Z}{a_0}\right)^3 \int e^{-2r_2Z/a_0} \frac{1}{|\mathbf{r}_1 - \mathbf{r}_2|} \, d^3r_2 \qquad \text{(A13.4}b\text{)}$$

The physical interpretation of this separation is clear: $U(r_1)$ represents the electrostatic potential at the radial position r_1 due to electron 2, after averaging (integrating) over the probability distribution for the position of electron 2. Then, Eq. (A13.4a) measures the potential energy of electron 1 as it moves in the potential created by electron 2.

First we evaluate Eq. (A13.4b) and introduce the dimensionless variables

$$\boldsymbol{\rho}_1 = \mathbf{r}_1 \frac{Z}{a_0} \qquad \boldsymbol{\rho}_2 = \mathbf{r}_2 \frac{Z}{a_0}$$

Thus

$$U(\rho_1) = \frac{e}{4\pi\varepsilon_0} \frac{1}{\pi} \frac{Z}{a_0} \int \rho_2^2 \, d\rho_2 \, d\Omega_2 \frac{1}{|\boldsymbol{\rho}_1 - \boldsymbol{\rho}_2|} e^{-2\rho_2} \qquad \text{(A13.5)}$$

In this integration $\boldsymbol{\rho}_1$ is a fixed vector both in magnitude and direction. We then divide the space in spherical shells of radius ρ_2 and thickness $d\rho_2$ as shown in Fig. A13.1. The volume of each shell is

$$dV(\rho_2) = 4\pi(\rho_2)^2 \, d\rho_2$$

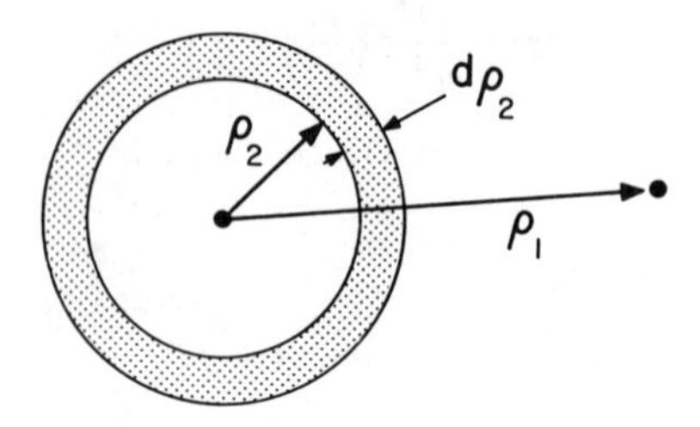

FIGURE A13.1. Radial coordinates for the integration of the matrix element of a two-electron atom. Here ρ_1 is assumed fixed and the integration is over the spherical shells at ρ_2.

We can recognize two cases:

(a) when $\rho_1 > \rho_2$; and
(b) when $\rho_1 < \rho_2$.

In the first case, the potential at ρ_1 due to a shell at ρ_2 is the same as if all the charge was at the origin. This follows from Gauss' law in view of the spherical symmetry of the charge distribution of the shell. Thus, for $\rho_1 > \rho_2$

$$dU(\rho_1) = \frac{e}{4\pi\varepsilon_0} \frac{Z}{a_0} \frac{4\pi}{\pi} (\rho_2)^2 \, d\rho_2 e^{-2\rho_2} \frac{1}{\rho_1} \qquad \text{(A13.6}a\text{)}$$

When $\rho_1 < \rho_2$, the potential inside a uniform spherical distribution of charge must be constant so that for $\rho_1 < \rho_2$

$$dU(\rho_1) = \text{const}$$

To determine the constant we note that $dU(\rho_1)$ must be continuous at $\rho_1 = \rho_2$. Thus for $\rho_1 < \rho_2$

$$dU(\rho_1) = \frac{e}{4\pi\varepsilon_0}\frac{Z}{a_0}\frac{4\pi}{\pi}(\rho_2)^2\, d\rho_2 e^{-2\rho_2}\frac{1}{\rho_2} \tag{A13.6b}$$

The integral of Eq. (13.5) can now be carried out in two parts

$$U(\rho_1) = \frac{e}{4\pi\varepsilon_0}\frac{4Z}{a_0}\left(\int_0^{\rho_1} e^{-2\rho_2}\rho_2\, d\rho_2 + \frac{1}{\rho_1}\int_{\rho_1}^{\infty} e^{-2\rho_2}(\rho_2)^2\, d\rho_2\right)$$

$$= \frac{e}{4\pi\varepsilon_0}\frac{Z}{a_0}\frac{1}{\rho_1}[1 - e^{-2\rho_1}(\rho_1 + 1)] \tag{A13.7}$$

Equation (A13.7) gives the electrostatic potential generated by electron 2 at the radial position ρ_1.

Finally, we average the position of electron 1 over the potential; that is, we perform the integral of Eq. (A13.4*a*). We have

$$\Delta E = \frac{e^2}{4\pi\varepsilon_0}\frac{Z}{a_0}\frac{1}{\pi}\int_0^{\infty}\frac{1}{\rho_1}(\rho_1)^2\, d\rho_1\, d\Omega_1 e^{-2\rho_1}[1 - e^{-2\rho_1}(\rho_1 + 1)]$$

$$= \frac{e^2}{4\pi\varepsilon_0}\frac{Z}{a_0}4\int_0^{\infty}\rho_1\, d\rho_1[e^{-2\rho_1} - e^{-4\rho_1}(1 + \rho_1)] \tag{A13.8}$$

Using

$$\int_0^{\infty} x^n e^{-\alpha x}\, dx = \frac{n!}{a^{n+1}}$$

we immediately obtain

$$\Delta E = \frac{e^2}{4\pi\varepsilon_0}\frac{Z}{a_0}[\tfrac{1}{4} - (\tfrac{1}{16} + \tfrac{1}{32})] = \tfrac{5}{8}\frac{e^2}{4\pi\varepsilon_0}\frac{Z}{a_0} \tag{A13.9}$$

This is the result that was used in Eq. (10.17*b*).

The procedure used here is restricted to spherically symmetric functions. It can be easily generalized to $l \neq 0$ cases by expanding $(|\mathbf{r}_1 - \mathbf{r}_2|)^{-1}$ in Legendre polynomials and using the addition theorem for the spherical harmonics. As is well known a different expansion is valid for the region $\rho_1 < \rho_2$ and for $\rho_1 > \rho_2$, just as in the example that we worked out. What is important is to appreciate the physical picture where we think of one electron as creating an average *electrostatic* potential and then evaluating the average energy of the second electron in that potential.

Appendix 14

Units and Physical Constants

A NOTE ON UNITS

As the reader must have noticed the discussion of atomic and nuclear phenomena is best carried out by using the units that are natural to this domain, rather than a specified set such as cgs or MKS units. We list below the units that are used throughout the text.

Length	cm
	F (Fermi) $1\ \mathrm{F} = 1\ \mathrm{fm} = 10^{-15}\ \mathrm{m} = 10^{-13}\ \mathrm{cm}$
	Å (Angstrom) $1\ \text{Å} = 10^{-10}\ \mathrm{m} = 10^{-8}\ \mathrm{cm}$
	a_0 (Bohr radius) $= \dfrac{4\pi\varepsilon_0\hbar^2}{m_e e^2} \simeq 0.53\times 10^{-8}\ \mathrm{cm}$
Energy	$\mathrm{eV} \simeq 1.6\times 10^{-19}\ \mathrm{J}$
	Ry (Rydberg) $= \dfrac{m_e e^4}{2\hbar^2}\dfrac{1}{(4\pi\varepsilon_0)^2} \simeq 13.6\ \mathrm{eV}$
Energy-length	$\hbar c \simeq 197.3\ \mathrm{MeV\text{-}F}$
Electric charge	$e \simeq 1.6\times 10^{-19}\ \mathrm{coul}$
	$\alpha = \dfrac{e^2}{4\pi\varepsilon_0}\dfrac{1}{\hbar c} \simeq \dfrac{1}{137}$
Cross section	mb (millibarn) $= 10^{-27}\ \mathrm{cm}^2$
Mass	
electron	$m_e c^2 \simeq 0.511\ \mathrm{MeV}$
proton	$m_p c^2 \simeq 938.3\ \mathrm{MeV}$
Wavelength	$\lambda = c/\nu = 2\pi(\hbar c/E)$
	$\lambda \simeq 12{,}400$ Å corresponds to $h\nu = 1\ \mathrm{eV}$
	$\dfrac{1}{\lambda} = \dfrac{\nu}{c}$ in cm^{-1} is referred to as wavenumbers
Temperature	$T = 11{,}600°\mathrm{K}$ corresponds to $kT = 1\ \mathrm{eV}$

Electric field	V/m
Magnetic field	T (Tesla)
	$pc = 0.03\ \mathrm{B}\rho$ (MeV, T, cm) (ρ is the radius of curvature—B magnetic field, p momentum)

NUMERICAL VALUES OF PHYSICAL CONSTANTS†

$c = 2.9979246 \times 10^{10}$ cm/sec (speed of light)

$e = 1.602189 \times 10^{-19}$ coul (magnitude of electron charge)

$h = 6.62618 \times 10^{-34}$ J-sec (Planck's constant)

$\hbar = h/2\pi = 1.054589 \times 10^{-34}$ J-sec

$= 6.58217 \times 10^{-16}$ eV-sec

$\hbar c = 197.328$ MeV-F

$$\alpha = \frac{e^2}{4\pi\varepsilon_0 \hbar c} = \frac{1}{137.0360} \text{ (fine-structure constant)}$$

$m_e = 0.511003\ \mathrm{MeV}/c^2$ (electron mass)

$= 9.10953 \times 10^{-28}$ g

$m_p = 938.279\ \mathrm{MeV}/c^2$ (proton mass)

$= 1.67265 \times 10^{-24}$ g

$k = 1.3806 \times 10^{-23}$ J/K (Boltzmann's constant)

$a_0 = 0.5291771 \times 10^{-8}$ cm (Bohr radius)

Ry $= 13.60580$ eV (Rydberg energy)

$$\mu_B = \frac{e\hbar}{2m_e c} = 5.78838 \times 10^{-9}\ \mathrm{eV/G}$$

$N_0 = 6.02205 \times 10^{23}$/mol (Avogadro's number)

$V_m = 22.4138$ liters/mol (grammolar volume for ideal gas at STP)

$\varepsilon_0 = 8.854 \times 10^{-12}$ coul/V-m (permittivity of free space)

$\mu_0 = 4\pi \times 10^{-7}$ V-sec/A-m (permeability of free space)

† From "Review of particle properties," *Phys. Lett.* **111B** (April 1982).

SUBJECT INDEX

Absorption of radiation, 193
Absorption spectrum of ammonia, 157
Accidental degeneracy, 319
Airy function, 343
Ammonia molecule, 152
Amplitude, 56
antisymmetric, 126, 427
superposition, 117
Analyzer of polarization, 100
Angular distribution, 510
Angular momentum
operator, 229, 604
commutation relations of, 230
composition of, 291, 293, 592
eigenvalues of, 280
representations of, 286
Schwinger's approach, 411
Annihilation operator, 371
Anomalous Zeeman effect, 106
Antineutrino, 111
Antisymmetric amplitude, 126, 427
Argand diagram, 552
Associated Laguerre polynomial, 395, 618
Associated Legendre polynomial, 291, 322
Average value, 41, 96
Azimuthal symmetry, 479

Balmer series, 321
Band structure of a crystal, 244, 253
Barrier penetration, 248
Basis states, 59
Beating, 35
Big-bang theory, 470
Binding energy, 421
Black body spectrum, 7, 465, 470
Block diagonal matrix, 476
Bohm-Aharonov effect, 421
Bohr radius, 326
Bohr theory, 320
Boltzmann distribution, 196, 465
Born approximation, 203, 540
Born, M., 204
Bose-Einstein statistics,106
Bosons, 126, 424
Bound states, 305, 363
Boundary conditions, 45, 239, 311
Box normalization, 571
Bra-ket notation, 57, 77
Bragg scattering, 132
Breit-Wigner formula, 530

Canonical variables, 224, 590
Cauchy's residue theorem, 538
Center of mass frame, 482
Center of mass motion, 306
Centrifugal barrier, 313
Charge conjugation, 173
Charmonium, 349
Clebsch-Gordan coefficients, 300, 593
Closure relation, 70
Column vector, 72
Compatible observables, 261
Complex conjugation, 73
Compton effect, 9
Compton wavelength, 409
Conservation laws, 264, 268
Constants of motion, 261
Coordinate representation, 218, 290
of Quantum Mechanical operators, 226, 229
Coordinate transformations, 73, 600
Correspondence principle, 40, 43, 111, 216
Cosmic background radiation, 470
Coulomb potential, 31, 201, 315
CP violation, 173
Creation operator, 371
Cross section
differential, 478
elastic, 528
inelastic, 528, 551
total, 528
Curvature, 241

Davisson-Germer experiment, 13
DeBroglie relations, 15
Debye-Scherrer pattern, 20
Decay matrix, 172
Decay of nucleus, 250
Decay width, 173
Degenerate states, 145

Delta function, 40, 564
representations of, 567
Density of states, 189
Determinant, 580
Deuteron, 407
Diatomic molecule, 381
Diffraction, by slit, 28
Diffraction, electron, 18
Diffraction grating, 29
Diffraction of neutrons, 119
Diffraction pattern, 13
Dirac equation, 331
Dirac identity, 177
Dirac, P.A.M., 57
Direct product of states, 123, 426
Discrete spectrum, 243
Dissociation energy, 460
Doppler shift, 470

Effective potential, 314
Effective range expansion, 506
Ehrenfest's theorem, 219, 231
Eigenstate, 79, 80, 587
Eigenvalue, 79, 585
Einstein, A., 8
Electric dipole moment, 159
Electron charge, 2
Electron configuration of elements, 434
Electron density, 48
Emission of radiation, 193
Emission spectrum, 438
Energy bands, 137
Energy density, 194, 468
Energy width, 212
Envelope, of wavepacket, 38
E.P.R. paradox, 98
Equation of motion, 136, 237
Exchange symmetry, 422
integral, 429, 445, 458
force, 446
Expansion coefficients, 60
Expectation values, 96

Fermi, E., 190, 250
Fermi-Dirac statistics, 106
Fermi energy, 47
Fermions, 53, 126, 424
Fermi's golden rule, 180, 190, 195
Fine-structure constant, 330
Fine-structure of atomic spectra, 332
Fine-structure splitting, 332
Flux, 248, 481
Free particle, 33, 570
in spherical coordinates, 492
Frequency, angular, 14
Fourier transform, 30, 201, 560

g-factor, 210
Gaussian function, 568
Gauss's law, 630
Generators of symmetry transformations, 272, 274
Good quantum number, 263
Grating, 29
Greens function, 535
advanced, 539
retarded, 539
stationary, 539
Ground state, 431
Groups, SU(2), 169
Group velocity, 39

Half-width, 190
Hamiltonian operator, 133
Heisenberg picture, 102, 232
Heisenberg, W., 1
Heitler-London, 455
Helicity, 63
Helium atom, 442
energy spectrum of, 448, 451
para and ortho, 446
wavefunctions of, 444
Hermite polynomials, 375, 613
Hermitian conjugation, 73
matrices, 79, 585
operators, 96, 274
Hilbert space, 62
Homonuclear molecules, 380, 436
Hulthen wavefunction, 418
Hydrogen atom, 32, 317
Hydrogen molecule, 452
Hyperfine splitting, 338
Hyperfine structure of atomic spectra, 332

Identical particles, 122, 421
interchange of, 124

Induced emission, 195
Inelastic scattering, 547
Inner product, 143
Integral equation, 536
Interference, 10, 117
Interpretation of quantum mechanics, 98
Invariance principles, 267
Inversion of coordinates, 99
Ionic binding, 459
Ionization potentials, 432
 of helium atom, 449
Isotope shift, 339
Isotopic spin, 173

Jackson, J.D., 198, 328
Jacobi identity, 301
Jacobian, 485

$K^\circ - \bar{K}^\circ$ system, 172
Kronecker symbol, 577

Laboratory frame, 482
Laguerre polynomials, 322, 613
Lamb shift, 337
Landau, L.D., 183
Laplace operator, 228, 535, 602
Larmor formula, 198
Larmor frequency, 132
Laser, neon, 50
Legendre equation, 292
Legendre polynomials, 291, 606
Lifetime, 198
Lifshitz, E.M., 183
Linear algebra, 56
Linear potential, 339
Linear superposition, 24, 69, 117
Linewidth, 213
Lippman-Schwinger equation, 535
Lorentizian curve, 219
Lorentz transformation, 113
Low energy n-p scattering, 504
Lyman series, 321

Magic number, 396
Magnetic moment, 3, 106
Magneton, Bohr, 4, 50
 nuclear, 107
Many-body system, 123, 421
Many-particle wavefunctions, 426
Mass spectroscopy, 408
Matrices
 block diagonal, 476
 diagonalization of, 585
 finding the eigenvalues of, 587
 hermitian, 79, 585
 multiplication of, 73, 578
 orthogonal, 74
 unitary, 74
Matrix algebra, 577
Matrix mechanics, 84
Mean square deviation, 225
Measurement process, 23, 67, 93, 126
Mercury, spectrum, 2
Meso-deuterium atom, 476
Mesons, 347
 μ, 112
 K, K°, 172
 ρ, 213
 π°, π^-, 98, 213
 $W^\pm$, 554
Microwave frequency, 193
Miller indices, 120
Molecular band spectra, 380, 387
Moment of inertia, 389, 440
Momentum representation, 69
Momentum space, 37
Momentum transfer, 201, 542
Monochromatic wave, 104

Neutrino, 111
Neutron, 51, 421, 504
Normal modes, 158
Normalization, 40
 of plane waves, 570
 of wave packets, 70
Nuclear force, 407
Nuclear magnetic resonance, 180, 204
Nucleus, 395
Number operator, 371
 representation, 374
Numerical values of physical constants, 634

Observables, 1, 93
Operators, 57
 eigenstates and eigenvalues of, 79
 hermitian, 96, 274

lowering, 171, 282, 366
permutation, 423
products of, 83
raising, 171, 282, 366
spherical coordinates, 600
spin exchange, 303
time translation, 134
unitary, 74, 87, 273
Optical theorem, 497, 552
Orthogonal basis, 59
matrix, 74
Orthonormality condition, 68
Overlap integral, 457, 629

Parity, 111, 265
operator, 267, 437
Partial wave analysis, 486, 497
Particle in a box, 43
Pauli exclusion principle, 47, 124, 428
Pauli spin matrices, 129, 167, 289
Pauli, W., 167
Penzias, A., 470
Periodic potential, 244
solutions, 246
Periodic table, 126, 421, 430
Permutation operator, 425
Perturbation theory, 140, 146
stationary states, 159
time-dependent, 185
Phase space, 48, 190
Phase velocity, 39
Photoelectric effect, 8
Photon, 9
Photon gas, 463
Planck, M., 7
Planck's constant, 7, 223
Planck's law, 197, 465
Poisson bracket, 223, 590
Poisson equation, 536
Polarization, linear, 65
circular, 63, 99
Position representation, 68, 218, 290
Positron, 111
Positronium, 476
Potential
Coulomb, 315
linear, 339
spherically symmetric, 309
square well, 399, 406
Precession, of spin, 5, 111
Probability amplitude, 23
current, 248
interpretation, 98
of transition, 184, 190
Process of measurement, 93
Projection operators, 83
Pseudoscalars, 99

Quadrupole moment tensor, 299
Quantization of energy, 240
Quantization volume, 193
Quantized observables, 1
Quantum chromodynamics, 352
Quantum conditions, 223
Quantum electrodynamics, 113, 337
Quarkonium, 346
Quarks, 151
charges and masses, 347
Quasi-stationary states, 195, 211, 533
energy width of, 211

Radial equation, Coulomb potential, 314
solutions, 319
Raman effect, 436
scattering, 439
spectra of molecules, 436, 440
Ramsauer-Townsend effect, 519, 530
Rayleigh, Lord, 489
Rayleigh-Jeans law, 7, 469
Rayleigh-Schrödinger perturbation, 145
Reflection coefficient, 250
Reflection operator, 269
Refraction, electron, 21
Refractive index, 21
Relativistic correction, 333
transformation, 114
Relaxation mechanism, 209
Representation, 57
of angular momentum operators, 286
of delta function, 567
momentum, 69
position, 68, 218, 290
of quantum-mechanical states, 61
Resonance condition, 195
peak, 529
scattering, 517, 526

Rodrigue's formula, 608
Rotation matrices, 298
Rotation-vibration states of molecules, 385
spectra of molecules, 386
Rotational state of ammonia, 158
Row vector, 75
Rutherford cross section, 203
Rydberg, 315

Scalar operators, 596
product, 75
Scattering
amplitude, 477
black disk, 553
Coulomb potential, 544
delta potential, 498
hard sphere, 502
identical particles, 521
length, 507
n-p, 504
proton, 15
static potential, 199
square-well potential, 500
Schrödinger, E., 84
Schrödinger equation, 237
Schrödinger picture, 102, 232
Screening, 433
Secular determinant, 156, 160, 586
Selection rules, 337, 350, 598
Separation of variables, 306
Shadow scattering, 553
Shell model of the nucleus, 395, 432
Similarity transformation, 168
Simple harmonic oscillator
in one dimension, 364
representations of, 371
in three dimensions, 388
Slater determinant, 428
Snell's law, 21
Spectral line, 3
Spherical Bessel function, 404, 491, 626
Spherical Hankel function, 404, 627
Spherical harmonics, 291, 491, 606, 610
Spherical Neumann function, 627
Spin angular momentum, 106
Spin exchange operator, 303
Spin-flip, 119
scattering, 515
Spin-orbit coupling, 329, 397, 449
Spin-statistics theorem, 126, 426
Splitting of degenerate states, 150
Spontaneous transition, 193
Square well potential, 47, 394, 406, 500
Stark effect, 360
State vector, 62
Stationary states, 104, 134
Statistical equilibrium, 195
Stepfunction, 569
Stern-Gerlach experiment, 3
Stimulated emission, 195
Stokes theorem, 122
Strangeness, 172
Superposition, of amplitudes, 35, 117
SU(2) group, 169
Symmetric amplitude, 126, 427
Symmetries, 264
in quantum mechanics, 268

Taylor expansion, 277
Thomas precession, 329
Time evolution operator, 134
Time independent perturbation, 140
of a two-level system, 159
Time reversal, 70, 86
Transformation, unitary, 86
matrix, 73, 76
Transition probability, 184, 190
Transitions, 179
Translation operator, 245
Transmission coefficient, 250
Transposed matrix, 73, 582
Trial wavefunction, 342
Tunnelling, 153, 247
Two-level systems, 134
examples of, 171
general description of, 167
H_2 molecule, 460
transitions in, 180

Uncertainty principle, 27, 224
Unitary transformation, 86
matrix, 74, 584
Units of physical constants, 633

Valence bond, 455
Van der Waals force, 474

Variational method, 339
application to linear potential, 342
application to quarkonium, 346
Vector operators, 597
Vector potential, 121
Vibrational states, 158, 382
Virial theorem, 235, 332, 379
Virtual states, 510

Wavefunction, 115, 238
Wave, harmonic, 14
Wave mechanics, 84
Wavenumber, 14
Wavepacket, 33
Wavepacket, Gaussian, 53
Wave, standing, 44
Weak interaction, 44, 554
Well, square, 47, 399, 406, 500
Wien's law, 7, 469

X-ray diffraction, 20

Yukawa potential, 545

Zeeman effect, 150, 329
Zero point energy, 368, 467